JACARANDA

MATHS QUEST 8

STAGE 4 NSW SYLLABUS | THIRD EDITION

T0342879

JACARANDA
MATHS QUEST 8

STAGE 4 NSW SYLLABUS | THIRD EDITION

BEVERLY LANGSFORD WILLING

CATHERINE SMITH

jacaranda
A Wiley Brand

Third edition published 2023 by
John Wiley & Sons Australia, Ltd
Level 4, 600 Bourke Street, Melbourne, Vic 3000

First edition published 2011
Second edition published 2014

Typeset in 10.5/13 pt TimesLTStd

ISBN: 978-0-7303-8686-5

Front cover images: © alexdndz/Shutterstock
© Marish/Shutterstock; © AlexanderTrou/Shutterstock
© Vector Juice/Shutterstock

Illustrated by various artists, diacriTech and Wiley Composition Services

Typeset in India by diacriTech

A catalogue record for this book is available from the National Library of Australia

NATIONAL LIBRARY OF AUSTRALIA

Printed in Singapore
M WEP217168 040823

The Publishers of this series acknowledge and pay their respects to Aboriginal Peoples and Torres Strait Islander Peoples as the traditional custodians of the land on which this resource was produced.

This suite of resources may include references to (including names, images, footage or voices of) people of Aboriginal and/or Torres Strait Islander heritage who are deceased. These images and references have been included to help Australian students from all cultural backgrounds develop a better understanding of Aboriginal and Torres Strait Islander Peoples' history, culture and lived experience.

It is strongly recommended that teachers examine resources on topics related to Aboriginal and/or Torres Strait Islander Cultures and Peoples to assess their suitability for their own specific class and school context. It is also recommended that teachers know and follow the guidelines laid down by the relevant educational authorities and local Elders or community advisors regarding content about all First Nations Peoples.

All activities in this resource have been written with the safety of both teacher and student in mind. Some, however, involve physical activity or the use of equipment or tools. **All due care should be taken when performing such activities.** To the maximum extent permitted by law, the author and publisher disclaim all responsibility and liability for any injury or loss that may be sustained when completing activities described in this resource.

The Publisher acknowledges ongoing discussions related to gender-based population data. At the time of publishing, there was insufficient data available to allow for the meaningful analysis of trends and patterns to broaden our discussion of demographics beyond male and female gender identification.

Contents

About this resource

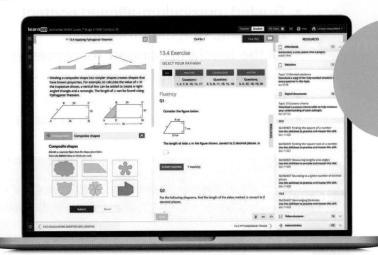

NEW FOR
2024 NSW SYLLABUS

JACARANDA
MATHS QUEST 8
NSW SYLLABUS
THIRD EDITION

Developed by teachers for students

Tried, tested and trusted. The third edition of the *Jacaranda Maths Quest series*, continues to focus on helping teachers achieve learning success for every student — ensuring no student is left behind, and no student is held back.

Because both what and how students learn matter

Learning is personal

Whether students need a challenge or a helping hand, you'll find what you need to create engaging lessons.

Whether in class or at home, students can get unstuck and progress! Scaffolded lessons, with detailed worked examples, are all supported by teacher-led video eLessons. Automatically marked, differentiated question sets are all supported by detailed worked solutions. And Brand-new Quick Quizzes support in-depth skill acquisition.

Learning is effortful

Learning happens when students push themselves. With learnON, Australia's most powerful online learning platform, students can challenge themselves, build confidence and ultimately achieve success.

Learning is rewarding

Through real-time results data, students can track and monitor their own progress and easily identify areas of strength and weakness.

And for teachers, Learning Analytics provide valuable insights to support student growth and drive informed intervention strategies.

Learn online with Australia's most

Everything you need for each of your lessons in one simple view

- Trusted, curriculum-aligned content
- Engaging, rich multimedia
- All the teaching-support resources you need
- Deep insights into progress
- Immediate feedback for students
- Create custom assignments in just a few clicks.

Practical teaching advice and ideas for each lesson provided in teachON

Teaching videos for all lessons

Reading content and rich media including embedded videos and interactivities

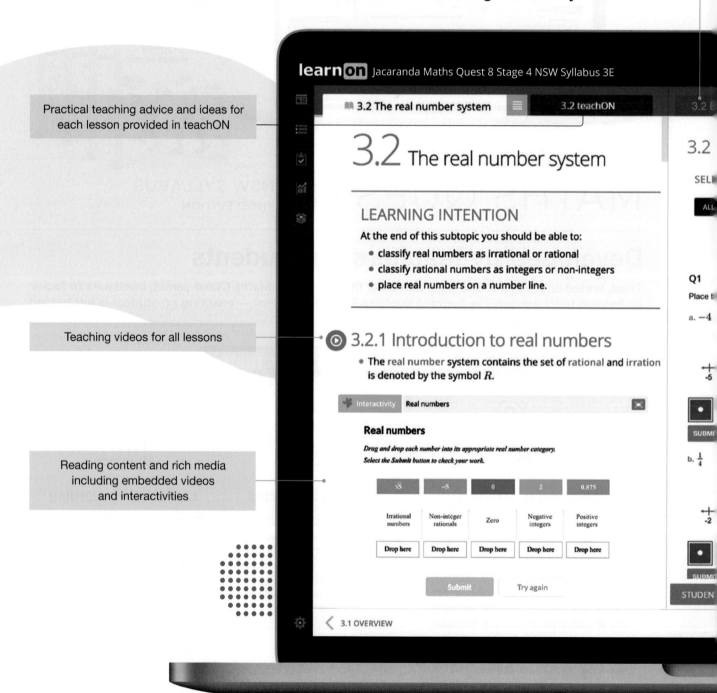

learn**On** Jacaranda Maths Quest 8 Stage 4 NSW Syllabus 3E

📖 3.2 The real number system ☰ 3.2 teachON 3.2

3.2 The real number system

LEARNING INTENTION

At the end of this subtopic you should be able to:
- classify real numbers as irrational or rational
- classify rational numbers as integers or non-integers
- place real numbers on a number line.

▶ 3.2.1 Introduction to real numbers

- The real number system contains the set of rational and irration is denoted by the symbol R.

Interactivity | Real numbers

Real numbers

Drag and drop each number into its appropriate real number category.
Select the Submit button to check your work.

| $\sqrt{5}$ | -5 | 0 | 2 | 0.875 |

| Irrational numbers | Non-integer rationals | Zero | Negative integers | Positive integers |
| Drop here | Drop here | Drop here | Drop here | Drop here |

Submit   Try again

< 3.1 OVERVIEW

3.2

SEL

ALL

Q1

Place t

a. -4

$\overset{+}{\underset{-5}{\mid}}$

☐

SUBMI

b. $\frac{1}{4}$

$\overset{+}{\underset{-2}{\mid}}$

☐

SUBMI

STUDEN

powerful learning tool, learnON

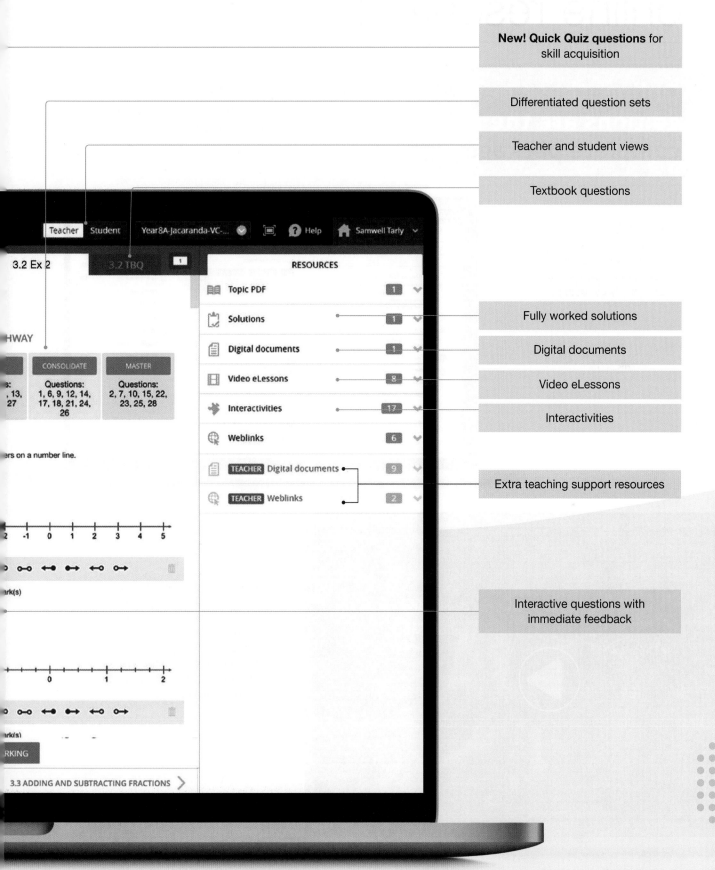

New! **Quick Quiz questions** for skill acquisition

Differentiated question sets

Teacher and student views

Textbook questions

Fully worked solutions

Digital documents

Video eLessons

Interactivities

Extra teaching support resources

Interactive questions with immediate feedback

Get the most from your online resources

Online, these new editions are the complete package

Trusted Jacaranda theory, plus tools to support teaching and make learning more engaging, personalised and visible.

Embedded interactivities and videos enable students to explore concepts and learn deeply by 'doing'.

New teaching videos for every lesson are designed to help students learn concepts by having a 'teacher at home', and are flexible enough to be used for pre- and post-learning, flipped classrooms, class discussions, remediation and more.

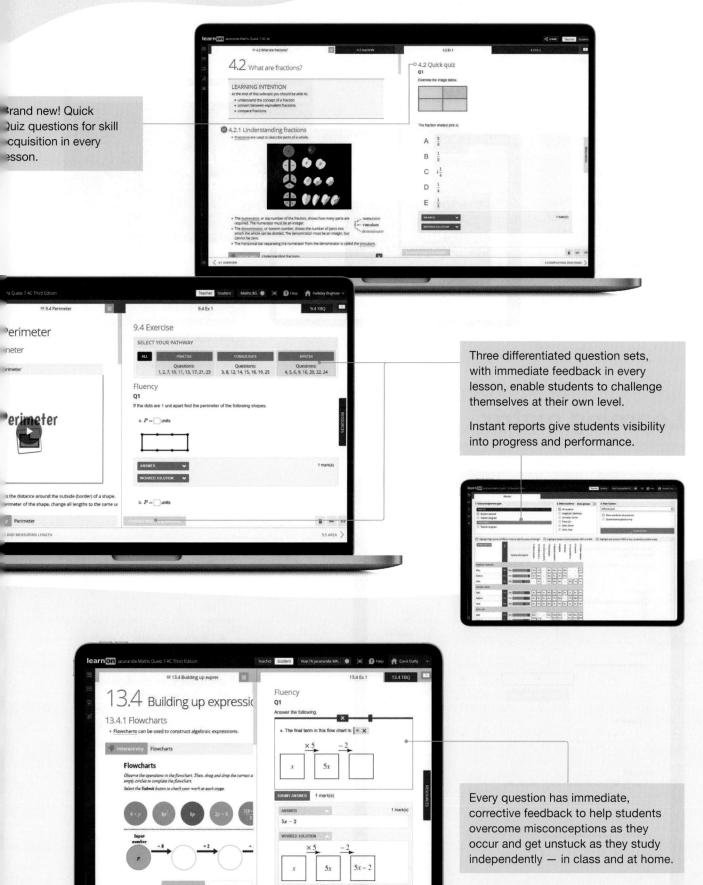

Brand new! Quick Quiz questions for skill acquisition in every lesson.

Three differentiated question sets, with immediate feedback in every lesson, enable students to challenge themselves at their own level.

Instant reports give students visibility into progress and performance.

Every question has immediate, corrective feedback to help students overcome misconceptions as they occur and get unstuck as they study independently — in class and at home.

NAPLAN Online Practice

Go online to complete practice NAPLAN tests. There are 6 NAPLAN-style question sets available to help you prepare for this important event. They are also useful for practising your Mathematics skills in general.

Also available online is a video that provides strategies and tips to help with your preparation.

Learning matrix

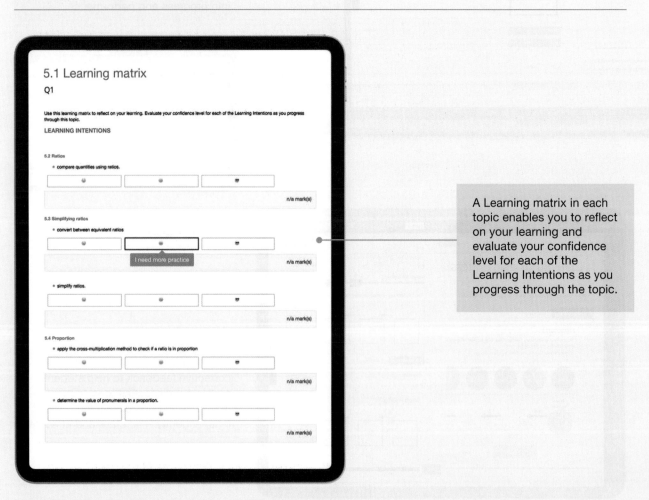

A Learning matrix in each topic enables you to reflect on your learning and evaluate your confidence level for each of the Learning Intentions as you progress through the topic.

A wealth of teacher resources

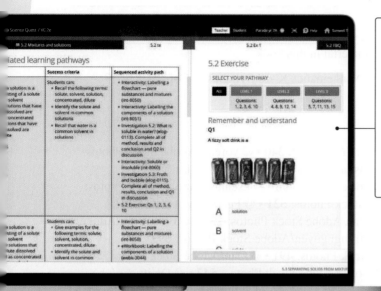

Enhanced teaching-support resources for every lesson, including:
- work programs and curriculum grids
- practical teaching advice
- three levels of differentiated teaching programs
- quarantined topic tests (with solutions)

Customise and assign

An inbuilt testmaker enables you to create custom assignments and tests from the complete bank of thousands of questions for immediate, spaced and mixed practice.

Reports and results

Data analytics and instant reports provide data-driven insights into progress and performance within each lesson and across the entire course.

Show students (and their parents or carers) their own assessment data in fine detail. You can filter their results to identify areas of strength and weakness.

Acknowledgements

The authors and publisher would like to thank the following copyright holders, organisations and individuals for their assistance and for permission to reproduce copyright material in this book.

Subject Outcomes, Objectives and Contents from the NSW Mathematics K–10 Syllabus © Copyright 2019 NSW Education Standards Authority.

Images

• © Maren Winter/Adobe Stock Photos: **45** • © Shutterstock: **76, 77, 332, 340** • © voren1/Adobe Stock Photos: **65** • © selensergen/Adobe Stock Photos: **22** • © Monkey Business/Adobe Stock Photos: **140** • © ramonefoster/Adobe Stock Photos: **62** • © elen31/Adobe Stock Photos: **206** • © kseniaso/Adobe Stock Photos: **329** • © Tiko/Adobe Stock Photos: **161** • © Drazen/Adobe Stock Photos: **521** • © Perytskyy/Adobe Stock Photos: **388** • © visoot/Adobe Stock Photos: **491** • © Christian/Adobe Stock Photos: **155** • © Elias Bitar/Adobe Stock Photos: **183** • © fizkes/Adobe Stock Photos: **316** • © trongnguyen/Adobe Stock Photos: **599** • © be free/Adobe Stock Photos: **102** • © Svitlana/Adobe Stock Photos: **491** • © One/Adobe Stock Photos: **500** • © Александр Суворов/Adobe Stock Photos: **549** • © daboost/Adobe Stock Photos: **547** • © Dmitry/Adobe Stock Photos: **23** • © Iryna/Adobe Stock Photos: **290** • © Maria Vitkovska/Adobe Stock Photos: **421** • © Roman Gorielov/Adobe Stock Photos: **120** • © arthurhidden/Adobe Stock Photos: **506** • © manassanant/Adobe Stock Photos: **14** • © Seventyfour/Adobe Stock Photos: **277** • © Jo Panuwat D/Adobe Stock Photos: **200** • © starsstudio/Adobe Stock Photos: **459** • © A.S./peopleimages.com/Adobe Stock Photos: **457** • © sacred art/Adobe Stock Photos: **266** • © Shipons Creative/Adobe Stock Photos: **141** • © bernardbodo/Adobe Stock Photos: **134** • © Still Horizon Studio/Adobe Stock Photos: **148** • © highwaystarz/Adobe Stock Photos: **395** • © WavebreakmediaMicro/Adobe Stock Photos: **163** • © Digital Vision: **235** • © Mark Dadswell/Getty Images: **206** • © Alamy Stock Photo: **193** • © 06photo/Shutterstock: **29, 470** • © 2happy/Shutterstock: **28** • © A G Baxter/Shutterstock: **552** • © A Sharma/Shutterstock: **142** • © aaronj9/Shutterstock: **430** • © AboutLife/Shutterstock: **64** • © Adam Calaitzis/Shutterstock: **276** • © Africa Studio/Shutterstock: **564** • © Aleksandr Petrunovskyi/Shutterstock: **386** • © Aleksandra Gigowska/Shutterstock: **28** • © AlenKadr/Shutterstock: **380** • © Alex Bogatyrev/Shutterstock: **515, 559** • © Alex Kravtsov/Shutterstock: **203** • © alexeisido/Shutterstock: **196** • © Ambrophoto/Shutterstock: **97** • © amphaiwan/Shutterstock: **368** • © Andrey_Popov/Shutterstock: **201, 321** • © Andriy Solovyov/Shutterstock: **182** • © Anna LoFi/Shutterstock: **205** • © antoniodiaz/Shutterstock: **157** • © ApoGapo/Shutterstock: **162** • © ArliftAtoz2205/Shutterstock: **139, 212** • © BONNINSTUDIO/Shutterstock: **382** • © canadastock/Shutterstock: **208** • © CapturePB/Shutterstock: **553** • © ChameleonsEye/Shutterstock: **135** • © charnsitr/Shutterstock: **162** • © chuyuss/Shutterstock: **472** • © CLS Digital Arts/Shutterstock: **232** • © Comaniciu Dan/Shutterstock: **162** • © conzorb/Shutterstock: **190** • © Curioso.Photography/Shutterstock: **147** • © Dejan Dundjerski/Shutterstock: **64** • © Denis Belitsky/Shutterstock: **213** • © Denphumi/Shutterstock: **40, 554** • © domnitsky/Shutterstock: **266** • © Dream79/Shutterstock: **161** • © Elena Veselova/Shutterstock: **175** • © ElenaGaak/Shutterstock: **309** • © Elnur/Shutterstock: **160** • © fotoknips/Shutterstock: **174** • © Fotokostic/Shutterstock: **193, 501** • © Frank Bach/Shutterstock: **219** • © Gena73/Shutterstock: **430** • © George Dolgikh/Shutterstock: **188** • © Gordon Bell/Shutterstock: **536** • © Ground Picture/Shutterstock: **468, 505** • © Heike Brauer/Shutterstock: **559** • © iurii/Shutterstock: **21** • © Izf/Shutterstock: **139** • © Jacek Chabraszewski/Shutterstock: **411, 436** • © jeffy11390/Shutterstock: **205** • © Jen Wolf/Shutterstock: **147** • © Jim Schwabel/Shutterstock: **544** • © Jiri Hera/Shutterstock: **295** • © Joshua Rainey Photography/Shutterstock: **221** • © Kam Hus/Shutterstock: **158** • © Kamila Starzycka/Shutterstock: **152** • © KKulikov/Shutterstock: **148** • © Kuchina/Shutterstock: **189** • © kurhan/Shutterstock: **514** • © Lane V. Erickson/Shutterstock: **337** • © Leszek Kobusinski/Shutterstock: **144** • © Lim Yong Hian/Shutterstock: **154** • © lzf/Shutterstock: **382** • © Makushin Alexey/Shutterstock: **143** • © Mamba Azul/Shutterstock: **339** • © Maridav/Shutterstock: **394** • © matimix/Shutterstock: **488** • © Maxx-Studio/Shutterstock: **297** • © Mega Pixel/Shutterstock: **244** • © Mikael Damkier/Shutterstock: Sonsedska Yuliia/Shutterstock, Andreina Nunez/Shutterstock, **173** • © Mitch Gunn/Shutterstock: **163, 203**

• © Monkey Business Images/Shutterstock: **158** • © MrJPEG/Shutterstock: **153** • © Neale Cousland/Shutterstock: **114, 181** • © Nednapa Sopasuntorn/Shutterstock: **422** • © nelson fontaine/Shutterstock: **331** • © Nerthuz/Shutterstock: **220** • © New Africa/Shutterstock: **157, 194** • © nikolpetr/Shutterstock: **437** • © nito/Shutterstock: **563** • © Oleksandr Lysenko/Shutterstock: **464** • © one photo/Shutterstock: **479** • © Osadchaya Olga/Shutterstock: **215** • © pathdoc/Shutterstock: **219** • © Paul Maguire/Shutterstock: **361** • © Phil Lowe/Shutterstock: **517** • © PhilipYb Studio/Shutterstock: **219** • © photastic/Shutterstock: **28** • © photokup/Shutterstock: **532** • © Phovoir/Shutterstock: **320** • © pio3/Shutterstock: **141** • © Pixel-Shot/Shutterstock: **299** • © PixieMe/Shutterstock: **512** • © Pixsooz/Shutterstock: **471** • © Pressmaster/Shutterstock: **112, 321** • © pzAxe/Shutterstock: **171** • © Rashevskyi Viacheslav/Shutterstock: **277** • © Rawpixel.com/Shutterstock: **275** • © robert_s/Shutterstock: **65** • © Rocksweeper/Shutterstock: **220** • © Ron Ellis/Shutterstock: **2** • © ronstik/Shutterstock: **198** • © Rudy Balasko/Shutterstock: **545** • © Ruth Black/Shutterstock: **92, 119** • © Sean Pavone/Shutterstock: **387** • © Sheila Fitzgerald/Shutterstock: **195** • © SherSS/Shutterstock: **205** • © Shots Studio/Shutterstock: **255** • © Shuang Li/Shutterstock: **488** • © Shutterstock/Andrey_Popov: **467** • © Shutterstock/baibaz: **168** • © Shutterstock/Brad J Mitchell: **581** • © Shutterstock/gornostay: **597** • © Shutterstock/Leah-Anne Thompso: **573** • © Shutterstock/Olya Maximenko: **328** • © Shutterstock/OmaPhoto: **572** • © Simone Andress/Shutterstock: **200** • © Skylines/Shutterstock: **436** • © solarseven/Shutterstock: **64** • © SpeedKingz/Shutterstock: **156, 506** • © sportpoint/Shutterstock: **558** • © STEKLO/Shutterstock: **146** • © Steve Green/Shutterstock: **548** • © Sun_Shine/Shutterstock: **149** • © SUPERGAL/Shutterstock: **557** • © suriya yapin/Shutterstock: **175** • © teez/Shutterstock: **241** • © Terence Wong/Shutterstock: **61** • © Tomsickova Tatyana/Shutterstock: **330** • © TY Lim/Shutterstock: **142** • © united photo studio/Shutterstock: **360** • © v74/Shutterstock: **150** • © Victor Moussa/Shutterstock: **540** • © Virrage Images/Shutterstock: **397** • © Vlad Teodor/Shutterstock: **309** • © Vlad1988/Shutterstock: **503** • © Wachiwit/Shutterstock: **471** • © wavebreakmedia/Shutterstock: **472** • © WHYFRAME/Shutterstock: **321** • © Woody Alec/Shutterstock: **128** • © yamix/Shutterstock: **153** • © Zety Akhzar/Shutterstock: **182** • © zhu difeng/Shutterstock: **146** • © 2020 Commonwealth of Australia as represented by the Department of Health and Aged Care: **207** • © aastock/Shutterstock: **85** • © alexdndz/Shutterstock, Marish/Shutterstock, AlexanderTrou/Shutterstock, Visual Generation/Shutterstock, Visual Generation/Shutterstock, **1, 33, 71, 127, 229, 287, 327, 393, 455, 531** • © ALEXEY GRIGOREV/Shutterstock: **10** • © Alhovik/Shutterstock: **485** • © Anastasios71/Shutterstock: **217** • © BananaStock: **480** • © Bohbeh/Shutterstock: **271** • © Brian A Jackson/Shutterstock: **512** • © burnel1/Shutterstock: **92** • © chrisdorney/Shutterstock: **254** • © Christophe BOISSON/Shutterstock: **554** • © Denton Rumsey/Shutterstock: **380** • © DibasUA/Shutterstock: **98, 117** • © Digital Stock: **235, 543** • © Double Brain/Shutterstock: **174, 184** • © Elena Burenkova/Shutterstock: **373** • © Evgeny Dubinchuk/Shutterstock: **541** • © Forefront Images/Shutterstock: **347** • © Galushko Sergey/Shutterstock: **585** • © hxdbzxy/Shutterstock: **154, 514** • © Hyung min Choi/Shutterstock: **195** • © James McDivitt/NASA: **552** • © Janaka Dharmasena/Shutterstock: **217** • © Janis Abolins/Shutterstock: **432** • © jannoon028/Shutterstock: **546** • © Jennifer Wright/John Wiley & Sons Australia: **385** • © Jethita/Shutterstock: **456** • © John Wiley & Sons Australia: **385, 508** • © karakotsya/Shutterstock: **273** • © Kari-Ann Tapp/John Wiley & Sons Australia: **86, 135, 421, 546** • © Kiev.Victor/Shutterstock: **271** • © kilukilu/Shutterstock: **359** • © Kitch Bain/Shutterstock: **72** • © klikkipetra/Shutterstock: **371** • © littlenySTOCK/Shutterstock: **9** • © LJSphotography/Alamy Stock Photo: **8** • © Macrovector/Shutterstock: **288** • © Malcolm Cross/John Wiley & Sons Australia: **240** • © Marcin Balcerzak/Shutterstock: **318** • © Margo Harrison/Shutterstock: **152** • © Marti Bug Catcher/Shutterstock: **360** • © Miceking/Shutterstock: **78** • © Michael D Brown/Shutterstock: **557** • © mkrol0718/Shutterstock: **22** • © mountain beetle/Shutterstock: **476** • © National Health and Medical Research Council: **484** • © Oleksiy Mark/Shutterstock: **488** • © oliveromg/Shutterstock: **113** • © Ollyy/Shutterstock: **116** • © phoelixDE/Shutterstock: **5** • © PhotoDisc: **338, 346, 379, 490, 514, 553** • © Photoongraphy/Shutterstock: **152** • © PictMotion/Shutterstock: **26** • © Renee Bryon/John Wiley & Sons Australia: **261** • © science photo/Shutterstock: **236** • © sirtravelalot/Shutterstock: **186** • © snake3d/Shutterstock: **203** • © Source: Adapted from healthdata.org: 2019: **485** • © Source: Wiley Art.: **337, 475, 487, 507** • © Source: © 2022 Commonwealth of Australia as represented by the Department of Health and Aged Care: **485** • © spass/Shutterstock: **58** • © steve estvanik/Shutterstock: **9** • © stevenku/Shutterstock: **234** • © Tartila/Shutterstock: **485** • © TeddyandMia/Shutterstock: **561** • © The Last Word/Shutterstock: **107** • © tirachard/Adobe Stock Photos: **519** • © tovovan/Shutterstock: **460** • © Uyanik/Shutterstock: **25** • © Vadim Sadovski/Shutterstock: **34**

Every effort has been made to trace the ownership of copyright material. Information that will enable the publisher to rectify any error or omission in subsequent reprints will be welcome. In such cases, please contact the Permissions Section of John Wiley & Sons Australia, Ltd.

NAPLAN practice

Go online to complete practice NAPLAN tests. There are 6 NAPLAN-style question sets available to help you prepare for this important event. They are also useful for practising your Mathematics skills in general.

Also available online is a video that provides strategies and tips to help with your preparation.

SET A
Calculator allowed

SET B
Non-calculator

SET C
Calculator allowed

SET D
Non-calculator

SET E
Calculator allowed

SET F
Non-calculator

1 Computation with integers

LESSON SEQUENCE

LESSON
1.1 Overview

Why learn this?

Integers are whole numbers that can be positive, negative or zero. You have been using integers all your life without even realising it. Every time you count from zero to ten or tell someone your age, you are using integers. Understanding integers is essential for dealing with numbers that you come across every day. Imagine you need to deliver something to number 30 in a particular street. A knowledge of integers will assist you to know if the house numbers are increasing or decreasing and which way you need to walk to find the house. Common uses of integers can be seen in sport scores, money transactions, heating and cooling appliances, and games. If you start taking notice, you will be amazed how often you use integers every day. Being able to add, subtract, multiply and divide integers is a critically important skill for everyday life and workplaces. Think about occupations in medicine, teaching, engineering, mechanics, hospitality, construction, design, agriculture, and sport. If you aspire to work in any of these fields, then being able to understand and compute integers will be crucial.

Hey students! Bring these pages to life online

▶ Watch videos

🧩 Engage with interactivities

A+ Answer questions and check solutions

Find all this and MORE in jacPLUS

Reading content and rich media, including interactivities and videos for every concept

Extra learning resources

Differentiated question sets

Questions with immediate feedback, and fully worked solutions to help students get unstuck

1. Determine the value of $18 - 15$.

2. In winter the midday temperature in Falls Creek is $4\,°C$ but, by midnight, the temperature drops by $6\,°C$. Calculate the temperature at midnight.

3. **MC** a. Select the highest number from the following options.

   A. -17 B. 0 C. -2 D. 10

   b. Select the lowest number from the following options.

   A. -17 B. 0 C. -2 D. 10

4. Determine the next three numbers in the following number sequence.

$$13,\ 10,\ 7,\ ...$$

5. Evaluate the following expressions.
   a. $5 + (-3)$
   b. $10 + (-18)$
   c. $-7 - 3$

6. Evaluate the following expressions.
   a. $5 - (-4)$
   b. $-3 - (-10)$
   c. $-6 - (-4)$

7. Evaluate the following.
   a. $-5 \times +3$
   b. -7×-4

8. **MC** Select the correct answer when evaluating $-(-5)^3$.

   A. -15 B. 15 C. 125 D. -125

9. Evaluate the following expressions when $p = 12$, $q = -4$ and $r = -5$.

   a. $\dfrac{p}{q}$

   b. $p + 2r - q$

10. Evaluate $36 + -2 \times -2 \div 4$.

11. Determine the missing number in the equation $-2 \times \underline{} \div -4 = -21$.

12. Write down the two possible values of m if $m^2 = 121$.

13. Determine the mean of $-6, 9, -15, 3$ and -1.

14. Two integers add to equal -3 and multiply to equal -54. Determine the answer if you divide the lower number by the higher number.

15. If $x + y + z = -5$, $\dfrac{x}{y} = -3$ and $x + z = -7$, determine the value of x.

LESSON
1.2 Adding and subtracting integers

LEARNING INTENTION

At the end of this lesson you should be able to:
- understand that integers can be negative, zero or positive
- understand that adding a negative integer is the same as subtracting a positive integer
- understand that subtracting a negative integer is the same as adding a positive integer
- add and subtract integers.

1.2.1 Integers

eles-3533

- **Integers** are positive whole numbers, negative whole numbers and zero. They can be represented on a number line.
- A group of integers is often referred to as the set Z.
$$Z = \{..., -4, -3, -2, -1, 0, 1, 2, 3, 4, ...\}$$
- Positive numbers and negative numbers have both **magnitude** (size or distance from 0) and **direction** (left or right of 0), and are often referred to as **directed numbers**.
- The number zero (0) is neither negative nor positive.

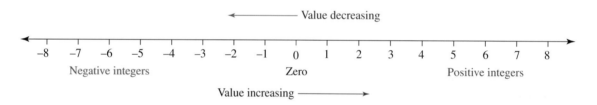

WORKED EXAMPLE 1 Representing words as integers

Write the integer suggested by each of the following descriptions.
a. The maximum temperature reached on a particular day at Mawson Station in Antarctica was 15 degrees Celsius below zero.
b. The roof of a building is 20 m above the ground.

THINK	WRITE
a. Numbers below zero are negative numbers.	**a.** −15 degrees Celsius
b. Numbers above zero are positive numbers.	**b.** +20 (or 20)

1.2.2 Addition of integers

eles-3534

- A number line can be used to add integers.
  - To add a positive integer, move to the right.
  - To add a negative integer, move to the left.

WORKED EXAMPLE 2 Adding integers using a number line

Use a number line to calculate the value of each of the following.

a. $-3 + (+2)$

b. $-3 + (-2)$

THINK

a. 1. Start at -3 and move 2 units to the right, as this is the addition of a positive integer.

WRITE

a.

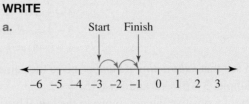

2. Write the answer.

$-3 + (+2) = -1$

b. 1. Start at -3 and move 2 units to the left, as this is the addition of a negative integer.

b.

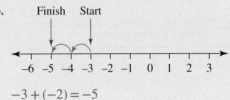

2. Write the answer.

$-3 + (-2) = -5$

DISCUSSION

Negative numbers are used to describe many real-life situations, including temperatures. What limitations would be placed on our ability to describe certain situations if we could not go below zero?

▶ 1.2.3 Subtraction of integers

eles-3535

- A number line can also be used to subtract integers.
- Consider the pattern:

$$3 - 1 = 2$$
$$3 - 2 = 1$$
$$3 - 3 = 0$$
$$3 - 4 = -1 \text{ and } 3 + (-4) = -1$$
$$3 - (-4) = 7 \text{ and } 3 + 4 = 7$$

- We can see that *subtracting a number* gives the same result as *adding its opposite*.
  For example, $3 - 5 = -2$ and $3 + (-5) = -2$.
- To subtract a positive integer, move to the left. This is the same as adding a negative integer.
- To subtract a negative integer, move to the right. This is the same as adding a positive integer.

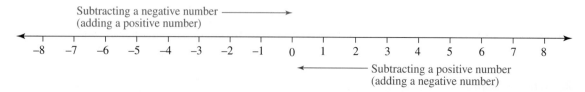

WORKED EXAMPLE 3 Subtracting integers using a number line

Use a number line to calculate the value of each of the following.

a. $-7 - (+1)$ b. $-2 - (-3)$

THINK   **WRITE**

a. 1. Subtracting an integer gives the same result   a. $-7 - (+1) = -7 + (-1)$
   as adding its opposite.

   2. Using a number line, start at -7 and move
   1 unit to the left.

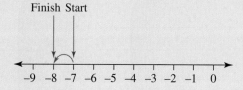

   3. Write the answer. $-7 - (+1) = -8$

b. 1. Subtracting an integer gives the same result   b. $-2 - (-3) = -2 + (+3)$
   as adding its opposite.

   2. Using a number line, start at -2 and move
   3 units to the right.

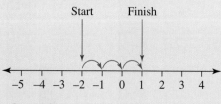

   3. Write the answer. $-2 - (-3) = 1$

Digital technology

To enter a negative number into a calculator, use the key marked ⊖. On a TI-30XB, this is positioned to the left of the ⏎ key. When pressed, this negative sign appears as a smaller, slightly raised dash compared to the subtraction symbol.

```
-2+1-(-5)              Math ▲


                           4
```

The expression $-2 + 1 - (-5)$ has been evaluated using a calculator. Observe whether the result is different with and without the brackets.

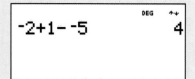

COMMUNICATING — COLLABORATIVE TASK: Walk the line

Equipment: A4 paper (if doing this activity outside, you will also need chalk and a pen).

1. Mark a number line from -12 to $+12$ on the floor using sheets of paper or on the ground using chalk.
2. Your teacher will call out a calculation, for example $7 - 13$. With a partner, find your starting point, in this case 7. Have one person walk the steps along the number line to the answer; in this case, walk 13 steps to the left, while the other remains at the starting point. Compare the start and finish positions.
3. Step 2 is repeated until each pair of students has solved a calculation.
4. As a pair, on an A4 sheet of paper, write an addition or subtraction question that can be solved using your number line. Write the answer and your names on the other side of the sheet. Hand your completed sheet to the teacher.
5. As a pair, you will now walk the line for the answer to another group's question.
6. Are there any relationships or shortcuts you can use when adding and subtracting positive and negative numbers on a number line?

Exercise 1.2 Adding and subtracting integers

learn on

1.2 Quick quiz on	1.2 Exercise

Individual pathways

■ PRACTISE	■ CONSOLIDATE	■ MASTER
1, 2, 3, 5, 8, 11, 15, 18, 21	4, 6, 9, 12, 14, 17, 19, 22	7, 10, 13, 16, 20, 23

Fluency

1. Select the integers from the following numbers.

$$3, \frac{1}{2}, -4, 201, 20.1, -4.5, -62, -3\frac{2}{5}$$

2. **WE1** Write an integer suggested by each of the following descriptions.
   a. A building lift has stopped five levels above the ground.
   b. A carpark is located on the fourth level below a building.
   c. The temperature is 23 °C.
   d. The bottom of Lake Eyre in South Australia is 15 metres below sea level.

For questions **3–10**, calculate the value of each of the expressions.

3. **WE2**
   a. $-3 + 2$
   b. $-7 + (-3)$
   c. $6 + (-7)$
   d. $-8 + (-5)$

4. a. $13 + (+6)$
   b. $12 + (-5)$
   c. $-25 + (+10)$
   d. $16 + (-16)$

5. **WE3**
   a. $7 - (+2)$
   b. $-18 - (+6)$
   c. $3 - (+8)$
   d. $11 - (+6)$

6. a. $17 - (-9)$
   b. $-28 - (-12)$
   c. $14 - (-8)$
   d. $-17 - (-28)$

7. a. $-31 + (-5)$
   b. $26 - (-10)$
   c. $-17 + (+3)$
   d. $28 - (-23)$

8. a. $17 - (+5)$
   b. $-13 - (-3)$
   c. $10 - (-3)$
   d. $-26 - (-15)$

9. a. $124 - (-26)$
   b. $-3 + (-4) - (-6)$
   c. $27 + (-5) - (-3)$
   d. $-10 + (+3) - (+6)$

10. a. $23 + (-15) - (-14)$
    b. $15 - (-4) + (-10)$
    c. $-37 - (-5) - (-10)$
    d. $-57 - (-18)$

Understanding

11. Complete the following table.

+	−8	+25	−18	+32
−6	$-8 + (-6) = -14$			
−13				
−16				
−19				

12. Complete the following table.

+	−11		+13	
	−16			
+17		36		
		18	12	
−28				−35

13. In a kitchen, some food is stored at $-18\,°C$ in a freezer and some at $4\,°C$ in the fridge. A roast is cooking in the oven at a temperature of $180\,°C$.

    a. Determine the difference in temperature between the food stored in the freezer and the food stored in the fridge.
    (*Hint:* difference = largest value − smallest value)
    b. Determine the difference in temperature between the food stored in the fridge and the roast cooking in the oven.
    c. Determine the difference in temperature between the food stored in the freezer and the roast cooking in the oven.

14. Calculate the difference between the two extreme temperatures recorded at Mawson Station in Antarctica in recent times.

Coldest = −31 °C
Warmest = 6 °C

15. Locate the button on your calculator that allows you to enter negative numbers. Use it to evaluate the following.

   a. $-458 + 157$
   b. $-5487 - 476$
   c. $-248 - (-658) - (-120)$
   d. $-42 + 57 - (-68) + (-11)$

16. Write out these equations, filling in the missing numbers.

   a. $-7 + \_\_ = 6$
   b. $8 + \_\_ = 12$
   c. $-15 - \_\_ = -26$
   d. $\_\_ - 13 + 21 = 79$

17. The following is from a homework sheet completed by a student in Year 8. Correct her work and give her a mark out of six. Make sure you include the correct answer if her answer is wrong.

   a. $-3 + (-7) = -10$
   b. $-4 - (-10) = -6$
   c. $-7 - 8 = 15$
   d. $9 - (-8) + (-7) = 10$
   e. $42 + 7 - (-11) = 60$
   f. $-17 + 4 - 8 = 21$

Communicating, reasoning and problem solving

18. Evaluate and compare the following pairs of expressions.

   a. $-4 + 1$ and $+1 - 4$
   b. $-7 + 5$ and $+5 - 7$
   c. $-8 + 3$ and $+3 - 8$
   d. What did you notice about the answers in parts a–c? A number line can be used to help you explain why this is the case.

19. Evaluate and compare the following pairs of expressions.

   a. $-2 + (-5)$ and $-(2 + 5)$
   b. $-3 + (-8)$ and $-(3 + 8)$
   c. $-7 + (-6)$ and $-(7 + 6)$
   d. What did you notice about the answers in question parts a–c? Explain why this is the case.

20. a. Explain positive and negative numbers to someone who does not know anything about them.
    b. Discuss strategies that you will use to remember how to add and subtract integers.

21. Insert the integers from −6 to +2 into the circles in the diagram shown, so that each line of three circles has a total of −3.

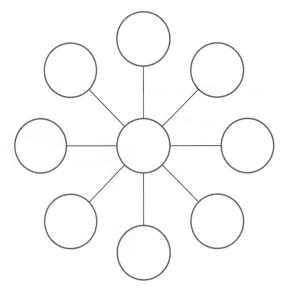

22. Igor visited the Macau Casino. The currency used in the casino is the Hong Kong dollar (HK$).

Arriving at the casino at 10 pm, Igor went straight to the baccarat table, where he won HK$270. The roulette wheel that he played next was a disaster, costing HK$340 in a very short space of time.

Igor moved to the 'vingt-et-un' table where he cleaned up, making HK$1450 after a run of winning hands.

Little did Igor know that when he sat down at the stud poker table it would signal the end of his night at the casino. He lost everything in one hand, HK$2750, not even leaving enough for a taxi to his hotel.

a. By writing a loss as a negative number and a win as a positive number, write a directed number sentence to represent this situation.
b. Determine how much money Igor had when he arrived at the casino.

23. The tip of this iceberg is 8 metres above sea level. If one-tenth of its total height is above the surface, represent the depth of its lowest point as an integer.

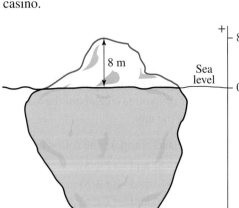

LESSON
1.3 Multiplying integers

> **LEARNING INTENTION**
>
> At the end of this lesson you should be able to:
> - understand that the product of two negative numbers is positive
> - understand that the product of a positive and a negative number is negative
> - multiply integers
> - evaluate simple indices and square roots.

▶ 1.3.1 Multiplication and powers

eles-3536

- Patterns in the answers in multiplication tables can be used to determine the product when two directed numbers are multiplied. Consider the following patterns.

$3 \times 3 = 9$	$-3 \times 3 = -9$
$3 \times 2 = 6$	$-3 \times 2 = -6$
$3 \times 1 = 3$	$-3 \times 1 = -3$

$3 \times -1 = -3$	$-3 \times -1 = 3$
$3 \times -2 = -6$	$-3 \times -2 = 6$
$3 \times -3 = -9$	$-3 \times -3 = 9$

- Looking closely at the signs of the answers in the table above, we can deduce the following rules when multiplying directed numbers.

> ## Determining the sign of the answer when multiplying integers
>
> - **When multiplying two integers with the same sign, the answer is positive.**
>
> $$+ \times + = +$$
> $$- \times - = +$$
>
> - **When multiplying two integers with different signs, the answer is negative.**
>
> $$+ \times - = -$$
> $$- \times + = -$$

WORKED EXAMPLE 4 Multiplying integers

Evaluate each of the following.
a. $-3 \times +7$ b. -8×-7

THINK	WRITE
a. The two numbers have different signs, so the answer is negative.	a. $-3 \times +7 = -21$
b. The two numbers have the same signs, so the answer is positive.	b. $-8 \times -7 = 56$ (or $+56$)

▶ 1.3.2 Powers and square roots of directed numbers

eles-3537

Powers

- Powers of a number give the number multiplied by itself multiple times.
  For example, $7^2 = 7 \times 7$ and $(-4)^3 = -4 \times -4 \times -4$
- When negative numbers are raised to a power, the sign of the answer will be:
  - positive if the power is even; for example, $(-3)^2 = -3 \times -3 = +9$.
  - negative if the power is odd; for example, $(-3)^3 = -3 \times -3 \times -3 = +9 \times -3 = -27$.

Squares

- A square number is any whole number multiplied by itself.
- All square numbers written in index form will have a power of 2.
- A square number can be illustrated by considering the area of a square with a whole number as its side length.
- Looking at the image shown, we can say that 5^2 or 25 is a square number since 5^2 or $25 = 5 \times 5$.

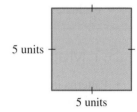

5 units

5 units

Square roots

- The square root of a number is a positive value that, when multiplied by itself, gives the original number.
- The symbol for the square root is $\sqrt{}$.
- Finding the square root of a number is the opposite of squaring the number.
  For example, if $5^2 = 25$, then $\sqrt{25} = 5$.
- Visually, the square root of a number is the side length of a square whose area is that number.
  For example, to determine $\sqrt{36}$ state the side length of a square whose area is 36.
  That is, $\sqrt{36} = 6$.

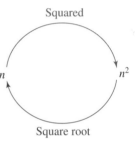

Squared

n n^2

Square root

WORKED EXAMPLE 5 Evaluating powers and square roots

Evaluate each of the following.

a. $(-5)^3$

b. **The square root of 64**

THINK

a. 1. Write the expression in expanded form.

2. Evaluate by working from left to right beginning with $-5 \times -5 = +25$.

b. Look for a positive number that, when squared, results in 64 ($8 \times 8 = 64$).

WRITE

a. $(-5)^3 = (-5) \times (-5) \times (-5)$

$= +25 \times (-5)$

$= -125$

b. $\sqrt{64} = 8$

DISCUSSION

Consider the outcome of squaring a negative number.

Investigate the value of $4^2, (-4)^2$ and -4^2. Do they give the same answer? Use the order of operations to help explain your reasoning.

Is it possible to determine the roots of negative numbers? Discuss what happens on the calculator if you try to evaluate the square root of a negative number.

 **Resources**

 Interactivity Multiplying integers (int-3704)

Exercise 1.3 Multiplying integers

learnon

1.3 Quick quiz on	**1.3 Exercise**

Individual pathways

■ PRACTISE	■ CONSOLIDATE	■ MASTER
1, 3, 6, 8, 12, 16, 18, 21	2, 4, 7, 9, 14, 15, 17, 19, 22	5, 10, 11, 13, 20, 23, 24

Fluency

1. **WE4** Evaluate each of the following.
   a. 3×4 b. 5×6 c. 7×2 d. 9×8

2. Evaluate each of the following.
   a. -2×5 b. 3×-8 c. -6×-7 d. 2×-13

3. Evaluate each of the following.
   a. -8×-6 b. -7×6 c. -10×75 d. -115×-10

4. Evaluate each of the following.
   a. -7×9 b. $+9 \times -8$ c. -11×-5 d. 150×-2

5. Use an appropriate method to evaluate the following.
   a. $-2 \times 5 \times -8 \times -10$
   b. $8 \times -1 \times 7 \times -2 \times 1$
   c. $8 \times -4 \times -1 \times -1 \times 6$
   d. $-3 \times -7 \times -2 \times -1 \times -1 \times -1$

6. Complete the following equations.
   a. $7 \times \underline{\quad} = -63$ b. $-3 \times \underline{\quad} = -21$ c. $16 \times \underline{\quad} = -32$ d. $\underline{\quad} \times -3 = 36$

7. Complete the following equations.
   a. $\underline{\quad} \times -9 = -72$ b. $\underline{\quad} \times -4 = 80$ c. $-10 \times \underline{\quad} = 60$ d. $-11 \times \underline{\quad} = 121$

8. **WE5a** Evaluate each of the following.
   a. $(-2)^3$ b. $(-3)^2$ c. $(-2)^4$ d. $(-3)^4$

9. Evaluate each of the following.
   a. $(-4)^2$ b. $(-5)^3$ c. $(-4)^4$ d. $(-5)^4$

10. If a negative number is raised to an even power, state whether the answer will be positive or negative.

11. If a negative number is raised to an odd power, state whether the answer will be positive or negative.

12. **WE5b** Evaluate the square root of each of the following numbers.
   a. 25   b. 81   c. 49   d. 121

Understanding

13. If $a = -2$, $b = -6$, $c = 4$ and $d = -3$, calculate the values of the following expressions.

    a. $a \times b \times c$ **b.** $a \times -b \times -d$ **c.** $b \times -c \times -d$

    d. $c \times -a \times -a$ **e.** $d \times -(-c)$ **f.** $a \times d \times b \times c^2$

14. For each of the following, write three possible sets of integers that will make the equation a true statement.

    a. $\_\_\_ \times \_\_\_ \times \_\_\_ = -12$ **b.** $\_\_\_ \times \_\_\_ \times \_\_\_ = 36$ **c.** $\_\_\_ \times \_\_\_ \times \_\_\_ = -36$

15. For each of the following, determine whether the result is a positive or negative value. You do not have to work out the value.

    a. $-25 \times 54 \times -47$ **b.** $-56 \times -120 \times -145$ **c.** $-a \times -b \times -c \times -d \times -e$

    Note: In part **c**, the pronumerals a, b, c, d and e are positive integers.

16. Use some examples to illustrate what happens when a number is multiplied by -1.

17. The notation $-(-3)$ is a short way of writing -1×-3.
    Write a similar expression for each of the following and then use an appropriate method to determine the answer.

    a. $-(-2)$ **b.** $-(+3)$ **c.** $-(-5)$

    d. $-(-(+5))$ **e.** $-(-(-7))$ **f.** $-(-(+4))$

Communicating, reasoning and problem solving

18. Explain, without calculating the result, why the expression below produces a negative result.
    $-2 \times -4 \times +3 \times -6 \times +4 \times +3$

19. For positive numbers, we can calculate any root of the number we like, including square root, cube root, fourth root and so on. Explain whether it is the same for negative numbers. Discuss whether we can calculate square roots, cube roots, fourth roots and so on for negative numbers. Use some examples to support your answer.

20. If the answer to $(-a)^n$ is negative, where a is an integer and n is a positive integer, establish whether a is positive or negative and whether n is odd or even. Give a reasoned explanation for your answer.

21. Eli and Luke are discussing the values of $(-5)^2$ and -5^2. Eli believes they give the same answer and Luke believes they do not. Explain who is correct.

22. If $a = -3$ and $b = -4$, evaluate $a^3 \times b^2$.

23. Evaluate $(-1)^n \times (-1)^{n+1}$ if:

    a. n is even   **b.** n is odd.

24. Consider a Mathematics test consisting of 30 multiple choice questions. A student scores 2 marks for a correct answer, -1 mark for an incorrect answer and zero marks for an unanswered question.

    Mary scores a total of 33 marks in the multiple choice section.

    Explain how she could have reached this total.

LESSON
1.4 Dividing integers

LEARNING INTENTION

At the end of this lesson you should be able to:
- understand that the quotient of two negative numbers is positive
- understand that the quotient of a positive and a negative number is negative
- divide integers.

▶ 1.4.1 Division of integers

eles-3538

- Division is the inverse or opposite operation of multiplication. We can use the multiplication facts for directed numbers to discover the division facts for directed numbers.

Multiplication fact	Division fact	Pattern
$4 \times 5 = 20$	$20 \div 5 = 4$ and $20 \div 4 = 5$	$\dfrac{\text{positive}}{\text{positive}} = \text{positive}$
$-4 \times -5 = 20$	$20 \div -5 = -4$ and $20 \div -4 = -5$	$\dfrac{\text{positive}}{\text{negative}} = \text{negative}$
$-4 \times 5 = -20$	$-20 \div 5 = -4$ and $-20 \div -4 = 5$	$\dfrac{\text{negative}}{\text{positive}} = \text{negative}$ and $\dfrac{\text{negative}}{\text{negative}} = \text{positive}$

Determining the sign of the answer when dividing integers

- **When dividing two integers with the same sign, the answer is positive.**

$$+ \div + = +$$
$$- \div - = +$$

- **When dividing two integers with different signs, the answer is negative.**

$$+ \div - = -$$
$$- \div + = -$$

- Remember that division statements can be written as fractions and then simplified.
  For example,

$$-12 \div -4 = \frac{-12}{-4}$$
$$= \frac{12 \times \cancel{-1}}{4 \times \cancel{-1}}$$
$$= \frac{3 \times \cancel{4}}{\cancel{4}}$$
$$= 3$$

WORKED EXAMPLE 6 Dividing two-digit integers

Evaluate each of the following.

a. $-56 \div 8$

b. $\dfrac{-36}{-9}$

THINK

a. The two numbers have different signs, so the answer is negative.

b. The two numbers have the same sign, so the answer is positive.

WRITE

a. $-56 \div 8 = -7$

b. $\dfrac{-36}{-9} = 4$

WORKED EXAMPLE 7 Dividing integers using long division

Evaluate the following.

a. $234 \div -6$

b. $-182 \div -14$

THINK

a. 1. Complete the division as if both numbers were positive numbers.

2. A positive number is divided by a negative number so the signs are different and therefore the sign of the answer is negative.

b. 1. Complete the division as if both numbers were positive numbers.

2. A negative number is divided by a negative number so the signs are the same and therefore the sign of the answer is positive.

WRITE

a. $$6\overline{)2\ 3^54}\ \ ^{39}$$

$234 \div -6 = -39$

b. $$14\overline{)1\ 8^42}\ \ ^{13}$$

$-182 \div -14 = 13$

COMMUNICATING — COLLABORATIVE TASK: Sign of the answer when dividing integers

Equipment: paper, pen

1. As a pair, select one of the divisions on the right.
2. Predict the sign of the answer to your division.
3. Draw a large number line, from –20 to +20, on your paper.
4. With your partner, evaluate your division and place its result in its correct position on the number line.
5. Check whether your prediction of the sign of the answer to your division is correct.
6. Work together as a class to check the rules 'When dividing two integers with the same sign, the answer is positive' and 'When dividing two integers with different signs, the answer is negative.'

$-35 \div 7$
$48 \div (-12)$
$(-24) \div (-6)$
$78 \div (-6)$
$-44 \div (-11)$
$36 \div (-3)$
$-36 \div (-9)$
$45 \div (-5)$
$51 \div (-17)$
$-57 \div (-3)$

 Resources

Interactivity Division of integers (int-3706)

Exercise 1.4 Dividing integers

learn

1.4 Quick quiz	1.4 Exercise

Individual pathways

■ PRACTISE	■ CONSOLIDATE	■ MASTER
1, 3, 5, 9, 12, 13, 16, 18, 21	2, 6, 8, 10, 14, 17, 19, 22	4, 7, 11, 15, 20, 23

Fluency

1. **WE6a** Evaluate the following.
   a. $-63 \div 9$ b. $8 \div -2$ c. $-8 \div 2$ d. $-6 \div -1$

2. Evaluate the following.
   a. $88 \div -11$ b. $0 \div -5$ c. $48 \div -3$ d. $-129 \div 3$

3. **WE6b** Evaluate each of the following.
   a. $\dfrac{-121}{-11}$ b. $\dfrac{-12}{3}$ c. $\dfrac{-36}{-12}$ d. $\dfrac{21}{-7}$

4. Evaluate the following.
   a. $-56 \div -7$ b. $184 \div -4$ c. $-55 \div -11$ d. $304 \div -8$

5. **WE7** Evaluate the following.
   a. $960 \div -8$ b. $-243 \div 9$ c. $-266 \div -7$ d. $-132 \div -4$

6. Evaluate the following.
   a. $-282 \div 6$ b. $1440 \div -9$ c. $324 \div -12$ d. $-3060 \div 17$

7. Evaluate the following.
   a. $-6000 \div -24$ b. $-2294 \div -37$ c. $4860 \div 15$ d. $-5876 \div -26$

Understanding

8. Write three different division statements, each of which has an answer of -8.

9. Copy and complete the following by placing the correct integer in the blank space.
   a. $-27 \div \underline{\quad} = -9$ b. $-68 \div \underline{\quad} = 34$ c. $72 \div \underline{\quad} = -8$ d. $-18 \div \underline{\quad} = -6$

10. Copy and complete the following by placing the correct integer in the blank space.
    a. $\underline{\quad} \div 7 = -5$ b. $\underline{\quad} \div -4 = -6$ c. $-132 \div \underline{\quad} = 11$ d. $-270 \div \underline{\quad} = 27$

11. Calculate the value of each of the following by working from left to right.
    a. $-30 \div 6 \div -5$ b. $-120 \div 4 \div -5$ c. $-800 \div -4 \div -5 \div 2$

12. If $a = -12$, $b = 3$, $c = -4$ and $d = -6$, calculate the value of each of the following expressions.
    a. $\dfrac{a}{c}$ b. $\dfrac{a}{b}$ c. $\dfrac{a}{d}$

13. If $a = -12$, $b = 3$, $c = -4$ and $d = -6$, calculate the value of each of the following expressions.

    a. $\dfrac{b}{c}$
           b. $\dfrac{b}{d}$
           c. $\dfrac{\left(\frac{a}{b}\right)}{d}$

14. If $a = -24$, $b = 2$, $c = -4$ and $d = -12$, calculate the value of each of the following expressions, by working from left to right.

    a. $a \div b \times c$
           b. $d \times c \div b \div c$
           c. $b \div c \div d \times a$

15. If $a = -24$, $b = 2$, $c = -4$ and $d = -12$, calculate the value of each of the following expressions, by working from left to right.

    a. $c \times a \div d \div b$
           b. $a \times b \div d \div d$
           c. $a \div d \times c \div b$

16. Copy and complete the following table. Divide the number on the top by the number in the left-hand column.

÷	+4	−10	+12	−8
−2	−2			
+7				
−3				
−10				

17. Copy and complete the following table. Divide the number on the top by the number in the left-hand column. Some of the integers are missing in the headings; you need to work these out, then fill in the remaining blank cells.

÷				−4
		−2		
−8	−4	3		
+6			−6	
				1

Communicating, reasoning and problem solving

18. $x \div y$ is equivalent to $\dfrac{x}{y}$.

    Use this information to simplify the following expressions. *Note:* 'Equivalent to' means 'equal to'.

    a. $x \div (-y)$
           b. $-x \div y$
           c. $-x \div (-y)$

19. The answer to $\dfrac{p \times q}{2 \times -5}$ is negative. Discuss what you can deduce about p and q.

20. The answer to $\dfrac{(-b)^3}{(-c)^4}$ is positive.

    a. Discuss what you can deduce about b.

    b. Write 3 examples to illustrate your deductions.

21. If $a = 2$ and $b = -6$, evaluate $\dfrac{(-a)^3}{(-b)^4}$.

22. Evaluate $\dfrac{(-1)^{n+1}}{(-1)^{n+2}}, n > 0$, if:

    a. $n = 1$
        b. $n = 2$
        c. n is even
        d. n is odd.

23. Evanka's last five scores in a computer game were $+6$, -9, -15, $+8$ and -4. Evaluate her average score.

LESSON
1.5 Order of operations with integers

LEARNING INTENTION

At the end of this lesson you should be able to:
- apply the order of operations to evaluate mathematical expressions.

▶ 1.5.1 Order of operations

eles-3539

- The order of operations is a set of mathematical rules used when working with directed numbers.

The order of operations

BIDMAS helps us to remember the correct order in which we should perform the various operations, working from left to right.

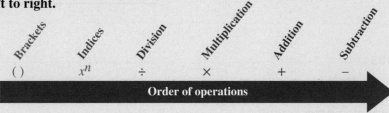

DISCUSSION

What effect do directed numbers have on the order of operations? Use some examples to help your explanation.

WORKED EXAMPLE 8 Using the order of operations

Calculate the value of each of the following.

a. $54 \div -6 + 8 \times -9 \div -4$

b. $-8 \div 2 + (-2)^3$

THINK	WRITE
a. 1. Write the expression.	a. $54 \div -6 + 8 \times -9 \div -4$
2. There are no brackets or powers, so working from left to right, complete all multiplication and division operations before any addition and subtraction.	
First operation: $54 \div -6 = -9$	$= -9 + 8 \times -9 \div -4$
Second operation: $8 \times -9 = -72$	$= -9 - 72 \div -4$
Third operation: $-72 \div -4 = +18$	$= -9 + 18$
Last operation: $-9 + 18 = 9$	$= 9$
3. Write the answer.	$54 \div -6 + 8 \times -9 \div -4 = 9$
b. 1. Write the expression.	b. $-8 \div 2 + (-2)^3$
2. Evaluate the cubed term.	$= -8 \div 2 + -8$
3. Complete the division.	$= -4 + -8$
4. Complete the addition.	$= -12$
5. Write the answer.	$-8 \div 2 + (-2)^3 = -12$

Digital technology

To calculate the square of a number, such as 8^2, press 8 and then press the $\boxed{x^2}$ key.

To calculate other powers such as $(-2)^3$, type the number including any brackets and then press the $\boxed{\wedge}$ or $x^{\square}$ key, followed by the power. *Note:* It is important to use brackets around powers when using digital technology otherwise you will receive a different answer.

The first screen demonstrates the calculation for Worked example 8b, and the second screen illustrates the importance of brackets

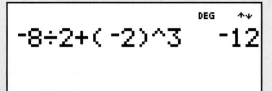

 Resources

 Video eLesson BIDMAS (eles-1883)

 Interactivity Order of operations (int-3707)

Exercise 1.5 Order of operations with integers

learn**on**

| **1.5 Quick quiz** on | **1.5 Exercise** |

Individual pathways

■ PRACTISE	■ CONSOLIDATE	■ MASTER
1, 3, 6, 8, 10, 13	2, 4, 7, 9, 11, 14	5, 12, 15

Fluency

1. **WE8a** Calculate the values of the following expressions.
   a. $-4-6-2$
   b. $-4\times2+1$
   c. $8\div(2-4)+1$
   d. $7-(3-1)+4$

2. Calculate the values of the following expressions.
   a. $6\times(4+1)$
   b. $-3-40\div8+2$
   c. $-4+5-6-7$
   d. $-5\times12+2$

3. Calculate the values of the following expressions.
   a. $12\div(2-4)-6$
   b. $13-(4-6)+2$
   c. $7\times(6+2)$
   d. $-6-36\div9+3$

4. Calculate the values of the following expressions.
   a. $-3+15-26-27$
   b. $-8\times11+12$
   c. $52\div(-9-4)-8$
   d. $23-(16-4)+7-3$

5. Calculate the values of the following expressions.

   a. $15 \times (-6 + 2)$ b. $-6 - 64 \div -16 + 8$

   c. $-3 \times -4 \times -1 \times 5$ d. $-6 \times (-13 + 5) + -4 + 2$

6. **WE8b** Evaluate each of the following.

   a. $-7 + 6 \times (-2)^3$ b. $(-9)^2 - 15 + 3$ c. $(-63 \div -7) \times (-3 + -2)^2$

7. Evaluate each of the following.

   a. $(-3)^3 - 3 \times -5$ b. $-5 \times -7 - [5 + (-8)^2]$ c. $[(-48 \div 8)^2 \times 36] \div -4$

Understanding

8. A class of Year 8 students were given the following question to evaluate.

$$4 + 8 \div -(2)^2 - 7 \times 2$$

   a. Several different answers were obtained, including -8, -12 and -17. Determine which one of these is the correct answer.

   b. Explain your answer by applying BIDMAS.

   c. Using only brackets, change the question in two ways so that the other two answers would be correct.

9. In a particular adventure video game, a player loses and gains points based on who or what they come in contact with during the game. See the list shown of the number of 'hit' points associated with each contact. Use the table to calculate the number of points the player has at the end of each round of the game.

Character	'Hit' points
Balrog	−100
Troll	−10
Orc	−5
Goblin	−2
Gnome	−1
Healing potion	+20
Cleric	+50

Round number	Points at the start of the round	Contacts during the round	Points at the end of the round
1	100	20 gnomes, 10 goblins and 3 healing potions	
2		3 gnomes, 5 goblins, 6 orcs and 5 healing potions	
3		3 orcs, 6 trolls and a cleric	
4		5 trolls, 1 balrog and a cleric	

10. Discuss the effect that directed numbers have on the order of operations.

Communicating, reasoning and problem solving

11. A viral maths problem posted on a social media site asks people to determine the answer to $6 \div 2(1 + 2)$. Most people respond with an answer of 1.

    a. Explain why the answer of 1 is incorrect and determine the correct answer.
    b. Insert an extra set of brackets in the expression so that 1 would be the correct answer.

12. Two numbers p and q have the same numerical value but the opposite sign; that is, one is positive and the other negative.

    a. If $-3 \times p + 4 \times q$ is positive, discuss what can be said about p and q.
    b. Test your answer to part **a** if the numerical values of p and q, written as $|p|$ and $|q|$, are both equal to 7.

13. Model each situation with integers, and then find the result.

    a. Jemma has $274 in the bank, and then makes the following transactions: 2 withdrawals of $68 each and 3 deposits of $50 each.
    b. If 200 boxes of apples were each 3 short of the stated number of 40 apples, evaluate the overall shortfall in the number of apples.
    c. A person with a mass of 108 kg wants to reduce their mass to 84 kg in 8 months. Determine the average mass reduction needed per month.

14. A classmate is recording the weather during July for a school project and wants your help to calculate the information. He records the following data for one week.

	Mon	Tue	Wed	Thu	Fri	Sat	Sun
Max (°C)	12.2	14.5	16.7	12.8	11.3	7.2	−0.3
Min (°C)	3.0	2.1	4.6	3.2	6.4	−2.9	−6.0

For the following, round all answers correct to 1 decimal place.

    a. Determine the difference between the lowest temperature and the highest temperature recorded during this week.

    b. Your classmate tries to predict temperatures and says that the minimum temperature $= \frac{1}{4} \times$ maximum temperature. He says this is the same as dividing by -4. Explain why this may not be correct.

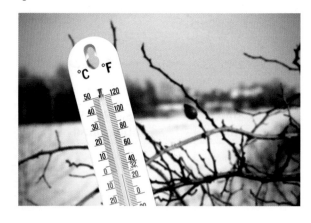

    c. He then tries to make another prediction, which involves taking the square root of the maximum temperature. Explain why this might not be a good idea.

    d. On the first Monday after this week, the temperature rises by 6.5 °C from Sunday's minimum temperature. It then drops by 3.2 °C overnight and rises by 8.9 °C on Tuesday. Evaluate the minimum temperature on Monday night and the maximum temperature on Tuesday.

e. His last prediction involves subtracting 6 from the maximum temperature, then dividing by 2 to predict the minimum temperature. Calculate the predicted values for the minimum temperature for each of his three methods, then copy and fill the provided table.

	Mon	Tue	Wed	Thu	Fri	Sat	Sun
Max (°C)	12.2	14.5	16.7	12.8	11.3	7.2	−0.3
Min (°C)	3.0	2.1	4.6	3.2	6.4	−2.9	−6.0
Max $\times \dfrac{1}{4}$							
$\sqrt{\text{Max}}$							
$\dfrac{(\text{Max} - 6)}{2}$							

f. Using your answer to **e**, discuss which method may be most accurate.

15. A frog and a snail are climbing an empty vertical pipe. The snail is 30 cm from the top of the pipe while the frog is 30 cm below the snail.

At the start of the first hour, the frog climbs 40 cm up the pipe and rests. The snail crawls 20 cm during this time, then also rests.

While they are resting, both the frog and the snail slip back down the pipe. The frog slips back 20 cm while the snail slips back 10 cm.

At the start of the second hour, the frog and the snail set off again and repeat the same process of climbing and resting until they reach the top of the pipe.

a. Explain which reaches the top of the pipe first, the frog or the snail.
b. Determine during which hour they reach the top of the pipe.

LESSON
1.6 Review

1.6.1 Topic summary

Integers

Integers are positive and negative whole numbers including 0.
- The integers −1, −2, −3, … are called negative integers.
- The integers 1, 2, 3, … are called positive integers.
- 0 is neither positive nor negative.

Addition of integers

A number line can be used to add integers.
- To add a positive integer, move to the right.
  e.g. $-4 + 1 = -3$
- To add a negative integer, move to the left.
  e.g. $5 + (-3) = 2$

Move right when adding a positive integer ⟶

⟵ Move left when adding a negative integer

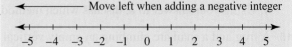

INTEGERS

Subtraction of integers

- Subtracting a number gives the same result as *adding its opposite*.

  e.g. $2 - (+5) = 2 + (-5) = -3$ or
  $-6 - (-11) = -6 + (+11) = 5$

Move right when subtracting a negative integer ⟶

⟵ Move left when subtracting a positive integer

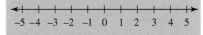

Division of integers

- When dividing integers with the *same sign*, the answer is *positive*.
  e.g. $-20 \div -4 = 5$ or $24 \div 6 = 4$
- When dividing integers with *different signs*, the answer is *negative*.
  e.g. $10 \div -5 = -2$ or $\dfrac{-35}{7} = -5$

Multiplication of integers

- When multiplying integers with the *same sign*, the answer is *positive*.
  e.g. $-4 \times -6 = 24$ or $7 \times 8 = 56$
- When multiplying integers with *different signs*, the answer is *negative*.
  e.g. $-5 \times 2 = -10$ or $9 \times -3 = -27$

Order of operations

- The order of operations is a set of rules we must follow so that we all have a common understanding of mathematical operations.
- The set order in which we calculate problems is:
  1. **Brackets**   ( ) or [ ]
  2. **Indices or roots** x^n or $\sqrt[n]{x}$
  3. **Division and Multiplication** (working left to right) ÷ or ×
  4. **Addition and Subtraction** (working left to right) + or −
- The acronym **BIDMAS** can be used to remember the correct order of operations.

Powers

- A negative number raised to an odd power will produce a negative number.
  e.g. $(-4)^3 = -64$
- A negative number raised to an even power will produce a positive number.
  e.g. $(-4)^2 = 16$

1.6.2 Project

Many board games engage the players in racing each other to the end of the board. Snakes and Ladders is an example of this style of game. You can climb ladders to get to the end quicker, but sliding down a snake means you get further away from the end.

You are going to make a board game that will help you to practise addition and subtraction of directed numbers. This game is played with two dice and is a race to the end of the number line provided on the board.

The diagram below shows part of a sample game, and can be used to explain the rules and requirements for your board game.

10	9	8	7	6
				5
				4

−2	−1	0	1	2	3
−3					

−4	−5	−6	−7	−8	−9	−10

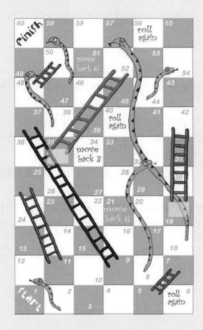

Two dice are to be used in this board game. Die 1 is labelled with N three times and P three times, and die 2 is labelled with the integers −1, −2, −3, +1, +2 and +3. (Stickers with these labels can be placed over two standard dice.)

- Rolling an N means you face the *negative* numbers; rolling a P means you face the *positive* numbers.
  - Imagine that you are at 0, and that you roll an N and +2. N means that you face the negative numbers; +2 means that you move forward 2 places.

- If you roll a P and +1, this means that you face the positive numbers and move forward 1 place.
- If you roll a P and −2, this means that you face the positive numbers and move *backward* 2 places.

1. For the three examples listed above, state which square you would end up in if you started at 0 each time.
2. Suppose that your first five turns at this game produced the following results on the dice:
   P and +1, P and +3, P and −2, N and −1, N and +2.
   If you started at 0, where did you end up after these five turns?

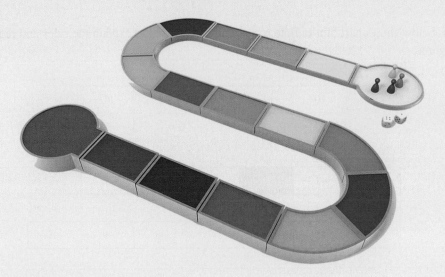

Your task is to design a board game similar to the one used in the example. The requirements for the game are listed below.

- The game is to be played with the two dice described earlier.
- Two or more players are required, taking turns to throw the dice.
- Start at 0 in the centre of the board. The race is on to get to either end of the board. The first person to reach an end is the winner.
- For 0 to be at the centre, your board will require an odd number of columns.
- At least 8 squares must have specific instructions — for example, 'go back 5 squares' or 'miss a turn'. You may even wish to include 'snakes and ladders' style obstacles.
- Use a sensible number of squares. If there are very few squares, the game will end too quickly; if there are many squares, they will be very small.

3. Work with a partner to make the two required dice and to design your board. Play with others to test the game and make necessary improvements if required. Be certain that all the requirements of the game are included.

 Resources

Interactivities Crossword (int-2723)

Sudoku puzzle (int-3182)

Fluency

1. State whether the following statement is true or false. The number -2.5 is called an integer.

2. State whether the following statement is true or false. $-6 < -2$

3. List the integers between -11 and -7.

4. Arrange these numbers in ascending order: $7, 0, -3, 10, -15$.

5. Calculate the value of each of the following.
   a. $-6 + (-8)$
   b. $16 - (-5)$
   c. $-3 - (+7) + (-2)$
   d. $-1 - (-5) - (+4)$

6. Write out the following equations and fill in the missing numbers.
   a. $7 - \underline{} = -14$
   b. $-19 + \underline{} = 2$
   c. $\underline{} - 13 - (-12) = 10$
   d. $-28 - \underline{} = -17$

Understanding

7. **MC** Select the correct statement from the following.
   A. Multiplying an even number of negative numbers together gives a negative answer.
   B. Squaring a negative number gives a negative answer.
   C. Dividing a negative number by another negative number gives a positive answer.
   D. Adding two negative numbers together gives a positive answer.

8. Evaluate each of the following.
   a. -12×-5
   b. $-(-10) \times 3 \times -2$
   c. $-24 \div -3$
   d. $-48 \div -4 \div -3$

9. Evaluate each of the following.
   a. $6 \times -3 \div -2$
   b. $-36 \div 3 \div -4 \times -9$
   c. $-8 \times -3 - (4 - -1) + -63 \div 7$
   d. $-9 + -9 \div -9 \times -9 - -9$

Communicating, reasoning and problem solving

10. Give an example of two numbers that fit each description that follows.
    If no numbers fit the description, explain why.
    a. Both the sum and the product of two numbers are negative.
    b. The sum of two numbers is positive and the quotient is negative.
    c. The sum of two numbers is 0 and the product is positive.

11. On a test, each correct answer scores 5 points, each incorrect answer scores -2 points and each question left unanswered scores 0 points.
    a. Suppose a student answers 16 questions correctly and 3 questions incorrectly, and does not answer 1 question. Write an expression for the student's score and determine the score.
    b. Suppose you answered all 20 questions on the test. What is the greatest number of questions you can answer incorrectly and still get a positive score? Explain your reasoning.

12. Write the following problem as an equation using directed numbers and determine the answer.

You have $25 and you spend $8 on lollies. You then spend another $6 on lunch.

A friend gives you $5 to buy lunch, which comes to only $3.50. You then find another $10 in your pocket and buy an ice-cream for $3.

Evaluate the amount of money left in total before you return your friend's change from lunch.

13. Write the following problem as an equation using directed numbers and determine the answer.
Two friends are on holiday; one decides to go skydiving and the other decides to go scuba diving. If the skydiving plane climbs to 4405 m above sea level, and the scuba diver goes to the ocean floor, which is 26 m below the surface, determine the vertical distance between the two friends.

14. You receive several letters in the mail: two gift cards worth $100 each, three bills worth $75 each and a voucher for $20. You want to evaluate the amount of money you end up with.
   a. Represent the situation using directed numbers.
   b. Solve the problem.

15. You earn $150 each time you work at the local races. If you work at three race meetings in one month, determine how much you earn that month.
   a. Represent the situation using directed numbers.
   b. Solve the problem.

16. For your birthday, you get three gift cards worth $40 each. Also, your brother gives you four movie vouchers worth $10 each. In total, determine how much more money you have after your birthday.
   a. Represent the situation using directed numbers.
   b. Solve the problem.

17. In science, directed numbers are often used to describe a direction or an increase or decrease in a measurement.

Directed numbers can describe the distance of an object from a reference point (known as the displacement, d, of the object). For example, if you are 200 km west of a town, and west is defined as a negative direction, you are -200 km from the town.

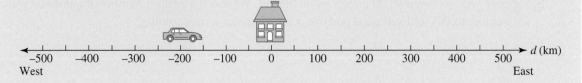

a. If a car travels 150 km in the easterly direction from −200 km, describe the displacement of the car from the town.

b. If a car travels from 300 km east of the town, describe the displacement of the car after it has travelled 450 km in the easterly direction.

18. Directed numbers can describe the direction in which an object is travelling. For example, travelling towards the east is often defined as the positive direction and towards the west as the negative direction. A car travelling west at 100 km/h goes at −100 km/h. Scientists use the term *velocity*, *v*, to mean a speed in a particular direction.

   a. If a car travels past a town at −100 km/h, determine where it will be in 2 hours' time.

   b. If a car goes past a town while travelling at −100 km/h, determine where the car was an hour ago.

on To test your understanding and knowledge of this topic, go to your learnON title at www.jacplus.com.au and complete the **post-test**.

Answers

Topic 1 Computation with integers

1.1 Pre-test

1. 3
2. $-2\,°C$
3. a. D   b. A
4. $4, 1, -2$
5. a. 2   b. -8 c. -10
6. a. 9   b. 7   c. -2
7. a. -15 b. 28
8. C
9. a. -3 b. 6
10. 37
11. -42
12. -11 and 11
13. -2
14. -1.5
15. -6

1.2 Adding and subtracting integers

1. $3, -4, 201, -62$
2. a. $+5$ b. -4 c. $+23$ d. -15
3. a. -1 b. -10 c. -1 d. -13
4. a. 19   b. 7   c. -15 d. 0
5. a. 5   b. -24 c. -5 d. 5
6. a. 26   b. -16 c. 22   d. 11
7. a. -36 b. 36   c. -14 d. 51
8. a. 12   b. -10 c. 13   d. -11
9. a. 150   b. -1 c. 25   d. -13
10. a. 22   b. 9   c. -22 d. -39

11.

+	−8	25	−18	32
−6	−14	19	−24	26
−13	−21	12	−31	19
−16	−24	9	−34	16
−19	−27	6	−37	13

12.

+	−11	19	13	−7
−5	−16	14	8	−12
17	6	36	30	10
−1	−12	18	12	−8
−28	−39	−9	−15	−35

13. a. $22\,°C$ b. $176\,°C$ c. $198\,°C$
14. $37\,°C$
15. a. -301 b. -5963 c. 530   d. 72
16. a. 13   b. 4   c. 11   d. 71
17. a. Correct   b. Incorrect; 6   c. Incorrect; -15
    d. Correct   e. Correct   f. Incorrect; -21

18. a. -3 b. -2
    c. -5 d. The answers are all the same.
19. a. $-2 + -5 = -7; -(2 + 5) = -7$
    b. $-3 + -8 = -11; -(3 + 8) = -11$
    c. $-7 + -6 = -13; -(7 + 6) = -13$
    d. For each pair of expressions, the answers are the same: each number is negative, so putting brackets around the sum and then attaching a negative sign gives the same outcome.
20. a. Individual responses will vary, but should include information about: the difference between positive and negative numbers including real-life examples; how using a number line could be used to help understand positive and negative numbers; common words used to indicate a positive or negative number; and where positive and negative numbers sit on a number line.
    b. Individual responses will vary, but should include tips about: which direction to move in on a number line when adding a positive integer, adding a negative integer, subtracting a positive integer and subtracting a negative integer. Your response may also include how these four different scenarios can be simplified into two groups.

21.

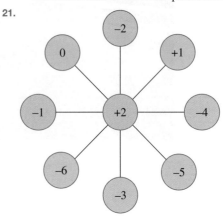

22. a. $270 - 340 + 1450 - 2750$
    b. HK$1370
23. -72 metres

1.3 Multiplying integers

1. a. 12   b. 30   c. 14   d. 72
2. a. -10 b. -24 c. 42   d. -26
3. a. 48   b. -42 c. -750 d. 1150
4. a. -63 b. -72 c. 55   d. -300
5. a. -800 b. 112   c. -192 d. 42
6. a. -9 b. 7   c. -2 d. -12
7. a. 8   b. -20 c. -6 d. -11
8. a. -8 b. 9   c. 16   d. 81
9. a. 16   b. -125 c. 256   d. 625
10. Positive
11. Negative
12. a. 5   b. 9   c. 7   d. 11
13. a. 48   b. -36 c. 72
    d. 16   e. -12 f. -576

14. There are many possible answers. A sample response is:
a. $-2 \times 2 \times 3$ b. $3 \times 3 \times 4$ c. $-2 \times 3 \times 6$

15. a. Positive   b. Negative   c. Negative

16. If a positive number is multiplied by -1, the number becomes negative.
If a negative number is multiplied by -1, the number becomes positive.

17. a. 2   b. -3 c. 5   d. 5   e. -7 f. 4

18. Because there is an odd number of negatives.

19. It is possible to find only odd-numbered roots for negative numbers. This means it is not possible to find the square root of a negative number. For example, $\sqrt[3]{-125} = -5$ but $\sqrt{-25}$ is not a real number.

20. $-a$ must be negative for the answer to be negative, so a must be positive. The power must be odd to give a negative result, so n is odd.

21. $(-5)^2 = 25$
$-5^2 = -25$
Luke is correct.

22. -432

23. a. -1 b. -1

24. One of the following: 17 correct, 1 incorrect; 18 correct, 3 incorrect; 19 correct, 5 incorrect; 20 correct, 7 incorrect; 21 correct, 9 incorrect

1.4 Dividing integers

1. a. -7 b. -4 c. -4 d. 6

2. a. -8 b. 0   c. -16 d. -43

3. a. 11   b. -4 c. 3   d. -3

4. a. 8   b. -46 c. 5   d. -38

5. a. -120 b. -27 c. 38   d. 33

6. a. -47 b. -160 c. -27 d. -180

7. a. 250   b. 62   c. 324   d. 226

8. There are many possible answers; examples include $-16 \div 2$, $-80 \div 10$, $-24 \div 3$.

9. a. 3   b. -2 c. -9 d. 3

10. a. -35 b. 24   c. -12 d. -10

11. a. 1   b. 6   c. -20

12. a. 3   b. -4 c. 2

13. a. $-\dfrac{3}{4}$ b. $-\dfrac{1}{2}$ c. $\dfrac{2}{3}$

14. a. 48   b. -6 c. -1

15. a. -4 b. $-\dfrac{1}{3}$ c. -4

16.

÷	4	−10	12	−8
−2	−2	5	−6	4
7	$\dfrac{4}{7}$	$\dfrac{-10}{7}$	$\dfrac{12}{7}$	$\dfrac{-8}{7}$
−3	$-\dfrac{4}{3}$	$\dfrac{10}{3}$	−4	$\dfrac{8}{3}$
−10	$-\dfrac{2}{5}$	1	$-\dfrac{6}{5}$	$\dfrac{4}{5}$

17.

÷	32	−24	−36	−4
12	$\dfrac{8}{3}$	−2	−3	$-\dfrac{1}{3}$
−8	−4	3	$\dfrac{9}{2}$	$\dfrac{1}{2}$
6	$\dfrac{16}{3}$	−4	−6	$-\dfrac{2}{3}$
−4	−8	6	9	1

18. a. $-\dfrac{x}{y}$ b. $-\dfrac{x}{y}$ c. $\dfrac{x}{y}$

19. They have the same sign: either both positive or both negative.

20. a. It is negative.
b. Sample responses can be found in the worked solutions in the online resources.

21. $-\dfrac{1}{162}$

22. a. -1 b. -1 c. -1 d. -1

23. -2.8

1.5 Order of operations with integers

1. a. -12 b. -7 c. -3 d. 9

2. a. 30   b. -6 c. -12 d. -58

3. a. -12 b. 17   c. 56   d. -7

4. a. -41 b. -76 c. -12 d. 15

5. a. -60 b. 6   c. -60 d. 46

6. a. -55 b. 69   c. 225

7. a. -12 b. -34 c. -324

8. a. -12
b. $4 + 8 \div -(2)^2 - 7 \times 2$
$= 4 + 8 \div -4 - 7 \times 2$
$= 4 + (-2) - 14$
$= -12$
c. $(4 + 8) \div -(2)^2 - 7 \times 2 = -17$
$4 + 8 \div (-(2))^2 - 7 \times 2 = -8$

9.

Round number	Points at the start of the round	Contacts during the round	Points at the end of the round
1	100	20 gnomes, 10 goblins and 3 healing potions	120
2	120	3 gnomes, 5 goblins, 6 orcs and 5 healing potions	177
3	177	3 orcs, 6 trolls and a cleric	152
4	152	5 trolls, 1 balrog and a cleric	52

10. Directed numbers are placed in brackets and applied first in operations. See the BIDMAS rule.

11. a. An answer of 1 is incorrect because the order of operations wasn't applied correctly.

$$6 \div 2(1 + 2) = 6 \div 2 \times 3$$
$$= 3 \times 3$$
$$= 9$$

The correct answer is 9.

b. $6 \div \left[2(1 + 2)\right] = 1$

12. a. $p < 0, q > 0$
b. $p = -7$ and $q = 7$: $-3 \times -7 + 4 \times 7 = 49$

13. a. $288   b. 600   c. 3 kg

14. a. $22.7\,°C$

b. Multiplying by $\dfrac{1}{4}$ is not the same as dividing by -4.

Multiplying by $\dfrac{1}{4}$ is the same as dividing by 4.

c. This is not a good idea because you cannot take the square root of a negative number, such as Sunday's maximum temperature.

d. Monday maximum $= 0.5\,°C$ and minimum $= -2.7\,°C$; Tuesday maximum $= 6.2\,°C$.

e.

	Mon	Tue	Wed	Thu	Fri	Sat	Sun
Max (°C)	12.2	14.5	16.7	12.8	11.3	7.2	−0.3
Min (°C)	3.0	2.1	4.6	3.2	6.4	−2.9	−6.0
Max $\times \dfrac{1}{4}$	3.1	3.6	4.2	3.2	2.8	1.8	−0.1
$\sqrt{\text{Max}}$	3.5	3.8	4.1	3.6	3.4	±2.7	N/A
$\dfrac{(\text{Max} - 6)}{2}$	3.1	4.3	5.4	3.4	2.7	0.6	−3.2

f. The method of multiplying the maximum temperature by $\dfrac{1}{4}$ gave the most accurate prediction of minimum temperature, but none of the methods is very accurate. Any prediction method should be tested over a number of weeks before deciding if it is accurate enough.

15. a. The frog and the snail get out of the pipe at exactly the same time.

b. They make it out of the pipe during the second hour.

Project

1. a. -2 b. 1   c. -2

2. 1

3. Designs will vary. Sample response is provided.
Instructions for making your own dice:
i. Print out as many copies of the die template as you need to make two dice that you consider adequate.
ii. Cut the die out along its outside border.
iii. Fold the die along each of the six sides (along the lines).

iv. With small pieces of clear tape, tape each edge to the adjacent edge. You should get a cube.

v. Roll the die to see if it works, then play the game. Your die may be a bit lopsided, but it should work. You might have to make several dice to get two that you consider adequate

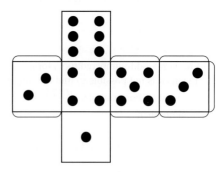

1.6 Review questions

1. False

2. True

3. $-10, -9, -8$

4. $-15, -3, 0, 7, 10$

5. a. -14 b. 21   c. -12 d. 0

6. a. 21   b. 21   c. 11   d. -11

7. C

8. a. 60   b. -60 c. 8   d. -4

9. a. 9   b. -27 c. 10   d. -9

10. a. There are many possible answers.
A correct answer will contain a positive and a negative number, with the negative number having a greater size than the positive number.

b. There are many possible answers.
A correct answer will contain one positive and one negative number, with the positive number having a greater size than the negative number.

c. Not possible

11. a. An expression for the student's score:
$16 \times 5 + 3 \times -2 + 1 \times 0 = 74$

b. 14

12. $19.50

13. $4431\,m$

14. a. $2 \times 100 - 3 \times 75 + 20$ b. $-$5

15. a. 3×150 b. $450

16. a. $3 \times 40 + 4 \times 10$ b. $160

17. a. $-50\,km$ b. $750\,km$

18. a. $-200\,km$ b. $100\,km$

2 Index laws

LESSON SEQUENCE

LESSON
2.1 Overview

Why learn this?

Indices (the plural of index) are an abbreviated way of expressing a number that has been multiplied by itself. They are also referred to as powers. For example, $4 \times 4 \times 4 \times 4 \times 4$ can be written in index form as 4^5. This is much simpler and neater than writing out all the repeated multiplication. Indices are very valuable in real life when dealing with very large or very small numbers, or with a number that is continually multiplied by itself. Indices are used in many parts of our modern technological world. For example, indices are used in computer game physics; investors use indices to track and measure share market growth; and engineers use indices to calculate the strength of materials used in buildings.

Biological scientists use indices to measure the growth and decay of bacteria. Indices are used by astrophysicists to calculate the distance, temperature and brightness of celestial objects. Did you know that the distance from Earth to the Sun is 150 000 000 km or 1.5×10^8 km? Using indices to represent a large number is helpful, as you do not have to write too many zeros. You are likely to come across indices in other subjects such as Science and Geography, so it is important that you understand what indices are, and that you can use them to perform simple calculations.

Hey students! Bring these pages to life online

▶ **Watch** videos

🧩 **Engage with** interactivities

A+ **Answer questions** and check solutions

Find all this and MORE in jacPLUS

Reading content and rich media, including interactivities and videos for every concept

Extra learning resources

Differentiated question sets

Questions with immediate feedback, and fully worked solutions to help students get unstuck

1. Write $3 \times 3 \times 3 \times 3 \times 3$ in index form.

2. Evaluate $3^2 \times 2^3$.

3. Simplify the fraction $\dfrac{42a^2}{48a^2}$.

4. Simplify $2c^3 \times 5c^4$.

5. **MC** From the list given, select the correct simplification of $\dfrac{c^{12}}{c^4}$.

   A. 3
   B. $\dfrac{c^3}{c^1}$
   C. c^3
   D. c^8

6. Simplify the fraction $\dfrac{18d^6}{24d^3}$.

7. Evaluate 7^0.

8. Show that $\dfrac{3m^5}{4m^3} \times \dfrac{8m^4}{12m^6} = \dfrac{1}{2}$.

9. Simplify the following expressions. Write your answers in simplified index form.
   a. $\left(7^5\right)^3$
   b. $(2w^5)^4$
   c. $(3p^5)^2 \times (2p^2)^0 \times 5p$

10. Simplify the following.

    a. $(3ab^2)^2 \times (5a^3b)$
    b. $\dfrac{(3a^2b)^2}{6ab}$

11. **MC** Select the correct simplification of $4f^3 + 3f^2 - 2f^3$.
    A. $5f^2$
    B. $3f^4$
    C. $3f^2 + 2$
    D. $2f^3 + 3f^2$

12. Simplify the following expression.

    $\left(\dfrac{3g^5}{4h^3}\right)^2$

13. Determine if $((2^3)^2)^3 = ((3^2)^3)^2$.

14. A cube's side length is written in index form as 7^3 cm. Write down the index form for the volume of that same cube.

15. The total surface area of a rectangular box is given by $2\,(lw + lh + wh)$, where l is the length, h is the height and w is the width of the box (all measured in cm).

    Given that $l = a^2b$, $w = ab^2$ and $h = a^2b^2$, write an expression for the surface area of the rectangular box, simplifying the expression as far as possible.

LESSON
2.2 Review of index form

LEARNING INTENTIONS

At the end of this lesson you should be able to:
- understand and apply index (exponent) notation
- identify base and power for a number
- write a term in factor form.

▶ 2.2.1 Index notation

eles-3548

- If a number or a variable is multiplied by itself several times, it can be written in a shorter form, which is referred to as **index** or **exponent notation**.
- A number expressed in index form has two parts:
  1. The base
  2. The power (also referred to as an index or exponent)
- The **base** tells us what number or variable is being multiplied.
- The **power** (index or exponent) tells us how many times the base will be written and multiplied by itself.
- Factor form is when all the multiplications are shown.
- When the answer corresponds to a number, it is called the **basic numeral**.
- Any number or variable that does not appear to have an index or power has an index of 1. For example, $2 = 2^1$ and $a = a^1$.
- Any non-zero number or variable raised to the power of zero is equal to 1. For example, $2^0 = 1$ and $a^0 = 1$.

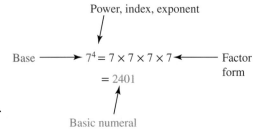

WORKED EXAMPLE 1 Identifying the base and power

State the base and power for each of the following.
a. 5^{14}
b. a^{35}

THINK	WRITE
a. 1. Write the number.	a. 5^{14}
2. Identify the base.	The base is 5.
3. Identify the power (the small number above and to the right of the base).	The power is 14.
b. 1. Write the number.	b. a^{35}
2. Identify the base.	The base is a.
3. Identify the power (the small number above and to the right of the base).	The power is 35.

WORKED EXAMPLE 2 Writing a number in factor form

Write the following in factor form.

a. 12^4

b. m^6

THINK	WRITE
a. 1. Write the number.	**a.** 12^4
2. The base is 12, so this is what will be multiplied. The power is 4, so this is how many times 12 should be multiplied.	$= 12 \times 12 \times 12 \times 12$
b. 1. Write the number.	**b.** m^6
2. The base is m, so this is what will be multiplied. The power is 6, so this is how many times m should be multiplied.	$= m \times m \times m \times m \times m \times m$

WORKED EXAMPLE 3 Expressing a number in index form

Express each of the following in index form.

a. $2 \times 5 \times 2 \times 2 \times 5 \times 2 \times 5$

b. $2 \times a \times 5 \times a \times a \times 3 \times b \times b \times b \times b \times b$

THINK	WRITE
a. 1. Write the numeric expression.	**a.** $2 \times 5 \times 2 \times 2 \times 5 \times 2 \times 5$
2. Collect the like terms together.	$= 2 \times 2 \times 2 \times 2 \times 5 \times 5 \times 5$
3. The number 2 has been multiplied by itself 4 times and the number 5 has been multiplied by itself 3 times.	$= 2^4 \times 5^3$
b. 1. Write the numeric expression.	**b.** $2 \times a \times 5 \times a \times a \times 3 \times b \times b \times b \times b \times b$
2. Collect the like terms together.	$= 2 \times 5 \times 3 \times a \times a \times a \times b \times b \times b \times b \times b$
3. The pronumeral a has been multiplied by itself 3 times and the pronumeral b has been multiplied by itself 5 times.	$= 30 \times a^3 \times b^5$
	$= 30\, a^3 b^5$

WORKED EXAMPLE 4 Expressing a number in factor form

Express $7 \times m^3 \times n^5$ in factor form.

THINK	WRITE
1. Write the numeric expression.	$7 \times m^3 \times n^5$
2. List the factors: 7 is written once, m is multiplied by itself 3 times, and n is multiplied by itself 5 times.	$= 7 \times m \times m \times m \times n \times n \times n \times n \times n$

⊳ 2.2.2 Prime factorisation using index notation

eles-3549

- Any composite number can be written as a product of its prime factors. Therefore, it can be written using index notation, as illustrated by the factor tree shown.

  Remember, a composite number is a number that can be divided by another whole number (for example 48). A prime number is a number that can only be divided by itself and 1 (for example 5). Remember that 1 is not a prime number.

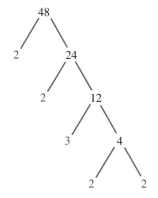

$$48 = 2 \times 2 \times 3 \times 2 \times 2$$
$$= 2^4 \times 3$$

WORKED EXAMPLE 5 Expressing numbers in index form using prime factorisation

Express 140 as a product of powers of prime factors using index notation.

THINK	WRITE
1. Express 140 as a product of a factor pair.	$140 = 14 \times 10$
2. Determine whether each number of the factor pair is prime. If the factors are prime, then no further calculations are required. If the factors are not prime, then each must be expressed as a product of another factor pair.	$140 = (2 \times 7) \times (2 \times 5)$ $\quad\ = 2 \times 7 \times 2 \times 5$ $\quad\ = 2 \times 2 \times 5 \times 7$
3. Write the answer.	$140 = 2^2 \times 5 \times 7$

COMMUNICATING — COLLABORATIVE TASK: Investigating powers

1. As a class, create a table on the board to show the numbers 1 to 10 raised to the power of 1 to 6.
2. Compare the results of raising 2 and 4 to different powers.
3. How can the table be used to show the value of different roots of numbers?
4. Investigate how the table would look if the numbers that were raised to different powers were unit fractions.
5. Investigate how the table would look if the numbers that were raised to different powers were decimals, such as 0.1, 0.2 and 0.3.
6. What is the difference between raising a number that is greater than 1 to a power and raising a number that is less than 1 to a power?

 Resources

⬥ **Interactivity** Review of index form (int-3708)

Exercise 2.2 Review of index form

2.2 Quick quiz on	2.2 Exercise

Individual pathways

■ PRACTISE	■ CONSOLIDATE	■ MASTER
1, 4, 7, 10, 13, 15, 19, 22, 25	2, 5, 8, 11, 14, 17, 20, 23, 26	3, 6, 9, 12, 16, 18, 21, 24, 27

Fluency

1. **WE1a** State the base and power for each of the following.
   a. 8^4
   b. 7^{10}
   c. 20^{11}

2. State the base and power for each of the following.
   a. 19^0
   b. 78^{12}
   c. 3^{100}

3. **WE1b** State the base and power for each of the following.
   a. c^{24}
   b. n^{36}
   c. d^{42}

4. Express the following in index form.
   a. $2 \times 2 \times 2 \times 2 \times 2 \times 2$
   b. $4 \times 4 \times 4 \times 4$
   c. $x \times x \times x \times x \times x \times x \times x$

5. Express the following in index form.
   a. $9 \times 9 \times 9$
   b. $11 \times l \times l \times l \times l \times l \times l \times l$
   c. $44 \times m \times m \times m \times m \times m$

6. **WE2a** Write the following in factor form.
   a. 4^2
   b. 5^4
   c. 7^5
   d. 6^3

7. **WE2b** Write the following in factor form.
   a. 3^6
   b. n^7
   c. a^4
   d. k^{10}

8. Express each of the following as a basic numeral.
   a. 3^5
   b. 4^4
   c. 2^8
   d. 11^3

9. Express each of the following as a basic numeral.
   a. 7^4
   b. 6^3
   c. 1^{10}
   d. 5^4

10. **MC** State what 6^3 means.
    A. 6×3
    B. $6 \times 6 \times 6$
    C. $3 \times 3 \times 3 \times 3 \times 3 \times 3$
    D. $6 + 6 + 6$

11. **MC** State what 3^5 means.
    A. 3×5
    B. 5×5
    C. $3 + 3 + 3 + 3 + 3$
    D. $3 \times 3 \times 3 \times 3 \times 3$

12. **WE3a** Express each of the following in index form.
    a. $6 \times 2 \times 2 \times 4 \times 4 \times 4 \times 4$
    b. $7 \times 7 \times 7 \times 7 \times 3 \times 3 \times 3 \times 3$
    c. $19 \times 19 \times 19 \times 19 \times 19 \times 2 \times 2 \times 2$
    d. $13 \times 13 \times 4 \times 4 \times 4 \times 4$

13. **WE3b** Express each of the following in index form.
    a. $66 \times p \times p \times m \times m \times m \times m \times m \times s \times s$
    b. $21 \times n \times n \times 3 \times i \times i \times i \times 6 \times r \times r \times r$
    c. $16 \times k \times e \times e \times e \times 12 \times p \times p$
    d. $11 \times j \times j \times j \times j \times j \times 9 \times p \times p \times l$

14. **WE4** Express each of the following in factor form.
    a. $15f^3 j^4$
    b. $7k^6 s^2$
    c. $4b^3 c^5$
    d. $19a^4 mn^3$

Understanding

15. **WE5** Using indices, express each of the following numbers as a product of its prime factors.

    a. 400   b. 225   c. 2000

16. Using indices, express each of the following numbers as a product of its prime factors.

    a. 64   b. 40   c. 36

17. Some basic numerals are written as the product of their prime factors. Identify each of these basic numerals.

    a. $2^3 \times 3 \times 5$ b. $2^2 \times 5^2$ c. $2^3 \times 3^3$

18. Some basic numerals are written as the product of their prime factors. Identify each of these basic numerals.

    a. $2^2 \times 7 \times 11$ b. $3^2 \times 5^2 \times 7$ c. $2^6 \times 5^4 \times 19$

19. Express each of the following numbers in index form with base 10.

    a. 10   b. 100   c. 1000   d. 1 000 000

20. Use your knowledge of place value to rewrite each of the following basic numerals in expanded form using powers of 10. The first number has been done for you.

	Basic numeral	Expanded form
	230	$2 \times 10^2 + 3 \times 10^1$
a.	500	
b.	470	
c.	2360	
d.	1980	
e.	5430	

21. Express each of the following as a basic numeral.

    a. $7 \times 10^4 + 5 \times 10^3$
    b. $3 \times 10^4 + 6 \times 10^2$
    c. $5 \times 10^6 + 2 \times 10^5 + 4 \times 10^2 + 8 \times 10^1$

Communicating, reasoning and problem solving

22. Explain how you will remember the meaning of base and index.

23. Explain what $a^3 b^4$ means. As part of your explanation, write $a^3 b^4$ as a basic numeral in factor form.

24. Explain which of the following scenarios results in a larger amount.
    - Having $1 000 000 deposited into your bank account immediately
    - Having 1 cent deposited into your bank account immediately, then having the amount in your account doubled each day for a period of 4 weeks

    Use calculations to help justify your response.

25. a, b and c are prime numbers.

    a. Write $8 \times a \times a \times b \times b \times b \times c \times c \times c \times c$ as a product of prime factors in index form.
    b. If $a = 2$, $b = 3$ and $c = 7$, calculate the value of the basic numeral represented by your answer in part a.

26. a. Rewrite the numbers 10 and 14 in expanded form with powers of 2 using 2^3, 2^2 and 2^1 as appropriate.
    b. Add the numbers in expanded form.
    c. Convert your answer for part b into a basic numeral.

27. **a.** Rewrite the numbers 140 and 680 in expanded form using powers of 10.
   b. Add the numbers in expanded form.
   c. Convert your answer for part b into a basic numeral.
   d. Try the question again with two numbers of your own. Choose numbers between 1000 and 10 000.
   e. What patterns do you notice?

LESSON
2.3 Multiplying powers

LEARNING INTENTION

At the end of this lesson you should be able to:
- apply the First Index Law to multiply terms that have the same base.

▶ 2.3.1 The First Index Law (multiplying numbers with the same base)

eles-3550

- Numbers in index form with the same base can be multiplied together by being written in factor form first.

  For example,

$$5^3 \times 5^2 = (5 \times 5 \times 5) \times (5 \times 5) = 5^5$$

$$\text{or } 4^4 \times 4^3 = (4 \times 4 \times 4 \times 4) \times (4 \times 4 \times 4) = 4^7$$

- By recognising a pattern, we can find a simpler, more efficient way to multiply numbers or variables in index form with the same base.
- The pattern observed from the above calculations is that when multiplying numbers or variables with the same base, we retain the base and add the powers together.

  For example,

$$5^3 \times 5^2 = 5^{3+2} \quad \text{and} \quad 4^4 \times 4^3 = 4^{4+3}$$
$$= 5^5 \qquad\qquad\qquad = 4^7$$

This pattern can be expressed as a general rule, as shown below.

First Index Law: Multiplying powers

When multiplying numbers in index form that have the same base, retain the base and add the powers.

$$a^m \times a^n = a^{m+n}$$

- If the variables in index form that are being multiplied have coefficients, the coefficients are multiplied together and the variables in index form are multiplied, and simplified using the First Index Law.

Applying the First Index Law to algebraic expressions

$$2a^4 \times 3a^5 \quad = \quad (2 \times 3) \quad \times \quad (a^4 \times a^5) \quad = \quad 6a^9$$

Coefficients   Coefficients multiplied   Variables multiplied

---

WORKED EXAMPLE 6 Simplifying using factor form

Simplify $2^3 \times 2^6$ by first writing it in factor form and then giving the answer in index form.

THINK	WRITE
1. Write the problem.	$2^3 \times 2^6$
2. Write it in factor form.	$= (2 \times 2 \times 2) \times (2 \times 2 \times 2 \times 2 \times 2 \times 2)$
3. Simplify by writing in index form.	$= 2^9$

WORKED EXAMPLE 7 Simplifying using the First Index Law

Simplify $7^4 \times 7 \times 7^3$, giving your answer in index form.

THINK	WRITE
1. Write the numeric expression.	$7^4 \times 7 \times 7^3$
2. Check all indices in the expression (the middle 7 has an index of 1).	$= 7^4 \times 7^1 \times 7^3$
3. Check if the bases are the same. They are all 7.	
4. Simplify by applying the First Index Law (add indices).	$= 7^{4+1+3}$
5. Write the answer.	$= 7^8$

WORKED EXAMPLE 8 Simplifying using the First Index Law

Simplify $5e^{10} \times 2e^3$.

THINK	WRITE
1. Write the algebraic expression.	$5e^{10} \times 2e^3$
2. The order is not important when multiplying, place the numbers first.	$= 5 \times 2 \times e^{10} \times e^3$
3. Multiply the constant terms.	$= 10 \times e^{10} \times e^3$
4. Check to see if the bases are the same. They are both e.	
5. Simplify by applying the First Index Law (add indices).	$= 10e^{10+3}$
6. Write the answer.	$= 10e^{13}$

⏵ 2.3.2 Multiplying expressions containing numbers with different bases

eles-3551

- When more than one variable is involved in multiplication, the First Index Law is applied to each variable separately.

WORKED EXAMPLE 9 Simplifying algebraic expressions

Simplify $7m^3 \times 3n^5 \times 2m^8 \times n^4$.

THINK	WRITE
1. Write the algebraic expression.	$7m^3 \times 3n^5 \times 2m^8 \times n^4$
2. The order is not important when multiplying, so place the numbers first and group the same variables together.	$= 7 \times 3 \times 2 \times m^3 \times m^8 \times n^5 \times n^4$
3. Simplify by multiplying the constant terms and applying the First Index Law to variables that are the same (add indices).	$= 42 \times m^{3+8} \times n^{5+4}$
4. Write the answer.	$= 42\,m^{11}n^9$

Exercise 2.3 Multiplying powers

learn on

2.3 Quick quiz on	2.3 Exercise

Individual pathways

■ PRACTISE	■ CONSOLIDATE	■ MASTER
1, 4, 7, 13, 15, 16, 19	2, 5, 8, 9, 11, 14, 17, 20	3, 6, 10, 12, 18, 21

Fluency

1. **WE6** Simplify the following expressions by first writing them in factor form and then giving the answer in index form.

   a. $3^7 \times 3^2$ b. $6^4 \times 6^3$ c. $10^6 \times 10^4$ d. $11^3 \times 11^3$

2. Simplify each of the following.

   a. $7^8 \times 7$ b. $2^{11} \times 2^3$ c. $5^2 \times 5^2$ d. $8^9 \times 8^2$

3. Simplify each of the following.

   a. $13^7 \times 13^8$ b. $q^{23} \times q^{24}$ c. $x^7 \times x^7$ d. $e \times e^3$

4. **WE7** Simplify each of the following, giving your answer in index form.

   a. $3^4 \times 3^6 \times 3^2$ b. $2^{10} \times 2^3 \times 2^5$ c. $5^4 \times 5^4 \times 5^9$ d. $6^8 \times 6 \times 6^2$

5. Simplify each of the following, giving your answer in index form.
   a. $10 \times 10 \times 10^4$
   b. $17^2 \times 17^4 \times 17^6$
   c. $p^7 \times p^8 \times p^7$
   d. $e^{11} \times e^{10} \times e^2$

6. Simplify each of the following, giving your answer in index form.
   a. $g^{15} \times g \times g^{12}$
   b. $e^{20} \times e^{12} \times e^6$
   c. $3 \times b^2 \times b^{10} \times b$
   d. $5 \times d^4 \times d^5 \times d^7$

7. **MC** Select which of the following is equal to $6 \times e^3 \times b^2 \times b^4 \times e$.
   A. $6b^6e^4$
   B. $6b^6e^3$
   C. $6b^9e$
   D. $6b^{10}e$

8. **MC** Select which of the following is equal to $3 \times f^2 \times f^{10} \times 2 \times e^3 \times e^8$.
   A. $32e^{11}f^{12}$
   B. $6e^{11}f^{12}$
   C. $6e^{23}f$
   D. $6e^{24}f^{20}$

9. **WE8** Simplify each of the following.
   a. $4p^7 \times 5p^4$
   b. $2x^2 \times 3x^6$
   c. $8y^6 \times 7y^4$

10. Simplify each of the following.
    a. $3p \times 7p^7$
    b. $12t^3 \times t^2 \times 7t$
    c. $6q^2 \times q^5 \times 5q^8$

Understanding

11. **WE9** Simplify each of the following.
    a. $2a^2 \times 3a^4 \times e^3 \times e^4$
    b. $4p^3 \times 2h^7 \times h^5 \times p^3$
    c. $2m^3 \times 5m^2 \times 8m^4$
    d. $2gh \times 3g^2h^5$
    e. $5p^4q^2 \times 6p^2q^7$

12. Simplify each of the following.
    a. $8u^3w \times 3uw^2 \times 2u^5w^4$
    b. $9dy^8 \times d^3y^5 \times 3d^7y^4$
    c. $7b^3c^2 \times 2b^6c^4 \times 3b^5c^3$
    d. $4r^2s^2 \times 3r^6s^{12} \times 2r^8s^4$
    e. $10h^{10}v^2 \times 2h^8v^6 \times 3h^{20}v^{12}$

13. Simplify each of the following.
    a. $3^x \times 3^4$
    b. $3^y \times 3^{y+2}$
    c. $3^{2y+1} \times 3^{4y-6}$
    d. $3^{\frac{1}{2}} \times 3^{\frac{2}{3}} \times 3^{\frac{3}{4}}$

14. Express the following basic numerals in index form: 9, 27 and 81.

15. Simplify each of the following expressions. (Give each answer in index form.)
    a. $3^4 \times 81 \times 9$
    b. $27 \times 3^n \times 3^{n-1}$

Communicating, reasoning and problem solving

16. Explain why $2^x \times 3^y$ does not equal $6^{(x+y)}$.

17. Follow the steps to write the expression indicated.
    Step 1: The prime number 5 is multiplied by itself n times.
    Step 2: The prime number 5 is multiplied by itself m times.
    Step 3: The answers from step 1 and step 2 are multiplied together.
    Explain how you arrive at your final answer. What is your answer?

18. The First Index Law can be applied only if the bases are the same. Why is that so? Give examples to help justify your response.

19. One grain of rice is placed on a chess board square, two grains of rice on the next square, four grains on the next square, eight grains on the next square and so on. Note that a chess board consists of 64 squares.

   a. Write the number of grains on the 10th square in index form.
   b. Write the number of grains on the nth square in index form.
   c. How many grains are on the 6th and 7th squares in total?
   d. How many grains are on the 14th and 15th squares in total? Write your answer in index form.
   e. Simplify your answer to part d by first taking out a common factor.

20. a. If $x^2 = x \times x$, what does $(x^3)^2$ equal?
    b. If the sides of a cube are 2^4 cm long, what is the volume of the cube in index form? (*Hint:* The volume of a cube of side length l cm is l^3 cm^3.)
    c. Evaluate the side length of a cube of volume 5^6 mm^3.
    d. Evaluate the side length of a cube of volume $(a^n)^{3p}$ mm^3.

21. If I square a certain number, then multiply the result by three times the cube of that number before adding 1, the result is 97. Determine the original number. Use algebra and show your working.

LESSON
2.4 Dividing powers

▶ 2.4.1 The Second Index Law (dividing numbers with the same base)
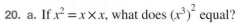
eles-3552

• Numbers in index form with the same base can be divided by first being written in factor form.

For example,

$$2^6 \div 2^4 = \frac{2 \times 2 \times 2 \times 2 \times 2 \times 2}{2 \times 2 \times 2 \times 2}$$
$$= \frac{\cancel{2} \times \cancel{2} \times \cancel{2} \times \cancel{2} \times 2 \times 2}{\cancel{2} \times \cancel{2} \times \cancel{2} \times \cancel{2}}$$
$$= 2 \times 2$$
$$= 2^2$$

• By recognising a pattern, a simpler, more efficient way to divide numbers or variables in index form with the same base can be found.
  The pattern observed from the above calculation is that when dividing numbers or variables with the same base, we retain the base and subtract the powers.

For example,

$$2^6 \div 2^4 = 2^{6-4} \qquad \text{or} \qquad 3^{10} \div 3^6 = 3^{10-6}$$
$$= 2^2 \qquad\qquad\qquad = 3^4$$

This pattern can be expressed as a general rule, as shown next.

Second Index Law: Dividing powers

When dividing numbers in index form with the same base, retain the base and subtract the powers.

$$a^m \div a^n = \frac{a^m}{a^n} = a^{m-n}$$

WORKED EXAMPLE 10 Simplifying using factor form

Simplify $\dfrac{5^{10}}{5^3}$ by first writing in factor form and then leaving your answer in index form.

THINK	WRITE
1. Write the numeric expression.	$\dfrac{5^{10}}{5^3}$
2. Change to factor form.	$= \dfrac{5 \times 5 \times 5 \times 5 \times 5 \times 5 \times 5 \times 5 \times 5 \times 5}{5 \times 5 \times 5}$
3. Cancel three 5s from the numerator and three 5s from the denominator.	$= \dfrac{5 \times 5 \times 5 \times 5 \times 5 \times 5 \times 5 \times \cancel{5} \times \cancel{5} \times \cancel{5}}{\cancel{5} \times \cancel{5} \times \cancel{5}}$ $= 5 \times 5 \times 5 \times 5 \times 5 \times 5 \times 5$
4. Write the answer in index form.	$= 5^7$

WORKED EXAMPLE 11 Simplifying using the Second Index Law

Simplify $d^{12} \div d^4$ using an index law.

THINK	WRITE
1. Write the algebraic expression and express it as a fraction.	$d^{12} \div d^4 = \dfrac{d^{12}}{d^4}$
2. Check to see if the bases are the same. They are both d.	
3. Simplify by using the Second Index Law (subtract indices).	$= d^{12-4}$
4. Write the answer in index form.	$= d^8$

▶ 2.4.2 Dividing algebraic terms containing coefficients

eles-3553

- When the variables in index form have numerical coefficients, we divide them as we would divide any other numbers and then apply the Second Index Law to the variables.
- In examples where the coefficients do not divide evenly, we simplify the fraction that is formed by them.
- When there is more than one variable involved in the division question, the Second Index Law is applied to each variable separately.

WORKED EXAMPLE 12 Simplifying when coefficients are present

Simplify $36d^7 \div 12d^3$, giving your answer in index form.

THINK	WRITE
1. Write the algebraic expression and express it as a fraction.	$36d^7 \div 12d^3 = \dfrac{36d^7}{12d^3}$
2. Divide the numbers (or coefficients).	$= \dfrac{3d^7}{d^3}$
3. Simplify by using the Second Index Law (subtract indices).	$= 3d^{7-3}$ $= 3d^4$

WORKED EXAMPLE 13 Simplifying using the First and Second index laws

Simplify $\dfrac{7t^3 \times 4t^8}{12t^4}$.

THINK	WRITE
1. Write the algebraic expression.	$\dfrac{7t^3 \times 4t^8}{12t^4}$
2. Multiply the numbers in the numerator and apply the First Index Law (add indices) in the numerator.	$= \dfrac{28t^{11}}{12t^4}$
3. Simplify the fraction formed and apply the Second Index Law (subtract indices).	$= \dfrac{7t^{11-4}}{3}$ $= \dfrac{7}{3}t^7$

on Resources

Interactivities Second Index Law (int-3711)
　　　　　　　　Dividing with coefficients (int-3712)

Exercise 2.4 Dividing powers

learn on

2.4 Quick quiz on	2.4 Exercise

Individual pathways

■ PRACTISE	■ CONSOLIDATE	■ MASTER
1, 2, 5, 8, 12, 15, 18, 21	3, 6, 9, 10, 13, 16, 19, 22	4, 7, 11, 14, 17, 20, 23

Fluency

1. **WE10** Simplify each of the following by first writing in factor form and leaving your answer in index form.

   a. $\dfrac{2^5}{2^2}$
   b. $\dfrac{7^7}{7^3}$
   c. $\dfrac{10^8}{10^5}$
   d. $\dfrac{9^4}{9^5}$

2. Simplify each of the following using the Second Index Law, leaving your answer in index form.

   a. $3^3 \div 3^2$
   b. $11^9 \div 11^2$
   c. $5^8 \div 5^4$
   d. $12^6 \div 12$

3. Simplify each of the following using the Second Index Law, leaving your answer in index form.

   a. $3^{45} \div 3^{42}$
   b. $13^{75} \div 13^{74}$
   c. $6^{23} \div 6^{19}$
   d. $\dfrac{10^{13}}{10^9}$

4. Simplify each of the following using the Second Index Law, leaving your answer in index form.

   a. $\dfrac{15^{456}}{15^{423}}$
   b. $\dfrac{h^{78}}{h}$
   c. $\dfrac{b^{77}}{b^7}$
   d. $\dfrac{f^{1000}}{f^{100}}$

5. **WE11** Simplify each of the following, giving your answer in index form.

   a. $3x^5 \div x^3$
   b. $6y^7 \div y^5$
   c. $8w^{12} \div w^5$
   d. $12q^{34} \div 4q^{30}$

6. Simplify each of the following, giving your answer in index form.

   a. $16f^{12} \div 2f^3$
   b. $100h^{100} \div 10h^{10}$
   c. $80j^{15} \div 20j^5$
   d. $\dfrac{45p^{14}}{9p^4}$

7. Simplify each of the following, giving your answer in index form.

   a. $\dfrac{48g^8}{6g^5}$
   b. $\dfrac{12b^7}{8b}$
   c. $\dfrac{81m^6}{18m^2}$
   d. $\dfrac{100n^{95}}{40n^5}$

8. **MC** Select which of the following is equal to $21r^{20} \div \left(14r^{10}\right)$.

   A. $7r^{10}$
   B. $\dfrac{3r^2}{2}$
   C. $7r^2$
   D. $\dfrac{3r^{10}}{2}$

9. **MC** Select which of the following is equal to $\left(\dfrac{2m^{33}}{16m^{11}}\right)$.

   A. $\dfrac{m^{22}}{8}$
   B. $\dfrac{8}{m^{22}}$
   C. $8m^{22}$
   D. $\dfrac{m^3}{8}$

10. **WE12** Simplify each of the following.

    a. $\dfrac{15p^{12}}{5p^8}$
    b. $\dfrac{18r^6}{3r^2}$
    c. $\dfrac{45a^5}{5a^2}$

11. Simplify each of the following.

    a. $\dfrac{60b^7}{20b}$
    b. $\dfrac{100r^{10}}{5r^6}$
    c. $\dfrac{9q^2}{q}$

Understanding

12. **WE13** Simplify each of the following.

    a. $\dfrac{8p^6 \times 3p^4}{16p^5}$

    b. $\dfrac{12b^5 \times 4b^2}{18b^2}$

    c. $\dfrac{25m^{12} \times 4n^7}{15m^2 \times 8n}$

13. Simplify each of the following.

    a. $\dfrac{27x^9y^3}{12xy^2}$

    b. $\dfrac{16h^7k^4}{12h^6k}$

    c. $\dfrac{12j^8 \times 6f^5}{8j^3 \times 3f^2}$

14. Simplify each of the following.

    a. $\dfrac{8p^3 \times 7r^2 \times 2s}{6p \times 14r}$

    b. $\dfrac{27a^9 \times 18b^5 \times 4c^2}{18a^4 \times 12b^2 \times 2c}$

    c. $\dfrac{81f^{15} \times 25g^{12} \times 16h^{34}}{27f^9 \times 15g^{10} \times 12h^{30}}$

15. Simplify each of the following.

    a. $2^{10} \div 2^p$

    b. $2^{7e} \div 2^{3e-4}$

    c. $\dfrac{5^{4x} \times 5^{3y}}{5^{2y} \times 5^x}$

    d. $\dfrac{3^{2-3m} \times 3^{7m}}{3^{5m} \times 3}$

16. Consider the fraction $\dfrac{8 \times 16 \times 4}{2 \times 32}$.

    a. Rewrite the fraction, expressing each basic numeral as a power of 2.
    b. Simplify by giving your answer:
       i. in index form
       ii. as a basic numeral.

17. Consider the fraction $\dfrac{6 \times 27 \times 36}{12 \times 81}$.

    a. Rewrite the fraction, expressing each basic numeral as the product of its prime factors.
    b. Simplify, giving the answer:
       i. in index form
       ii. as a basic numeral.

Communicating, reasoning and problem solving

18. In a Maths test, one of the questions was:

    Simplify the following expression $\dfrac{12^x}{3^y}$.

    Sam, who is a student in this class, wrote his answer as $4^{(x-y)}$.
    Explain whether Sam's answer is correct.

19. Step 1: The prime number 3 is multiplied by itself p times.
    Step 2: The prime number 3 is multiplied by itself q times.
    Step 3: The answer from step 1 is divided by the answer from step 2.
    State your answer. Explain how you arrived at your final answer.

20. Using two different methods for simplifying the expression $\dfrac{y^4}{y^6}$, show that $y^{-2} = \dfrac{1}{y^2}$.

21. By considering $a^p \div a^p$, show that any base raised to the power of zero equals 1.
    (*Hint:* Any number or variable, except zero, raised to the power of zero is equal to 1. For example, $2^0 = 1$ and $a^0 = 1$.)

22. By considering $m^6 \div m^7$, show that $m^{-1} = \dfrac{1}{m}$.

23. I cube a certain number, then multiply the result by six. I now divide the result by a certain number to the power of five. The result is 216. Determine the number. Show your working.

LESSON
2.5 The zero index

▶ 2.5.1 The Third Index Law (the power of zero)

eles-3554

- Consider the following two methods of simplifying $2^3 \div 2^3$.

Method 1	Method 2
$2^3 \div 2^3 = \dfrac{2 \times 2 \times 2}{2 \times 2 \times 2}$ $= \dfrac{8}{8}$ $= 1$	$2^3 \div 2^3 = \dfrac{2^3}{2^3}$ $= 2^{3-3}$ $= 2^0$ $= 1$

- As the two results must be the same, 2^0 must equal 1.
- Another way to establish the meaning of the zero index is to consider the following pattern of numbers:

3^4	3^3	3^2	3^1	3^0	3^{-1}	3^{-2}	3^{-3}	3^{-4}
$3 \times 3 \times 3 \times 3$	$3 \times 3 \times 3$	3×3	3	?	$\dfrac{1}{3}$	$\dfrac{1}{3 \times 3}$	$\dfrac{1}{3 \times 3 \times 3}$	$\dfrac{1}{3 \times 3 \times 3 \times 3}$

$\div 3 \qquad \div 3 \qquad \div 3 \quad ? \quad ? \quad \div 3 \qquad \div 3 \qquad \div 3$

As each consecutive number in this pattern is found by dividing the previous number by 3, it follows that 3^0 must equal 1.

> **Third Index Law: The zero index**
>
> **Any base, excluding zero, that is raised to the power of zero is equal to 1.**
>
> $$a^0 = 1 \text{ where } a \neq 0$$

- Any non-zero numerical or algebraic expression that is raised to the power of zero is equal to 1.
  For example,

$$(2^2 \times 3)^0 = 1 \text{ or } (2abc^2)^0 = 1$$

WORKED EXAMPLE 14 Simplifying using the Third Index Law

Determine the value of 15^0.

THINK	WRITE
1. Write the numeric expression.	15^0
2. Any non-zero base with an index of zero is equal to 1 (Third Index Law).	$= 1$

WORKED EXAMPLE 15 Simplifying using the Third Index Law

Determine the value of $(25 \times 36)^0$.

THINK	WRITE
1. Write the numeric expression.	$(25 \times 36)^0$
2. Everything within the brackets has an index of zero; therefore, according to the Third Index Law, it is equal to 1.	$= 1$

WORKED EXAMPLE 16 Simplifying using the Third Index Law

Determine the value of $19e^5 a^0$.

THINK	WRITE
1. Write the algebraic expression.	$19e^5 a^0$
2. The base a has a power of zero, so it is equal to 1.	$= 19e^5 \times 1$
3. Simplify and write the answer.	$= 19e^5$

WORKED EXAMPLE 17 Simplifying using various index laws

Simplify $\dfrac{6m^3 \times 11m^{14}}{3m^{10} \times 2m^7}$.

THINK	WRITE
1. Write the algebraic expression.	$\dfrac{6m^3 \times 11m^{14}}{3m^{10} \times 2m^7}$
2. Multiply the constant terms and apply the First Index Law (add indices) in both the numerator and denominator.	$= \dfrac{66m^{3+14}}{6m^{10+7}}$ $= \dfrac{66m^{17}}{6m^{17}}$
3. Divide the constant terms and simplify using the Second Index Law (subtract indices).	$= 11m^{17-17}$ $= 11m^0$
4. The base m has a power of zero, so it is equal to 1 (Third Index Law).	$= 11 \times 1$
5. Simplify and write the answer.	$= 11$

Digital technology

Scientific calculators can evaluate expressions in index form where the base is a number.

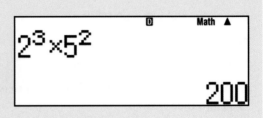

Exercise 2.5 The zero index

learn on

2.5 Quick quiz   **on**

2.5 Exercise

Individual pathways

■ PRACTISE	■ CONSOLIDATE	■ MASTER
1, 4, 7, 10, 13, 15, 18	2, 5, 8, 11, 14, 16, 19	3, 6, 9, 12, 17, 20

Fluency

1. **WE14** Determine the value of each of the following.

   a. 16^0 b. 44^0 c. f^0 d. h^0

2. **WE15** Determine the value of each of the following.

   a. $(23 \times 8)^0$ b. $7^0 \times 6^0$ c. $(35z^4)^0$ d. $(12w^7)^0$

3. Determine the value of each of the following. Check your answers to parts a and b with a calculator.

   a. 4×3^0 b. $9^0 + 11$ c. $c^0 - 10$ d. $3p^0 + 19$

4. **WE16** Determine the value of each of the following.

   a. $12m^3k^0$ b. $7c^0 + 14m^0$ c. $32g^0 + 40h^0$ d. $\dfrac{8k^0}{7j^0}$

5. Determine the value of each of the following.

   a. $\dfrac{16t^0}{8y^0}$ b. $\dfrac{6b^2 \times 5c^0}{3s^0}$ c. $\dfrac{4d^0 \times 9p^2}{12q^0}$ d. $\dfrac{3p \times 4d^0}{2z^0 \times 6p}$

6. Evaluate each of the following.

   a. $e^{10} \div e^{10}$ b. $a^{12} \div a^{12}$ c. $(4b^3)^0 \div (4b^3)^0$

7. Evaluate each of the following.

   a. $84f^{11} \div 12f^{11}$ b. $30z^9 \div 10z^9$ c. $99t^{13} \div 33t^{13}$

Understanding

8. Simplify each of the following.

   a. $\dfrac{21p^4}{21p^4}$

   b. $\dfrac{40f^{33}}{10f^{33}}$

   c. $\dfrac{54p^6q^8}{27p^6q^8}$

   d. $\dfrac{16p^{11}q^{10}}{8p^2q^{10}}$

9. Simplify each of the following.

   a. $\dfrac{24a^9e^{10}}{16a^9e^6}$

   b. $\dfrac{x^4y^2z^{11}}{x^4yz^{11}}$

   c. $\dfrac{7i^7m^6r^4}{21i^7m^3r^4}$

   d. $\dfrac{3c^5d^3l^9}{12c^2d^3l^9}$

10. **MC** You are told that there is an error in the statement $3p^7q^3r^5s^6 = 3p^7s^6$. Determine what the left-hand side should be changed to in order to make the statement correct.

    A. $\left(3p^7q^3r^5s^6\right)^0$

    B. $\left(3p^7\right)^0q^3r^5s^6$

    C. $3p^7\left(q^3r^5s^6\right)^0$

    D. $3p^7\left(q^3r^5\right)^0s^6$

11. **MC** You are told that there is an error in the statement $\dfrac{8f^6g^7h^3}{6f^4g^2h} = \dfrac{8f^2}{g^2}$. Determine what the left-hand side should be changed to

    should be changed to in order to make the statement correct.

    A. $\dfrac{8f^6\left(g^7h^3\right)^0}{(6)^0f^4g^2(h)^0}$

    B. $\dfrac{8\left(f^6g^7h^3\right)^0}{\left(6f^4g^2h\right)^0}$

    C. $\dfrac{8\left(f^6g^7\right)^0h^3}{\left(6f^4\right)^0g^2h}$

    D. $\dfrac{8f^6g^7h^3}{\left(6f^4g^2h\right)^0}$

12. **MC** Select which of the following is equal to $\dfrac{6k^7m^2n^8}{4k^7\left(m^6n\right)^0}$.

    A. $\dfrac{3}{2}$

    B. $\dfrac{3n^8}{2}$

    C. $\dfrac{3m^2}{2}$

    D. $\dfrac{3m^2n^8}{2}$

13. **WE17** Simplify each of the following.

    a. $\dfrac{2a^3 \times 6a^2}{12a^5}$

    b. $\dfrac{3c^6 \times 6c^3}{9c^9}$

    c. $\dfrac{5b^7 \times 10b^5}{25b^{12}}$

    d. $\dfrac{8f^3 \times 3f^7}{4f^5 \times 3f^5}$

    e. $\dfrac{9k^{12} \times 4k^{10}}{18k^4 \times k^{18}}$

14. Simplify each of the following.

    a. $\dfrac{2h^4 \times 5k^2}{20h^2 \times k^2}$

    b. $\dfrac{p^3 \times q^4}{5p^3}$

    c. $\dfrac{m^7 \times n^3}{5m^3 \times m^4}$

    d. $\dfrac{8u^9 \times v^2}{2u^5 \times 4u^4}$

    e. $\dfrac{9x^6 \times 2y^{12}}{3y^{10} \times 3y^2}$

Communicating, reasoning and problem solving

15. Explain why $x^0 = 1$.

16. Simplify $\dfrac{2^0x^2}{2^2x^0}$, explaining each step of your method.

17. If $a^{\frac{1}{2}}$ is equivalent to $\sqrt{a}$, show what $a^{-\frac{1}{2}}$ is equivalent to.

18. Use indices, and multiplication and division, to set up four expressions that simplify to y^5. At least one of your four expressions must involve the use of the Third Index Law.

19. I raise a certain number to the power of three, then multiply the answer by three to the power of zero. I then multiply the result by the certain number to the power of four and divide the answer by three times the certain number to the power of seven. If the final answer is three multiplied by the certain number squared, find the certain number. Show each line of your method.

20. A Mathematics class is asked to simplify $\dfrac{64x^6y^6z^3}{16x^2y^6z^3}$. Peter's answer is $4x^3$. Explain why this is incorrect, pointing out Peter's error. Determine the correct answer. Identify another source of possible error involving indices.

LESSON
2.6 Raising powers

LEARNING INTENTIONS

At the end of this lesson you should be able to:
- apply the Fourth Index Law when raising a power to another power by multiplying the indices
- combine different index laws to simplify expressions.

▶ 2.6.1 The Fourth Index Law (raising a power to another power)

eles-3555

- The expression $\left(2^3\right)^4$ is an example of a power term $\left(2^3\right)$ raised to another power (4).
- Raising a power term to another power is a variation of the First Index Law.

For example,

$$\left(2^3\right)^4 = 2^3 \times 2^3 \times 2^3 \times 2^3$$
$$= 2^{3+3+3+3}$$
$$= 2^{12}$$

can be simplified to

$$\left(2^3\right)^4 = 2^{3\times4}$$
$$= 2^{12}$$

$$\left(a^2\right)^3 = a^2 \times a^2 \times a^2$$
$$= a^{2+2+2}$$
$$= a^6$$

can be simplified to

$$\left(a^2\right)^3 = a^{2\times3}$$
$$= a^6$$

- The pattern observed from these calculations is that when raising a power to another power, we retain the base and multiply the powers.
- This pattern can be expressed as a general rule, as shown below.

Fourth Index Law: Raising powers

When raising a power to another power, retain the base and multiply the powers.

$$(a^m)^n = a^{m \times n}$$

- **Every number and variable inside the brackets should have its index multiplied by the power outside the brackets. That is:**

$$(a \times b)^m = a^m \times b^m$$

$$\left(\frac{a}{b}\right)^m = \frac{a^m}{b^m}$$

(**These are sometimes called the Fifth Index Law and the Sixth Index Law.**)

- **Every number or variable inside the brackets must be raised to the power outside the brackets. For example:**

$$(3 \times 2)^4 = 3^4 \times 2^4 \quad \text{and} \quad (2a^4)^3 = 2^3 \times a^{4 \times 3}$$
$$= 8a^{12}$$

WORKED EXAMPLE 18 Simplifying using the Fourth Index Law

Simplify the following, leaving answers in index form.

a. $\left(7^4\right)^8$

b. $\left(\dfrac{3^2}{5^3}\right)^3$

THINK	WRITE
a. 1. Write the numeric expression.	a. $\left(7^4\right)^8$
2. Use the Fourth Index Law (retain the base and multiply the indices).	$= 7^{4 \times 8}$
3. Simplify and write the answer.	$= 7^{32}$
b. 1. Write the numeric expression.	b. $\left(\dfrac{3^2}{5^3}\right)^3$
2. Use the Fourth Index Law (retain the base and multiply the indices) in both the numerator and denominator.	$= \dfrac{3^{2 \times 3}}{5^{3 \times 3}}$
3. Simplify (here the bases are different and therefore the indices cannot be subtracted) and write the answer.	$= \dfrac{3^6}{5^9}$

WORKED EXAMPLE 19 Simplifying using the Fourth Index Law

Simplify $\left(2b^5\right)^2 \times \left(5b^8\right)^3$.

THINK	WRITE
1. Write the numeric expression.	$\left(2b^5\right)^2 \times \left(5b^8\right)^3$
2. Simplify using the Fourth Index Law (retain the base and multiply the indices).	$= 2^{1\times2} \times b^{5\times2} \times 5^{1\times3} b^{8\times3}$ $= 2^2 b^{10} \times 5^3 b^{24}$
3. Evaluate the coefficients.	$= 4b^{10} \times 125b^{24}$ $= 500b^{10} \times b^{24}$
4. Simplify using the First Index Law (add indices) and write the answer.	$= 500b^{34}$

WORKED EXAMPLE 20 Simplifying algebraic expressions using the Fourth Index Law

Simplify $\left(\dfrac{2a^5}{d^2}\right)^3$.

THINK	WRITE
1. Write the algebraic expression.	$\left(\dfrac{2a^5}{d^2}\right)^3$
2. Simplify using the Fourth Index Law (retain the base and multiply the indices) for each term inside the brackets.	$= \dfrac{2^{1\times3} a^{5\times3}}{d^{2\times3}}$ $= \dfrac{2^3 a^{15}}{d^6}$
3. Calculate the coefficient and write the answer. The bases, a and d, are different and therefore the indices cannot be subtracted here.	$= \dfrac{8a^{15}}{d^6}$

 Resources

◆ **Interactivity** Fourth Index Law (int-3715)

Exercise 2.6 Raising powers

learn on

2.6 Quick quiz   on   2.6 Exercise

Individual pathways

■ PRACTISE	■ CONSOLIDATE	■ MASTER
1, 4, 7, 10, 13, 16, 19, 20, 21, 26	2, 5, 8, 11, 14, 17, 22, 23, 27	3, 6, 9, 12, 15, 18, 24, 25, 28

Fluency

1. **WE18** Simplify each of the following, leaving your answers in index form.

   a. $\left(3^2\right)^3$ b. $\left(6^8\right)^{10}$ c. $\left(11^{25}\right)^4$ d. $\left(5^{12}\right)^{12}$ e. $\left(\left(3^2\right)^2\right)^2$

2. Simplify each of the following, leaving your answers in index form.

   a. $\left(3^2 \times 10^3\right)^4$ b. $\left(13 \times 17^3\right)^5$ c. $\left(\dfrac{3^3}{2^2}\right)^{10}$ d. $\left(3w^9 q^2\right)^4$ e. $\left(\dfrac{7e^5}{r^2 q^4}\right)^2$

3. **WE19** Simplify each of the following.

   a. $\left(p^4\right)^2 \times \left(q^3\right)^2$ b. $\left(r^5\right)^3 \times \left(w^3\right)^3$ c. $\left(b^5\right)^2 \times \left(n^3\right)^6$

4. Simplify each of the following.

   a. $\left(j^6\right)^3 \times \left(g^4\right)^3$ b. $\left(q^2\right)^2 \times \left(r^4\right)^5$ c. $\left(h^3\right)^8 \times \left(j^2\right)^8$

5. Simplify each of the following.

   a. $\left(f^4\right)^4 \times \left(a^7\right)^3$ b. $\left(t^5\right)^2 \times \left(u^4\right)^2$ c. $\left(i^3\right)^5 \times \left(j^2\right)^6$

6. Simplify each of the following.

   a. $\left(2^3\right)^4 \times \left(2^4\right)^2$ b. $\left(t^7\right)^3 \times \left(t^3\right)^4$ c. $\left(a^4\right)^0 \times \left(a^3\right)^7$

7. Simplify each of the following.

   a. $\left(b^6\right)^2 \times \left(b^4\right)^3$ b. $\left(e^7\right)^8 \times \left(e^5\right)^2$ c. $\left(g^7\right)^3 \times \left(g^9\right)^2$

8. Simplify each of the following.

   a. $\left(3a^2\right)^4 \times \left(2a^6\right)^2$ b. $\left(2d^7\right)^3 \times \left(3d^2\right)^3$ c. $\left(10r^{12}\right)^4 \times \left(2r^3\right)^2$

9. **MC** Select which of the following is equal to $\left(p^7\right)^2 \div p^2$.

   A. p^7 B. p^{12} C. p^{16} D. $p^{4.5}$

10. **MC** Select which of the following is equal to $\dfrac{\left(w^5\right)^2 \times \left(p^7\right)^3}{\left(w^2\right)^2 \times \left(p^3\right)^5}$.

    A. $w^2 p^6$ B. $(wp)^6$ C. $w^{14} p^{36}$ D. $w^2 p^2$

11. **MC** Select which of the following is equal to $\left(r^6\right)^3 \div \left(r^4\right)^2$.

    A. r^3 B. r^4 C. r^8 D. r^{10}

Understanding

12. Simplify each of the following.

    a. $\left(a^3\right)^4 \div \left(a^2\right)^3$ b. $\left(m^8\right)^2 \div \left(m^3\right)^4$ c. $\left(n^5\right)^3 \div \left(n^6\right)^2$ d. $\left(b^4\right)^5 \div \left(b^6\right)^2$

13. Simplify each of the following.

   a. $\left(f^7\right)^3 \div \left(f^2\right)^2$ b. $\left(g^8\right)^2 \div \left(g^5\right)^2$ c. $\left(p^9\right)^3 \div \left(p^6\right)^3$ d. $\left(y^4\right)^4 \div \left(y^7\right)^2$

14. Simplify each of the following.

   a. $\dfrac{\left(c^6\right)^5}{\left(c^5\right)^2}$ b. $\dfrac{\left(f^5\right)^3}{\left(f^2\right)^4}$ c. $\dfrac{\left(k^3\right)^{10}}{\left(k^2\right)^8}$ d. $\dfrac{\left(p^{12}\right)^3}{\left(p^{10}\right)^2}$

15. **WE20** Simplify each of the following.

   a. $\left(\dfrac{3b^4}{d^3}\right)^2$ b. $\left(\dfrac{5h^{10}}{2j^2}\right)^2$ c. $\left(\dfrac{2k^5}{3t^8}\right)^3$

16. Simplify each of the following.

   a. $\left(\dfrac{7p^9}{8q^{22}}\right)^2$ b. $\left(\dfrac{5y^7}{3z^{13}}\right)^3$ c. $\left(\dfrac{4a^3}{7c^5}\right)^4$

17. Simplify each of the following using the index laws.

   a. $g^3 \times 2g^5$ b. $2p^6 \times 4p^2$ c. $\left(w^3\right)^6$ d. $12x^6 \div (2x)$ e. $\left(2d^3\right)^2$

18. Simplify each of the following using the index laws.

   a. $5a^6 \times 3a^2 \times a^2$ b. $15s^8 \div \left(5s^2\right)$ c. $4bc^6 \times 3b^3 \times 5c^2$

   d. $\dfrac{14x^8}{7x^4}$ e. $\left(f^4 g^3\right)^2$

19. Simplify each of the following using the index laws.

   a. $\dfrac{16u^6 v^5}{6u^3 v}$ b. $x^2 y^4 \times xy^3$ c. $5a^6 b^2 \times a^2 \times 3ab^3$

   d. $x^2 y^4 \div \left(xy^3\right)$ e. $\left(4p^2 q^5\right)^3$

Communicating, reasoning and problem solving

20. Explain why $\left(\left(a^b\right)^c\right)^0 = 1$.

21. a. Simplify $\left(4^3\right)^2$, leaving your answer in index form.
    b. Use a calculator to determine the value of your answer to part **a**.
    c. Use a calculator to determine the value of $\left(4^3\right)^2$.

22. Simplify each of the following, giving your answer in index form. Justify your answer in each case.

   a. $\left(w^3\right)^4 \div w^2$ b. $\dfrac{4x^5 \times 3x}{2x^4}$ c. $\left(2a^3\right)^2 \times 3a^5$

   d. $12x^6 \times 2x \div \left(3x^5\right)$ e. $2d^3 + d^2 + 5d^3$

23. A Mathematics class is asked to simplify $\left(r^4\right)^3 \div \left(r^3\right)^2$. Karla's answer is r.
    Explain why Karla's answer is incorrect and identify her error.
    State the correct answer.

24. Simplify each of the following, giving your answer in index form. Justify your answer in each case.

a. $\dfrac{\left(2k^3\right)^2}{4k^4}$

b. $\dfrac{4p^5}{p^4 \times 6p}$

c. $15s^8t^3 \div \left(5s^2t^2\right) \times 2st^4$

d. $12b^4c^6 \div \left(3b^3\right) \div \left(4c^2\right)$

e. $\left(f^4g^3\right)^2 - fg^3 \times f^7g^3$

25. Simplify each of the following, giving your answer in index form. Justify your answer in each case.

a. $\dfrac{\left(3p^3\right)^2 \times 4p^7}{2\left(p^4\right)^3}$

b. $2\left(x^2y\right)^4 \times 8xy^3$

c. $24x^2y^4 \div \left(12xy^3\right) - xy$

26. Using BIDMAS, calculate $2^{\left(2^{\left(2^2\right)}\right)}$ and $\left(\left(2^2\right)^2\right)^2$, and show that they are not equal. You can use your calculator to assist you.

27. a. Simplify the following, leaving numbers in index form.
$$3^{2^2},\ 2^{2^{2^2}},\ 2^{3^3},\ 5^2,\ 2^{5^2},\ 2^{2^5}$$
b. Arrange the numbers in ascending order.

28. a. Identify as many different expressions as possible that when raised to a power will result in $16x^8y^{12}$.

b. Identify as many different expressions as possible that when raised to a power will result in $3^{12n}a^{6n}b^{12n}$.

LESSON
2.7 Review

2.7.1 Topic summary

Index (or exponent) notation

- Index (or exponent) notation is a short way of writing a repeated multiplication.
  e.g. $2 \times 2 \times 2 \times 2 \times 2 \times 2$ can be written as 2^6, which is read as '2 to the power of 6'.
- The base is the number that is being repeatedly multiplied, and the index is the number of times it is multiplied.
  $$2^6 = 2 \times 2 \times 2 \times 2 \times 2 \times 2$$
  $$= 64$$
- In the above example, the number 64 is called a basic numeral.

INDEX LAWS

First Index Law

- When numbers with the same base are multiplied, keep the base the same and add the powers.
  $$a^m \times a^n = a^{m+n}$$
  e.g. $x^4 \times x^3 = x^7$
  $3x^2 \times 5x^4 = 15x^6$

Fourth Index Law

- When a power is raised to another power, keep the base the same and multiply the powers.
  $$\left(a^m\right)^n = a^{m \times n}$$
  e.g. $\left(x^3\right)^4 = x^{12}$

Second Index Law

- When numbers with the same base are divided, keep the base the same and subtract the powers.
  $$a^m \div a^n = a^{m-n}$$
  e.g. $x^7 \div x^4 = x^3$
  $20x^6 \div 12x^2 = \dfrac{20x^6}{12x^2}$
  $= \dfrac{5x^4}{3}$

Fifth and Sixth Index Laws

- Every term inside brackets must be raised to the power outside the brackets.
  $$(a \times b)^m = a^m \times b^m$$
  $$\left(\frac{a}{b}\right)^m = \frac{a^m}{b^m}$$
  e.g. $(2a)^5 = 2^5 \times a^5$
  $= 32a^5$
  $\left(\dfrac{ab}{3x}\right)^4 = \dfrac{a^4 b^4}{3^4 x^4}$

Third Index Law

- Any term (excluding 0) raised to the power of 0 is equal to 1.
  $$a^0 = 1$$
  e.g. $(2a)^0 = 1$
  $(2x^2 \times 5a^3)^0 = 1$

2.7.2 Project

Scientific notation and standard form

Scientists work with many extremely large (and small) numbers, which are not easy to use in their basic numeral form. For example, the distance to the nearest star outside the solar system, Proxima Centauri, is 40 000 000 000 000 000 m, and the radius of a hydrogen atom is 0.000 000 000 025 m.

Such numbers can look a little clumsy. Counting the zeros can be hard on the eye, and it's easy to miss one. Furthermore, your calculator would not be able to fit all the digits on its screen!

Scientists use powers of 10 in a number system called **scientific notation** or **standard form**. They have also come up with prefixes that stand for certain powers of 10. There is a prefix for every third power.

Work with a partner and use the internet to complete the following table, which shows the scientific notation prefixes and abbreviations for a wide range of numbers.

Your calculator will accept very large or very small numbers when they are entered because it uses scientific notation.

Scientific notation	Basic numeral	Name	SI prefix	SI symbol
1.0×10^{12}	1 000 000 000 000		tera	
1.0×10^{9}		Billion		
1.0×10^{6}	1 000 000		mega	M
1.0×10^{3}				
	100	Hundred	hecto	
1.0×10^{1}			deca	da
1.0×10^{-1}	0.1			
1.0×10^{-2}	0			
1.0×10^{-3}	0.001	Thousandth		
1.0×10^{-6}		Millionth	micro	μ
1.0×10^{-9}			nano	
1.0×10^{-12}	0.000 000 000 001	Trillionth		p

Note: SI is the abbreviation for International System of Units.

Use the following steps to write the number 825 460 in scientific notation.

Step 1: Place a decimal point so that the number appears to be between 1 and 10, that is a single digit then a decimal point.

8.254 60

Step 2: Count how many decimal places the decimal point is from its old position. (*Note:* For whole numbers, this is at the right-hand end of the number.) In this case, it is five places away.

$$8.254\,60$$

Step 3: Multiply the number in step 1 by the power of 10 equal to the number of places in step 2.

$$8.254\,60 \times 10^5$$

Note: If your number was made smaller in step 1, multiply it by a positive power to increase it to its true value. If your number was made larger in step 1, multiply it by a negative power to reduce it to its true value.

Proxima Centauri, near the Southern Cross, is the closest star to Earth and is 4.2 light-years away. A light-year is the distance that light travels in 1 year. Light travels at 300 000 kilometres per second.

1. Write 300 000 km/s in scientific notation.
2. Calculate the distance travelled by light in 1 minute.
3. Determine the distance travelled by light in 1 hour.
4. Calculate the distance travelled by light in 1 day.
5. Multiply your answer in question **4** by 365.25 to find the length of a light-year in kilometres. (Why do we multiply by 365.25?) Write this distance in scientific notation.
6. Calculate the distance from Earth to Proxima Centauri in kilometres.
7. Evaluate the distance from Earth to some other stars in both light-years and kilometres.
8. If light takes 500 seconds to travel from the Sun to Earth, evaluate the distance from Earth to the Sun in kilometres. Express your answer in scientific notation.

 Resources

Interactivities Crossword (int-2620)
Sudoku puzzle (int-3183)

Exercise 2.7 Review questions

learn on

Fluency

1. State the base for each of the following.
   a. 5^{10} b. 9^4 c. x^8 d. w^7

2. State the power or index for each of the following.
   a. 11^6 b. 23^5 c. C^{17} d. L^{100}

3. Write the following in index form.
   a. $7 \times 7 \times 7 \times 7$ b. $3 \times 3 \times 3 \times 3 \times 3 \times 3 \times 3$
   c. $m \times m \times m \times m \times m$ d. $k \times k \times k \times n \times n \times n \times n \times n$

4. Write each of the following as a basic numeral.
   a. 6^2 b. 8^2 c. 3^4 d. 2^7 e. 5^3

5. Evaluate each of the following.
   a. $7^2 - 4^2$

   b. $9^2 + 3^3 - 5^2$

6. Simplify each of the following.
   a. $3^5 \times 3^6$
   b. $10^{11} \times 10^4$
   c. $7^3 \times 7^6$
   d. $j^4 \times j^6 \times j^9$

7. Simplify each of the following.
   a. $t^4 \times t^5 \times t$
   b. $2z^5 \times 6z \times z$
   c. $5w^3 \times 7w^{12} \times w^{14}$
   d. $2e^2p^3 \times 6e^3p^5$

8. Simplify each of the following.
   a. $6^5 \div 6^2$
   b. $12^{10} \div 12$
   c. $5^{24} \div 5^{14}$
   d. $2^6 \div 2^2$

9. Simplify each of the following.
   a. $\dfrac{3^{20}}{3^{11}}$
   b. $\dfrac{m^{99}}{m^{66}}$
   c. $\dfrac{p^{15}}{p}$
   d. $\dfrac{h^7 \times h^{11}}{h^5}$

Understanding

10. Simplify each of the following.
    a. $\dfrac{L^6 \times L^2 \times L^4}{L^8}$
    b. $\dfrac{y^5 \times y^7 \times y^2}{y^8}$
    c. $\dfrac{a^7 \times a \times a^5}{a^3 \times a^6}$
    d. $\dfrac{c^4 \times c^2 \times c \times c^7}{c^3 \times c^8 \times c^4}$

11. Simplify the following.
    a. 4^0
    b. $r^4 s^0 u^9$
    c. 1966^0
    d. m^0

12. Simplify the following.
    a. $d^2 e^6 f^0$
    b. zb^0
    c. $7w^0$
    d. $8q^0 - 2q^0$

13. Simplify the following.
    a. $4s^0 + 60t^0$
    b. $v^0 w^5$
    c. $\left(x^3 y^6\right)^0$
    d. klm^0

14. Raise each of the following to the given power.
    a. $\left(2^4\right)^3$
    b. $\left(6^9\right)^2$
    c. $\left(7^4\right)^{10}$
    d. $\left(n^{21}\right)^6$

15. Raise each of the following to the given power.
    a. $\left(r^{16} i^{12}\right)^2$
    b. $\left(b^2 d^8\right)^{20}$
    c. $\left(2pm^3\right)^3$
    d. $\left(9wz^4\right)^2$

16. **MC** Select which of the following is equal to $\left(\dfrac{4b^4}{d^2}\right)^3$.

    A. $\dfrac{4b^3}{d^3}$
    B. $\dfrac{12b^{12}}{d^6}$
    C. $\dfrac{64b^{12}}{d^6}$
    D. $\dfrac{64b^7}{d^5}$

Communicating, reasoning and problem solving

17. a. Evaluate each of the following.
    i. $(-1)^1$
    ii. $(-1)^2$
    iii. $(-1)^3$
    iv. $(-1)^4$
    v. $(-1)^5$
    vi. $(-1)^6$
    b. Use your answers to part **a** to complete the following sentence:
    If negative one is raised to an even power, the result is ____; if it is raised to an odd power, the result is ____.
    c. Consider the expression $(-1)^k + (-1)^l$.
    Determine all possible values of the above expression. Specify the values of k and l for which each result occurs.

▶

18. At 9 am there were 10 bacteria in a Petri dish.

   a. If the number of bacteria doubles every minute, evaluate how many bacteria were in the Petri dish after:

   i. 1 minute   **ii.** 2 minutes

   iii. 3 minutes   **iv.** 10 minutes.

   b. Develop the rule that connects the number of bacteria, N, and the time, t (in minutes), after 9 am.

   c. Use your answer to part **b** to calculate the number of bacteria in the dish at 10 am.

   Give your answer in index form (do not evaluate).

19. Lena receives an email containing a chain letter. She is asked to forward this letter to 5 friends (or else she will have a lot of bad luck!).

   Lena promptly sends 5 letters as instructed. (Let's call this the *first round* of letters.)

   Each of Lena's 5 friends also sends 5 letters. (Call this the *second round* of letters.)

   a. Determine the number of letters sent in the second round. Give your answer:

   i. as a basic numeral   **ii.** in index form.

   b. Determine the number of letters that would be sent in the third round. Give your answer:

   i. as a basic numeral   **ii.** in index form.

   c. Assuming the chain is not broken and each recipient sent out 5 letters, determine the total number of letters sent in the first four rounds. Give your answer:

   i. as a basic numeral   **ii.** in index form.

20. Nathan is considering participating in the Premier's Reading Challenge. He decided to test himself first by trying to read a 400-page book in 6 days.

   Nathan read 7 pages on day 1.

   After performing some basic arithmetic computations, he realised that he needed to increase that amount to be able to finish the book on time.

   Nathan decided to double the number of pages read every day.

   a. Determine the number of pages that Nathan read on:

   i. day 2   **ii.** day 3.

   b. Develop the formula connecting the number of pages P read per day and the number of days d.

   c. Use your answer to part **b** to find the number of pages Nathan will read on day 6.

   d. Show, with mathematical calculations, whether Nathan will finish the book in six days if he continues according to plan.

21. Alex bought a second-hand car for \$25 000. Each year the car depreciates by 20% (i.e. each year it loses 20% of its value).

   a. Calculate the value of the car at the end of the first year.

   b. Evaluate the value of the car at the end of the second year.

   c. The value, V, of the car can be found using the formula $V = 25\,000 \times 0.8^t$, where t is the number of years after purchase.

   Explain the meaning of the numbers 25 000 and 0.8.

   d. Create a table and record the results for $t = 1$ to $t = 5$.

   e. Alex decided that he will sell his car when its value falls below \$5000.

   Determine after how many years the value of his car will fall below \$5000.

22. The number, E, of employees in a large firm grows according to the rule $E = 60 \times 1.15^t$, where t is the number of years from the year 2018.
   a. Determine the number of people the firm employed in the year 2018.
   b. Determine the number of employees there were in:
      i. 2019
      ii. 2020.
   c. Determine the number of years taken for the number of employees to exceed 200.

23. Four rabbits were accidentally introduced to a small island. The population of rabbits doubled every 4 weeks.
   a. Determine the number of rabbits on the island:

      i. 8 weeks later
      ii. 24 weeks later
      iii. 1 year later.

   b. After 1 year, to cope with the rabbit problem, some foxes were brought to the island.
      As a result, the population of rabbits started declining by 10% each week.
      After the foxes had been brought in, determine how many rabbits were left after:

      i. 1 week
      ii. 2 weeks
      iii. 10 weeks.

24. There were 50 bacteria of Type X and 30 bacteria of Type Y in a Petri dish. The number of bacteria of Type X doubles every 4 hours; the number of bacteria of Type Y quadruples every 6 hours.
   Evaluate the total number of bacteria in the dish after:
   a. 12 hours
   b. 1 day
   c. 2 days.

25. A basic numeral can be expressed in standard form (also called scientific notation) by being written as a number between 1 and 10 multiplied by a power of 10.
   For example, the number 4000 in standard form is written as 4.0×10^3.
   For each of the following situations, express the basic numeral in standard form.
   a. A company declares an annual profit of 3 billion dollars.
   b. The diameter of Earth is (approximately) 12 750 km.
   c. The half-life of a certain radioactive element is 5 000 000 years.
   d. Light travels at a speed of 300 000 km/s.

26. a. Express the basic numerals 4, 8 and 16 as powers of 2.
   b. Use your answers from part a to simplify the following expression.
   $$\frac{4^x \times 8^y}{16}$$

27. If $a^2 = 7$, determine the value of:
   a. a^{4+1}
   b. $2a^6$
   c. $3a^6 - 4a^4$.

28. A rubber ball is dropped from a balcony that is 10 m above the ground. The ball bounces to $\frac{3}{4}$ of its previous height after each bounce.
   a. Determine the greatest height of the ball above the ground after:
      i. 1 bounce
      ii. 3 bounces
      iii. 5 bounces.
   b. Determine when the height of the ball above the ground will be less than 1 m.

   on To test your understanding and knowledge of this topic, go to your learnON title at www.jacplus.com.au and complete the **post-test**.

Answers

Topic 2 Index laws

2.1 Pre-test

1. 3^5

2. 72

3. $\dfrac{7}{8}$

4. $10\,c^7$

5. D

6. $\dfrac{3d^3}{4}$

7. 1

8. $\dfrac{3m^5}{4m^3} \times \dfrac{8m^4}{12m^6} = \dfrac{3m^5}{4m^3} \times \dfrac{2m^4}{3m^6}$

   $\qquad = \dfrac{3m^5 \times 2m^4}{4m^3 \times 3m^6}$

   $\qquad = \dfrac{6m^9}{12m^9}$

   $\qquad = \dfrac{6}{12}$

   $\qquad = \dfrac{1}{2}$

9. a. 7^{15}

   b. $16w^{20}$

   c. $45p^{11}$

10. a. $45a^5b^5$

    b. $\dfrac{3a^3b}{2}$

11. D

12. $\dfrac{9g^{10}}{16h^6}$

13. $\left(\left(2^3 \right)^2 \right)^3 = 2^{18} = 262\,144$

    $\left(\left(3^2 \right)^3 \right)^2 = 3^{12} = 531\,441$

    Therefore, they are not equal.

14. 7^9

15. $2\left(a^3b^3 + a^4b^3 + a^3b^4 \right)$

2.2 Review of index form

1. a. Base $= 8$; power $= 4$

   b. Base $= 7$; power $= 10$

   c. Base $= 20$; power $= 11$

2. a. Base $= 19$; power $= 0$

   b. Base $= 78$; power $= 12$

   c. Base $= 3$; power $= 100$

3. a. Base $= c$; power $= 24$

   b. Base $= n$; power $= 36$

   c. Base $= d$; power $= 42$

4. a. 2^6 b. 4^4 c. x^5

5. a. 9^3 b. $11l^7$ c. $44m^5$

6. a. 4×4 b. $5 \times 5 \times 5 \times 5$

   c. $7 \times 7 \times 7 \times 7 \times 7$ d. $6 \times 6 \times 6$

7. a. $3 \times 3 \times 3 \times 3 \times 3 \times 3$

   b. $n \times n \times n \times n \times n \times n \times n$

   c. $a \times a \times a \times a$

   d. $k \times k \times k \times k \times k \times k \times k \times k \times k$

8. a. 243   b. 256

   c. 256   d. 1331

9. a. 2401   b. 216

   c. 1   d. 625

10. B

11. D

12. a. $2^2 \times 4^4 \times 6$ b. $3^4 \times 7^4$

    c. $2^3 \times 19^5$ d. $4^4 \times 13^2$

13. a. $66m^5p^2s^2$ b. $378i^3n^2r^3$

    c. $192e^3kp^2$ d. $99j^5lp^2$

14. a. $15 \times f \times f \times f \times j \times j \times j \times j$

    b. $7 \times k \times k \times k \times k \times k \times k \times s \times s$

    c. $4 \times b \times b \times b \times c \times c \times c \times c \times c$

    d. $19 \times a \times a \times a \times a \times m \times n \times n \times n$

15. a. $400 = 2^4 \times 5^2$ b. $225 = 3^2 \times 5^2$

    c. $2000 = 2^4 \times 5^3$

16. a. $64 = 2^6$ b. $40 = 2^3 \times 5$ c. $36 = 2^2 \times 3^2$

17. a. 120   b. 100   c. 216

18. a. 308   b. 1575   c. 760 000

19. a. 10^1 b. 10^2 c. 10^3 d. 10^6

20. a. 5×10^2

    b. $4 \times 10^2 + 7 \times 10^1$

    c. $2 \times 10^3 + 3 \times 10^2 + 6 \times 10^1$

    d. $1 \times 10^3 + 9 \times 10^2 + 8 \times 10^1$

    e. $5 \times 10^3 + 4 \times 10^2 + 3 \times 10^1$

21. a. 75 000   b. 30 600   c. 5 200 480

22. The base is the number being multiplied and the index represents how many times the base should be multiplied by itself.

23. Factors multiplied together in 'shorthand' form:
    $a \times a \times a \times b \times b \times b \times b$

24. The second option is better.

25. a. $2^3a^2b^3c^4$ b. 2 074 464

26. a. $1 \times 2^3 + 1 \times 2^1$, $1 \times 2^3 + 1 \times 2^2 + 1 \times 2^1$

    b. $2 \times 2^3 + 1 \times 2^2 + 2 \times 2^1 = 1 \times 2^4 + 1 \times 2^2 + 1 \times 2^2$

    $\qquad = 1 \times 2^4 + 2 \times 2^2$

    $\qquad = 1 \times 2^4 + 1 \times 2^3$

    c. 24

27. a. $1 \times 10^2 + 4 \times 10^1$, $6 \times 10^2 + 8 \times 10^1$

    b. $7 \times 10^2 + 12 \times 10^1 = 8 \times 10^2 + 2 \times 10^1$

    c. 820

d. Personal response required. For example, $3160 + 4550$.
$$3160 = 3 \times 10^3 + 1 \times 10^2 + 6 \times 10^1$$
$$4550 = 4 \times 10^3 + 5 \times 10^2 + 5 \times 10^1$$
$$3 \times 10^3 + 1 \times 10^2 + 6 \times 10^1 + 4 \times 10^3 + 5 \times 10^2 + 5 \times 10^1$$
$$= 7 \times 10^3 + 6 \times 10^2 + 11 \times 10^1$$
$$= 7000 + 600 + 110$$
$$= 7710$$

e. Sample responses can be found in the worked solutions in the online resources.

2.3 Multiplying powers

1. **a.** 3^9 **b.** 6^7
   c. 10^{10} **d.** 11^6

2. **a.** 7^9 **b.** 2^{14}
   c. 5^4 **d.** 8^{11}

3. **a.** 13^{15} **b.** q^{47}
   c. x^{14} **d.** e^4

4. **a.** 3^{12} **b.** 2^{18}
   c. 5^{17} **d.** 6^{11}

5. **a.** 10^6 **b.** 17^{12}
   c. p^{22} **d.** e^{23}

6. **a.** g^{28} **b.** e^{38}
   c. $3b^{13}$ **d.** $5d^{16}$

7. A

8. B

9. **a.** $20p^{11}$ **b.** $6x^8$ **c.** $56y^{10}$

10. **a.** $21p^8$ **b.** $84t^6$ **c.** $30q^{15}$

11. **a.** $6a^6e^7$ **b.** $8h^{12}p^6$ **c.** $80m^9$
    d. $6g^3h^6$ **e.** $30p^6q^9$

12. **a.** $48u^9w^7$ **b.** $27d^{11}y^{17}$ **c.** $42b^{14}c^9$
    d. $24r^{16}s^{18}$ **e.** $60h^{38}v^{20}$

13. **a.** 3^{x+4} **b.** 3^{2y+2}
    c. 3^{6y-5} **d.** $3^{\frac{23}{12}}$

14. $9 = 3^2$; $27 = 3^3$; $81 = 3^4$

15. **a.** 3^{10} **b.** 3^{2n+2}

16. The bases, 2 and 3, are different. The index laws do not apply.

17. Step 1: 5^n
    Step 2: 5^m
    Step 3: 5^{n+m}

18. Adding the powers when multiplying terms with the same base gives an equivalent answer to evaluating the indices separately. Trying to add powers when the bases are different does not result in an equivalent answer to evaluating the indices separately.

19. **a.** 2^9 **b.** 2^{n-1}
    c. $2^5 + 2^6 = 96$ **d.** $2^{13} + 2^{14}$
    e. $3(2^{13})$

20. **a.** x^6 **b.** 2^{12}
    c. 5^2 **d.** $(a^n)^p = a^{np}$

21. 2

2.4 Dividing powers

1. **a.** 2^3 **b.** 7^4
   c. 10^3 **d.** $\dfrac{1}{9}$

2. **a.** 3   **b.** 11^7
   c. 5^4 **d.** 12^5

3. **a.** 3^3 **b.** 13
   c. 6^4 **d.** 10^4

4. **a.** 15^{33} **b.** h^{77}
   c. b^{70} **d.** f^{900}

5. **a.** $3x^2$ **b.** $6y^2$
   c. $8w^7$ **d.** $3q^4$

6. **a.** $8f^9$ **b.** $10h^{90}$
   c. $4j^{10}$ **d.** $5p^{10}$

7. **a.** $8g^3$ **b.** $\dfrac{3b^6}{2}$
   c. $\dfrac{9m^4}{2}$ **d.** $\dfrac{5n^{90}}{2}$

8. D

9. A

10. **a.** $3p^4$ **b.** $6r^4$ **c.** $9a^3$

11. **a.** $3b^6$ **b.** $20r^4$ **c.** $9q$

12. **a.** $\dfrac{3p^5}{2}$ **b.** $\dfrac{8b^5}{3}$ **c.** $\dfrac{5m^{10}n^6}{6}$

13. **a.** $\dfrac{9x^8y}{4}$ **b.** $\dfrac{4hk^3}{3}$ **c.** $3f^3j^5$

14. **a.** $\dfrac{4p^2rs}{3}$ **b.** $\dfrac{9a^5b^3c}{2}$ **c.** $\dfrac{20f^6g^2h^4}{3}$

15. **a.** 2^{10-p} **b.** 2^{4e+4} **c.** 5^{3x+y} **d.** 3^{1-m}

16. **a.** $\dfrac{2^3 \times 2^4 \times 2^2}{2^1 \times 2^5}$
    b. **i.** 2^3 **ii.** 8

17. **a.** $\dfrac{2 \times 3 \times 3^3 \times 2^2 \times 3^2}{2^2 \times 3 \times 3^4}$
    b. **i.** $2^1 \times 3^1$ **ii.** 6

18. Sam's answer is incorrect. The bases, 12 and 3, are different. The laws of indices do not apply.

19. Step 1: 3^p
    Step 2: 3^q
    Step 3: $3^p \div 3^q = 3^{p-q}$

20. Use the factor form method of simplification and the Second Index Law (as the other method) to deduce that $y^{-2} = \dfrac{1}{y^2}$.

21. $a^p \div a^p = 1$ and $a^p \div a^p = a^{p-p} = a^0 = 1$

22. $m^6 \div m^7 = m^{6-7} = m^{-1}$ and $m^6 \div m^7 = \dfrac{m^6}{m^7} = \dfrac{1}{m}$; thus, $\dfrac{1}{m} = m^{-1}$.

23. $\pm \dfrac{1}{6}$

2.5 The zero index

1. a. 1   b. 1   c. 1   d. 1
2. a. 1   b. 1   c. 1   d. 1
3. a. 4   b. 12   c. -9 d. 22
4. a. $12m^3$ b. 21   c. 72   d. $\dfrac{8}{7}$
5. a. 2   b. $10b^2$ c. $3p^2$ d. 1
6. a. 1   b. 1   c. 1
7. a. 7   b. 3   c. 3
8. a. 1   b. 4   c. 2   d. $2p^9$
9. a. $\dfrac{3e^4}{2}$ b. y c. $\dfrac{m^3}{3}$ d. $\dfrac{c^3}{4}$
10. D
11. A
12. D
13. a. 1   b. 2   c. 2   d. 2   e. 2
14. a. $\dfrac{h^2}{2}$ b. $\dfrac{q^4}{5}$ c. $\dfrac{n^3}{5}$ d. v^2 e. $2x^6$
15. Any base raised to the power of 0 equals 1.
16. $2^0 = 1,\ x^0 = 1 \Rightarrow \dfrac{2^0 x^2}{2^2 x^0} = \dfrac{x^2}{2^2} = \dfrac{x^2}{4}$
17. $a^{-\frac{1}{2}} = \dfrac{1}{\sqrt{a}}$
18. Four sample expressions are: $\dfrac{y^{20}}{y^{15}}$, $\dfrac{6x^2 y \times 5xy^5}{15x^3 \times 2y}$, $\dfrac{(6xy)^0}{6} \times 6y^{\frac{7}{2}} \times y^3,\ 3x^0 + y^5 - 3$
19. $\pm \dfrac{1}{3}$
20. Peter has treated the different occurrences of x incorrectly. He has calculated $x^{6 \div 2}$ instead of x^{6-2}.
    The answer is $4x^4$.
    If 64 and 16 are converted to base 2 or base 4, and the indices are divided, not subtracted, an error will occur.

2.6 Raising powers

1. a. 3^6 b. 6^{80} c. 11^{100}
   d. 5^{144} e. 3^8
2. a. $3^8 \times 10^{12}$ b. $13^5 \times 17^{15}$ c. $\dfrac{3^{30}}{2^{20}}$
   d. $3^4 w^{36} q^8$ e. $\dfrac{7^2 e^{10}}{r^4 q^8}$
3. a. $p^8 q^6$ b. $r^{15} w^9$ c. $b^{10} n^{18}$
4. a. $j^{18} g^{12}$ b. $q^4 r^{20}$ c. $h^{24} j^{16}$
5. a. $f^{16} a^{21}$ b. $t^{10} u^8$ c. $i^{15} j^{12}$

6. a. 2^{20} b. t^{33} c. a^{21}
7. a. b^{24} b. e^{66} c. g^{39}
8. a. $324a^{20}$ b. $216d^{27}$ c. $40\,000r^{54}$
9. B
10. B
11. D
12. a. a^6 b. m^4 c. n^3 d. b^8
13. a. f^{17} b. g^6 c. p^9 d. y^2
14. a. c^{20} b. f^7 c. k^{14} d. p^{16}
15. a. $\dfrac{9b^8}{d^6}$ b. $\dfrac{25h^{20}}{4j^4}$ c. $\dfrac{8k^{15}}{27t^{24}}$
16. a. $\dfrac{49p^{18}}{64q^{44}}$ b. $\dfrac{125y^{21}}{27z^{39}}$ c. $\dfrac{256a^{12}}{2401c^{20}}$
17. a. $2g^8$ b. $8p^8$ c. w^{18}
    d. $6x^5$ e. $4d^6$
18. a. $15a^{10}$ b. $3s^6$ c. $60b^4 c^8$
    d. $2x^4$ e. $f^8 g^6$
19. a. $\dfrac{8u^3 v^4}{3}$ b. $x^3 y^7$ c. $15a^9 b^5$
    d. xy e. $64p^6 q^{15}$
20. Any base, even a complex one like this, raised to the power of 0 equals 1.
21. a. 4^6 b. 4096   c. 4096
22. a. w^{10} b. $6x^2$ c. $12a^{11}$
    d. $8x^2$ e. $7d^3 + d^2$
23. For the first term, Karla has added the powers instead of multiplying, which is incorrect. She has multiplied the powers in the second term, which is correct according to the index laws.
    $$r^7 \div r^6 = r^1 = r$$
    The correct answer is r^6.
    $$\dfrac{(r^4)^3}{(r^3)^2} = \dfrac{r^{12}}{r^6} = r^6$$
24. a. k^2 b. $\dfrac{2}{3}$ c. $6s^7 t^5$
    d. bc^4 e. 0
25. a. $18p$ b. $16x^9 y^7$ c. xy
26. $2^{\left(2^{\left(2^2\right)}\right)} = 2^{2^4} = 2^{16} = 65\,536;$

    $2^{(2^{(2^2)})}$ ■   _{Hin} ▲

    65536

    $\left(\left(2^2\right)^2\right)^2 = \left((4)^2\right)^2 = 16^2 = 256\ (= 2^8)$

    $((2^2)^2)^2$ ■   _{Hin} ▲

    256

27. a. $3^4; 2^{16}; 2^{27}; 5^2; 2^{25}; 2^{32}$

 b. $5^2; 3^{2^2}; 2^{2^{2^2}}; 2^{5^2}; 2^{3^3}; 2^{2^5}$

28. a. $\left(16x^8y^{12}\right)^1, \left(4x^4y^6\right)^2, \left(2x^2y^3\right)^4$

 b. $\left(3^{12n}a^{6n}b^{12n}\right)^1, \left(3^{12}a^6b^{12}\right)^n, \left(3^6a^3b^6\right)^{2n}, \left(3^{4n}a^{2n}b^{4n}\right)^3,$
 $\left(3^4a^2b^4\right)^{3n}, \left(3^2a^1b^2\right)^{6n}, \left(3^{6n}a^{3n}b^{6n}\right)^2, \left(3^{2n}a^nb^{2n}\right)^6$

Project

See the table at the bottom of the page.*

1. 3.0×10^5 km/s

2. 1.8×10^7 km/min

3. 1.08×10^9 km/h

4. 2.592×10^{10} km/day

5. 9.47×10^{12} km. On average, there are 365.25 days in 1 year.

6. 3.98×10^{13} km

7. Answers will vary. A sample response is given here. Alpha Centauri A and B are roughly 4.35 light-years and 4.12×10^{13} km away.

8. 1.5×10^8 km

2.7 Review questions

1. a. 5 **b.** 9 **c.** x **d.** w

2. a. 6 **b.** 5 **c.** 17 **d.** 100

3. a. 7^4 **b.** 3^7 **c.** m^5 **d.** k^3n^5

4. a. 36 **b.** 64 **c.** 81 **d.** 128
 e. 125

5. a. 33 **b.** 83

6. a. 3^{11} **b.** 10^{15} **c.** 7^9 **d.** j^{19}

7. a. t^{10} **b.** $12z^7$ **c.** $35w^{29}$ **d.** $12e^5p^8$

8. a. 6^3 **b.** 12^9 **c.** 5^{10} **d.** 2^4

9. a. 3^9 **b.** m^{33} **c.** p^{14} **d.** h^{13}

10. a. L^4 **b.** y^6 **c.** a^4 **d.** $\dfrac{1}{c}$

11. a. 1 **b.** r^4u^9 **c.** 1 **d.** 1

12. a. d^2e^6 **b.** z **c.** 7 **d.** 6

13. a. 64 **b.** w^5 **c.** 1 **d.** kl

14. a. 2^{12} **b.** 6^{18} **c.** 7^{40} **d.** n^{126}

15. a. $i^{24}r^{32}$ **b.** $b^{40}d^{160}$ **c.** $8p^3m^9$ **d.** $81w^2z^8$

16. C

17. a. i. -1 **ii.** 1 **iii.** -1
 iv. 1 **v.** -1 **vi.** 1

 b. Positive one; negative one

 c. -2 if k and l are both odd; 0 if one of the powers is odd and one is even; 2 if both k and l are even.

18. a. i. 20 **ii.** 40 **iii.** 80 **iv.** 10 240

 b. $N = 10 \times 2^t$

 c. 10×2^{60}

19. a. i. 25 **ii.** 5^2

 b. i. 125 **ii.** 5^3

 c. i. 780 **ii.** $5 + 5^2 + 5^3 + 5^4$

20. a. i. 14 **ii.** 28

 b. $P = 7 \times 2^{d-1}$

 c. 224

 d. Yes, the total for 6 days is 441, which is more than 400.

21. a. $20\,000

 b. $16\,000

 c. 25 000 represents the purchase price of the car; 0.8 means 80% (expressed as a decimal) — this is the portion of the value that the car retains after each year.

 d. Create a table recording the value of the car for $t = 1$ up to $t = 5$, use the formula $V = \$25\,000 \times 0.8^t$

---

*

Scientific notation	Basic numeral	Name	SI prefix	SI symbol
1.0×10^{12}	1 000 000 000 000	Trillion	tera	T
1.0×10^9	1 000 000 000	Billion	giga	G
1.0×10^6	1 000 000	Million	mega	M
1.0×10^3	1000	Thousand	kilo	k
1.0×10^2	100	Hundred	hecto	h
1.0×10^1	10	Ten	deca	da
1.0×10^{-1}	0.1	Tenth	deci	d
1.0×10^{-2}	0.01	Hundredth	centi	c
1.0×10^{-3}	0.001	Thousandth	milli	m
1.0×10^{-6}	0.000 001	Millionth	micro	μ
1.0×10^{-9}	0.000 000 001	Billionth	nano	n
1.0×10^{-12}	0.000 000 000 001	Trillionth	pico	p

t	1	2	3	4	5
Value of car	$20 000	$16 000	$12 800	$10 240	$8 192

e. After 8 years

22. a. 60
   b. i. 69   ii. 79
   c. 9 years

23. a. i. 16   ii. 256   iii. 32 768
   b. i. 29 491   ii. 26 542   iii. 11 425

24. a. 880   b. 10 880   c. 2 170 880

25. a. 3.0×10^9 b. 1.275×10^4
   c. 5.0×10^6 d. 3.0×10^5

26. a. 2^2, 2^3, 2^4
   b. $2^{2x+3y-4}$

27. a. $49\sqrt{7}$ b. 686   c. 833

28. a. i. 7.5 m   ii. 4.218 75 m   iii. 2.373 05 m
   b. After 9 bounces

3 Real numbers

LESSON SEQUENCE

LESSON
3.1 Overview

Why learn this?

Whole numbers cannot be used to describe everything. Real numbers include positive and negative whole numbers (integers) and decimal numbers (or fractions). Imagine you have 1 apple to share between 2 people. What portion would each person get? This is impossible to describe using whole numbers, as each person would receive one-half. This value can be expressed using a fraction $\left(\dfrac{1}{2}\right)$ or a decimal number (0.5). Real numbers are used every day. Think about how you could count dollars and cents if you did not understand decimals. How could you follow a recipe or share something equally among friends if you did not understand fractions? Fractions and decimals are used for telling the time, calculating a discount on a sale item, measuring height and determining statistics for a sports match.

All occupations require a good knowledge of calculating real numbers. A builder needs to use decimals to measure things accurately, nurses and doctors use decimals to monitor blood pressure and administer the correct medication, and accountants use decimals when preparing tax returns and calculating profits and losses. A good understanding of adding, subtracting, multiplying and dividing real numbers will be crucial for your everyday life!

Hey students! Bring these pages to life online

▶ Watch videos

Engage with interactivities

A+ Answer questions and check solutions

Find all this and MORE in jacPLUS ▶

Reading content and rich media, including interactivities and videos for every concept

Extra learning resources

Differentiated question sets

Questions with immediate feedback, and fully worked solutions to help students get unstuck

1. Answer the following.

   a. Simplify $\dfrac{25}{30}$.

   b. Write the following fraction as a mixed number expressed in simplest form: $\dfrac{13}{5}$.

2. Evaluate the following fraction calculations, giving your answers in simplified fraction form.

   a. $\dfrac{3}{4} + \dfrac{1}{7}$

   b. $\dfrac{4}{5} - \dfrac{1}{2}$

3. Evaluate the following fraction calculations, giving your answers in simplified fraction form.

   a. $\dfrac{3}{4} \times \dfrac{5}{6}$

   b. $\dfrac{1}{6} \div \dfrac{1}{2}$

4. State whether the following is True or False: $\dfrac{7}{2}$ is a rational number.

5. Evaluate the following.
   a. $1.05 + 2.006$
   c. $2.16 - 1.847$

   b. 24.5×3.6
   d. $15.769 \div 1.3$

6. Write the ratio $\dfrac{4}{7} : \dfrac{6}{7}$ in its simplest form.

7. By rounding to the first digit, estimate the answer to $631 \div 19$.

8. Evaluate the following fraction calculations. Write the answers as mixed numbers in simplified form.

   a. $2\dfrac{1}{2} - 1\dfrac{1}{3}$

   b. $2\dfrac{1}{4} \times 1\dfrac{3}{7}$

9. Determine the value of the following fraction calculation, giving your answer in simplified form.

   $$\left(-1\dfrac{1}{3}\right) \div \left(-\dfrac{2}{3}\right)$$

10. Convert the following decimals to fractions in their simplest form.
    a. 3.14

    b. 0.625

11. **MC** From the following list, select the correct decimal for $\dfrac{2}{7}$.

    A. $0.\dot{2}86\,27\dot{2}$ 　　　 **B.** $0.\dot{2}\dot{8}$ 　　　 **C.** $0.\dot{2}85\,71\dot{4}$ 　　　 **D.** $0.\dot{2}67\,71\dot{4}$

12. By rounding the numerator to a multiple of the denominator, provide a whole number estimate of the value of the fraction $\dfrac{661}{50}$.

▶

13. **MC** The answer to $\dfrac{378 \times 490}{42 \times 5}$ has the digits 8, 8, 2.

By using any estimation method, choose the correct answer to the calculation from the given possibilities.

    A. 8882   **B.** 882   **C.** 88.2   **D.** 8.82

14. The area of a triangle is given as $A = \dfrac{1}{2}bh$ where b is the base length and h is the height of the triangle.

The area of a particular triangle is $6\dfrac{1}{4}$ cm^2 and the base of the triangle measures $1\dfrac{2}{3}$ cm.

Calculate the height of the triangle as a simplified mixed fraction.

15. Amira goes shopping and spends 20% of her money on a pair of shorts. She spends another one-third of her money on a present for her mum, and a quarter of her money on lunch out with her friends. She is left with $26. Evaluate how much Amira spent on the shorts.

LESSON
3.2 The real number system

LEARNING INTENTION

At the end of this lesson you should be able to:
- classify real numbers as irrational or rational
- classify rational numbers as integers or non-integers
- place real numbers on a number line.

3.2.1 Introduction to real numbers

eles-3576

- The **real number** system contains the set of **rational** and **irrational** numbers. It is denoted by the symbol R.

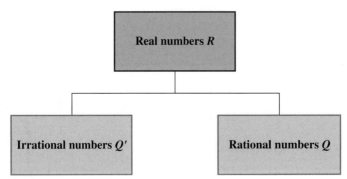

Rational numbers

- A rational number is any number that can be expressed as a ratio (or fraction) of two whole numbers in the form $\dfrac{a}{b}$, where a and b are whole numbers and $b \neq 0$. For example, $\dfrac{3}{5}$ is a rational number.
- The set of rational numbers is denoted by the symbol Q.
- All integers are rational numbers, as they can be expressed in the form $\dfrac{a}{b}$.

For example, $6 = \dfrac{6}{1}$ and $-3 = \dfrac{-3}{1}$.

- Terminating decimals and recurring decimals are also rational numbers.

  For example, $0.4 = \dfrac{4}{10} = \dfrac{2}{5}$ is a terminating decimal and $0.\dot{3} = 0.3333\ldots = \dfrac{1}{3}$ is a recurring decimal.

Irrational numbers

- Irrational numbers are numbers that *cannot* be expressed in the form $\dfrac{a}{b}$, where a and b are whole numbers and $b \neq 0$. Examples include $\sqrt{2}$, $\sqrt{7}$ and $2\sqrt{3}$.
- Irrational numbers can be expressed as decimals that do not terminate or repeat in any pattern.
- π (pi) is a special irrational number that relates the diameter of a circle to its circumference. It can be expressed as a decimal that never terminates or repeats in any pattern: $\pi = 3.141\,592\ldots$
- The value of π has been calculated to millions of decimal places by computers, yet it has still been found to be non-terminating and non-recurring.
- While there is no official symbol for the set of irrational numbers, we will use the symbol Q'. That is, the set of irrational numbers contains the numbers that are not rational.

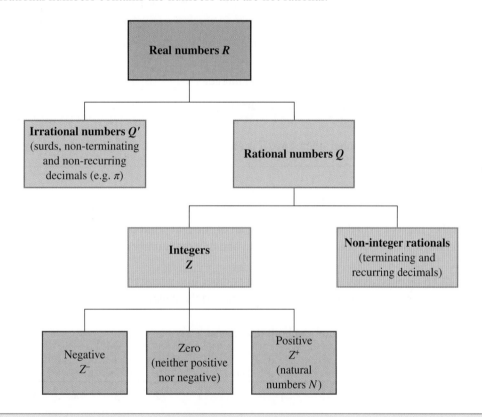

Digital technology

Your calculator can be useful in helping decide whether a number is rational or irrational.

If a number can be expressed as a fraction in the form $\dfrac{a}{b}$ where a and b are whole numbers and $b \neq 0$, then it is rational.

Most scientific calculators can convert a number into a fraction. Depending on the brand of calculator used, this may appear as f ◁▷ d, S ⇔ D or similar.

The calculator screens show how a recurring decimal can be converted to a fraction, while π and $\sqrt{5}$ cannot. This helps confirm whether a number is rational or irrational.

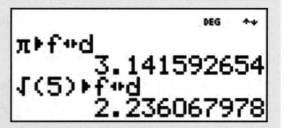

WORKED EXAMPLE 1 Classifying real numbers

Classify the following numbers as irrational, non-integer rational, or integer.

a. **1.2**   b. **−21**   c. $\sqrt{5}$ d. $\sqrt{4}$

THINK

a. 1.2 is a terminating decimal; therefore, it is rational. 1.2 is not an integer.

b. −21 is a negative whole number; therefore, it is an integer.

c. $\sqrt{5} = 2.360\,679\,\ldots$ is a non-terminating, non-recurring decimal; therefore, it is irrational.

d. $\sqrt{4} = 2$ is a positive whole number; therefore, it is an integer.

WRITE

a. Non-integer rational

b. Integer

c. Irrational

d. Integer

3.2.2 The real number line

eles-3577

- The real number line is a visual way of displaying all of the real numbers and their values.
- Any point on the real number line is a real number, and any real number can be placed on the real number line.
- The **origin**, often labelled as O, is the point on the number line where the number 0 sits.
- The positive real numbers sit to the right of the origin, and the negative real numbers sit to the left of the origin.
- The real number line extends infinitely in both directions.

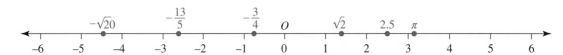

WORKED EXAMPLE 2 Placing real numbers on a number line

Place the following numbers on a real number line.
You may use a calculator to determine the values where required.

a. π b. $\sqrt{9}$ c. $-\dfrac{9}{5}$ d. $-\sqrt{2}$

THINK

a. 1. Use a calculator to determine the approximate value of π.

   2. Place π on the number line.

WRITE

a. $\pi \approx 3.14$

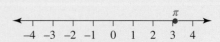

b. 1. Determine the value of $\sqrt{9}$.

2. Place $\sqrt{9}$ on the number line.

b. $\sqrt{9} = 3$

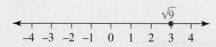

c. 1. Determine the value of $-\dfrac{9}{5}$.

2. Place $-\dfrac{9}{5}$ on the number line.

c. $-\dfrac{9}{5} = -1.8$

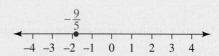

d. 1. Use a calculator to determine the approximate value of $-\sqrt{2}$.

2. Place $-\sqrt{2}$ on the number line.

d. $-\sqrt{2} \approx -1.41$

---

on Resources

Interactivities Real numbers (int-3717)
The number line (int-3720)

Exercise 3.2 The real number system

learnon

3.2 Quick quiz on   3.2 Exercise

Individual pathways

■ PRACTISE	■ CONSOLIDATE	■ MASTER
1, 5, 9	2, 4, 6, 10, 11	3, 7, 8, 12

Fluency

1. **WE2** Place the following numbers on a number line.

 a. -4 b. $\dfrac{1}{4}$ c. $\sqrt{5}$ d. -3.6

2. Place the following numbers on a number line.

 a. 1.9 b. $-\dfrac{2}{4}$ c. 2.4 d. $2\sqrt{2}$

3. Place the following numbers on a number line.

 a. $\sqrt{10}$ b. -3.8 c. $\dfrac{48}{12}$ d. $\dfrac{-5}{-2}$

4. Identify which of the following numbers sits furthest to the right on a number line.

$$-1.8, \quad 5.3, \quad 5\frac{5}{10}, \quad \frac{-12}{-2}, \quad \sqrt{30}, \quad (-2)^2$$

Understanding

5. **WE1** Classify the following numbers as irrational, non-integer rational, or integer.

   a. 7
   b. $-\frac{1}{3}$
   c. $\sqrt{3}$
   d. $\frac{9}{2}$

6. Classify the following numbers as irrational, non-integer rational, or integer.

   a. $-\sqrt{5}$
   b. $\frac{35}{7}$
   c. $2\sqrt{3}$
   d. $-\frac{1}{13}$

7. Classify the following numbers as irrational, non-integer rational, or integer.

   a. $\sqrt{100}$
   b. -3.75
   c. $\sqrt{6}$
   d. $1.\dot{1}$

8. Which of the following numbers are rational?

$$9.17, \quad \frac{\sqrt{2}}{2}, \quad 10, \quad 7.\dot{3}\dot{4}, \quad \frac{\sqrt{36}}{3}, \quad -2, \quad \frac{\pi}{3}$$

Communicating, reasoning and problem solving

9. Explain why 0 is a rational number.

10. The inclusion of zero in the set of natural numbers has been the centre of an ongoing mathematical discussion. The natural numbers are the numbers that we use to count things. Explain whether zero be included in the set of natural numbers.

11. A farmer creates a paddock that is square in shape and has an area of 5 km^2. Determine the perimeter of the paddock. Show your working.

12. If you multiply an irrational number by a rational number, is the product rational or irrational? Explain your reasoning.

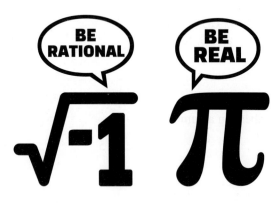

LESSON
3.3 Adding and subtracting fractions

LEARNING INTENTION

At the end of this lesson you should be able to:
- recognise and create equivalent fractions
- add and subtract fractions.

3.3.1 Equivalent fractions

eles-3578

- A fraction has two parts. The top part is called the **numerator** and the bottom part is called the **denominator**.
- The horizontal bar separating the numerator from the denominator is called the **vinculum**.

$$\frac{2}{3} \leftarrow \text{Numerator}$$
$$\leftarrow \text{Denominator}$$

Numerator • how many parts are required
Denominator • the number of equal parts the whole has been divided into
• describes the size of the parts

This line (the vinculum) means 'divide'.

- **Equivalent fractions** are fractions that are equal in value. That is, they have the same ratio between the numerator and denominator; for example, $\frac{4}{6} = \frac{2}{3}$.

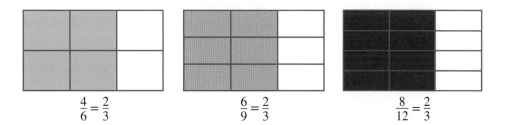

$$\frac{4}{6} = \frac{2}{3} \qquad \frac{6}{9} = \frac{2}{3} \qquad \frac{8}{12} = \frac{2}{3}$$

Equivalent fractions

Equivalent fractions can be produced by multiplying or dividing the numerator and denominator by the same number.

For example:

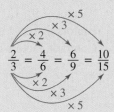

$$\frac{2}{3} = \frac{4}{6} = \frac{6}{9} = \frac{10}{15} \qquad \frac{30}{40} = \frac{15}{20} = \frac{3}{4}$$

WORKED EXAMPLE 3 Writing equivalent fractions

Fill in the missing numbers on the following equivalent fractions.

a. $\dfrac{3}{4} = \dfrac{6}{-}$

b. $\dfrac{5}{6} = \dfrac{-}{18}$

c. $\dfrac{2}{5} = \dfrac{20}{-}$

d. $\dfrac{6}{9} = \dfrac{-}{3}$

THINK	WRITE
a. The numerator has been multiplied by 2; therefore, the denominator must also be multiplied by 2.	a. $4 \times 2 = 8$ $\dfrac{3}{4} = \dfrac{6}{8}$
b. The denominator has been multiplied by 3; therefore, the numerator must also be multiplied by 3.	b. $5 \times 3 = 15$ $\dfrac{5}{6} = \dfrac{15}{18}$
c. The numerator has been multiplied by 10; therefore, the denominator must also be multiplied by 10.	c. $5 \times 10 = 50$ $\dfrac{2}{5} = \dfrac{20}{50}$
d. The denominator has been divided by 3; therefore, the numerator must also be divided by 3.	d. $6 \div 3 = 2$ $\dfrac{6}{9} = \dfrac{2}{3}$

▶ 3.3.2 Simplifying fractions

eles-3579

- Fractions can be simplified if the numerator and denominator share a common factor.
- If the **highest common factor** (HCF) between the numerator and denominator is 1, then the fraction is in its simplest form.

Simplifying fractions

To simplify a fraction, divide the numerator and denominator by their highest common factor.

WORKED EXAMPLE 4 Simplifying fractions

Write $\dfrac{9}{12}$ in simplest form.

THINK	WRITE
1. Determine the common factors of the numerator and denominator.	Factors of 9: ①, ③, 9 Factors of 12: ①, 2, ③, 4, 6, 12 Common factors of 9 and 12: 1, 3
2. Divide the numerator and denominator by their highest common factor.	$\dfrac{9}{12} = \dfrac{9 \div 3}{12 \div 3} = \dfrac{3}{4}$

ⓑ 3.3.3 Adding and subtracting fractions

eles-3580

- When adding or subtracting fractions, the denominators *must* be the same.
- If denominators are different, convert to equivalent fractions with the **lowest common denominator** (LCD). The LCD is the smallest multiple of all the denominators in a set of fractions.

Adding and subtracting fractions with same denominators

To add or subtract fractions with the same denominator, perform the required operation on the numerators.

For example:

$$\frac{1}{7} + \frac{3}{7} = \frac{1+3}{7} = \frac{4}{7}$$

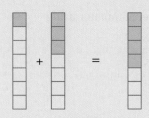

$$\frac{8}{9} - \frac{5}{9} = \frac{8-5}{9} = \frac{3}{9}$$

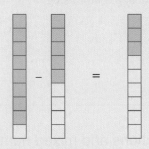

Adding and subtracting fractions with different denominators

To add or subtract fractions with different denominators, convert both fractions to equivalent fractions with the same lowest common denominator (LCD).

For example:

$$\frac{1}{2} + \frac{1}{4} = \frac{2}{4} + \frac{1}{4} = \frac{3}{4}$$

$$\frac{2}{3} - \frac{1}{2} = \frac{4}{6} - \frac{3}{6} = \frac{1}{6}$$

WORKED EXAMPLE 5 Adding and subtracting proper fractions

Evaluate the following, giving your answers in simplest form.

a. $\dfrac{1}{8} + \dfrac{5}{8}$

b. $\dfrac{1}{2} - \dfrac{1}{3}$

c. $\dfrac{3}{4} + \dfrac{1}{6}$

THINK

a. 1. As the denominators are equal, add the numerators and leave the denominator unchanged.

   2. Simplify the fraction.

b. 1. Determine the lowest common denominator.

   2. Write both fractions with the same denominator using equivalent fractions.

   3. Subtract the numerators.

c. 1. Determine the lowest common denominator.

   2. Write both fractions with the same denominator using equivalent fractions.

   3. Add the numerators.

WRITE

a. $\dfrac{1}{8} + \dfrac{5}{8} = \dfrac{6}{8}$

$= \dfrac{3}{4}$

b. The lowest common denominator of 2 and 3 is 6.

$\dfrac{1}{2} - \dfrac{1}{3} = \dfrac{1 \times 3}{2 \times 3} - \dfrac{1 \times 2}{3 \times 2}$

$= \dfrac{3}{6} - \dfrac{2}{6}$

$= \dfrac{1}{6}$

c. The lowest common denominator of 4 and 6 is 12.

$\dfrac{3}{4} + \dfrac{1}{6} = \dfrac{3 \times 3}{4 \times 3} + \dfrac{1 \times 2}{6 \times 2}$

$= \dfrac{9}{12} + \dfrac{2}{12}$

$= \dfrac{11}{12}$

Converting between improper fractions and mixed numbers

- An **improper fraction** has a numerator greater than the denominator; for example, $\dfrac{7}{3}$.

- A **mixed number** contains a whole number part and a proper fraction part; for example, $2\dfrac{1}{3}$.

- Mixed numbers can be expressed as improper fractions, and improper fractions can be expressed as mixed numbers.

$2\dfrac{1}{3} = 1 + 1 + \dfrac{1}{3} =$ $= \dfrac{7}{3}$

WORKED EXAMPLE 6 Converting between improper fractions and mixed numbers

a. **Express** $3\dfrac{2}{3}$ **as an improper fraction.**

b. **Express** $\dfrac{23}{5}$ **as a mixed number.**

THINK

a. 1. Write the mixed number as the sum of the whole number and the fraction.

WRITE

a. $3\dfrac{2}{3} = 3 + \dfrac{2}{3}$

2. Express the whole number as an improper fraction with a denominator of 3.

$$= \frac{9}{3} + \frac{2}{3}$$

3. Perform the addition of the numerators.

$$= \frac{11}{3}$$

b. 1. Write the improper fraction.

b. $\frac{23}{5} = 23 \div 5$

2. Determine how many times the denominator can be divided into the numerator and what the remainder is.

$$= 4 \text{ remainder } 3$$

3. Write the answer.

$$= 4\frac{3}{5}$$

Adding and subtracting improper fractions and mixed numbers

- To add or subtract improper fractions, ensure the fractions share a common denominator, then perform the addition or subtraction on the numerators.
- To add or subtract mixed numbers, first convert the mixed numbers to improper fractions, then solve as usual.
- An alternative method for adding and subtracting mixed numbers is to add or subtract the whole number part first, and then the fractions.

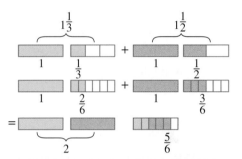

DISCUSSION

Why are mixed numbers preferred to improper fractions when answering worded questions?

WORKED EXAMPLE 7 Adding and subtracting improper fractions and mixed numbers

Evaluate the following.

a. $\frac{11}{5} + \frac{27}{10}$

b. $3\frac{1}{12} - 1\frac{5}{12}$

c. $2\frac{2}{3} + 3\frac{1}{2}$

THINK

WRITE

a. 1. Write the question.

a. $\frac{11}{5} + \frac{27}{10}$

2. Determine the lowest common denominator.

$$\text{LCD}(5, 10) = 10$$

3. Write each fraction with the same denominator using equivalent fractions.

$$= \frac{11 \times 2}{5 \times 2} + \frac{27}{10}$$

$$= \frac{22}{10} + \frac{27}{10}$$

4. Write the answer.

$$= \frac{49}{10}$$

b. 1. Write the question.

b. $3\frac{1}{12} - 1\frac{5}{12}$

2. Convert the mixed numbers to improper fractions.

$$= \frac{37}{12} - \frac{17}{12}$$

3. Perform the subtraction and then simplify the answer. $= \dfrac{20}{12}$

$= \dfrac{5}{3}$

4. Write the answer as a mixed number if appropriate. $= 1\dfrac{2}{3}$

c. 1. Write the question. **c.** $2\dfrac{2}{3} + 3\dfrac{1}{2}$

2. Change each mixed number to an improper fraction. $= \dfrac{8}{3} + \dfrac{7}{2}$

3. Write both fractions with the same denominator using equivalent fractions. $= \dfrac{8 \times 2}{3 \times 2} - \dfrac{7 \times 3}{2 \times 3}$

$= \dfrac{16}{6} + \dfrac{21}{6}$

4. Add the fractions. $= \dfrac{37}{6}$

5. Write the answer as a mixed number if appropriate. $= 6\dfrac{1}{6}$

---

on Resources

▶ **Video eLesson** Addition and subtraction of fractions (eles-1862)

🧩 **Interactivities** Addition and subtraction of proper fractions (int-3718)
Addition and subtraction of mixed numbers (int-3719)
Addition and subtraction of fractions (int-3721)

Exercise 3.3 Adding and subtracting fractions

learn on

3.3 Quick quiz on	3.3 Exercise

Individual pathways

■ PRACTISE	■ CONSOLIDATE	■ MASTER
1, 4, 5, 6, 12, 13, 16, 20, 23	2, 7, 9, 10, 11, 14, 15, 17, 19, 21, 24	3, 8, 18, 22, 25

Fluency

1. **WE3** Fill in the numbers missing from the following equivalent fractions.

a. $\dfrac{1}{6} = \dfrac{5}{-}$ **b.** $\dfrac{2}{7} = \dfrac{}{28}$ **c.** $\dfrac{5}{8} = \dfrac{35}{-}$ **d.** $\dfrac{-12}{-} = \dfrac{-4}{11}$

2. Fill in the numbers missing from the following equivalent fractions.

a. $\dfrac{1}{9} = \dfrac{}{81}$ **b.** $\dfrac{-6}{40} = \dfrac{}{20}$ **c.** $\dfrac{36}{-} = \dfrac{3}{8}$ **d.** $-\dfrac{13}{18} = -\dfrac{26}{-}$

3. Fill in the numbers missing from the following equivalent fractions.

a. $\dfrac{-}{6} = \dfrac{-45}{18}$

b. $\dfrac{42}{30} = \dfrac{-7}{-}$

c. $\dfrac{88}{48} = \dfrac{-}{6}$

d. $\dfrac{132}{144} = \dfrac{-}{-12}$

4. WE4 Write the following fractions in simplest form.

a. $\dfrac{2}{10}$

b. $\dfrac{81}{90}$

c. $\dfrac{21}{24}$

d. $\dfrac{63}{72}$

5. WE5a Solve the following, giving your answers in simplest form.

a. $\dfrac{1}{5} + \dfrac{4}{5}$

b. $\dfrac{2}{8} + \dfrac{3}{8}$

c. $\dfrac{3}{17} + \dfrac{6}{17}$

d. $\dfrac{21}{27} - \dfrac{16}{27}$

6. Solve the following, giving your answers in simplest form.

a. $\dfrac{2}{5} + \dfrac{1}{4}$

b. $\dfrac{3}{4} + \dfrac{5}{8}$

c. $\dfrac{6}{10} - \dfrac{2}{5}$

d. $\dfrac{9}{14} - \dfrac{2}{7}$

7. WE5b Evaluate the following, giving your answers in simplest form.

a. $\dfrac{3}{4} + \dfrac{5}{6}$

b. $\dfrac{9}{10} - \dfrac{1}{3}$

c. $\dfrac{2}{5} + \dfrac{3}{4}$

d. $\dfrac{8}{9} - \dfrac{3}{4}$

8. Evaluate the following, giving your answers in simplest form.

a. $\dfrac{11}{13} - \dfrac{2}{3}$

b. $\dfrac{2}{5} + \dfrac{6}{11}$

c. $\dfrac{1}{5} - \dfrac{2}{17}$

d. $\dfrac{19}{21} - \dfrac{3}{5}$

9. WE6a Express the following as improper fractions.

a. $3\dfrac{4}{7}$

b. $4\dfrac{12}{13}$

c. $5\dfrac{2}{5}$

d. $9\dfrac{5}{8}$

10. WE6b Express the following as mixed numbers in their simplest form.

a. $\dfrac{16}{3}$

b. $\dfrac{52}{6}$

c. $\dfrac{25}{4}$

d. $\dfrac{42}{35}$

Understanding

11. Arrange the following set in ascending order.

$$\dfrac{5}{4}, \quad 1\dfrac{13}{24}, \quad \dfrac{11}{8}, \quad \dfrac{17}{12}, \quad \dfrac{39}{24}, \quad 1\dfrac{6}{12}$$

12. Nafisa eats $\dfrac{5}{8}$ of a block of chocolate for afternoon tea and $\dfrac{3}{8}$ of the block after dinner.

Determine how much of the block Nafisa has eaten altogether.

For questions **13–19**, evaluate, giving your answers in simplest form.

13. a. $\dfrac{3}{15} + \dfrac{11}{15} - \dfrac{2}{15}$

b. $\dfrac{8}{25} + \dfrac{34}{50} - \dfrac{7}{25}$

c. $\dfrac{21}{30} + \dfrac{5}{6} + \dfrac{9}{10}$

14. WE7a

a. $\dfrac{13}{8} - \dfrac{5}{4}$

b. $\dfrac{23}{7} + \dfrac{3}{8}$

c. $\dfrac{-16}{9} - \dfrac{4}{5} + \dfrac{5}{3}$

15. a. $\dfrac{27}{18} - \dfrac{31}{9}$

b. $-\dfrac{12}{5} + \dfrac{7}{3} + \dfrac{-11}{6}$

c. $-\dfrac{34}{10} - \dfrac{-21}{5} + \dfrac{15}{6}$

16. **WE7b**

  a. $-2\dfrac{3}{5} - 4\dfrac{1}{5}$

  b. $6\dfrac{7}{9} - 3\dfrac{5}{9}$

  c. $8\dfrac{4}{5} - 4\dfrac{1}{5}$

17. **WE7c**

  a. $6\dfrac{1}{4} + 3\dfrac{1}{6}$

  b. $12\dfrac{2}{5} + 8\dfrac{7}{9}$

  c. $4\dfrac{3}{4} - 5\dfrac{1}{6} + 3\dfrac{3}{12}$

18. a. $\dfrac{9}{4} - 2\dfrac{1}{16} + \dfrac{-13}{8}$

  b. $\dfrac{-3}{8} - 4\dfrac{1}{3} + \dfrac{51}{12}$

  c. $3\dfrac{11}{20} + \dfrac{-10}{3} - 2\dfrac{2}{5}$

19. Seven bottles of soft drink were put out onto the table at a birthday party.
    Calculate the amount of soft drink that was left over after $5\dfrac{2}{9}$ bottles were consumed.

Communicating, reasoning and problem solving

20. In a class, $\dfrac{1}{3}$ of the students ride their bikes to school, $\dfrac{1}{4}$ catch the bus and the rest get a lift. Determine what fraction of the class gets a lift to school. Show your working.

21. Frank has a part-time job at the local newsagency. If he spends $\dfrac{1}{3}$ of his pay on comic books and $\dfrac{2}{5}$ on lollies, evaluate what fraction of his pay is left over. Show your working.

22. A Year 8 class organised a cake stall to raise some money. They had 10 whole cakes to start with. If they sold $2\dfrac{3}{4}$ cakes at recess and then $5\dfrac{7}{8}$ cakes at lunch time, determine the number of cakes that were left over. Show your working.

23. You need to fill a container with exactly $2\dfrac{5}{12}$ cups of water. If you have cups that measure $\dfrac{1}{4}, \dfrac{1}{3}, \dfrac{1}{2}$ and 1 cup of water, determine the quickest way to measure this amount.

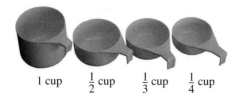

1 cup $\tfrac{1}{2}$ cup $\tfrac{1}{3}$ cup $\tfrac{1}{4}$ cup

24. A tray used to bake a muesli bar slice holds 3 cups of ingredients.
    For the recipe shown, determine whether or not the mixture will fit in the tray.

25. Polly and Neda had divided up some coins. Neda was upset as Polly had more coins than Neda. Polly said, 'Here's one-third of my coins.' Neda was moved by Polly's generosity and gave back one-half of her total. Polly gave her one-quarter of her new total and an extra coin.

  a. Assuming that Polly started with x coins and Neda started with y coins, show that after their final exchange of coins Polly had
  $\dfrac{15x + 9y - 24}{24}$ coins and Neda had $\dfrac{9x + 15y + 24}{24}$ coins.

  b. Show that the total number of coins Polly and Neda had between them after their final exchange was the same total with which they started.

Muesli bar slice

Ingredients:
$\tfrac{1}{2}$ cup of dried fruit
$\tfrac{3}{4}$ cup of grated apple
$1\tfrac{1}{4}$ cups of muesli
$\tfrac{2}{3}$ cup of apple juice
1 tablespoon $\left(\tfrac{2}{25} \text{ cup}\right)$ of canola margarine

LESSON
3.4 Multiplying and dividing fractions

LEARNING INTENTION

At the end of this lesson you should be able to:
- multiply and divide fractions.

▶ 3.4.1 Multiplication of fractions

eles-3581

- When multiplying fractions, the denominators do not have to be the same.

Multiplying fractions

To multiply fractions, simply multiply the numerators and then multiply the denominators.

$$\frac{1}{5} \times \frac{3}{4} = \frac{1 \times 3}{5 \times 4}$$

$$= \frac{3}{20}$$

- Mixed numbers must be converted into improper fractions before multiplying.
- A numerator and denominator can be simplified by dividing by a common factor prior to the multiplication;

  for example, $\dfrac{8^2}{9} \times \dfrac{5}{4^1} = \dfrac{2 \times 5}{9 \times 1} = \dfrac{10}{9}$.

- The word *of* is often used in practical applications of fraction multiplication. It can be replaced with a multiplication sign to evaluate expressions.

WORKED EXAMPLE 8 Multiplying fractions

Evaluate $\dfrac{2}{5} \times -\dfrac{5}{8}$.

THINK	WRITE
1. Write the expression and cancel the common factors in numerators and denominators.	$\dfrac{2^1}{5^1} \times -\dfrac{5^1}{8^4}$
2. Multiply the numerators and then multiply the denominators. *Note:* positive × negative = negative	$= \dfrac{1}{1} \times -\dfrac{1}{4}$
3. Write the answer.	$= -\dfrac{1}{4}$

▶ 3.4.2 Division of fractions

eles-3582

- The reciprocal of a number is 1 divided by the number. That is, $\dfrac{1}{\text{number}}$.

  For example, the reciprocal of 8 is $\dfrac{1}{8}$.

- To determine the reciprocal of a fraction, simply flip the whole fraction.
  For example, the reciprocal of $\frac{3}{8}$ is $\frac{8}{3}$.
- To determine the reciprocal of a mixed number, express it as an improper fraction first, then flip it.
- Reciprocals are used when dividing fractions. To divide fractions, multiply by the reciprocal.

Dividing fractions

To divide two fractions, multiply by the reciprocal.

$$\frac{5}{6} \div \frac{1}{3} = \frac{5}{6} \times \frac{3}{1} = \frac{15}{6} = \frac{5}{2}$$

This process can be remembered easily by the saying KEEP, CHANGE, FLIP.

Keep the first fraction the same.

Change the division sign into a multiplication sign.

Flip the second fraction.

$$\frac{5}{6} \qquad \div \qquad \frac{1}{3}$$

$$\text{KEEP} \quad \text{CHANGE} \quad \text{FLIP}$$

$$\frac{5}{6} \qquad \times \qquad \frac{3}{1}$$

WORKED EXAMPLE 9 Dividing fractions

Evaluate $-\frac{3}{4} \div -1\frac{1}{2}$.

THINK	WRITE
1. Write the question.	$-\frac{3}{4} \div -1\frac{1}{2}$
2. Change any mixed numbers into improper fractions first.	$= -\frac{3}{4} \div -\frac{3}{2}$
3. Convert the question into a multiplication problem by: • keeping the leftmost fraction the same • changing the ÷ sign to a × sign • flipping (taking the reciprocal of) the second fraction.	$= -\frac{3}{4} \times -\frac{2}{3}$
4. Multiply the fractions by multiplying numerators together, then multiplying the denominators together. *Note:* negative × negative = positive	$= \frac{3 \times 2}{4 \times 3}$ $= \frac{6}{12}$
5. Simplify the fraction by dividing by the highest common factor. HCF $= 6$	$= \frac{1}{2}$
6. Write the answer.	$-\frac{3}{4} \div -1\frac{1}{2} = \frac{1}{2}$

COMMUNICATING — COLLABORATIVE TASK: Simple finite continued fractions

A simple finite continued fraction is an expression in the form $a + \cfrac{1}{b + \cfrac{1}{c + \cfrac{1}{d + \cfrac{1}{e + \cfrac{1}{\ldots}}}}}$, where a, b, c, d, e

etc. are integers.

Any rational number can be represented by a simple finite continued fraction.

For instance, $\dfrac{11}{38} = \cfrac{1}{3 + \cfrac{1}{2 + \cfrac{1}{5}}}$, and $\dfrac{38}{11} = 3 + \cfrac{1}{2 + \cfrac{1}{5}}$

But how can those integers a, b, c etc. be found?

Euclid's algorithm, which can be used to compute the greatest common divisor (GCD) of two numbers, can be used.

Algorithm

Let a and b be two positive integers, with $a > b$.

Step 1: Divide a by b and determine the reminder r. If $r = 0$, then the GCD is b.

Step 2: If $r \neq 0$, then set $a = b$ and $b = r$ and go back to step 1.

Using $a = 38$ and $b = 11$:

Step 1

$$38 = 3 \text{ times } 11 + 5$$

The reminder is $5 \neq 0$, so we move to step 2.

Step 2

Now $a = 11$ and $b = 5$.

Repeat step 1 with $a = 11$ and $b = 5$.

Step 1

$$11 = 2 \times 5 + 1$$

The reminder is $1 \neq 0$, so we move to step 2.

Step 2

Now $a = 5$ and $b = 1$

Repeat step 1 with $a = 5$ and $b = 1$.

Step 1

$$5 = 5 \times 1 + 0$$

The reminder is zero, so the algorithm terminates.

In this case, the GCD of 11 and 38 is 1.

What interests us in the case of simple finite continued fractions are the quotients for each step, **3**, **2**, and **5**, which are, in order, the values of the integers $a, b,$ and c.

Equipment: paper, pen, calculator

1. As a pair, select one of the 12 fractions on the right.
2. Work together as a class to determine if, given the values of the numerator and denominator, it is easy to determine whether a chosen fraction will be represented by

$$a + \cfrac{1}{b + \cfrac{1}{c}} \quad \text{or by} \quad \cfrac{1}{a + \cfrac{1}{b + \cfrac{1}{c}}}$$

3. With your partner, use Euclid's algorithm and the examples provided to determine the values of the integers a, b and c.
4. Use your findings to determine the continued fraction representation of the reciprocal of your chosen fraction.
5. Swap your findings with another group and use a calculator to check their answer.

$\dfrac{31}{37}$	$\dfrac{37}{31}$
$\dfrac{25}{31}$	$\dfrac{37}{31}$
$\dfrac{25}{31}$	$\dfrac{31}{25}$
$\dfrac{24}{7}$	$\dfrac{7}{24}$
$\dfrac{18}{7}$	$\dfrac{7}{18}$
$\dfrac{5}{9}$	$\dfrac{9}{5}$
$\dfrac{519}{64}$	$\dfrac{64}{519}$

 Resources

▶ **Video eLesson** Multiplication and division of fractions (eles-1867)

Interactivities Multiplication of fractions (int-3722)

Multiplication and division of negative fractions (int-3723)

Exercise 3.4 Multiplying and dividing fractions

learn on

3.4 Quick quiz	**3.4 Exercise**

Individual pathways

■ PRACTISE	■ CONSOLIDATE	■ MASTER
1, 2, 4, 6, 8, 12, 14, 15, 18, 21	3, 9, 10, 13, 16, 19, 22	5, 7, 11, 17, 20, 23, 24

Fluency

1. **WE8** Solve the expressions.

   a. $\dfrac{3}{4} \times \dfrac{1}{2}$ b. $\dfrac{1}{8} \times \dfrac{1}{7}$ c. $\dfrac{1}{2} \times \dfrac{5}{6}$ d. $\dfrac{5}{7} \times \dfrac{1}{3}$

2. Evaluate the expressions.

   a. $\dfrac{2}{5} \times \dfrac{3}{5}$ b. $\dfrac{3}{7} \times \dfrac{7}{9}$ c. $\dfrac{5}{8} \times \dfrac{11}{20}$ d. $\dfrac{5}{6} \times \dfrac{3}{10}$

3. Solve the expressions.

   a. $\dfrac{11}{20} \times \dfrac{2}{3}$ b. $\dfrac{1}{3} \times \dfrac{3}{5}$ c. $\dfrac{2}{3} \times \dfrac{9}{10}$ d. $\dfrac{6}{7} \times \dfrac{14}{15}$

4. Solve the expressions.

   a. $-\dfrac{1}{2} \times \dfrac{1}{3}$ b. $-\dfrac{3}{4} \times -\dfrac{1}{5}$ c. $\dfrac{1}{3} \times -\dfrac{3}{4}$ d. $-\dfrac{2}{3} \times 7$

5. Evaluate the expressions.

   a. $-\dfrac{3}{4} \times \dfrac{5}{6}$ b. $-\dfrac{8}{9} \times 1\dfrac{3}{4}$ c. $-\dfrac{5}{6} \times \dfrac{3}{10}$ d. $-3\dfrac{1}{7} \times -\dfrac{7}{8}$

6. **WE9** Solve the expressions.

   a. $\dfrac{1}{3} \div \dfrac{1}{2}$ b. $\dfrac{7}{8} \div \dfrac{3}{2}$ c. $\dfrac{2}{5} \div \dfrac{1}{4}$ d. $\dfrac{4}{14} \div \dfrac{1}{3}$

7. Evaluate the expressions.

   a. $\dfrac{3}{4} \div \dfrac{7}{8}$ b. $\dfrac{12}{15} \div \dfrac{4}{3}$ c. $\dfrac{1}{5} \div \dfrac{10}{12}$ d. $\dfrac{5}{6} \div \dfrac{8}{9}$

Understanding

8. Solve the expressions.

   a. $3\dfrac{1}{2} \times 1\dfrac{3}{5}$ b. $1\dfrac{2}{10} \times 1\dfrac{1}{5}$ c. $3\dfrac{2}{4} \times 2\dfrac{1}{2}$ d. $2\dfrac{2}{3} \times 1\dfrac{1}{2}$

9. Evaluate the expressions.

   a. $6 \times 2\dfrac{1}{6}$ b. $1\dfrac{3}{5} \times \dfrac{5}{8}$ c. $5\dfrac{3}{4} \times 2\dfrac{2}{5}$ d. $4\dfrac{3}{4} \times 2\dfrac{1}{2}$

10. Solve the expressions.

    a. $1\dfrac{6}{10} \div 1\dfrac{3}{5}$ b. $3\dfrac{5}{7} \div 2\dfrac{1}{6}$ c. $1\dfrac{5}{7} \div \dfrac{1}{3}$ d. $1\dfrac{1}{6} \div \dfrac{2}{1}$

11. Evaluate the expressions.

    a. $1\dfrac{1}{3} \div \dfrac{5}{6}$ b. $3\dfrac{1}{2} \div 1\dfrac{3}{5}$ c. $10\dfrac{4}{5} \div 2\dfrac{1}{2}$ d. $7\dfrac{8}{9} \div 7\dfrac{1}{2}$

12. Solve the expressions.

    a. $-\dfrac{1}{5} \div \dfrac{1}{2}$ b. $\dfrac{2}{3} \div -\dfrac{3}{4}$

    c. $\dfrac{3}{2} \div -4$ d. $-\dfrac{7}{4} \div -\dfrac{2}{1}$

13. Evaluate the expressions.

    a. $-\dfrac{1}{8} \div \dfrac{3}{4}$ b. $-2\dfrac{1}{4} \div -\dfrac{1}{2}$

    c. $2\dfrac{2}{3} \div -1\dfrac{1}{9}$ d. $-\dfrac{3}{5} \div 2\dfrac{5}{8}$

14. Determine $\dfrac{3}{4}$ of 16.

15. An assortment of 75 lollies is to be divided evenly among 5 children.

   a. Determine the fraction of the total number of lollies that each child will receive.

   b. Calculate the number of lollies each child will receive.

16. Solve the following by applying BIDMAS.

   a. $-\dfrac{2}{3} + \dfrac{1}{6} \times -\dfrac{2}{5}$

   b. $1\dfrac{1}{2} \times -\dfrac{5}{6} \div \dfrac{4}{7}$

   c. $-\dfrac{7}{8} \div -1\dfrac{3}{4} - \dfrac{1}{2}$

17. Evaluate the following by applying BIDMAS.

   a. $\left(\dfrac{2}{5} - \dfrac{6}{7}\right) \times -3\dfrac{1}{3}$

   b. $\left(-1\dfrac{1}{2} - 3\dfrac{4}{5}\right) \div \dfrac{3}{5}$

   c. $\dfrac{9}{10} \times -\dfrac{5}{3} \div \left(1\dfrac{2}{7} - 2\dfrac{1}{2}\right)$

Communicating, reasoning and problem solving

18. Sam has been collecting caps from all around the world. If he has a total of 160 caps and $\dfrac{1}{4}$ of them are from the USA, determine how many non-USA caps he has.
   Show your working.

19. Year 8's cake stall raised $120. If they plan to give $\dfrac{1}{4}$ to a children's charity and $\dfrac{2}{3}$ to a charity for the prevention of cruelty to animals, determine how much each group will receive and how much is left over.
   Show your working.

20. In the staff room there is $\dfrac{7}{8}$ of a cake left over from a meeting. If 14 members of staff would all like a piece, evaluate the fraction of the original cake they will each receive.
   Show how you reached your answer.

21. You come across a fantastic recipe for chocolate pudding that you want to try, but it says that the recipe feeds five people. Explain what you would do to modify the recipe so that it feeds only one person.

22. A wealthy merchant died and left 17 camels to be shared among his three children. The eldest was to have half of the camels, the second child one-third, and the third child one-ninth. 'It's not possible!' protested the eldest. A wise man lent his camel to the children, raising the total to 18. The eldest child then took half (nine camels); the second child took one-third (six camels); and the youngest child took one-ninth (two camels). The wise man then departed on his own camel and everyone was happy.

    a. Explain what the eldest child meant by 'It's not possible!'

    b. Calculate the value of $\dfrac{1}{2} + \dfrac{1}{3} + \dfrac{1}{9}$ and comment on the relationship between the numerator, the denominator and the number of camels.

23. The distance between Sydney and Melbourne is approximately 878 km. The XPT train takes 11 hours to travel from Sydney to Melbourne. The driving time by car is $8\dfrac{3}{4}$ hours, and it takes $1\dfrac{1}{2}$ hours to fly.

    If the average speed is found by dividing the total distance by the travelling time, calculate the average speed when travelling from Sydney to Melbourne by:

    a. train
    b. car
    c. plane.

24. Fractions of the type $\dfrac{1}{a + \dfrac{1}{b + \dfrac{1}{c}}}$ are called finite simple continued fractions (see the collaborative task on continued fractions in section 3.4.2).

    Write $\dfrac{7}{30}$ as a continued fraction. Evaluate the values of the pronumerals a, b and c.

Chocolate pudding
(serves 5)

Ingredients:

$1\dfrac{1}{2}$ cups plain flour

1 teaspoon baking soda

$\dfrac{1}{2}$ cup sugar

1 teaspoon vanilla essence

$1\dfrac{1}{2}$ cups melted chocolate

$\dfrac{1}{3}$ cup cocoa powder

$\dfrac{1}{2}$ cup canola oil

$\dfrac{1}{2}$ cup water

1 cup milk

LESSON
3.5 Terminating and recurring decimals

LEARNING INTENTION

At the end of this lesson you should be able to:
- convert a fraction into a decimal
- use correct notation to write a recurring decimal
- convert a decimal into a fraction.

⏵ 3.5.1 Fractions to decimals

eles-3583

- Numbers that can be written as fractions are rational numbers.
- To write a fraction as a decimal, divide the numerator by the denominator.
- Fractions expressed as decimals are either **terminating decimals**, which have a fixed number of decimal places, or **recurring decimals**, which have an infinite number of decimal places.
- Recurring decimals with one recurring digit are written with a dot above the recurring digit.

- Recurring decimals with more than one recurring digit are written with a line above the recurring digits:

$$0.333333\ldots = 0.\dot{3}$$

$$0.484\,848\ldots = 0.\overline{48}$$

(Sometimes two dots are used instead of a line, with the dots shown above the first and last digits of the repeating pattern; for example, $0.484\,848\ldots = 0.\dot{4}\dot{8}$).

WORKED EXAMPLE 10 Converting fractions to decimals

Convert the following fractions to decimals. State whether the decimal is a recurring decimal or a terminating decimal.

a. $\dfrac{1}{5}$ b. $4\dfrac{5}{12}$ c. $-\dfrac{5}{8}$ d. $\dfrac{2}{3}$

THINK	WRITE
a. 1. Write the question.	a. $\dfrac{1}{5}$
2. Rewrite the question using division.	$= 1 \div 5$
3. Divide, adding zeros as required.	$\begin{array}{r} 2.0 \\ 5\overline{)1.0} \end{array}$
4. Write the answer.	$\dfrac{1}{5} = 0.2$ This is a terminating decimal.
b. 1. Write the question.	b. $4\dfrac{5}{12}$
2. Convert the mixed number to an improper fraction.	$= \dfrac{53}{12}$
3. Rewrite the question using division.	$= 53 \div 12$
4. Since 12 is not a power of 2 or 5, it will be a recurring decimal.	
5. Divide, adding zeros until a pattern occurs.	$\begin{array}{r} 4.\ \ 4\ 1\ 6\ 6\ 6 \\ 12\overline{)53.{}^{5}0^{2}0^{8}0^{8}0^{8}0} \end{array}$
6. Write the answer with a dot above the recurring number.	$4\dfrac{5}{12} = 4.41\dot{6}$ This is a recurring decimal.
c. 1. The size of the decimal will be the same as for $\dfrac{5}{8}$.	c. $\dfrac{5}{8}$
2. Rewrite the question using division.	$= 5 \div 8$
3. $8 = 2^3$, so an exact answer will be found. Divide, adding zeros as required.	$\begin{array}{r} 0.\ 6\ 2\ 5 \\ 8\overline{)5.{}^{5}0^{2}0^{4}0} \end{array}$
4. Write the answer, remembering that the original fraction was negative.	$-\dfrac{5}{8} = -0.625$ This is a terminating decimal.

d. 1. Rewrite the question using division.	d.	$\dfrac{2}{3} = 2 \div 3$
2. Since 3 is not a power of 2 or 5, it will be a recurring decimal.		$\dfrac{0.\ 6\ 6\ 6}{3\overline{)2.{}^{2}0{}^{2}0{}^{2}0}}$
3. Write the answer, remembering to place a dot above the recurring number.		$\dfrac{2}{3} = 0.\dot{6}$
		This is a recurring decimal.

3.5.2 Decimals to fractions

eles-3584

- When changing a decimal to a fraction, rewrite the decimal as a fraction with the same number of zeros in the denominator as there are decimal places in the question. Simplify the fraction by cancelling.

WORKED EXAMPLE 11 Converting decimals to fractions

Convert the following decimals to fractions in simplest form.

a. 0.25 b. 1.342 c. −0.8

THINK	WRITE
a. 1. Write the question.	a. 0.25
2. Rewrite as a fraction with the same number of zeros in the denominator as there are decimal places in the question. Simplify the fraction by cancelling.	$= \dfrac{\cancel{25}^{1}}{\cancel{100}^{4}}$
3. Write the answer.	$= \dfrac{1}{4}$
b. 1. Write the question.	b. 1.342
2. Rewrite the decimal in expanded form.	$= 1 + 0.342$
3. Write as a mixed number with the same number of zeros in the denominator as there are decimal places in the question and cancel.	$= 1 + \dfrac{\cancel{342}^{171}}{\cancel{1000}^{500}}$
4. Write the answer.	$= 1\dfrac{171}{500}$
c. 1. Write the question.	c. -0.8
2. Rewrite as a fraction with the same number of zeros in the denominator as there are decimal places in the question. Simplify the fraction by cancelling.	$= -\dfrac{\cancel{8}^{4}}{\cancel{10}^{5}}$
3. Write the answer.	$= -\dfrac{4}{5}$

Equipment: paper, pen

1. In pairs, look for a pattern in the following conversions:

$0.\dot{3} = \dfrac{3}{9}$	$0.\dot{7} = \dfrac{7}{9}$	$1.\dot{5} = 1\dfrac{5}{9}$
$0.\overline{13} = \dfrac{13}{99}$	$0.\overline{67} = \dfrac{67}{99}$	$4.\overline{88} = 4\dfrac{88}{99}$
$0.\overline{137} = \dfrac{137}{999}$	$0.\overline{871} = \dfrac{871}{999}$	$1.\overline{712} = 1\dfrac{712}{999}$
$0.\overline{7615} = \dfrac{7615}{9999}$	$0.\overline{1432} = \dfrac{1432}{9999}$	$6.\overline{1151} = 6\dfrac{1151}{9999}$

2. With your partner, write a short description of the pattern you observed.
3. Swap your description with another group and provide feedback on their description. Is it concise, correct and clear?
4. Work together as a class to create a short explanation (less than 100 words) on how to easily convert a recurring fraction to a decimal.
5. As a class, discuss whether you can adapt this for any type of recurring fraction.
   For instance, can you write $1.8\dot{3}$ as a fraction?
   Hint: You might want to consider the following.

$$1.8\dot{3} = 1.8 + 0.0\dot{3} = \frac{9}{5} + \frac{0.\dot{3}}{10}$$

Exercise 3.5 Terminating and recurring decimals

learnon

3.5 Quick quiz on	**3.5 Exercise**

Individual pathways

■ PRACTISE	■ CONSOLIDATE	■ MASTER
1, 3, 6, 9, 10, 14, 17	2, 4, 7, 11, 12, 15, 18	5, 8, 13, 16, 19

Fluency

1. **WE10a** Convert the following fractions to decimals, giving exact answers or using the correct notation for recurring decimals where appropriate.

   a. $\dfrac{4}{5}$ b. $\dfrac{1}{4}$ c. $\dfrac{3}{4}$ d. $\dfrac{7}{4}$ e. $\dfrac{2}{3}$

2. Convert the following fractions to decimals, giving an exact answer or using the correct notation for recurring decimals where appropriate.

   a. $\dfrac{5}{12}$ b. $\dfrac{9}{11}$ c. $\dfrac{21}{22}$ d. $\dfrac{13}{6}$ e. $\dfrac{7}{15}$

3. **WE10b** Convert the following mixed numbers to decimal numbers, giving exact answers or using the correct notation for recurring decimals where appropriate.
   Check your answers using a calculator.

   a. $6\dfrac{1}{2}$
   b. $1\dfrac{3}{4}$
   c. $3\dfrac{2}{5}$
   d. $8\dfrac{4}{5}$
   e. $6\dfrac{3}{4}$

4. Convert the following mixed numbers to decimal numbers, giving exact answers or using the correct notation for recurring decimals where appropriate.
   Check your answers using a calculator.

   a. $12\dfrac{9}{10}$
   b. $5\dfrac{2}{3}$
   c. $11\dfrac{11}{15}$
   d. $1\dfrac{5}{6}$
   e. $4\dfrac{1}{3}$

5. **WE10c** Convert the following fractions to decimal numbers, giving exact answers or using the correct notation for recurring decimals where appropriate.

   a. $-\dfrac{4}{15}$
   b. $-\dfrac{7}{9}$
   c. $-1\dfrac{5}{6}$
   d. $-5\dfrac{8}{9}$
   e. $-3\dfrac{1}{7}$

6. **WE11** Convert the following decimal numbers to fractions in simplest form.

   a. 0.4
   b. 0.8
   c. 1.2
   d. 3.2
   e. 0.56

7. Convert the following decimal numbers to fractions in simplest form.

   a. 0.75
   b. 1.30
   c. 7.14
   d. 4.21
   e. 10.04

8. Convert the following decimal numbers to fractions in simplest form.

   a. 7.312
   b. 9.940
   c. 84.126
   d. 73.90
   e. 0.0042

9. Of the people at a school social, $\dfrac{3}{4}$ were boys. Write this fraction as a decimal number.

Understanding

10. On a recent science test, Katarina answered all questions correctly, including the bonus question, and her score was $\dfrac{110}{100}$. Calculate this as a decimal value.

11. Alison sold the greatest number of chocolates in her scouting group. She sold $\dfrac{5}{9}$ of all chocolates sold by the group. Write this as a decimal number, correct to 2 decimal places.

12. Alfonzo ordered a pizza to share with three friends, but he ate 0.6 of it. Calculate the fraction that was left for his friends.

13. Using examples, explain the difference between rational and irrational numbers.

Communicating, reasoning and problem solving

14. By converting to decimal fractions, arrange the following in order from lowest to highest, showing all of your working.

$$\frac{6}{10}, \quad \frac{3}{4}, \quad \frac{2}{3}, \quad \frac{5}{8}, \quad \frac{6}{7}$$

15. To change $0.\overline{14}$ from a recurring decimal to a fraction, put the digits after the decimal place (14) on the numerator. The denominator has the same number of digits as the numerator (2), with all of these digits being 9. Thus, $0.\overline{14} = \frac{14}{99}$.
    a. Write $0.\dot{7}$ as a fraction, showing all of your working.
    b. Write $0.\overline{306}$ as a fraction, showing all of your working.
    c. Write $2.\dot{2}$ as a mixed number, showing all of your working.

16. Use examples to show your reasoning for each of the following questions.
    a. When you add two recurring decimals, do you always get a recurring decimal?
    b. When you add a recurring decimal and a terminating decimal, do you get a recurring decimal?
    c. When you subtract two recurring decimals, do you get a recurring decimal?
    d. When you subtract a recurring decimal from a terminating decimal, do you get a recurring decimal?
    e. When you subtract a terminating decimal from a recurring decimal, do you get a recurring decimal?

17. Robin Hoot, the captain of a motley pirate crew known throughout the Seven Seas as 'The Ferry Men', kept $0.\dot{3}$ of all stolen loot for himself. The rest was split between his crew.
    a. If there were 20 Ferry Men, determine the fraction of the loot that each Ferry Man received.
    b. Little George, one of the Ferry Men, received 15 gold doubloons. Evaluate the amount of loot altogether.
    c. If there were p pirates not including Robin Hoot, and each of them received d doubloons, determine the total amount of loot.

18. The pig ate $\frac{2}{5}$ of the chocolate cake the animals had stolen from Mrs Brown's windowsill where it had been cooling. The goat and the cow ate $0.\dot{2}$ and 0.35 of the cake respectively.
    a. If the duck and the turkey finished the rest, having equal shares, determine the fraction that the duck ate.
    b. If the duck and the turkey had r grams of the cake each, determine how much of the cake the pig had.

19. James had 18 litres of water shared unequally between three buckets.
    Then he:
    1. poured three-quarters of the water in bucket 1 into bucket 2
    2. poured half the water that was now in bucket 2 into bucket 3
    3. poured a third of the water that was now in bucket 3 into bucket 1.

    After the pouring, all the buckets contained equal amounts of water. Evaluate the amount of water contained by each bucket at the start.

LESSON
3.6 Adding and subtracting decimals

LEARNING INTENTION

At the end of this lesson you should be able to:
- add and subtract decimals.

▶ 3.6.1 Addition of positive decimals

eles-3585

- When adding decimals, the decimal points must *always* align vertically so that the place values of the numbers being added are aligned.
- The place values are summarised in the following table:

Thousands	Hundreds	Tens	Units	·	Tenths	Hundredths	Thousandths	Ten thousandths
1000	100	10	1	·	$\dfrac{1}{10}$	$\dfrac{1}{100}$	$\dfrac{1}{1000}$	$\dfrac{1}{10\,000}$

- When decimals with different numbers of decimal places are added or subtracted, trailing zeros can be written so that both decimals have the same number of decimal places.

WORKED EXAMPLE 12 Adding decimal numbers

Calculate $3.586 + 4.1 + 2.07$.

THINK	WRITE
1. Set out the addition in vertical columns. Line up the decimal points so that the digits of the same place value are underneath each other.	$\begin{array}{r} 3.586 \\ 4.1 \\ +2.07 \\ \hline \end{array}$
2. Fill in the smallest place values with trailing zeros so that both numbers have the same number of decimal places.	$\begin{array}{r} 3.586 \\ 4.100 \\ +2.070 \\ \hline \end{array}$
3. Add the digits in each place value, working from right to left and carrying any tens over to the next place value column.	$\begin{array}{r} \overset{1}{}3.586 \\ 4.100 \\ +2.070 \\ \hline 9.756 \\ \hline \end{array}$

▶ 3.6.2 Subtraction of positive decimals

eles-3586

- Decimals can be subtracted using a similar method to addition.
- If the digit in the number being subtracted is greater than the digit it is being subtracted from, you will need to 'borrow' from the next available place value.

WORKED EXAMPLE 13 Subtracting decimal numbers

Calculate $658.59 - 248.258$.

THINK	WRITE
1. Set out the subtraction in vertical columns. Line up the decimal points and make sure that the place values are underneath each other. Fill in the smallest place values with trailing zeros so that both numbers have the same number of decimal places.	$\begin{array}{r} 658.590 \\ -248.258 \\ \hline \end{array}$
2. Subtract the digits as you would subtract whole numbers, working from right to left. Write the decimal point in the answer directly below the decimal points in the question.	$\begin{array}{r} {}^{8\,10} \\ 658.59\cancel{0} \\ -248.258 \\ \hline 410.332 \end{array}$

3.6.3 Addition and subtraction of positive and negative decimals

eles-3587

- To add two numbers with the same sign, add and keep the sign.

WORKED EXAMPLE 14 Adding two negative numbers

Calculate $-3.64 + (-2.9)$.

THINK	WRITE
1. Set the addition out in vertical columns. Line up the decimal points so that the digits of the same place value are underneath each other.	$\begin{array}{r} -3.64 \\ +-2.90 \\ \hline \end{array}$
2. As the two numbers have the same sign (−), add them together and keep the sign in the answer.	$\begin{array}{r} {}^{1} \\ -3.64 \\ +-2.90 \\ \hline -6.54 \end{array}$

- To add two numbers with different signs, remove the signs and subtract the smaller number from the larger one.
- The sign of the answer will be the sign of the larger number.

WORKED EXAMPLE 15 Adding numbers with different signs

Calculate $-5.7 + 1.63$.

THINK	WRITE
1. As the numbers have different signs, remove the signs and subtract the smaller one from the larger one. This makes it easier to subtract. We will re-insert the negative once we calculate the answer.	$\begin{array}{r} 5.70 \\ -1.63 \\ \hline \end{array}$

2. Evaluate the subtraction.

$$\begin{array}{r} {}^{6}5.\overset{1}{7}\overset{0}{\cancel{0}} \\ -\ 1.63 \\ \hline 4.07 \end{array}$$

3. As the larger number was negative, the answer must be negative. In other words, re-insert the negative into the answer.

$-5.7 + 1.63 = -4.07$

 Resources

Exercise 3.6 Adding and subtracting decimals

learn

3.6 Quick quiz	3.6 Exercise

Individual pathways

■ PRACTISE	■ CONSOLIDATE	■ MASTER
1, 3, 4, 6, 8, 13, 15, 16, 19	2, 5, 9, 11, 14, 17, 20	7, 10, 12, 18, 21

Fluency

1. **WE12** Solve the following.
   a. $8.3 + 4.6$
   b. $16.45 + 3.23$
   c. $13.06 + 4.2$

2. Evaluate the following.
   a. $7.9 + 12.4$
   b. $128.09 + 4.35$
   c. $5.308 + 33.671 + 3.74$

3. Solve the following.
   a. $0.93 + 4.009 + 1.3$
   b. $56.830 + 2.504 + 0.1$
   c. $25.3 + 89 + 4.087 + 7.77$

4. **WE13** Evaluate the following.
   a. $4.56 - 2.32$
   b. $19.97 - 12.65$
   c. $124.99 - 3.33$

5. Evaluate the following.
   a. $63.872 - 9.051$
   b. $43.58 - 1.25$
   c. $87.25 - 34.09$

6. Solve the following.
   a. $125.006 - 0.04$
   b. $35 - 8.97$
   c. $42.1 - 9.072$

7. **MC** The difference between 47.09 and 21.962 is:
   A. 26.93
   B. 25.932
   C. 26.128
   D. 25.128

8. **MC** The sum of 31.5 and 129.62 is:
   A. 98.12
   B. 161.12
   C. 150.12
   D. 444.62

Understanding

9. **WE14** Calculate the following.

   a. $-0.4 + (-0.5)$ b. $-3.4 - 7.17$ c. $-0.6 - 0.72$

10. Calculate the following.

    a. $-2.6 - 1.7$ b. $-132.6 - 31.21$ c. $-0.021 + (-0.97)$

11. **WE15** Calculate the following.

    a. $0.2 + (-0.9)$ b. $-5.41 + 2.87$ c. $1.2 + (-4.06)$

12. Calculate the following.

    a. $-0.8 + 0.23$ b. $335 + (-411.2)$ c. $-3.2 - (-0.65)$

13. On a recent shopping trip, Salmah spent the following amounts: $45.23, $102.78, $0.56 and $8.65.

    a. Calculate the amount she spent altogether.
    b. If Salmah started with $200.00, calculate the amount she had left after her trip.

14. Dagmar is in training for the athletic carnival. The first time she ran the 400 m, it took her 187.04 seconds. After a week of intensive training, she had reduced her time to 145.67 seconds.
    Determine by how many seconds, correct to 2 decimal places, she had cut her time.

15. Kathie runs each morning before school. On Monday she ran 1.23 km, on Tuesday she ran 3.09 km, she rested on Wednesday, and on both Thursday and Friday she ran 2.78 km.
    Calculate the number of kilometres she ran for the week.

Communicating, reasoning and problem solving

16. Students in a school are planning to raise money for a charity by holding running days for different year levels. They hope to run a total of 1946 km. The Year 7s ran 278.2 km, the Year 8s ran 378.4 km and the Year 9s ran 526 km. Determine how many more kilometres the students must run to reach their goal.

17. During 2019, international visitors to Australia spent $45.2 billion. During 2022, after the global travel restrictions due to the pandemic, international visitors spent $9.7 billion. Evaluate by how much spending by international visitors increased or decreased from 2019 to 2022. Show your working.

18. A group of bushwalkers (Alex, Brett, Alzbeta and Mia) were staying at Port Stephens. From their camp at Hawks Nest, they wanted to walk 8.2 km to Barry Creek.
    At the 5-km mark, Mia developed blisters because she had unsuitable footwear on, and was then accompanied back 3.8 km by Alex. There they met another walking group who had bandaids with them. After fixing Mia's feet, Mia and Alex decided to walk back past their Hawks Nest camp and continue on another 2.1 km until they came to Myall Lakes National Park. Drawing a sketch of the situation might be useful.

    a. Evaluate the distance Alex and Mia had to walk to get back to Hawks Nest after Mia fixed her blisters.
    b. Determine the distance between Barry Creek and Myall Lakes National Park.
    c. If Alex and Mia had decided to meet Alzbeta and Brett at Barry Creek instead of going to Myall Lakes National Park, determine how far they would have walked in total to get to Barry Creek.
       Justify your answer.

19. When decimals are being added, give reasons why the decimal points are lined up one underneath the other.

20. **a.** A friend incorrectly rounded 23.925 38 to 2 decimal places as 23.92. Discuss where your friend made a mistake.

   Explain how to correctly round to 2 decimal places.

   b. Your friend started working at a milk bar and was trying to round numbers in his head. A customer's purchase came to $12.97 and the customer gave your friend $15.00.

   Your friend realised that the change would come to $2.03. Because 3 is less than 5, your friend rounded it down and gave the person $2.00 back.

   Explain why the customer might discuss their change with your friend.

Rounding guidelines for cash transactions

Final cash amount	Round to:
1 and 2 cents	nearest 10
3 and 4 cents	nearest 5
6 and 7 cents	nearest 5
8 and 9 cents	nearest 10

Source: www.accc.gov.au

21. By writing each of these recurring decimals correct to 4 decimal places and showing your working, evaluate:

   a. $0.\dot{3} + 0.\dot{5}$

   b. $0.\dot{3} + 0.\dot{8}$

   c. $0.\dot{3} + 0.\dot{7}$

   d. $0.\dot{4} + 0.\overline{53}$

   e. $0.\dot{7} + 0.\dot{6}$

   f. $2.\overline{534} - 0.\overline{12}$

LESSON
3.7 Multiplying and dividing decimals

LEARNING INTENTION

At the end of this lesson you should be able to:
- multiply and divide decimals.

3.7.1 Multiplication of positive decimals

eles-3588

- To multiply decimals, ignore the decimal point and multiply as you would whole numbers. Count the number of digits after the decimal point in each of the multiplying numbers. Adding them together gives the number of decimal places in the answer.
- It is a good idea to estimate the answer to make sure that your answer makes sense.

WORKED EXAMPLE 16 Multiplying decimal numbers

Calculate the value of 34.6 × 0.74, giving an exact answer.

THINK	WRITE
1. Estimate the answer by rounding the numbers to the first digit.	Estimate $= 30 \times 0.7$ $= 21$
2. Set out the question as you would for whole numbers, temporarily ignoring the decimal places. Multiply the numbers as you would multiply whole numbers.	$$\begin{array}{ccccc} & & \overset{3}{3} & \overset{1}{4} & \overset{4}{6} \\ \times & & & 7 & 4 \\ \hline & 1 & 3 & 8 & 4 \\ 2 & 4 & 2 & 2 & 0 \\ \hline 2 & 5 & 6 & 0 & 4 \\ \hline \end{array}$$
3. 34.6 has 1 decimal place and 0.74 has 2 decimal places. Therefore, there will be 3 decimal places in the answer.	$34.6 \times 0.74 = 25.604$

▶ 3.7.2 Division of positive decimals

eles-3589

- When dividing decimals, make sure that the divisor (the number you are dividing by) is a whole number.
- If the divisor is not a whole number, make it a whole number by either:
  - writing the question as a fraction and multiplying the numerator and denominator by an appropriate power of 10, or
  - multiplying the dividend and divisor by an appropriate power of 10.
  - Once the divisor is a whole number, divide the numbers as usual, making sure that the decimal point in the answer is directly above the decimal point in the question.
- Extra zeros can be placed after the decimal point in the dividend if needed.

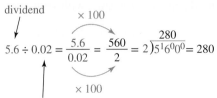

$$5.6 \div 0.02 = \frac{5.6}{0.02} = \frac{560}{2} = 2\overline{)5^16^00^0} = 280$$

WORKED EXAMPLE 17 Dividing decimal numbers

Calculate:

a. $54.6 \div 8$ 　　　　　　　　　　　　　　**b.** $89.356 \div 0.06$

Round your answers to 2 decimal places.

THINK	WRITE
a. 1. Estimate the answer by rounding the numbers to the first digit.	**a.** Estimate $= \dfrac{50}{8}$ $= 6\dfrac{1}{4}$
2. Write the question as shown, adding extra zeros to the dividend. Write the decimal point in the answer directly above the decimal point in the question and divide as usual.	$\dfrac{6.\ 8\ 2\ 5}{8\overline{)54.^66^20^40}}$
3. Write the question and answer, rounded to the required number of decimal places.	$54.6 \div 8 = 6.83$ (to 2 decimal places)
b. 1. Estimate the answer by rounding the numbers to the first digit.	**b.** Estimate $= \dfrac{90}{0.06}$ $= 1500$
2. Write the question.	$89.356 \div 0.06$
3. Multiply both parts by an appropriate multiple of 10 so that the divisor is a whole number (in this case, 100).	$= (89.356 \times 100) \div (0.06 \times 100)$ $= 8935.6 \div 6$
4. Divide, adding extra zeros to the dividend. Write the decimal point in the answer directly above the decimal point in the question and divide as for short division.	$\dfrac{1\ 4\ 8\ 9.\ 2\ 6\ 6}{6\overline{)8^29^53^55.^16^40^40}}$
5. Write the question and answer, rounded to the required number of decimal places.	$89.356 \div 0.06 = 1489.27$ (to 2 decimal places)

1. Form pairs and write five questions involving times tables, but change them slightly so that one or both of the numbers are decimals, for example 5×0.9.
2. Swap your questions with another pair and solve their questions. As a group of four, discuss the answers and any difficulties that you may have had.
3. In your original pairs, write five questions using division based on times tables. Again, ensure that one or both of the numbers are decimals.
4. This time, swap your questions with a different pair and solve their questions.
5. As a class, discuss how your answers change when dividing and multiplying by a decimal rather than a whole number.

3.7.3 Multiplication and division of positive and negative decimals

eles-3590

- When positive and negative decimals are being multiplied and divided, the rules for multiplying and dividing integers apply.

Determining the sign of an answer

- **When multiplying or dividing numbers with the *same* sign, the result is *positive*.**
- **When multiplying or dividing numbers with *different* signs, the result is *negative*.**

WORKED EXAMPLE 18 Multiplying positive and negative decimals

Simplify -3.8×0.05.

THINK	WRITE
1. Estimate the answer by rounding the numbers to the first digit.	$\text{Estimate} = -4 \times 0.05$ $\qquad\qquad = -0.2$
2. Ignore the decimal points and multiply as for positive whole numbers.	$\begin{array}{r} {}^{4}3\,8 \\ \times\ 5 \\ \hline 190 \end{array}$
3. Count the number of decimal places in the question and insert the decimal point. The final answer should have 3 decimal places. The signs are different, so insert a negative sign in the answer.	$-3.8 \times 0.05 = -0.190$ $\qquad\qquad\qquad = -0.19$

WORKED EXAMPLE 19 Dividing positive and negative decimals

Determine the quotient of $-0.015 \div -0.4$, giving an exact answer.

THINK	WRITE
1. Estimate the answer by rounding the numbers to the first digit.	$\text{Estimate} = \dfrac{-0.02}{-0.4}$ $\qquad\qquad = 0.05$
2. Write the question.	$-0.015 \div -0.4$

3. Multiply both parts by 10 to produce a whole number divisor.

$= -(0.015 \times 10) \div -(0.4 \times 10)$

$= -0.15 \div -4$

4. Divide until an exact answer is achieved or until a recurring pattern is evident.

$$\begin{array}{r} 0.0\ 3\ 7\ 5 \\ 4\overline{)0.1\ 5^3 0^2 0} \end{array}$$

5. The signs are the same, so the answer is positive.

$-0.015 \div -0.4 = 0.0375$

 Resources

▶ **Video eLessons** Multiplication of decimals (eles-2311)
Division of decimals (eles-1877)

✦ **Interactivities** Multiplication of positive decimals (int-3726)
Division of positive decimals (int-3727)
Division of negative decimals (int-3728)

Exercise 3.7 Multiplying and dividing decimals

learn

| 3.7 Quick quiz   on | 3.7 Exercise |

Individual pathways

■ PRACTISE	■ CONSOLIDATE	■ MASTER
1, 3, 4, 8, 12, 14, 16	2, 5, 6, 10, 15, 17, 18	7, 9, 11, 13, 19, 20

Fluency

1. **WE16** Calculate the following, giving exact answers.
   a. 6.2×0.8
   b. 7.9×1.2
   c. 109.5×5.6
   d. 5.09×0.4

2. Calculate the following, giving exact answers.
   a. 65.7×3.2
   b. 32.76×2.4
   c. 123.97×4.7
   d. 3.4×642.1

3. Calculate the following, giving exact answers.
   a. 576.98×2
   b. 0.6×67.9
   c. 23.4×6.7
   d. 52.003×12

4. **WE17a** Calculate the following. Give answers rounded to 2 decimal places.
   a. $43.2 \div 7$
   b. $523.9 \div 4$
   c. $6321.09 \div 8$
   d. $2104 \div 3$

5. Calculate the following. Give answers rounded to 2 decimal places.
   a. $286.634 \div 3$
   b. $76.96 \div 12$
   c. $27.8403 \div 11$
   d. $67.02 \div 9$

6. **WE17b** Calculate the following. Give answers rounded to 2 decimal places where appropriate.
   a. $53.3 \div 0.6$
   b. $960.43 \div 0.5$
   c. $3219.09 \div 0.006$
   d. $478.94 \div 0.016$

7. Calculate the following. Give answers rounded to 2 decimal places where appropriate.
   a. $25.865 \div 0.004$
   b. $26.976 \div 0.0003$
   c. $12.00053 \div 0.007$
   d. $35.064 \div 0.005$

8. Estimate the following by rounding each number to the first digit. Check your estimates with a calculator and comment on their accuracy.
   a. 5.1×13.4
   b. $73.8 \div 11.4$
   c. $4.9 \div 0.13$
   d. 46.2×0.027

9. Estimate the following by rounding each number to the first digit. Check the accuracy of your estimates by using a calculator.

   a. 0.14×984 **b.** $405 \div 36.15$ **c.** 17.9×4.97 **d.** $0.58 \div 0.0017$

Understanding

10. WE18 Solve the following.

    a. 0.3×-0.2 **b.** $(-0.3)^2$ **c.** -0.8×0.9 **d.** $(-0.6)^2$

11. Evaluate the following.

    a. 4000×-0.5 **b.** -0.02×-0.4 **c.** -4.9×0.06 **d.** $(0.2)^2 \times -40$

12. WE19 Determine the quotient of each of the following, giving an exact answer.

    a. $-8.4 \div 0.2$ **b.** $0.15 \div -0.5$ **c.** $-15 \div 0.5$ **d.** $0.049 \div -0.07$

13. Determine the quotient of each of the following, giving an exact answer.

    a. $-0.0036 \div 0.06$ **b.** $270 \div -0.03$ **c.** $0.8 \div -0.16$ **d.** $(1.2)^2 \div 0.04$

14. Solve the following, giving answers rounded to 1 decimal place.
    (*Hint:* Pay attention to the order of operations.)

    a $4.6 \times 2.1 + 1.2 \times 3.5$ **b** $5.9 \times 1.8 - 2.4 \times 3.8$

15. Evaluate the following, giving answers rounded to 1 decimal place.
    (*Hint:* Pay attention to the order of operations.)

    a. $6.2 + 4.5 \div 0.5 - 7.6$ **b.** $11.4 - 7.6 \times 1.5 + 2$

Communicating, reasoning and problem solving

16. A group of 21 Year 8 students were going on an excursion to the planetarium. If the total cost was $111.30, determine the amount paid by each student.

17. Determine the decimal halfway between 2.01 L and 2.02 L.

18. A shoebox with a mass of 150 g measures 30.0 cm by 16.5 cm along the bottom, with a height of 10.3 cm.

    a. The volume of the shoebox is found by multiplying its three linear dimensions together. Evaluate its volume.

    b. Density is defined as mass divided by volume. Evaluate the density of the shoebox in g/cm^3.

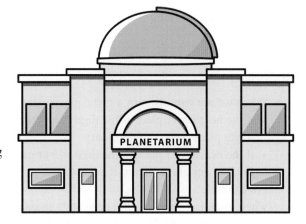

19. A square of grass has a side length of 3.4 metres. There is a path 1.2 metres wide around the outside of the square of grass.

    a. Calculate the area of the square of grass.
    b. Determine the area of the path, explaining your method.

20. Speed equals the distance travelled divided by the time taken.

    a. If Septimus Squirrel travels 27.9 metres in 8.4 seconds and Fortescue Fieldmouse travels 18.7 metres in 6.4 seconds, determine which of them travels at the greater speed. Show full working.
    b. If she is travelling at 5 metres per second, determine how long (in hours) it will take Genevieve Goose to travel round the equator, assuming the circumference of Earth at the equator is approximately 40 075 kilometres. Give your answer to 1 decimal place.

    Hint: To convert seconds to hours, divide time in seconds by 3600.

LESSON
3.8 Estimation

LEARNING INTENTION

At the end of this lesson you should be able to:
- determine an estimate to a problem using different rounding techniques.

▶ 3.8.1 Estimation

eles-3591

- Sometimes an **estimate** of the answer is all that is required (an estimate is an answer close to the actual answer, but found using easier numbers).

WORKED EXAMPLE 20 Calculating an estimate

Marilyn and Kim disagree about the answer to the following calculation:

$7.3 + 7.1 + 6.9 + 6.8 + 7.2 + 7.3 + 7.4 + 6.6$. **Marilyn says the answer is 56.6, but Kim thinks it is 46.6. Obtain an estimate for the calculation and determine who is correct.**

THINK	WRITE
1. Carefully analyse the values and devise a method to estimate the total.	Each of the values can be approximated to 7 and there are eight values.
2. Perform the calculation using the rounded numbers.	$7 \times 8 = 56$
3. Answer the question.	Marilyn is correct, because the approximate value is very close to 56.6.

Rounding to

- When we are **rounding to** a given place value:
  - if the next lower place value digit is less than 5, leave the place value digit as it is, and add zeros to all lower place values, if necessary
  - if the next lower place value digit is 5 or greater, increase the given place value digit by 1 and add zeros to all lower place values, if necessary.

    For example:

    when rounding 25 354 to the nearest thousand, the result is 25 000

    when rounding 25 354 to the nearest hundred, the result is 25 400.

Rounding up to

- When **rounding up**, the digit in the desired place value is increased by 1 regardless of the digits in the lower place positions (as long as they are not all zeros).
- Zeros are added to the lower place positions to retain the place value.

    For example:

    when rounding up 3420 to the nearest thousand, the result is 4000

    when rounding up 3420 to the nearest hundred, the result is 3500.

Rounding down to

- When **rounding down**, all digits following the desired place value are replaced by zeros, leaving the given place value unchanged.

    For example, when rounding down 635 to the nearest ten, the result is 630.

Consider the number 39 461 and perform the following operations.
a. Round to the nearest thousand.
b. Round up to the nearest hundred.
c. Round down to the nearest ten.

THINK	WRITE
a. 1. Consider the digit in the thousands place position and the digit in the next lower place position.	**a.** The digit 9 lies in the thousands position. The digit 4, which is less than 5, lies in the hundreds position.
2. Write the answer, adding the required number of zeros.	The number 39 461 rounded to the nearest thousand is 39 000.
b. 1. Consider the digit in the hundreds position and all following digits.	**b.** The digit 4 lies in the hundreds position. As the digits in the lower place positions are not all zeros, round this up to 5, and the lower place positions will all become 0.
2. Write the answer, adding the required number of zeros.	The number 39 461 rounded up to the nearest hundred is 39 500.
c. 1. Consider the digit in the tens position and all following digits.	**c.** The digit 6 lies in the tens position. As the digit in the lower place position is not zero, round this down to 6, and the lower place position will become 0.
2. Write the answer, adding the required number of zeros.	The number 39 461 rounded down to the nearest ten is 39 460.

3.8.2 Estimating by rounding to the first digit

eles-3592

- When estimating answers to calculations, sometimes it is simplest to round all numbers in the calculation to the first digit and then perform the operation.

Provide an estimate to the following calculations by first rounding each number to its first digit. Check your estimate with a calculator.

Comment on the accuracy of your estimate.

a. $394 + 76 - 121$

b. $\dfrac{692 \times 32}{19 \times 87}$

THINK	WRITE
a. 1. Round each of the numbers to the first digit.	**a.** Rounded to the first digit, 394 becomes 400, 76 becomes 80 and 121 becomes 100.
2. Perform the calculation using the rounded numbers.	$394 + 76 - 121 \approx 400 + 80 - 100$ ≈ 380

3. Check using a calculator. Comment on how the rounded result compares with the actual answer.	Using a calculator, the result is 349. The estimate compares well to the actual (calculator) value.
b. 1. Round each of the numbers to the first digit.	**b.** Rounded to the first digit, 692 becomes 700, 32 becomes 30, 19 becomes 20 and 87 becomes 90.
2. Perform the calculation using the rounded numbers.	$$\frac{692 \times 32}{19 \times 87} \approx \frac{\cancel{700}^{35} \times \cancel{30}^{1}}{\cancel{20}_{1} \times \cancel{90}_{3}}$$ $$\approx \frac{35}{3}$$ $$\approx 12$$
3. Check using a calculator. Comment on how the rounded result compares with the actual answer.	Using a calculator, the result is 13.4 (rounded to 1 decimal place). The estimate is very close to the actual (calculator) value.

▶ 3.8.3 Estimating by rounding the dividend to a multiple of the divisor

- To make division easier, the dividend can be rounded to a multiple of the divisor. For example, in $20\,532 \div 7$, $20\,532$ (the dividend) could be rounded to $21\,000$. Because we know that 21 is a multiple of 7, we could see, through mental approximation, that the answer is close to 3000 (the exact answer is 2933).

WORKED EXAMPLE 23 Rounding the dividend to a multiple of the divisor

Provide estimates for the calculation $\dfrac{537}{40}$ by:

a. rounding the dividend up to the nearest hundred
b. rounding the dividend to the nearest ten
c. rounding the dividend to a multiple of the divisor.

THINK	WRITE
a. 1. Round the dividend up to the nearest hundred.	**a.** 537 rounded up to the nearest hundred is 600.
2. Perform the division. Write the estimation.	$$\frac{537}{40} \approx \frac{\cancel{600}^{15}}{\cancel{40}_{1}}$$ $$\approx 15$$
b. 1. Round the dividend to the nearest ten.	**b.** 537 rounded to the nearest ten is 540.
2. Perform the division. Write the estimation.	$$\frac{537}{40} \approx \frac{\cancel{540}^{27}}{\cancel{40}^{2}}$$ $$\approx 13.5$$
c. 1. Round the dividend to a multiple of the divisor.	**c.** 520 is a multiple of 40.
2. Perform the division. Write the estimation.	$$\frac{537}{40} \approx \frac{\cancel{520}^{13}}{\cancel{40}_{1}}$$ $$\approx 13$$

---

110 Jacaranda Maths Quest 8 Stage 4 NSW Syllabus Third Edition

- Different methods will give slightly different estimates.

WORKED EXAMPLE 24 Using an estimation technique

The exact answer to $\dfrac{132 \times 77}{55}$ has the digits 1848. Use any estimation technique to locate the position of the decimal point.

THINK	WRITE
1. Round each of the numbers to the first digit.	Rounded to the first digit, 132 becomes 100, 77 becomes 80 and 55 becomes 60.
2. Perform the calculation using the rounded numbers and write the estimate, ignoring the decimal.	$\dfrac{132 \times 77}{55} \approx \dfrac{100 \times \overset{4}{\cancel{80}}}{\underset{3}{\cancel{60}}}$ $\approx \dfrac{400}{3}$ ≈ 133
3. Use the estimate obtained to locate the position of the decimal point. Write the correct answer.	The estimate gives an answer between 100 and 200. This indicates that the decimal point should be between the last two digits. The correct answer is 184.8.

 Resources

Interactivities Rounding (int-3730)
Rounding to the first digit (int-3731)

Exercise 3.8 Estimation

learn**on**

3.8 Quick quiz **on**	3.8 Exercise

Individual pathways

■ PRACTISE	■ CONSOLIDATE	■ MASTER
1, 2, 5, 10, 13, 16	3, 4, 6, 12, 14, 17	7, 8, 9, 11, 15, 18

Fluency

1. **WE20** Marilyn and Kim disagree about the answer to the following calculation:

$$8.6 + 9.2 + 8.7 + 8.8 + 8.9 + 9.3 + 9.4 + 8.6$$

Marilyn says the answer is 81.5, but Kim thinks it is 71.5. Obtain an estimate for the calculation and determine who is correct.

2. **WE21** For each of the following numbers:

   i. round to the first digit
   ii. round up to the first digit
   iii. round down to the first digit.

     a. 239   b. 4522   c. 21   d. 53 624   e. 592   f. 1044

3. Round each of the numbers in question **2** down to the nearest ten.

4. Round each of the numbers in question **2** up to the nearest hundred.

5. **WE22** Determine an estimate for each of the following by rounding each number to the first digit.

   a. $78 \div 21$ b. $297 + 36$
   c. $587 - 78$ d. $235 + 67 + 903$
   e. $1256 - 678$

6. Determine an estimate for each of the following by rounding each number to the first digit.

   a. 789×34 b. 56×891
   c. $1108 \div 53$ d. $345 + 8906 - 23 + 427$
   e. $907 \div 88$

7. Determine an estimate for each of the following by rounding each number to the first digit.

   a. $326 \times 89 \times 4$ b. $2378 \div 109$
   c. $7 \times 211 - 832$ d. $977 \div 10 \times 37$
   e. $(12\,384 - 6910) \times (214 + 67)$

8. **WE23** Provide estimates for each of the following by first rounding the dividend to a multiple of the divisor.

   a. $35\,249 \div 9$ b. $2396 \div 5$
   c. $526\,352 \div 7$ d. $145\,923 \div 12$
   e. $92\,487 \div 11$ f. $5249 \div 13$

9. **WE24** Use any of the estimation techniques to locate the position of the decimal point in each of the following calculations. The correct digits for each one are shown in brackets.

   a. $\dfrac{369 \times 16}{288}$ (205)   b. $\dfrac{42\,049}{14 \times 20}$ (150 175)   c. $\dfrac{99 \times 270}{1320}$ (2025)

   d. $\dfrac{285 \times 36}{16 \times 125}$ (513)   e. $\dfrac{256 \times 680}{32 \times 100}$ (544)   f. $\dfrac{7290 \times 84}{27 \times 350}$ (648)

Understanding

10. If 127 people came to a school social and each paid \$5 admission, determine an estimate for the amount of money collected.

11. Estimate the whole numbers between which each of the following will lie.

　　a. $\sqrt{20}$　　　　　**b.** $\sqrt{120}$　　　　　**c.** $\sqrt{180}$　　　　　**d.** $\sqrt{240}$

12. Complete the table below with the rounded question, the estimated answer and the exact answer. Use rounding to the first digit. The first one has been completed.

	Question	Rounded question	Estimated answer	Exact answer
a.	789×56	800×60	48 000	44 184
b.	$124 \div 5$			
c.	$678 + 98 + 46$			
d.	235×209			
e.	$7863 - 908$			
f.	63×726			
g.	$39\,654 \div 227$			
h.	$1809 - 786 + 467$			
i.	$21 \times 78 \times 234$			
j.	$942 \div 89$			
k.	$\dfrac{492 \times 94}{38 \times 49}$			
l.	$\dfrac{54\,296}{97 \times 184}$			

Communicating, reasoning and problem solving

13. Consider the multiplication 18×44.

　　a. Round both numbers to the first digit and then complete the multiplication.

　　b. Multiply 18 by 44 and then round the answer to the first digit.

　　c. Compare your answers to parts **a** and **b**. Discuss what you notice. Explain whether this is the case for all calculations.

14. Determine an approximate answer to each of the worded problems below by rounding to the first digit. Remember to write each answer in a sentence.

　　a. A company predicted that it would sell 13 cars in a month at \$28 999 each. About how much money would they take in sales?

　　b. A tap was leaking 8 mL of water each hour. Calculate approximately how many millilitres of water would be lost if the tap was allowed to leak for 78 hours.

　　c. The Year 8 cake stall sold 176 pieces of cake for 95 cents each. Calculate an estimate for the amount of money they made.

　　d. Steven swam 124 laps of a 50-m pool and, on average, each lap took him 47 seconds. If he swam non-stop, determine approximately how many seconds he was swimming for.

　　e. An audience of 11 784 people attended a recent Kylie concert at Rod Laver Arena and paid \$89 each for their tickets. Calculate the approximate amount of money taken at the door.

　　f. A shop sold 4289 articles at \$4.20 each. Estimate the amount of money that was paid altogether.

　　g. On Clean Up Australia Day, 19 863 people volunteered to help. If they each picked up 196 pieces of rubbish, estimate how many pieces of rubbish were collected altogether.

15. If two numbers are being multiplied, explain how you can predict the size of the rounding error relative to their respective rounding-off points.

16. Sports commentators often estimate the crowd size at sports events. They estimate the percentage of the seats that are occupied, and then use the venue's seating capacity to estimate the crowd size.

   a. The photograph below shows the crowd at an AFL match between Collingwood and Essendon at the Melbourne Cricket Ground. Use the information given with the photograph to calculate the estimated number of people in attendance.
   b. Choose two well-known sports venues and research the maximum seating capacity for each.
   c. If you estimate that three out of every four seats are occupied at each of the venues chosen in part b for particular sports events, determine how many people are in attendance at each venue.

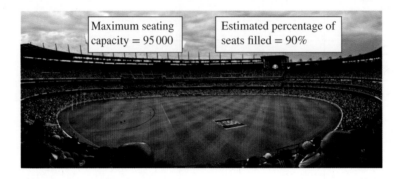

Maximum seating capacity = 95 000

Estimated percentage of seats filled = 90%

17. For each of the following multiplications, calculate the product when only:
   i. the first number is rounded to the nearest 10
   ii. the second number is rounded to the nearest 10.
      a. 18×45 b. 18×46 c. 18×47 d. 18×48 e. 18×49
      f. Describe any patterns that you notice in your answers to parts a–e.

18. Explore the rounding error patterns that occur in calculating the following multiplications when only:
   i. the first number is rounded to the nearest 10
   ii. the second number is rounded to the nearest 10.
      a. 15×44 b. 16×44 c. 17×44

LESSON
3.9 Review

3.9.1 Topic summary

The real number system

- The set of real numbers, R, contains rational numbers, Q, and irrational numbers, Q'.
- Within the rational numbers there are integers, Z, and non-integer numbers.
- Integers can be broken up into positive numbers, Z^+, negative numbers, Z^-, and zero.

REAL NUMBERS

Estimation

- Estimating the answer before doing any calculations helps you decide whether your answer is accurate.
- An effective way to determine an estimate is to round the numbers to the first digit and then do the calculations.
- Other estimation techniques, such as rounding to the nearest 10 or 100, can also be used.

Terminating and recurring decimals

- Terminating decimals have a fixed number of decimal places.
  e.g. 0.5, 0.25, 7.125
- Recurring decimals have an infinite (never-ending) number of decimal places.
  e.g. $\frac{1}{3} = 0.333\,333\ldots$
- Recurring decimals that have one recurring digit are written with a dot above the recurring digit.
  e.g. $\frac{1}{3} = 0.\dot{3}$
- Recurring decimals with more than one digit are written with a line above the recurring digits.
  e.g. $\frac{1}{7} = 0.\overline{142\,857}$

Operations with decimals

- When adding or subtracting decimals, first align the decimal points. Then add as you would for whole numbers, working from right to left.
- When multiplying decimals, ignore the decimal point and multiply as if the numbers are integers. The number of decimal places in the answer is equal to the total number of decimal places in the two numbers being multiplied.
  e.g. 2.1×0.35 (3 decimal places in total)
  $\rightarrow 21 \times 35 = 735$
  $\therefore 2.1 \times 0.35 = 0.735$
- When dividing decimals, the divisor must be a whole number. To convert, multiply both the divisor and dividend by a power of 10. Then divide as for whole numbers, ensuring the decimal point in the answer is placed directly above the decimal point in the question.

Fractions

- A fraction represents part of a whole.
- Equivalent fractions are fractions in which the numerators and denominators are different but in the same ratio.
  e.g. $\frac{3}{12} = \frac{1}{4}$
- Fractional answers must always be given in simplified form. To simplify a fraction, divide the numerator and denominator by the same number.

Operations with fractions

- When adding or subtracting fractions, the denominators must be the same. Convert to the lowest common denominator if needed, then add or subtract the numerators only.
- When multiplying fractions, multiply the numerators together and multiply the denominators together, then simplify if possible.
  e.g. $\frac{3}{4} \times \frac{2}{5} = \frac{6}{20} = \frac{3}{10}$
- When dividing fractions, follow the **Keep, Change, Flip** method, then multiply the fractions together.

3.9.2 Project

A growing nation

Details about a population are collected in a census. Australia's most recent census took place in August 2021, on a day known as 'census day'. The information provided by the population is collected and then analysed by the Australian Bureau of Statistics (ABS) over a period of 2 years. After this, the information is released to the public. The following table displays some selected characteristics for Australia based on the information collected in the 2021 census.

Characteristics	Total persons
Males in Australia	12 545 154
Females in Australia	12 877 635
People aged 15 years and over	20 784 780
People aged 65 years and over	4 378 088
People born in Australia	17 019 815
People who speak English only	18 303 662

1. What was the total population of Australia on census day in 2021?
2. According to the values in the table, how many people were not born in Australia?
3. Express the male population as a proportion of the entire population. Give your answer as a decimal. (Divide the male population by the total population.) Repeat this for the female population.
4. How many people were aged 14 years and younger?

The following table shows the same characteristics as the first table, but it relates to New South Wales only.

Characteristics	Total persons
Males in New South Wales	3 984 166
Females in New South Wales	4 087 995
People aged 15 years and over	6 602 162
People aged 65 years and over	1 424 145
People born in Australia	5 206 544
People who speak English only	5 457 982

5. What was the population of New South Wales on census day in 2021?
6. Express both the male and female population of New South Wales as a proportion of the total population of the state.
7. How does the proportion of males and females in New South Wales compare with the proportion of males and females in the entire Australian population?
8. On a separate page, compare the other characteristics in the table with those for the entire population.

Australia's population has increased since Federation in 1901. The following table shows some of the information collected in the 1901 census.

Characteristics	Total persons
Males in Australia	1 977 928
Females in Australia	1 795 873
People born in Australia	2 908 303

9. How much did Australia's population grow in the 120 years since 1901?
10. Compare the proportion of people born in Australia at the 1901 census with the 2021 census.
11. The Australian Bureau of Statistics website (www.abs.gov.au) can be used to investigate other characteristics from the 2021 census.
    Use the ABS website to investigate another characteristic from the 2021 census.
    State your findings and include all working to justify your conclusions.

Exercise 3.9 Review questions

learn on

Fluency

1. **MC** The numbers 0.68 and $\sqrt{7}$ can be classified as:
   A. irrational and non-integer rational respectively.
   B. integer and irrational respectively.
   C. rational and non-integer rational respectively.
   D. non-integer rational and irrational respectively.

2. Solve the following.
   a. $\dfrac{2}{3} + \dfrac{6}{7}$
   b. $\dfrac{3}{5} + 4\dfrac{1}{2}$
   c. $2\dfrac{3}{4} - 1\dfrac{1}{8}$
   d. $\dfrac{5}{6} + \dfrac{3}{12} + \dfrac{4}{15}$

3. Evaluate the following.
   a. $\dfrac{127}{64} - \dfrac{5}{8} + 2\dfrac{3}{4}$
   b. $2\dfrac{1}{2} + 3\dfrac{1}{2} - 1\dfrac{3}{5}$
   c. $-1\dfrac{19}{60} + \dfrac{1}{4}$
   d. $-\dfrac{3}{5} - \dfrac{7}{10}$

4. Solve the following.
   a. $\dfrac{2}{5} \times \dfrac{7}{8}$
   b. $\dfrac{3}{4} \div \dfrac{7}{8}$
   c. $\dfrac{22}{6} \times \dfrac{8}{11}$
   d. $4\dfrac{1}{3} \times 9\dfrac{1}{2}$

5. Evaluate the following.

   a. $7\dfrac{1}{5} \div \dfrac{8}{20}$　　　b. $\dfrac{9}{4} \div 8\dfrac{1}{2}$　　　c. $-\dfrac{7}{8} \times \dfrac{5}{14}$　　　d. $-2\dfrac{3}{4} \div -\dfrac{3}{8}$

6. Solve the following.

   a. $2.4 + 3.7$　　　b. $11.62 - 4.89$　　　c. $12.04 + 2.9$　　　d. $5.63 - 0.07$

7. Evaluate the following.
   a. $34.2 - 4.008$　　　b. $34.09 + 1.2$　　　c. $-2.48 + 1.903$　　　d. $-1.63 - 2.54$

8. Evaluate the following, rounding answers to 2 decimal places where appropriate.

   a. 432.9×2　　　b. 78.02×3.4　　　c. $543.7 \div 0.12$　　　d. $9.65 \div 1.1$

9. Evaluate the following, rounding answers to 2 decimal places where appropriate.

   a. 923.06×0.00045　　b. $74.23 \div 0.0007$　　c. $0.08 \div -0.4$　　d. $-1.02 \div -0.5$

Understanding

10. Convert the following decimals to fractions in simplest form.
    a. 0.7　　　b. 0.45　　　c. 1.85　　　d. 2.4

11. Convert the following fractions to decimals, giving exact answers or using the correct notation for recurring decimals where appropriate.

    a. $\dfrac{3}{2}$　　　b. $\dfrac{14}{25}$　　　c. $\dfrac{8}{75}$　　　d. $\dfrac{137}{6}$

12. For each of the following numbers:
    i. round to the first digit
    ii. round up to the first digit
    iii. round down to the first digit.
    a. $39\,260$　　　b. 222　　　c. 3001

13. Provide estimates for each of the following by first rounding the dividend to a multiple of the divisor.
    a. $809 \div 11$　　　b. $7143 \div 9$　　　c. $13\,216 \div 12$

14. The answer to $\dfrac{99 \times 1560}{132 \times 312}$ contains the digits 375, in that order. Use an estimating technique to determine the position of the decimal point and write the true answer.

15. Use your estimation skills to determine approximate answers for the following.
    a. 306×12
    b. $268 + 3075 + 28 + 98\,031$
    c. $4109 \div 21$
    d. $19\,328 - 4811$

Communicating, reasoning and problem solving

16. In order to raise money for charity, a Year 8 class organised a cake stall. Starting the day with 9 whole cakes, they sold $2\frac{1}{4}$ cakes at recess and $4\frac{5}{8}$ at lunchtime.

    a. Determine the number of cakes left over.
    b. If there were 20 students in the class, explain whether they would be able to share the leftover cake equally if it had been cut up into eighths.
    c. The cake stall raised $150. If they plan to give $\frac{1}{5}$ to the Red Cross and $\frac{2}{3}$ to World Vision, determine how much each group will receive and how much is left over.

17. Identify the fractions equivalent to $\frac{3}{4}$ that have the following properties.

    a. The difference between the denominator and the numerator is 7.
    b. The product of the denominator and numerator is 300.

18. At Teagan's farm, there are 24 horses. One-sixth of them are brown and one-quarter of them are black. If half of the remaining horses are chestnut and the other half grey, evaluate the number of grey horses.

19. On the diagram shown, EFGH is half the area of ABCD, JKLM is half the area of EFGH and NPQR is half the area of JKLM. Determine what fraction of ABCD is NPQR.

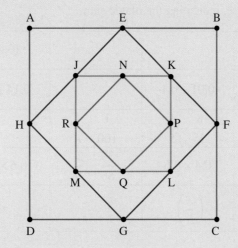

20. The five points shown on the number line are evenly spaced. Evaluate the value for B.

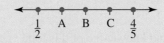

21. You and a friend have decided to test the combinations of cordial and water shown in the following table. The mixture with the highest cordial content will be the one you and your friend will give to 180 Year 8 students.

Mix A	Mix B
2 cups cordial 3 cups water	1 cup cordial 2 cups water
Mix C	**Mix D**
5 cups cordial 9 cups water	3 cups cordial 5 cups water

a. Explain which mix will be used.

b. If each student will drink $\dfrac{1}{2}$ of a cup each, determine how much you will need to make.

22. The skin of a banana weighs about $\dfrac{1}{8}$ the mass of the whole banana. The cost of a bundle of bananas was $5.12. They were peeled and the fruit was found to weigh 7.6 kg in total. Evaluate the price/kg for whole bananas, including fruit and skin.
Use mathematical reasoning to justify your answer.

23. James took a loan from a bank to pay a debt of $14 200. He pays $670.50 per month for 2 years. Evaluate the interest paid by James in total.

24. Magic squares show a grid of numbers that have the same sum horizontally, vertically and diagonally. It is not necessary for the numbers to be integers.
Complete this magic square, indicating the magic sum.

1.7×0.2			$(0.4)^2$	$\dfrac{1}{10} + \dfrac{1}{5}$
$\dfrac{19}{25} - \dfrac{3}{10}$	$0.01 \div 0.1$		$0.14 \div 0.5$	$\dfrac{2^3}{5^2}$
	$\dfrac{8}{25} - \dfrac{1}{5}$	$\dfrac{1}{100} + \dfrac{1}{4}$		$\dfrac{33}{100} \div \dfrac{3}{4}$
	0.4×0.6	$\dfrac{39}{50} - \dfrac{2}{5}$	0.6×0.7	
$\dfrac{11}{100} \div \dfrac{1}{2}$	$\left(\dfrac{3}{5}\right)^2$			$\dfrac{17}{25} - \dfrac{1}{2}$

 To test your understanding and knowledge of this topic, go to your learnON title at www.jacplus.com.au and complete the **post-test**.

Answers

Topic 3 Real numbers

3.1 Pre-test

1. a. $\dfrac{5}{6}$ b. $2\dfrac{3}{5}$

2. a. $\dfrac{25}{28}$ b. $\dfrac{3}{10}$

3. a. $\dfrac{5}{8}$ b. $\dfrac{1}{3}$

4. True

5. a. 3.056   b. 88.20
   c. 0.313   d. 12.13

6. $2 : 3$

7. 30

8. a. $1\dfrac{1}{6}$ b. $3\dfrac{3}{14}$

9. 2

10. a. $\dfrac{157}{50}$ b. $\dfrac{5}{8}$

11. C

12. 13

13. B

14. $7\dfrac{1}{2}$ cm

15. $24

3.2 The real number system

1. a.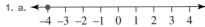

   b. $\dfrac{1}{4}$

   c. $\sqrt{5}$

   d. -3.6

2. a. 1.9

   b. $-\dfrac{2}{4}$

   c. 2.4

   d. $2\sqrt{2}$

3. a.  $\sqrt{10}$

b. -3.8

c. $\dfrac{48}{12}$

d. $\dfrac{-5}{2}$

4. $\dfrac{-12}{-2}$ (6)

5. a. Integer   b. Non-integer rational
   c. Irrational   d. Non-integer rational

6. a. Irrational   b. Integer
   c. Irrational   d. Non-integer rational

7. a. Integer   b. Non-integer rational
   c. Irrational   d. Non-integer rational

8. $9.17, 10, 7.\dot{3}\dot{4}, \dfrac{\sqrt{36}}{3}, -2$

9. 0 can be expressed as a ratio of 2 integers. For example, $0 = \dfrac{0}{1}$.

10. Answers will vary. Example: zero should not be included in the set of natural numbers because it holds no counting value and would not be used.

11. $4\sqrt{5}$

12. Irrational. An irrational number has an infinite, non-repeating decimal representation. Multiplying this by a rational number will not change the fact that it is infinite and non-repeating.

3.3 Adding and subtracting fractions

1. a. $\dfrac{1}{6} = \dfrac{5}{30}$ b. $\dfrac{2}{7} = \dfrac{8}{28}$
   c. $\dfrac{5}{8} = \dfrac{35}{56}$ d. $\dfrac{-12}{33} = \dfrac{-4}{11}$

2. a. $\dfrac{1}{9} = \dfrac{9}{81}$ b. $\dfrac{-6}{40} = \dfrac{-3}{20}$
   c. $\dfrac{36}{96} = \dfrac{3}{8}$ d. $-\dfrac{13}{18} = -\dfrac{26}{36}$

3. a. $\dfrac{-15}{6} = \dfrac{-45}{18}$ b. $\dfrac{42}{30} = \dfrac{-7}{-5}$
   c. $\dfrac{88}{48} = \dfrac{11}{6}$ d. $\dfrac{132}{144} = \dfrac{-11}{-12}$

4. a. $\dfrac{1}{5}$ b. $\dfrac{9}{10}$ c. $\dfrac{7}{8}$ d. $\dfrac{7}{8}$

5. a. 1   b. $\dfrac{5}{8}$ c. $\dfrac{9}{17}$ d. $\dfrac{5}{27}$

6. a. $\dfrac{13}{20}$ b. $\dfrac{11}{8}$ c. $\dfrac{1}{5}$ d. $\dfrac{5}{14}$

7. a. $\dfrac{19}{12}$ b. $\dfrac{17}{30}$ c. $\dfrac{23}{20}$ d. $\dfrac{5}{36}$

8. a. $\dfrac{7}{39}$ b. $\dfrac{52}{55}$ c. $\dfrac{7}{85}$ d. $\dfrac{32}{105}$

9. a. $\dfrac{25}{7}$ b. $\dfrac{64}{13}$ c. $\dfrac{27}{5}$ d. $\dfrac{77}{8}$

10. a. $5\dfrac{1}{3}$ b. $8\dfrac{2}{3}$ c. $6\dfrac{1}{4}$ d. $1\dfrac{1}{5}$

11. $\dfrac{5}{4}, \dfrac{11}{8}, \dfrac{17}{12}, 1\dfrac{6}{12}, 1\dfrac{13}{24}, \dfrac{39}{24}$

12. All of the block

13. a. $\dfrac{4}{5}$ b. $\dfrac{18}{25}$ c. $\dfrac{73}{30}$

14. a. $\dfrac{3}{8}$ b. $\dfrac{205}{56}$ c. $-\dfrac{41}{45}$

15. a. $-\dfrac{35}{18}$ b. $-\dfrac{19}{10}$ c. $\dfrac{33}{10}$

16. a. $-6\dfrac{4}{5}$ b. $3\dfrac{2}{9}$ c. $4\dfrac{3}{5}$

17. a. $9\dfrac{5}{12}$ b. $21\dfrac{8}{45}$ c. $2\dfrac{5}{6}$

18. a. $-\dfrac{23}{16}$ b. $-\dfrac{11}{24}$ c. $-\dfrac{131}{60}$

19. $1\dfrac{7}{9}$

20. $\dfrac{5}{12}$

21. $\dfrac{4}{15}$

22. $1\dfrac{3}{8}$

23. 1 cup $+\dfrac{1}{2}$ cup $+\dfrac{1}{4}$ cup $+$ two $\dfrac{1}{3}$ cups

24. It will not fit.

25. Sample responses can be found in the worked solutions in the online resources.

3.4 Multiplying and dividing fractions

1. a. $\dfrac{3}{8}$ b. $\dfrac{1}{56}$ c. $\dfrac{5}{12}$ d. $\dfrac{5}{21}$

2. a. $\dfrac{6}{25}$ b. $\dfrac{1}{3}$ c. $\dfrac{11}{32}$ d. $\dfrac{1}{4}$

3. a. $\dfrac{11}{30}$ b. $\dfrac{1}{5}$ c. $\dfrac{3}{5}$ d. $\dfrac{4}{5}$

4. a. $-\dfrac{1}{6}$ b. $\dfrac{3}{20}$ c. $-\dfrac{1}{4}$ d. $-\dfrac{14}{3}$

5. a. $-\dfrac{5}{8}$ b. $-\dfrac{14}{9}$ c. $-\dfrac{1}{4}$ d. $\dfrac{11}{4}$

6. a. $\dfrac{2}{3}$ b. $\dfrac{7}{12}$ c. $\dfrac{8}{5}$ d. $\dfrac{6}{7}$

7. a. $\dfrac{6}{7}$ b. $\dfrac{3}{5}$ c. $\dfrac{6}{25}$ d. $\dfrac{15}{16}$

8. a. $5\dfrac{3}{5}$ b. $1\dfrac{11}{25}$ c. $8\dfrac{3}{4}$ d. 4

9. a. 13 b. 1 c. $13\dfrac{4}{5}$ d. $11\dfrac{7}{8}$

10. a. 1 b. $\dfrac{12}{7}$ c. $\dfrac{36}{7}$ d. $\dfrac{7}{12}$

11. a. $\dfrac{8}{5}$ b. $\dfrac{35}{16}$ c. $\dfrac{108}{25}$ d. $\dfrac{142}{135}$

12. a. $-\dfrac{2}{5}$ b. $-\dfrac{8}{9}$ c. $-\dfrac{3}{8}$ d. $\dfrac{7}{8}$

13. a. $-\dfrac{1}{6}$ b. $\dfrac{9}{2}$ c. $-\dfrac{12}{5}$ d. $-\dfrac{8}{35}$

14. 12

15. a. $\dfrac{1}{5}$ b. 15

16. a. $-\dfrac{11}{15}$ b. $-\dfrac{35}{16}$ c. 0

17. a. $\dfrac{32}{21}$ b. $-\dfrac{53}{6}$ c. $\dfrac{21}{17}$

18. 120

19. Children's charity: $30
    Cruelty to animals charity: $80
    Left over: $10

20. $\dfrac{1}{16}$

21. Divide all of the ingredients' quantities by 5.

22. a. It is not possible to evenly divide 17 by 2, 3 or 9.

   b. $\dfrac{1}{2} + \dfrac{1}{3} + \dfrac{1}{9} = \dfrac{17}{18}$
    The numerator is the number of camels given away and the denominator is the number of camels required for this situation to be possible.

23. a. $79\dfrac{9}{11}$ km/h   b. $100\dfrac{12}{35}$ km/h   c. $585\dfrac{1}{3}$ km/h

24. $a = 4, b = 3, c = 2$

3.5 Terminating and recurring decimals

1. a. 0.8 b. 0.25 c. 0.75
   d. 1.75 e. $0.\dot{6}$

2. a. $0.41\dot{6}$ b. $0.\overline{81}$ c. $0.9\overline{54}$
   d. $2.1\dot{6}$ e. $0.4\dot{6}$

3. a. 6.5 b. 1.75 c. 3.4
   d. 8.8 e. 6.75

4. a. 12.9 b. $5.\dot{6}$ c. $11.7\dot{3}$
   d. $1.8\dot{3}$ e. $4.\dot{3}$

5. a. $-0.2\dot{6}$ b. $-0.\dot{7}$ c. $-1.8\dot{3}$
   d. $-5.\dot{8}$ e. $-3.\overline{142\,857}$

6. a. $\dfrac{2}{5}$ b. $\dfrac{4}{5}$ c. $1\dfrac{1}{5}$

   d. $3\dfrac{1}{5}$ e. $\dfrac{14}{25}$

7. a. $\dfrac{3}{4}$ b. $1\dfrac{3}{10}$ c. $7\dfrac{7}{50}$

   d. $4\dfrac{21}{100}$ e. $10\dfrac{1}{25}$

8. a. $7\dfrac{39}{125}$ b. $9\dfrac{47}{50}$ c. $84\dfrac{63}{500}$

   d. $73\dfrac{9}{10}$ e. $\dfrac{21}{5000}$

9. 0.75

10. 1.1

11. 0.56

12. $\dfrac{2}{5}$

13. Sample responses can be found in the online resources.

14. $\dfrac{6}{10}, \dfrac{5}{8}, \dfrac{2}{3}, \dfrac{3}{4}, \dfrac{6}{7}$

15. a. $\dfrac{7}{9}$ b. $\dfrac{34}{111}$ c. $2\dfrac{2}{9}$

16. a. No, the sum of two recurring decimals may be an exact

   number; for example, $\dfrac{1}{3} + \dfrac{2}{3} = \dfrac{3}{3} = 1$.

   b. Yes, the recurring part is maintained; for example,

   $\dfrac{1}{3} + \dfrac{1}{2} = \dfrac{5}{6} = 0.8\dot{3}$.

   c. No, it could result in zero; for example, $\dfrac{1}{3} - \dfrac{1}{3} = 0$.

   d. Yes, the recurring part is maintained.

   e. Yes, the recurring part is maintained.

17. a. $\dfrac{1}{30}$

   b. 450 gold doubloons

   c. $1.5pd$

18. a. $\dfrac{1}{72}$ b. $\dfrac{144r}{5}g$

19. Bucket 1: 12 L; bucket 2: 3 L; bucket 3: 3 L

3.6 Adding and subtracting decimals

1. a. 12.9   b. 19.68   c. 17.26

2. a. 20.3   b. 132.44   c. 42.719

3. a. 6.239   b. 59.434   c. 126.157

4. a. 2.24   b. 7.32   c. 121.66

5. a. 54.821   b. 42.33   c. 53.16

6. a. 124.966   b. 26.03   c. 33.028

7. D

8. B

9. a. -0.9 b. -10.57 c. -1.32

10. a. -4.3 b. -163.81 c. -0.991

11. a. -0.7 b. -2.54 c. -2.86

12. a. -0.57 b. -76.2 c. -2.55

13. a. $157.22   b. $42.78

14. 41.37 seconds

15. 9.88 km

16. 763.4 km

17. $35.5 billion

18. a. 1.2 km   b. 10.3 km   c. 15.8 km

19. When adding and subtracting decimals, the decimal points must be lined up underneath each other to ensure digits from the same place value are added or subtracted.

20. a. The third decimal place is 5, and so should be rounded up. Your friend made the mistake of rounding down. 23.925 rounded to 2 decimal places is 23.93.

   b. $2.03 should be rounded to the nearest 5 cents, but your friend rounded to the nearest 10 cents. $2.03 rounded to the nearest 5 cents is $2.05.

21. a. 0.8889   b. 1.2222   c. 1.1111
   d. 0.9798   e. 1.4444   f. 2.4133

3.7 Multiplying and dividing decimals

1. a. 4.96   b. 9.48   c. 613.2   d. 2.036

2. a. 210.24   b. 78.624   c. 582.659   d. 2183.14

3. a. 1153.96   b. 40.74   c. 156.78   d. 624.036

4. a. 6.17   b. 130.98   c. 790.14   d. 701.33

5. a. 95.54   b. 6.41   c. 2.53   d. 7.45

6. a. 88.83   b. 1920.86   c. 536 515   d. 29 933.75

7. a. 6466.25   b. 89 920   c. 1714.36   d. 7012.80

8. a. 50   b. 7   c. 50   d. 1.5

9. a. 100   b. 10   c. 100   d. 300

10. a. -0.06 b. 0.09   c. -0.72 d. 0.36

11. a. -2000 b. 0.008   c. -0.294 d. -1.6

12. a. -42 b. -0.3 c. -30 d. -0.7

13. a. -0.06 b. -9000 c. -5 d. 36

14. a. 13.9   b. 1.5

15. a. 7.6   b. 2.0

16. $5.30

17. 2.015

18. a. 5098.5 cm^3 b. 0.0294 g/cm^3

19. a. 11.56 m^2 b. 22.08 m^2

20. a. Septimus Squirrel   b. 2226.4 hours

3.8 Estimation

1. Estimate: 72. Kim is correct.

2. a. i. 200   ii. 300   iii. 200
   b. i. 5000   ii. 5000   iii. 4000

c. i. 20   ii. 30   iii. 20

d. i. 50 000   ii. 60 000   iii. 50 000

e. i. 600   ii. 600   iii. 500

f. i. 1000   ii. 2000   iii. 1000

3. a. 230   b. 4520   c. 20
   d. 53 620   e. 590   f. 1040

4. a. 300   b. 4600   c. 100
   d. 53 700   e. 600   f. 1100

5. a. 4   b. 340   c. 520
   d. 1170   e. 300

6. a. 24 000   b. 54 000   c. 20
   d. 9680   e. 10

7. a. 108 000   b. 20   c. 600
   d. 4000   e. 810 000

8. a. 4000   b. 500   c. 70 000
   d. 12 000   e. 8000   f. 400

9. a. 20.5   b. 150.175   c. 20.25
   d. 5.13   e. 54.4   f. 64.8

10. $500

11. a. 4 and 5   b. 10 and 11   c. 13 and 14
    d. 15 and 16

12. See the table at the foot of the page.*

13. a. 800
    b. 800
    c. Does not apply to all calculations; for example, 13×27.

14. a. $300 000
    b. 800 mL
    c. $200
    d. 5000 seconds
    e. $900 000
    f. $16 000
    g. 4 000 000 pieces of litter

15. You can predict how large a rounding error will be by calculating the % error for each number being multiplied, and then multiplying these % errors to calculate the final % error.

16. a. 85 500
    b. Sample responses can be found in the worked solutions in the online resources.
    c. Sample responses can be found in the worked solutions in the online resources.

17. a. i. 900   ii. 900
    b. i. 920   ii. 900
    c. i. 940   ii. 900
    d. i. 960   ii. 900
    e. i. 980   ii. 900
    f. When the first number is rounded, the answers increase. When the second number is rounded, the answers remain constant.

18. a. i. 880
       ii. 600
    b. i. 880
       ii. 640
    c. i. 880
       ii. 680
    The error increases if the smaller number is rounded. The error decreases when the rounded number is closer to the actual number.

Project

1. 25 422 789
2. 8 402 974
3. Male: 0.493; female: 0.507
4. 4 638 009
5. 8 072 161
6. Male: 0.494; female: 0.506

*12.

		Question	Simplified question	Estimated answer	Exact answer
a.		789×56	800×60	48 000	44 184
b.		$124 \div 5$	$100 \div 5$	20	24.8
c.		$678 + 98 + 46$	$700 + 100 + 50$	850	822
d.		235×209	200×200	40 000	49 115
e.		$7863 - 908$	$8000 - 900$	7100	6955
f.		63×726	60×700	42 000	45 738
g.		$39 654 \div 227$	$40 000 \div 200$	200	174.69
h.		$1809 - 786 + 467$	$2000 - 800 + 500$	1700	1490
i.		$21 \times 78 \times 234$	$20 \times 80 \times 200$	320 000	383 292
j.		$942 \div 89$	$900 \div 90$	10	10.58
k.		$\dfrac{492 \times 94}{38 \times 49}$	$\dfrac{500 \times 100}{40 \times 50}$	25	24.84
l.		$\dfrac{54 296}{97 \times 184}$	$\dfrac{50 000}{100 \times 200}$	2.5	3.04

7. The proportion of males and females in NSW is almost the same as the proportion across Australia.

8. The proportion of people who speak only English in NSW is significantly lower than in Australia as a whole. NSW also has a slightly lower proportion of people who were born in Australia and a slightly higher proportion of people aged 65 years or older.

9. Australia's population grew by 21 648 from 1901 to 2021.

10. In 1901, the proportion of people born in Australia was 0.771.
    In 2021, the proportion of people born in Australia was 0.669.
    The proportion of people born in Australia in 1901 is greater than the proportion born in Australia in 2021.

11. Answers will vary with each individual's research.

3.9 Review questions

1. D

2. a. $1\frac{11}{21}$ b. $5\frac{1}{10}$ c. $1\frac{5}{8}$ d. $1\frac{7}{20}$

3. a. $4\frac{7}{64}$ b. $4\frac{2}{5}$ c. $-1\frac{1}{15}$ d. $-1\frac{3}{10}$

4. a. $\frac{7}{20}$ b. $\frac{6}{7}$ c. $\frac{8}{3}$ d. $\frac{247}{6}$

5. a. 18 b. $\frac{9}{34}$ c. $-\frac{5}{16}$ d. $\frac{22}{3}$

6. a. 6.1 b. 6.73 c. 14.94 d. 5.56

7. a. 30.192 b. 35.29 c. −0.577 d. −4.17

8. a. 865.8 b. 265.27 c. 4530.83 d. 8.77

9. a. 0.42   b. 106 042.86
   c. −0.20   d. 2.04

10. a. $\frac{7}{10}$ b. $\frac{9}{20}$ c. $\frac{37}{20}$ d. $\frac{12}{5}$

11. a. 1.5 b. 0.56 c. $0.10\dot{6}$ d. $22.8\dot{3}$

12. a. i. 40 000   ii. 40 000   iii. 30 000
    b. i. 200   ii. 300   iii. 200
    c. i. 3000   ii. 4000   iii. 3000

13. a. 70   b. 800   c. 1100

14. 3.75

15. a. 3000 b. 103 330 c. 200 d. 15 000

16. a. $2\frac{1}{8}$
    b. No, they will be 3 pieces short.
    c. Red Cross: $30
       World Vision: $100
       Left over: $20

17. a. $\frac{21}{28}$ b. $\frac{15}{20}$

18. 7

19. NPQR is $\frac{1}{8}$ the area of ABCD.

20. $\frac{13}{20}$

21. a. Mix A
    b. 90 cups (36 cups of cordial, 54 cups of water)

22. $0.60 per kg

23. $1892

24. The magic sum is 1.3.

0.34	0.48	0.02	0.16	0.30
0.46	0.1	0.14	0.28	0.32
0.08	0.12	0.26	0.4	0.44
0.2	0.24	0.38	0.42	0.06
0.22	0.36	0.5	0.04	0.18

4 Percentages

LESSON
4.1 Overview

Why learn this?

Percentages are used to describe many different aspects of information and even have their own symbol: %. One per cent means one-hundredth; therefore, 1% means one per hundred, 10% means ten per hundred and 50% means 50 per hundred. Percentages can be used as an alternative to decimals and fractions. We can write *one-half* as a decimal (0.5), a fraction $\left(\dfrac{1}{2}\right)$ and a percentage (50%).

Why do we have so many ways of writing the same number? Depending on the context, it may be easier to use a certain form. Percentages are commonly used in finance and shopping. It is easier to express an interest rate as 5% rather than 0.05 or $\dfrac{1}{20}$, and easier to say that items are discounted by 70% rather than by 0.7 or $\dfrac{7}{10}$. When you see an interest rate of 5% (5 per hundred), you can easily calculate that for every $100 you will earn $5 in interest.

You will see percentages used for discounts at shops, interest rates for bank accounts and loans, rates of property growth or loss, statistics for sports matches, data used in the media, and company statements about profit and loss. Understanding percentages will help you deal with your own finances and make decisions regarding your income once you are working.

Hey students! Bring these pages to life online

▶ **Watch videos**

🧩 **Engage with interactivities**

A+ **Answer questions and check solutions**

Find all this and MORE in jacPLUS ▶

Reading content and rich media, including interactivities and videos for every concept

Extra learning resources

Differentiated question sets

Questions with immediate feedback, and fully worked solutions to help students get unstuck

1. Calculate 5% of $150.

2. **MC** Which of the following is the correct simplified fraction of 35%?

   A. $\dfrac{7}{10}$ B. $\dfrac{35}{100}$ C. $\dfrac{7}{20}$ D. $\dfrac{35}{1}$

3. Arrange the following numbers in ascending order: $\dfrac{3}{5}$, 22%, $\dfrac{1}{4}$, 0.31, 111%.

4. **MC** Select the correct percentage of 22 g in 1 kg.
   A. 22.2%   B. 22%   C. 2.2%   D. 0.22%

5. Calculate 23% of $80.

6. Tennis equipment at a sports shop is reduced by 15% for an end-of-financial-year sale. A racket has an original price of $90. Calculate the new sale price.

7. A cricket bat is reduced from $400 to $380. Calculate the percentage discount.

8. **MC** To calculate an 8% increase of an amount, what number do you multiply the original amount by?
   A. 108   B. 8   C. 1.8   D. 1.08

9. William is 55 years old and was born in Scotland. He lived in England for 45% of his life and in Australia for 11 years, and the rest of his life was spent in Scotland. Determine how long he lived in Scotland for. Write the answer in years and months.

10. **MC** When the original price of an item is multiplied by 0.78, what percentage has the item increased or decreased by?
    A. Increased by 78%   B. Decreased by 78%
    C. Decreased by 0.22%   D. Decreased by 22%

11. In an auction, an apartment originally priced at $2 750 000 sells for $2 820 000. Calculate the percentage profit made on the sale. Write the answer to 2 decimal places.

12. **MC** The cost of a sofa, including GST, is $890. What would be the cost of the sofa before GST?
    A. $801.10   B. $809.09   C. $801   D. $809.10

13. An item is reduced by 10%, and then increased by 11.$\dot{1}$%. This takes the item back to its original price (to the nearest cent). True or False?

14. The price of a car is reduced by 10% three weeks in a row. Calculate the percentage drop in price by the end of the third week. Write the answer to the nearest whole number.

15. The UK pound (£) can be exchanged for 1.6 Australian dollars (A$). The New Zealand dollar (NZ$) can be exchanged for 0.92 Australian dollars. A Toyota Yaris (excluding GST/VAT) costs £9400, NZ$16 000 and A$13 900. VAT (the UK equivalent of GST) is 20%. GST in New Zealand is 15%. GST in Australia is 10%. In which of the countries is Toyota Yaris the cheapest, including GST/VAT?

LESSON
4.2 Percentages

▶ 4.2.1 Writing percentages in different ways

eles-3618

- The term **per cent** means 'per hundred'.
- The symbol for percentage is %. For example, 60% (60 per cent) means 60 parts out of 100.
- A quantity can be expressed in different ways using percentages, fractions and decimals.

Expressing percentages as fractions or decimals

To convert a percentage to a fraction or a decimal, divide by 100.

$$60\% = \frac{60}{100} = 0.60$$

WORKED EXAMPLE 1 Converting percentages to fractions and decimals

Convert the following percentages to fractions and then decimals.
a. 67%   b. 55%

THINK	WRITE
a. 1. To convert to a fraction, write the percentage, then change it to a fraction with a denominator of 100.	a. $67\% = \frac{67}{100}$
2. To convert 67% to a decimal, think of it as 67.0%, then divide it by 100 by moving the decimal point 2 places to the left.	$67\% = 0.67$
b. 1. To convert 55% to a fraction, write the percentage, then change it to a fraction by adding a denominator of 100.	b. $55\% = \frac{55}{100}$
2. The fraction is not in simplest form, so cancel by dividing the numerator and the denominator by 5.	$55\% = \frac{55}{100} = \frac{11}{20}$
3. To convert 55% to a decimal, think of it as 55.0%, then divide it by 100 by moving the decimal point 2 places to the left.	$55\% = 0.55$

- There are a number of common percentages, and their fraction and decimal equivalents, with which you should be familiar.

Percentage	Fraction	Decimal
$8\frac{1}{3}\%$	$\frac{1}{12}$	$0.08\dot{3}$
10%	$\frac{1}{10}$	0.1
$11\frac{1}{9}\%$	$\frac{1}{9}$	$0.\dot{1}$
12.5%	$\frac{1}{8}$	0.125
$16\frac{2}{3}\%$	$\frac{1}{6}$	$0.1\dot{6}$
25%	$\frac{1}{4}$	0.25
$33\frac{1}{3}\%$	$\frac{1}{3}$	$0.\dot{3}$
50%	$\frac{1}{2}$	0.5
$90\frac{10}{11}\%$	$\frac{10}{11}$	$0.\overline{90}$
100%	1	1

- When converting a fraction or decimal to a percentage, do the inverse of dividing by 100; that is, multiply by 100.

Converting fractions or decimals to percentages

To express a fraction or decimal as a percentage, multiply by 100%.

$$\frac{1}{2} = \frac{1}{2} \times 100\%$$
$$= 50\%$$

Digital technology

Scientific calculators have a % button that can be used to compute calculations involving percentages.

Percentages can be converted into decimals and fractions. Most calculators have a button that converts decimals and fractions. Depending on the brand of calculator, this may appear as S ⇔ d, f ◁▷ d or similar.

```
55%                    ⒟    Math  ▲

                           0.55
86%                    ⒟    Math  ▲

                             43
                             ──
                             50
```

- The easiest method of comparing percentages, fractions and decimals is to convert all of them to their decimal form and use place values to compare them.

WORKED EXAMPLE 2 Comparing fractions, decimals and percentages

Place the following quantities in ascending order, and then place them on a number line.

$$45\%, \frac{7}{10}, 0.36, 80\%, 2\frac{1}{2}, 110\%, 1.54$$

THINK	WRITE
1. Convert all of the quantities into their decimal equivalents.	$0.45, 0.7, 0.36, 0.80, 2.5, 1.10, 1.54$
2. Place them in ascending order.	$0.36, 0.45, 0.7, 0.80, 1.10, 1.54, 2.5$
3. Place them in ascending order in their original form.	$0.36, 45\%, \frac{7}{10}, 80\%, 110\%, 1.54, 2\frac{1}{2}$
4. Draw a number line from 0 to 3, with increments of 0.25.	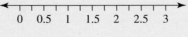
5. Place the numbers on the number line.	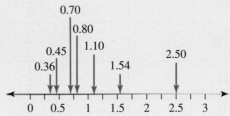

4.2.2 Percentage increases and decreases

eles-3619

- Percentage increases and decreases can be used to calculate and compare prices, markups, discounts, population changes, company profits and many other quantities.
- To calculate a percentage increase or decrease, calculate the net increase or decrease and then express it as a percentage of the initial value.

> **Calculating a percentage change**
>
> $$\text{Percentage change} = \frac{\text{increase or decrease in quantity}}{\text{original quantity}} \times \frac{100\%}{1}$$

Note: Percentage increases of more than 100% are possible; for example, the increase from 3 to 7.5 is an increase of 150%.

WORKED EXAMPLE 3 Calculating percentage increase

Calculate the percentage increase when a shop owner marks up a $50 item to $70.

THINK	WRITE
1. The quantity has increased, so calculate the difference between $50 and $70.	Increase = $70 − $50 = $20
2. The percentage increase can be calculated by creating the fraction 20 out of 50 and then multiplying by 100%.	Percentage increase = $\frac{20}{50} \times 100\%$ = 40%
3. Write the answer.	The percentage increase is 40%.

Calculate the percentage decrease, rounded to 2 decimal places, when the population of a town falls from 62 000 people to 48 000 people.

THINK	WRITE
1. The difference between 62 000 and 48 000 is 14 000.	Decrease $= 62\,000 - 48\,000$ $= 14\,000$
2. The percentage decrease can be calculated by creating the fraction 14 000 out of 62 000 and then multiplying by 100%.	Percentage decrease $= \dfrac{14\,000}{62\,000} \times 100\%$ $= 22.58\%$
3. Write the answer.	The percentage decrease is 22.58%.

Resources

Video eLesson Decimals, fractions and percentages (eles-1868)

Interactivities Percentages, fractions and decimals (int-3741)

Percentage increase and decrease (int-3742)

Exercise 4.2 Percentages

learnon

4.2 Quick quiz on	**4.2 Exercise**

Individual pathways

■ PRACTISE 1, 3, 5, 8, 10, 14, 17	■ CONSOLIDATE 2, 4, 7, 9, 15	■ MASTER 6, 11, 12, 13, 16

Fluency

1. **WE1** Convert the following percentages to fractions and then to decimals.

   a. 24%　　　　　　b. 13%　　　　　　c. 1.5%　　　　　　d. 250%

2. Convert the following percentages to fractions and then to decimals.

   a. 47%　　　　　　b. 6.6%　　　　　　c. 109.8%　　　　　d. 10.02%

3. Express the following percentages as fractions in simplest form.

   a. 20%　　　　　　b. 35%　　　　　　c. 61%　　　　　　d. 105%

4. Express the following percentages as fractions in simplest form.

   a. 11%　　　　　　b. 82%　　　　　　c. 12.5%　　　　　　d. 202%

5. Express the following decimals as percentages.

   a. 0.15　　　　　　b. 0.85　　　　　　c. 3.10　　　　　　d. 0.024

6. Express the following fractions as percentages. Round your answer to 2 decimal places where appropriate.

   a. $\dfrac{7}{8}$
   b. $\dfrac{3}{5}$
   c. $\dfrac{5}{6}$
   d. $2\dfrac{1}{3}$

Understanding

7. **WE2** For the following sets of numbers, place the numbers in ascending order and then on a number line.

   a. 1.6, 25%, $\dfrac{7}{8}$, 75%, 10%, $3\dfrac{1}{2}$, 2.4
   b. $3\dfrac{4}{5}$, 330%, 4.5%, 150%, 3, $2\dfrac{1}{3}$, 2.8

8. **WE3** Calculate the percentage increase when 250 increases to 325.

9. **WE4** Calculate the percentage decrease, rounded to 2 decimal places, when the population of fish in a pond decreases from 1500 to 650.

10. Express $120 as a percentage of $400.

11. In a library, there are 24 children, 36 women and 42 men. Calculate the percentage of women visiting the library.
    Give your answer rounded to 2 decimal places.

12. During a sale, a jacket originally priced at $79.99 is decreased in price to $55.99. Calculate the percentage decrease.

Communicating, reasoning and problem solving

13. A group of students was practising their basketball free throws. Each student had four shots and the results are displayed in the table.

Free throw results	Number of students	Percentage of students
No shots in	3	
One shot in, three misses	11	
Two shots in, two misses	10	
Three shots in, one miss	4	
All shots in	2	

   a. Identify how many students participated in the game.
   b. Complete the table to show the percentage of students for each result.
   c. Calculate how many students made exactly 25% of their shots.
   d. Calcuate what percentage of students made less than 50% of their shots.

14. The price of entry into a theme park has increased by 10% every year since the theme park opened. If the latest price rise increased the ticket cost to $8.80, explain how to determine the price of a ticket 2 years ago. Show your calculations in your explanation.

15. Survey your classmates on the brand of mobile phone that they have. Present your results in a table showing each brand of phone as a percentage, fraction and decimal of the total number of phones.

16. The table shows the percentage of households with 0 to 5 children. Calculate:

    a. the percentage of households that have 6 or more children
    b. the percentage of households that have fewer than 2 children
    c. the fraction of households that have no children
    d. the fraction of households that have 1, 2 or 3 children.

Number of children	Percentage (%)
0	56
1	16
2	19
3	6
4	2
5	1

17. Use the bunch of flowers shown to answer these questions.

    a. Calculate the percentage of the flowers that are yellow.
    b. What fraction of the flowers are pink?
    c. Write two of your own questions and swap with a classmate.

LESSON
4.3 Finding percentages of an amount

LEARNING INTENTIONS

At the end of this lesson you should be able to:
- calculate percentages of an amount
- increase or decrease a value by a percentage.

▶ 4.3.1 Calculating percentages of an amount

eles-3621

- As percentages can't be used directly in calculations, they must be converted into fractions or decimals.
- Percentages of an amount can be determined using calculations with either fractions or decimals.

Digital technology

The percentage button and the multiplication symbol can be used to help determine percentages of an amount.

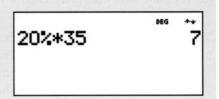

Using decimals

- To calculate a percentage of an amount using decimals, follow these steps:
    1. Write the percentage as a decimal.
    2. Change 'of' to × (multiplication).
    3. Multiply.

WORKED EXAMPLE 5 Calculating the percentage of an amount using decimals

Of the 250 students selected at random to complete a survey, 16% were in Year 11. Calculate how many of the students were in Year 11.

THINK	WRITE
1. Decide what percentage of the total is required. Write an expression to find the percentage of the total.	16% of 250
2. Write the percentage as a decimal. Change 'of' to ×.	$= 0.16 \times 250$
3. Multiply.	$= 40$
4. Answer the question by writing a sentence.	40 of the 250 students were in Year 11.

Using fractions

- To calculate a percentage of an amount, follow these steps:
    1. Write the percentage as a fraction with a denominator of 100.
    2. Change 'of' to ×.
    3. Write the amount as a fraction over 1 if it is not already a fraction.
    4. Cancel.
    5. Perform the multiplication.
    6. Simplify.

Calculate 20% of 35.

THINK	WRITE
1. Write the question.	20% of 35
2. Write the percentage as a fraction with a denominator of 100, change 'of' to '×', write the amount as a fraction over 1 and cancel.	$= \dfrac{20}{\cancel{100}^{20}} \times \dfrac{\cancel{35}^{7}}{1}$
3. Cancel again.	$= \dfrac{\cancel{20}^{1}}{\cancel{20}^{1}} \times \dfrac{7}{1}$
4. Multiply numerators and multiply denominators.	$= \dfrac{7}{1}$
5. Simplify by dividing the numerator by the denominator.	$= 7$
6. Answer the question.	20% of 35 is 7.

▶ 4.3.2 Increasing or decreasing a quantity by $x\%$

eles-3622

- To increase a quantity by $x\%$, multiply it by $(100 + x)\%$. For example, to increase 70 by 30%, multiply it by 130%.
- To decrease a quantity by $x\%$, multiply it by $(100 - x)\%$. For example, to decrease 70 by 40%, multiply it by 60%.

Note: Convert the percentage to a decimal or fraction before multiplying.

a. A newborn baby weighed 3.5 kg. After 1 month the baby's weight had increased by 20%. Calculate the weight of the baby after 1 month.

b. Carlos went for a run on Tuesday evening and ran for 10.2 km. When he next went for a run on Thursday evening, he ran 15% less than he did on Tuesday. Calculate how far he ran on Thursday.

THINK	WRITE
a. 1. Add the percentage increase to 100%.	a. $100\% + 20\% = 120\%$
2. Express the percentage as a fraction and multiply by the amount to be increased.	$\dfrac{120}{100} \times 3.5 = \dfrac{\cancel{120}^{12}}{\cancel{100}^{20}} \times \dfrac{\cancel{35}^{7}}{\cancel{10}^{1}}$ $= \dfrac{\cancel{12}^{3}}{\cancel{20}^{5}} \times \dfrac{7}{1}$ $= \dfrac{21}{5}$ $= 4.2 \text{ kg}$
3. Write the answer.	The weight of the baby after 1 month is 4.2 kg.

b. 1. Subtract the percentage decrease from 100%.   **b.** $100\% - 15\% = 85\%$

2. Express the percentage as a fraction and multiply by the amount to be increased.

$$\frac{85}{100} \times 10.2 = \frac{\cancel{85}^{17}}{\cancel{100}^{50}} \times \frac{\cancel{102}^{51}}{\cancel{10}^{2}}$$

$$= \frac{867}{100}$$

$$= 8.67 \text{ km}$$

3. Write the answer.   Carlos ran 8.67 km on Tuesday.

 Resources

▷ **Video eLesson** Percentages of an amount (eles-1882)

🧩 **Interactivity**   Percentage of an amount (int-3743)

Exercise 4.3 Finding percentages of an amount   **learn**

4.3 Quick quiz on	**4.3 Exercise**

Individual pathways

■ PRACTISE	■ CONSOLIDATE	■ MASTER
1, 3, 5, 8, 9, 12, 14, 18, 20, 23, 25, 28	2, 4, 6, 10, 15, 16, 19, 21, 26, 29	7, 11, 13, 17, 22, 24, 27, 30

Fluency

1. **WE5&6** Calculate the following.

  a. 50% of 20   **b.** 20% of 80   **c.** 5% of 60   **d.** 10% of 30

2. Calculate the following.

  a. 31% of 300   **b.** 40% of 15   **c.** 12% of 50   **d.** 35% of 80

3. Calculate the following.

  a. 70% of 110   **b.** 52% of 75   **c.** 90% of 70   **d.** 80% of 5000

4. Calculate the following.

  a. 44% of 150   **b.** 68% of 25   **c.** 24% of 175   **d.** 38% of 250

5. Calculate the following.

  a. 95% of 200   **b.** 110% of 50   **c.** 150% of 8   **d.** 125% of 20

6. Calculate the following.

  a. 66% of 20   **b.** 2% of 95   **c.** 55% of 45   **d.** 15% of 74

7. Calculate the following.

    a. 95% of 62 **b.** 32% of 65 **c.** 18% of 80 **d.** 82% of 120

8. **MC** 60% of 30 is:

    A. $19\dfrac{4}{5}$ **B.** $\dfrac{31}{5}$ **C.** 186 **D.** 18

9. Calculate the following, rounding answers to the nearest 5 cents.

    a. 1% of $268 **b.** 1% of $713 **c.** 1% of $573 **d.** 1% of $604

10. Calculate the following, rounding answers to the nearest 5 cents.

    a. 1% of $19.89 **b.** 1% of $429.50 **c.** 1% of $4.25 **d.** 1% of $6.49

11. Calculate the following, rounding answers to the nearest 5 cents.

    a. 1% of $9.99 **b.** 1% of $0.24 **c.** 1% of $0.77 **d.** 1% of $1264.37

12. Calculate the following, rounding answers to the nearest 5 cents.

    a. 22% of $10 **b.** 13% of $14 **c.** 35% of $210 **d.** 12% of $150

13. Calculate the following, rounding answers to the nearest 5 cents.

    a. 2% of $53 **b.** 7% of $29 **c.** 45% of $71.50 **d.** 33% of $14.50

Understanding

14. Wiradjuri country is the largest indigenous nation in New South Wales. It covers 16% of the of the total land area. If New South Wales has a total land area of $800\,000\,\text{km}^2$, determine the land area of Wiradjuri country.

15. In a survey, 40 people were asked if they liked or disliked Vegemite. Of the people surveyed, 5% said they disliked Vegemite. Calculate how many people:

    a. disliked Vegemite
    b. liked Vegemite.

16. **WE7** The grocery bill for Mika's shopping was $250. The following week, Mika spent 7% more on his groceries. How much did he spend in the following week?

17. A long-distance runner completed a 15-kilometre run in 120 minutes. The next time she ran 15 kilometres, she reduced her time by 5%. How fast did she complete the 15 kilometres on the second occasion?

18. Maria is buying a new set of golf clubs. The clubs are marked at $950, but if Maria pays cash, the shop will take 10% off the marked price. How much will the clubs cost if Maria pays cash?

19. When you multiply a quantity by 0.77, determine by what percentage you are decreasing the quantity.

20. Increase the following quantities by the given percentages.

    a. 33 kg by 10%
    b. 50 lb by 20%
    c. 83 cm by 100%

21. Decrease the following quantities by the given percentages.

    a. 25 kg by 10%

    b. 40 km by 20%

    c. $96 by 90%

22. Ninety per cent of students at a school were present for school photographs. If the school has 1100 students, calculate how many were absent on the day the photographs were taken.

23. Jim can swim 50 m in 31 seconds. If he improves his time by 10%, calculate Jim's new time.

24. Thirty-two thousand four hundred people went to the SCG to watch a Sydney versus Collingwood football match. Of the crowd, 42% went to the game by car and 55% caught public transport.
Calculate how many people:

    a. arrived by car

    b. caught public transport.

Communicating, reasoning and problem solving

25. When I am 5% older than I am now, I will be 21 years old. Calculate how old I am now.

26. The price of bread has increased to 250% of its price 20 years ago. If a loaf of bread costs $2.00 now, determine how much it would have cost 20 years ago.

27. My mother is three times older than I am. My sister is 75% of my age, and 12% of my grandfather's age. My father is 40, 4 years older than my mother.
Determine the ages of my sister and my grandfather.

28. In a Maths competition, the top 8% of students across the state achieve a score of 40 or more out of a possible 50.

    a. In a school where 175 students have entered the Maths competition, calculate how many scores higher than 40 you would expect.

    b. In one school, there were 17 scores of 40 or more, and 204 scores that were less than 40. Compare the results to determine whether the students performed better than the state average.

29. Broadcasting regulations specify that 55% of television programs shown between 6 pm and midnight must be Australian content and that, between 6 pm and midnight, there should be no more than 13 minutes per hour of advertising.
Calculate:

    a. how many minutes of advertising are allowed between 6 pm and midnight

    b. for how many minutes programs are screened between 6 pm and midnight

    c. the maximum percentage of time spent screening advertising

    d. how many minutes of Australian content must be screened between 6 pm and midnight.

30. I am 27 years old and have lived in Australia for 12 years. If I continue to live in Australia, calculate how old I will be when the number of years I have lived here is 75% of my age.

LESSON
4.4 Discount

▶ 4.4.1 Applying discount

eles-3623

- A discount is a reduction in price, commonly used by businesses aiming to clear out old stock or attract new customers.
- There are two types of discounts:
  - A fixed price discount is a set amount (in dollars) that a product is discounted by.
  - A percentage discount is a discount that is a set percentage of the product's price.

Calculating discount

In general, if an $r\%$ discount is applied:

$$\text{Discount} = \frac{r}{100} \times \text{original price}$$

Calculating selling price of a discounted item

- **Method 1**
  Use the percentage remaining after the percentage discounted has been subtracted from 100%; that is, if an item for sale has a 10% discount, then the price must be 90% of the marked price.

WORKED EXAMPLE 8 Calculating the price of a discounted item

Calculate the sale price on a pair of shoes marked $95 if a 10% discount is given.

THINK	WRITE
1. Determine the percentage of the marked price that is paid, by subtracting the percentage discount from 100%.	$100\% - 10\% = 90\%$
2. Calculate the sale price of the shoes.	90% of $\$95 = 0.9 \times \$95$ $= \$85.50$
3. Write the answer in a sentence.	The sale price of the shoes is $85.50.

- **Method 2**

  The new sale price of the item can be solved by calculating the amount of the discount, then subtracting the discount from the marked price.

  Alternative solution to Worked example 8:

$$\text{Discount} = 10\% \text{ of } \$95.00$$
$$= \$9.50$$
$$\text{Sale price} = \text{marked price} - \text{discount}$$
$$= \$95.00 - \$9.50$$
$$= \$85.50$$

WORKED EXAMPLE 9 Calculating discount and sale price

Peddles is a bicycle store that has offered a discount of 15% on all goods.
Determine:
a. the cash discount allowed on a bicycle costing $260
b. the sale price of the bicycle.

THINK

a. Calculate the discount, which is 15% of the marked price.

b. 1. To calculate the sale price, subtract the discount from the marked price.

   2. Write the answer in a sentence.

WRITE

a. $\text{Discount} = 15\% \text{ of } \$260$
$$= 0.15 \times \$260$$
$$= \$39$$
The cash discount allowed is $39.

b. $\text{Sale price} = \text{marked price} - \text{discount}$
$$= \$260 - \$39$$
$$= \$221$$

The sale price of the bicycle is $221.

Calculating the percentage discount

- When given the original and the discounted prices, the percentage discount can be determined.

Calculating percentage discount

To calculate the percentage discount, write the discounted amount as a percentage of the original price.

$$\text{Percentage discount} = \frac{\text{discounted amount}}{\text{original price}} \times \frac{100}{1}\%$$

WORKED EXAMPLE 10 Calculating percentage discount

At Peddles, the price of a bicycle is reduced from $260 to $200. Calculate the percentage discount.

THINK	WRITE
1. Calculate the amount of the discount.	Discount = $260 − $200 = $60

2. Write the discount as a percentage of the original price.	Percentage discount = $\dfrac{60}{260} \times 100\%$
	        = 23.0769...%
	        ≈ 23%

3. Write the answer in a sentence.	The percentage discount is about 23%.

Note: Just as some items are on sale or reduced, some items or services increase in price.

Examples are the cost of electricity or a new model of car.

We calculate the percentage increase in price in a similar way to the percentage decrease. However, instead of subtracting the discounted value, we add the extra percentage value.

COMMUNICATING — COLLABORATIVE TASK: Let's go shopping!

Equipment: sales catalogues from nearby shops, paper, pen, calculator

Part A

1. As a class, brainstorm percentage discounts that you see advertised in sales. Pick three common ones.
2. Each person should think of an item they want to buy and its current price. A volunteer might like to draw a table on the board with the column headings 'Item' and then the three common percentage discounts.
3. Each person should then calculate the new prices of the selected item, assuming the discount shown on the board. Repeat this process for the item listed underneath yours.
4. As a class, fill in the table and discuss the results.

Part B

1. Work in groups of three or four. Select a page from one of the sales catalogues and calculate the percentage discount on *five* items.
2. Discuss the results as a class. How would you calculate the average percentage discount shown on the items in the catalogue?

---

on Resources

Interactivities Selling price (int-3745)
Discount (int-3744)

Exercise 4.4 Discount

4.4 Quick quiz on	4.4 Exercise

Individual pathways

■ PRACTISE	■ CONSOLIDATE	■ MASTER
1, 3, 5, 8, 10, 13, 17, 19, 23, 27	2, 4, 6, 11, 12, 14, 18, 21, 24, 25, 28	7, 9, 15, 16, 20, 22, 26, 29

Fluency

1. Calculate the discount on each of the items in the table, using the percentage shown.

	Item	Marked price	Discount
a.	Smart watch	$210	20%
b.	Skateboard	$185	25%

2. Calculate the discount on each of the items in the table, using the percentage shown.

	Item	Marked price	Discount
a.	Mobile phone	$330	15%
b.	Tennis racquet	$190	40%

3. Without using a calculator, calculate the percentage discount of the following.

	Marked price	Discount
a.	$100	$10
b.	$250	$125

4. Without using a calculator, calculate the percentage discount of the following.

	Marked price	Discount
a.	$90	$30
b.	$80	$20

5. **WE8** Calculate the sale price of each item with the following marked prices and percentage discounts.

	Marked price	Discount
a.	$1000	15%
b.	$250	20%

6. Calculate the sale price of each item with the following marked prices and percentage discounts.

	Marked price	Discount
a.	$95	12%
b.	$156	$33\frac{1}{3}\%$

7. Calculate the sale price of each item with the following marked prices and percentage discounts.

	Marked price	Discount
a.	$69.95	$7\frac{1}{2}\%$
b.	$345	30%

8. Determine the percentage discount given on the items shown in the table. Round to the nearest per cent.

	Original price	Selling price
a.	$25	$15
b.	$100	$72

9. Determine the percentage discount given on the items shown in the table. Round to the nearest per cent.

	Original price	Selling price
a.	$69	$50
b.	$89.95	$70

Understanding

10. Decrease the following amounts by the percentages given.
    a. $50 by 10%
    b. $90 by 50%
    c. $45 by 20%

11. A tablet computer that usually sells for $599 was advertised with a saving of $148. Calculate the percentage discount being offered. Round to the nearest per cent.

12. The following items are all discounted.

$380	$450	$260	$600
25% discount	20% discount	$33\frac{1}{3}\%$ discount	15% discount

    a. Compare the values of the discounts to decide which item had the largest dollar discount.
    b. Identify which items have the same dollar discount.
    c. Calculate the difference between the largest and the smallest dollar discounts.
    d. If the surfboard had a discount of 20%, would $470 be enough to buy it?

13. **WE9** A sale discount of 20% was offered by the music store Solid Sound. Calculate:
    a. the cash discount allowed on a $350 sound system
    b. the sale price of the system.

14. Fitness trackers are advertised at $69.95, less 10% discount. Calculate the sale price.

15. A store-wide clearance sale advertised 15% off everything.

    a. Determine the selling price of a pair of jeans marked at $49.
    b. If a camera marked at $189 was sold for $160.65, determine whether the correct percentage was deducted.

16. T-shirts are advertised at $15.95 less 5% discount. Calculate the cost of five T-shirts.

17. **WE10** Calculators were advertised at $20, discounted from $25. What percentage discount was given?

18. CDs normally selling for $28.95 were cleared for $23.95. Calculate the percentage discount given (correct to 1 decimal place).

19. At a sale, Negar bought a $120 jacket for $48. What percentage of the original price did she save?

20. Arjun bought a mobile phone priced $199.95 and signed up for a 1-year plan. He received a 10% discount on the telephone and a 15% discount on the $75 connection fee.
    How much did Arjun pay altogether (correct to the nearest 5 cents)?

21. Alannah bought two hairdryers for $128 each. She sold one at a loss of 5% and the other for a profit of 10%.

    a. Determine the selling price of each.
    b. Will she have made a profit or a loss?

22. **MC** Kristen's car insurance was $670, but she had a 'no claim bonus' discount of 12%. Which of the following will not give the amount she must pay?

    A. First calculate 12% of $670 and add your answer to $670.
    B. Calculate $(88 \div 100) \times 670$.
    C. Find 88% of $670.
    D. First calculate 12% of $670, and subtract your answer from $670.

Communicating, reasoning and problem solving

23. Is there a difference between 75% off $200 and 75% of $200? Explain.

24. Concession movie tickets sell for $12.00 each, but if you buy 4 or more you get $1.00 off each ticket. What percentage discount is this (correct to 2 decimal places)? Show your working.

25. Henry buys a computer priced at $1060, but with a 10% discount. Sancha finds the same computer selling at $840 plus a tax of 18%.
    Who has the better price? Explain.

26. You are in a surf shop and you hear 'For today only: take fifty percent off the original price and then a further forty percent off that.' You hear a customer say 'This is fantastic! You get ninety percent off the original price!'
    Is this statement correct? Explain why.

27. What would you multiply the original prices of items by to get their new prices with:

    a. a 35% discount   b. an 11% increase   c. a 6% discount   d. a 100% increase?

28. A student was completing a discount problem where she needed to calculate a 25% discount on $79. She misread the question and calculated a 20% discount to get $63.20.
    She then realised her mistake and took a further 5% from $63.20. Is this the same as taking 25% off $79?
    Use calculations to support your answer.

29. **a.** At the local market there is a 'buy two, get one free' offer on handmade soaps. Explain what percentage discount this is equivalent to.

b. At a rival market there is a 'buy one, get another half price' offer on soaps.
Explain whether this deal is the same, better or worse than the discount offered in part **a**.

LESSON
4.5 Profit and loss

LEARNING INTENTIONS

At the end of this lesson you should be able to:
- calculate profit from cost price and selling price
- calculate the selling price of an item from cost price and profit/loss
- calculate the cost price of an item from selling price and profit/loss.

▶ 4.5.1 Cost prices and selling prices

eles-3624

- Overhead costs are not directly linked to a specific product, but are required to sell products. These include staff wages, rent, store improvements, electricity and advertising.
- The **cost price** of a product is the total price that a business pays for the product including overhead costs.
- The **selling price** is the price that a customer buys a product for.
- **Profit** is the amount of money made on a sale. It is the difference between the total of the retailer's costs (cost price) and the price for which the goods actually sell (selling price).

> ### The profit equation
>
> $$\text{Profit} = \text{selling price} - \text{cost price}$$
>
> *Note:* **If the profit is negative, it's said that a loss has been made.**

Calculating the selling price

- The selling price of an item can be calculated by using the information of percentage profit or loss.

> ### Calculating the selling price from percentage profit or loss
>
> The following equations can be used to determine the selling price of an item, given the cost price and the percentage profit or loss.
>
> Selling price = (100% + percentage profit) × cost price
> Selling price = (100% − percentage loss) × cost price

WORKED EXAMPLE 11 Calculating selling price given percentage profit

Ronan operates a sports store at a fixed profit margin of 65%. Calculate how much he would sell a pair of running shoes for, if they cost him $40.

THINK

1. Determine the selling price by first adding the percentage profit to 100%, then determining this percentage of the cost price.

2. Write the answer in a sentence.

WRITE

Selling price = (100 + 65)% of $40
$\qquad$ = 165% of $40
$\qquad$ = 1.65 × $40
$\qquad$ = $66

The running shoes would sell for $66.

WORKED EXAMPLE 12 Calculating selling price given percentage loss

David bought a surfboard for $300 and sold it at a 20% loss a year later. Calculate the selling price.

THINK

1. Determine the selling price by first subtracting the percentage loss from 100%, then determining this percentage of the cost price.

2. Write the answer in a sentence.

WRITE

Selling price = (100 − 20)% of $300
$\qquad$ = 80% of $300
$\qquad$ = 0.80 × $300
$\qquad$ = $240

David sold the surfboard for $240.

- Profit or loss is usually calculated as a percentage of the cost price.

Percentage profit on cost price

$$\text{Percentage profit} = \frac{\text{profit}}{\text{cost}} \times 100\%$$

$$\text{Percentage loss} = \frac{\text{loss}}{\text{cost}} \times 100\%$$

WORKED EXAMPLE 13 Calculating profit as a percentage of the cost price

A music store buys records at $15 each and sells them for $28.95 each. Calculate the percentage profit made on the sale of a record.

THINK

1. Calculate the profit on each record: profit = selling price − cost price.

2. Calculate the percentage profit: $\frac{\text{profit}}{\text{cost}} \times 100\%$.

3. Write the answer in a sentence, rounding to the nearest per cent if applicable.

WRITE

Profit = $28.95 − $15
= $13.95

Percentage profit = $\frac{13.95}{15} \times 100\%$
= 93%

The profit is 93% of the cost price.

- Modern accounting practice favours calculating profit or loss as a percentage of the selling price. This is because commissions, discounts, taxes and other items of expense are commonly based on the selling price.

Percentage profit on selling price

$$\text{Percentage profit} = \frac{\text{profit}}{\text{selling price}} \times 100\%$$

$$\text{Percentage loss} = \frac{\text{loss}}{\text{selling price}} \times 100\%$$

Calculating the cost price

- If you are given the selling price and the percentage profit or loss, you can work backwards to calculate the cost price.

> **Cost price**
>
> Cost price = selling price − profit
> or
> = selling price + loss

WORKED EXAMPLE 14 Calculating cost price

A fashion store sells a pair of jeans for $180. If they made a percentage profit of 80% of the selling price, determine the cost price of the pair of jeans.

THINK

1. Enter the given information into the percentage profit selling price formula.

2. Rearrange the formula to make profit the subject.

3. Complete the calculation to determine the profit.

4. Subtract the profit from the selling price to determine the cost price.

WRITE

$$\text{Percentage profit} = \frac{\text{profit}}{\text{selling price}} \times 100\%$$

$$80\% = \frac{\text{profit}}{180} \times 100\%$$

$$\frac{80}{100} = \frac{\text{profit}}{180}$$

$$\frac{80}{100} \times 180 = \text{profit}$$

$$\text{Profit} = \frac{80}{100} \times 180$$

$$\text{Profit} = \$144$$

$$\text{Cost price} = \$180 - \$144$$
$$= \$36$$

 Resources

Interactivity Profit and loss (int-3746)

| 4.5 Quick quiz   on | 4.5 Exercise |

Individual pathways

■ PRACTISE	■ CONSOLIDATE	■ MASTER
1, 3, 6, 8, 11, 13, 16, 18, 21	2, 4, 7, 9, 12, 14, 17, 19, 22	5, 10, 15, 20, 23, 24

Assume percentage profit or loss is calculated on the cost price unless otherwise stated.

Fluency

1. Calculate the profit or loss for each of the following.

	Cost price	Selling price
a.	$15	$20
b.	$40	$50

2. Calculate the profit or loss for each of the following.

	Cost price	Selling price
a.	$52	$89.90
b.	$38.50	$29.95

3. **WE11&12** Calculate the selling price of each of the following.

	Cost price	%	Profit/loss
a.	$18	40%	profit
b.	$116	25%	loss

4. Calculate the selling price of each of the following.

	Cost price	%	Profit/loss
a.	$1300	30%	profit
b.	$213	75%	loss

5. Calculate the selling price of each of the following.

	Cost price	%	Profit/loss
a.	$699	$33\frac{1}{3}$	profit
b.	$5140	7%	loss

6. **WE14** Calculate the cost price of the following.

	Selling price	Percentage profit of selling price
a.	$80	55%
b.	$125	90%

7. Calculate the cost price of the following.

	Selling price	Percentage profit of selling price
a.	$3500	24%
b.	$499.95	35%

Understanding

8. **WE13** A restored motorbike was bought for $350 and later sold for $895.
   a. Calculate the profit.
   b. Calculate the percentage profit. Give your answer correct to the nearest whole number.

9. A music store sold a drum kit for $480. If they made a percentage profit of 75% of the selling price, determine the cost price of the drum kit.

10. James's Secondhand Bookshop buys secondhand books for $4.80 and sells them for $6.00.
    a. What is the ratio of the profit to the cost price?
    b. What is the percentage profit on the cost price?
    c. What is the ratio of the profit to the selling price?
    d. What is the percentage profit on the selling price?
    e. Discuss how the answers to parts **a** and **b** are related.

11. A retailer bought a laptop for $1200 and advertised it for $1525.
    a. Calculate the profit.
    b. Calculate the percentage profit (to the nearest whole number) on the cost price.
    c. Calculate the percentage profit (to the nearest whole number) on the selling price.
    d. Compare the differences between the answers to parts **b** and **c**.

12. Rollerblades bought for $139.95 were sold after six months for $60.
    a. Calculate the loss.
    b. Calculate the percentage loss. Give your answer to the nearest whole number.

13. Calculate the selling price for each item.
    a. Jeans costing $20 are sold with a profit margin of 95%.
    b. A soccer ball costing $15 is sold with a profit margin of 80%.
    c. A sound system costing $499 is sold at a loss of 45%.
    d. A skateboard costing $30 is sold with a profit margin of 120%.

14. Determine the cost price for the following items.
    a. A diamond ring sold for $2400 with a percentage profit of 60% of the selling price.
    b. A cricket bat sold for $69 with a percentage profit of 25% of the selling price.
    c. A 3-seater sofa sold for $1055 with a percentage profit of 35% of the selling price.

15. A fruit-and-vegetable shop bought 500 kg of tomatoes for $900 and sold them for $2.80 per kg.

   a. What is the profit per kilogram?
   b. Calculate the profit as a percentage of the cost price (round to 1 decimal place).
   c. Calculate the profit as a percentage of the selling price (round to 1 decimal place).
   d. Compare the answers to parts **b** and **c**.

16. Sonja bought an old bike for $20. She spent $47 on parts and paint and renovated it. She then sold it for $115 through her local newspaper. The advertisement cost $10.

   a. What were her total costs?
   b. What percentage profit (to the nearest whole number) did she make on costs?
   c. What percentage profit (to the nearest whole number) did she make on the selling price?

17. **MC** A clothing store operates on a profit margin of 150%. The selling price of an article bought for p is:

   A. $151p$
   B. $150p$
   C. $2.5p$
   D. $1.5p$

Communicating, reasoning and problem solving

18. A fruit-and-vegetable retailer buys potatoes by the tonne (1 tonne is 1000 kg) for $180 and sells them in 5-kg bags for $2.45. What percentage profit does he make (to the nearest whole number)? Show your working.

19. What discount can a retailer offer on her marked price of $100 so that she ends up selling at no profit and no loss, if she had initially marked her goods up by $50? Justify your answer.

20. Two business partners bought a business for $158 000 and sold it for $213 000. The profit was to be shared between the two business partners in the ratio of 3 : 2.
What percentage share does each person receive?
How much does each receive?

21. To produce a set of crockery consisting of a dinner plate, soup bowl, bread plate and coffee mug, the costs per item are $0.98, $0.89, $0.72 and $0.69 respectively.
These items are packaged in boxes of 4 sets and sell for $39.
If a company sells 4000 boxes in a month, what is its total profit?

22. Copy and complete the table below.

Cost per item	Items sold	Sale price	Total profit
$4.55	504	$7.99	
$20.00		$40.00	$8040.00
$6.06	64 321		$225 123.50
	672	$89.95	$28 425.60

23. The method used to calculate profits can make a difference when comparing different profits.

Cost = $20.00
Price = $120.00

Cost = $26 500.00
Price = $32 000.00

Cost = $1.00 (homemade)
Price = $3.50

   a. **i.** Describe the profits on each of the items above as a raw amount.
      ii. List the items from largest profit to smallest profit.
     iii. Discuss whether this is a fair method of comparing the profits.
   b. **i.** Express the profit on each of the items as a percentage of its cost.
      ii. List the items from largest profit to smallest profit.
     iii. Discuss whether this is a fair method of comparing the profits.
   c. **i.** Express the profit on each of the items as a percentage of its price.
      ii. List the items from largest profit to smallest profit.
     iii. Discuss whether this is a fair method of comparing the profits.

24. Max bought a car for $6000.00. He sold it to Janine for 80% of the price he paid for it. Janine sold it to Jennifer at a 10% loss. Jennifer then sold it to James for 75% of the price she paid. What did James pay for the car?
What was the total percentage loss on the car from Max to James?

LESSON
4.6 Goods and Services Tax (GST)

LEARNING INTENTIONS

At the end of this lesson you should be able to:
- understand what GST is
- calculate prices before and after GST.

▶ 4.6.1 Investigating GST

eles-3625

- **GST** is a tax imposed by the Australian federal government on goods and services. (As with all taxes, there are exemptions, but these will not be considered here.)
  - Goods: A tax of 10% is added to new items that are purchased, such as petrol, clothes and some foods.
  - Services: A tax of 10% is added to services that are paid for, such as work performed by plumbers, painters and accountants.

Calculating the amount of GST

The amount of GST on an item can be determined by dividing by 10 if the price is pre-GST, or by dividing by 11 if the price is inclusive of GST.

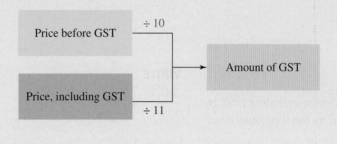

WORKED EXAMPLE 15 Calculating the amount of GST

A packet of potato chips costs $1.84 before GST.
Calculate:
a. the GST charged on the packet of chips
b. the total price the customer has to pay, if paying with cash.

THINK	WRITE
a. 1. GST is 10%. Calculate 10% of 1.84.	a. 10% of $1.84 = $\dfrac{\$1.84}{10}$ or $0.184
2. Write the answer in a sentence.	The GST charged on the packet of chips is $0.18 (rounded).
b. 1. Total equals pre-GST price plus GST.	b. $1.84 + $0.18 = $2.02
2. Write the answer in a sentence.	The total price the customer has to pay is $2.00 (rounded down by the seller).

- To calculate the pre-GST cost, when the total you are given includes GST, divide the GST-inclusive amount by 110 and multiply by 100. This is equivalent to dividing by 1.1.

Calculating the cost without GST

$$\text{Cost without GST} = \frac{\text{cost with GST}}{1.1}$$

WORKED EXAMPLE 16 Calculating prices before GST

A plumber's hourly charge includes GST. If she worked for 5 hours and the total bill including GST was $580, calculate her hourly price before GST.

THINK	WRITE
1. Calculate the hourly price including GST by dividing the total bill by the total number of hours (5).	$\dfrac{\$580}{5} = \$116$
2. Calculate the hourly price excluding GST.	110% of Pre-GST hourly rate $= \$116$ Pre-GST hourly rate $= \dfrac{\$116}{1.1}$ $= \$105.45$
3. Write the answer.	The plumber's hourly rate is $105.45 before GST.

COMMUNICATING — COLLABORATIVE TASK: Interpreting receipts

Collect some receipts from a variety of different shops. Look for the section of the receipt that details the GST information. Are any items on your receipts exempt from GST? As a class, collate your findings and determine any similarities in the GST-exempt items/services.

 Resources

 Interactivity Goods and Services Tax (int-3748)

Exercise 4.6 Goods and Services Tax (GST)

learnon

4.6 Quick quiz on	4.6 Exercise

Individual pathways

■ PRACTISE	■ CONSOLIDATE	■ MASTER
1, 3, 6, 8, 13, 16, 17, 20	2, 4, 7, 9, 10, 14, 18, 21	5, 11, 12, 15, 19, 22

Fluency

1. Explain GST in your own words.

2. Does GST apply below? Answer yes or no for each example.
   - a. Petrol
   - b. A lawyer's fee
   - c. Hotel accommodation
   - d. Lounge room carpet
   - e. Floor tiling
   - f. Wages at a fast-food restaurant

3. **WE15** The pre-GST price of a packet of laundry powder is $4.50.

   a. Calculate the GST on the laundry powder.
   b. Calculate the total price including GST.

4. The pre-GST price of a tin of peaches is $2.12.

   a. Calculate the GST on the tin of peaches.
   b. Calculate the total price including GST.

5. The pre-GST price of 1 kg of jellybeans is $3.85.

   a. Calculate the GST on the jellybeans.
   b. Calculate the total price including GST.

6. **WE16** The prices of the following items are inclusive of GST. Calculate the pre-GST price of each.

   a. 1 kg of apples at $3.85
   b. A basketball that costs $41.80

7. The prices of the following items are inclusive of GST. Calculate the pre-GST price of each.

   a. 5 kg of potatoes at $6.50
   b. A couch that costs $730

8. Millie buys a pack of batteries and pays 25 cents GST. How much did she pay in total for the batteries?

9. A new bicycle costs $450, including GST. How much is the GST?

Understanding

10. The telephone company Ringtel charges home customers $42.50 per month plus $0.24 per local call. Determine the monthly phone bill, including GST, if a customer makes 51 local calls in a month.

11. All car rental agencies use similar charging plans. Drivo charges $44 per day plus $0.47 per kilometre travelled. A customer wishes to rent a car for four days and travels 1600 km. Calculate the customer's total bill, including GST.

12. Expresso is a company that operates in the 'we-visit-you' car repair business. It charges $85 per hour plus a flat $40 visiting fee.

    a. Set up an expression, including GST, for the cost of a repair that takes t hours.
    b. If the repair takes 3 hours and 30 minutes, determine the final cost.

13. A company that installs floor tiles charges $35 per square metre for the actual tiles, and a fee of $100 plus $10 per square metre to install the tiles in a home. Let the area of the floor to be tiled be $x \, \text{m}^2$.

    a. Determine an expression, including GST, that represents the total cost of tiling in terms of x.
    b. What would be the total cost for a $20 \, \text{m}^2$ floor?

14. To buy my new super-dooper mobile phone outright I must pay $30 per month, including GST, for 3 years. How much GST will I pay?

15. In the United Kingdom a similar tax, called the value-added tax or VAT, is levied at 20%. If I paid £67 for a jumper purchased in a shop on Bond Street, London:

    a. how much VAT did I pay
    b. what was the pre-VAT price of the jumper?

16. In New Zealand GST is levied at 15% of the purchase price of goods.
    If I buy a pair of jeans and pay NZ$12 in GST, calculate the total price I paid for the jeans in NZ dollars.

Communicating, reasoning and problem solving

17. Explain what the terms *inclusive of GST* and *exclusive of GST* mean.

18. Explain why the amount of GST on an item is not equivalent to 10% of the GST-inclusive price.

19. Explain why the pre-GST price of an item is not equivalent to 10% off the GST-inclusive price.

20. The GST rate is 10%. This means that when a business sells something or provides a service, it must charge an extra $\frac{1}{10}$ of the price or cost. That extra money then must be sent to the tax office.

    For example, an item that would otherwise be worth $100 now has GST of $10 added, so the price tag will show $110.
    The business will then send that $10 to the tax office, along with all the other GST it has collected on behalf of the government.

    a. Suppose a shopkeeper made sales totalling $15 400.
       Determine how much of that amount is GST.
    b. Explain whether there is a number they can quickly divide by to calculate the GST.

21. Taking GST to be 10%, calculate:

    a. the GST payable on an item whose pre-GST price is P, and the price payable
    b. the pre-GST price of an item that costs A, and how much GST would need to be paid.

22. In the country Snowdonia, GST is 12.5%. Kira has purchased a new hairdryer that cost 111 kopeks including GST. There are 100 plens in 1 kopek.

    a. Calculate how much GST Kira paid.
    b. If 1 Australian dollar = 2 kopeks, calculate how much GST Kira would have paid if she had purchased the hairdryer in Melbourne, where GST is currently 10%.

LESSON
4.7 Review

4.7.1 Topic summary

PERCENTAGES

Discount

- A discount is a reduction in price.

$$\text{Percentage discount} = \frac{\text{discount}}{\text{original selling price}} \times \frac{100\%}{1}$$

- The new sale price can be obtained either by subtracting the discount amount from the original price, or by calculating the remaining percentage. e.g. A 10% discount means there is 90% remaining of the original selling price.

Percentages

- The term 'per cent' means 'per hundred'.
- Percentages can be converted into fractions and decimals by dividing the percentage by 100.

$$\text{e.g. } 77\% = \frac{77}{100} = 0.77$$

- To convert fractions and decimals into percentages, multiply by 100.
- Percentage increase/decrease

$$= \frac{\text{amount of increase/decrease}}{\text{original amount}} \times \frac{100}{1}\%$$

- To find the percentage of an amount, convert the percentage to a fraction or decimal and then multiply.

$$\text{e.g. } 25\% \text{ of } 48 = \frac{25}{100} \times \frac{48}{1} = 12$$

- To increase a quantity by $x\%$, multiply the quantity by $(100 + x)\%$.
- To decrease a quantity by $x\%$, multiply the quantity by $(100 - x)\%$.

Profit and loss

- A profit means the selling price > cost price, i.e. money has been made.

    Profit = selling price − cost price

- A loss means the selling price < cost price, i.e. money has been lost.

    Loss = cost price − selling price

Percentage profit and loss is usually calculated based on the cost price.

$$\text{Percentage profit/loss} = \frac{\text{profit or loss}}{\text{cost price}} \times \frac{100\%}{1}$$

Good and Services Tax (GST)

- GST is the Goods and Services Tax. This is a 10% tax added by the government to the cost of many items and services.

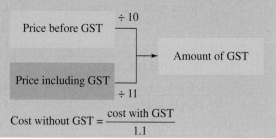

$$\text{Cost without GST} = \frac{\text{cost with GST}}{1.1}$$

4.7.2 Project

The composition of gold in jewellery

You may be aware that most gold jewellery is not made of pure gold. The materials used in jewellery are usually alloys, or mixtures of metals. The finest gold used in jewellery is 24 carat and is known as fine gold. Gold in this form is very soft and is easily scratched. Most metals will form an alloy with gold; silver, copper and zinc are commonly used in jewellery making. Other metals may be used to create coloured gold.

A table of the composition of some of the common gold alloys used in jewellery is shown below.

Gold name	Composition	
Gold (24 carat)	Gold	100%
Yellow gold (22 carat)	Gold	91.67%
	Silver	5%
	Copper	2%
	Zinc	1.33%
Pink gold (18 carat)	Gold	75%
	Copper	20%
	Silver	5%
Rose gold (18 carat)	Gold	75%
	Copper	22.25%
	Silver	2.75%
Red gold (18 carat)	Gold	75%
	Copper	25%
White gold (18 carat)	Gold	75%
	Palladium	10%
	Nickel	10%
	Zinc	5%
Grey-white gold (18 carat)	Gold	75%
	Iron	17%
	Copper	8%
Green gold (18 carat)	Gold	75%
	Silver	20%
	Copper	5%
Blue gold (18 carat)	Gold	75%
	Iron	25%
Purple gold	Gold	80%
	Aluminium	20%

Use the table to answer the following questions.

1. Study the table and list the metals used to create the alloys of gold mentioned.
2. A particular rose-gold bracelet weighs 36 grams. Calculate the masses of the various components in the bracelet.
3. How much more gold would there be in a yellow-gold bracelet of the same mass? What fraction is this of the mass of the bracelet?
4. Pink, rose and red gold all contain 75% gold. In addition, they each contain copper, and pink and rose gold also contain silver. Describe the effect you feel the composition of the alloy has on the colour of the gold.
5. Why does white gold not contain any copper?
6. Compare the composition of the alloys in red gold and blue gold.
7. Twenty-four-carat gold is classed as 100% gold. On this basis, an alloy of gold containing 75% gold has a carat value of 18 carat. Note this fact in the table above. Purple gold is 80% gold. What would its carat value be?
8. Just as there are various qualities of gold used in jewellery making, the same is true of silver jewellery. Sterling silver, which is commonly used, is actually not pure silver.
   Find out about the composition of silver used in jewellery making. Write a short report on your findings on a separate sheet of paper.

Exercise 4.7 Review questions

learn

Fluency

1. Calculate these amounts.
   a. $2.45 + $13.20 + $6.05   **b.** $304.60 − $126.25   **c.** $9.65 × 7

2. What is $65.50 ÷ 11? (Round your answer to the nearest 5 cents.)

3. Yindi purchased a handbag for $250 and later sold it on eBay for $330.
   a. Calculate the percentage profit on the cost price.
   b. Calculate the percentage profit on the selling price.
   c. Compare the answers to parts **a** and **b**.

4. William owns a hairdressing salon and raises the price of men's haircuts from $26.50 to $29.95 $29.95. Determine the percentage by which he increased the price of men's haircuts. Give your answer correct to the nearest per cent.

5. A discount of 18% on a tennis racquet reduced its price by $16.91. Calculate the sale price.

6. A washing machine bought for $129 was later sold for $85. What percentage loss was made on the sale?

7. Calculate the percentage profit on a sound bar purchased for $320 and later sold for $350.

8. A 15% discount reduced the price of a basketball by $4.83. What was the original price?

Understanding

9. Tim works in a sports shop. He purchased wholesale golf shirts for $55 each. If he made 163% profit, determine the sale price of the golf shirts.

10. The profit on a gaming console is $240. If this is 60% of the cost price, calculate:

    a. the cost price   **b.** the selling price.

11. A music store sells an acoustic guitar for $899. If the cost price of the guitar is $440, determine the profit the store makes, taking GST into account.

12. A company made a profit of $238 000. This represents a 10% profit increase compared to the previous year. Determine last year's profit.

13. A camping goods shop operates on a profit margin of 85%. Calculate how much the shop would have paid for a sleeping bag that sells for $89.95.

Communicating, reasoning and problem solving

14. After a 5% discount, a telephone bill is $79.50. Calculate the original amount of the bill.

15. Pablo spent $82.20 at the supermarket. If 15% of this was spent on tomatoes priced at $3 per kilogram, determine the weight of tomatoes purchased.

16. You buy ten pairs of headphones for a total of $150. Determine the price you should sell six pairs for if you wish to make a profit of 25% on each pair.

17. An art dealer sold two paintings at an auction. The first painting sold for $7600, making a 22% loss on its cost. The second painting sold for $5500, making a profit of 44%.
    Explain whether the art dealer made an overall profit or loss.

18. Jacques' furniture shop had a sale with $\frac{1}{3}$ off the usual price of lounge suites. If the original price of a suite was \$5689 including GST, determine the sale price including GST.

19. Goods listed at \$180 were discounted by 22%.
 a. Calculate the sale price.
 b. If they had sold for \$100, determine what the percentage discount would have been.

20. Steve Smith buys a cricket bat for \$85, signs it and donates it for an auction.

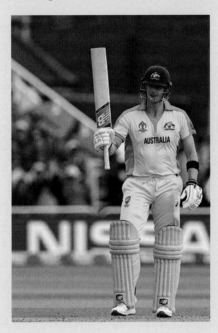

If it sells for \$500, calculate:
 a. the percentage increase in the bat's value
 b. the dollar value of the signature.

21. Andrew buys a pair of jeans for \$59.95. The original price tag was covered by a 30% sticker but the sign on top of the rack said 'Additional 15% off already reduced prices'.
 a. Calculate the original price of the jeans. Give your answer correct to the nearest 5 cents.
 b. Determine what percentage of the original cost Andrew ended up saving.

22. Café Noir charges a 1% levy on the bill for trading on Sundays. If the final bill is \$55.55, determine the original price, taking into account that the levy has been charged and then 10% GST has been added.

on To test your understanding and knowledge of this topic, go to your learnON title at www.jacplus.com.au and complete the **post-test**.

Answers

Topic 4 Percentages

4.1 Pre-test

1. $7.50
2. C
3. $22\%, \frac{1}{4}, 0.31, \frac{3}{5}, 111\%$
4. C
5. $18.40
6. $76.50
7. 5%
8. D
9. 19 years 3 months
10. D
11. 2.55%
12. B
13. True
14. 27%
15. Australia

4.2 Percentages

1. a. 0.24   b. 0.13   c. 0.015   d. 2.5
2. a. 0.47   b. 0.066   c. 1.098   d. 0.1002
3. a. $\frac{1}{5}$ b. $\frac{7}{20}$ c. $\frac{61}{100}$ d. $\frac{21}{20}$
4. a. $\frac{11}{100}$ b. $\frac{41}{50}$ c. $\frac{1}{8}$ d. $\frac{101}{50}$
5. a. 15%   b. 85%   c. 310%   d. 2.4%
6. a. 87.5%   b. 60%   c. 83.33%   d. 233.33%
7. a. $10\%, 25\%, 75\%, \frac{7}{8}, 1.6, 2.4, 3\frac{1}{2}$

   b. $4.5\%, 150\%, 2\frac{1}{3}, 2.8, 3, 330\%, 3\frac{4}{5}$
8. 30%
9. 56.67%
10. 30%
11. 35.29%
12. 30%
13. a. 30

    b.

Free throw results	Number of students	Percentage of students
No shots in	3	10%
One shot in, three misses	11	36.6%
Two shots in, two misses	10	33.3%
Three shots in, one miss	4	13.3%
All shots in	2	6.6%

c. 11

d. $46.\dot{6}\%$

14. $7.27
15. Answers will vary. To calculate the percentage, fraction and decimal of a particular brand, the number of phones of that brand needs to be divided by the total number of phones.
16. a. 0%   b. 72%   c. $\frac{14}{25}$ d. $\frac{41}{100}$
17. a. 38%   b. $\frac{8}{21}$

4.3 Finding percentages of an amount

1. a. 10   b. 16   c. 3   d. 3
2. a. 93   b. 6   c. 6   d. 28
3. a. 77   b. 39   c. 63   d. 4000
4. a. 66   b. 17   c. 42   d. 95
5. a. 190   b. 55   c. 12   d. 25
6. a. 13.2   b. 1.9   c. 24.75   d. 11.1
7. a. 58.9   b. 20.8   c. 14.4   d. 98.4
8. D
9. a. $2.70   b. $7.15   c. $5.75   d. $6.05
10. a. $0.20   b. $4.30   c. $0.05   d. $0.05
11. a. $0.10   b. $0.00   c. $0.00   d. $12.65
12. a. $2.20   b. $1.80   c. $73.50   d. $18.00
13. a. $1.05   b. $2.05   c. $32.20   d. $4.80
14. $128\,000\,\text{km}^2$
15. a. 2   b. 38
16. $267.50
17. 114 minutes
18. $855
19. 23%
20. a. 36.3 kg   b. 60 lb   c. 166 cm
21. a. 22.5 kg   b. 32 km   c. $9.60
22. 110
23. 27.9 seconds
24. a. 13 608 people   b. 17 820 people
25. 20 years old
26. $0.80
27. Sister: 9 years old; grandfather: 75 years old
28. a. 14

    b. 7.69% of students achieved a score of 40 or more, which is just below the state average.
29. a. 78 minutes   b. 282 minutes
    c. $21.\dot{6}\%$ d. 155.1 minutes
30. 60 years old

4.4 Discount

1. a. $42   b. $46.25
2. a. $49.50   b. $76
3. a. 10%   b. 50%

4. a. $33\frac{1}{3}\%$ b. 25%

5. a. $850   b. $200

6. a. $83.60   b. $104

7. a. $64.70   b. $241.50

8. a. 40%   b. 28%

9. a. 28%   b. 22%

10. a. $45   b. $45   c. $36

11. 25%

12. a. Mobile phone $95   b. Surfboard and bike
    c. $8.35   d. No

13. a. $70   b. $280

14. $62.96

15. a. $41.65   b. Yes

16. $75.76

17. 20%

18. 17.3%

19. 60%

20. $243.70

21. a. $121.60; $140.80   b. Profit

22. A

23. Yes, there is a difference in the meanings. 75% off $200 = $150 off the price, so you would pay only $50.
    75% of $200 = $150, i.e. $\frac{3}{4}$ of $200

24. $1.00/$12.00 × 100% = 8.33%, so this is an 8.33% discount.

25. Henry pays $954; Sancha pays $991.20. Henry has the best buy.

26. No, the statement is not correct. For example, if you have a cost of $100, a 50% discount = $50 and a 40% discount (on that $50) = $20.
    Total discount = $70; this represents a 70% discount, not 90%.

27. a. 65%   b. 111%   c. 94%   d. 200%

28. 95% of $63.20 = $60.05; 75% of $79 = $59.25. The two methods calculate percentages of different amounts so they result in different answers.

29. a. 33.33%
    b. This deal is worse than the deal offered in part a as it is equivalent to only a 25% discount; however, it should be used if you only want to buy two soaps.

4.5 Profit and loss

1. a. $5 profit   b. $10 profit

2. a. $37.90 profit   b. $8.55 loss

3. a. $25.20   b. $87

4. a. $1690   b. $53.25

5. a. $932   b. $4780.20

6. a. $36.00   b. $12.50

7. a. $2660   b. $324.95

8. a. $545   b. 156%

9. $120

10. a. 1 : 4
    b. 25%
    c. 1 : 5
    d. 20%
    e. The ratio of the profit to the cost price as a fraction is the same as the percentage profit on the cost price.

11. a. $325
    b. 27%
    c. 21%
    d. The percentage profit is greater on the cost price.

12. a. $79.95   b. 57%

13. a. $39   b. $27   c. $274.45   d. $66

14. a. $960   b. $51.75   c. $685.75

15. a. $1.00 profit per kg
    b. 55.6%
    c. 35.7%
    d. The percentage profit is greater on the cost price.

16. a. $77   b. 49%   c. 33%

17. C

18. 172%

19. 50%

20. 60%, 40%; $33 000, $22 000

21. $103 520

22.

Cost per item	Items sold	Sale price	Total profit
$4.55	504	$7.99	$1733.76
$20.00	402	$40.00	$8040.00
$6.06	64 321	$9.56	$225 123.50
$47.65	672	$89.95	$28 425.60

23. a. i. $100, $5500, $2.50
    ii. Car, shoes, cookies
    iii. Not fair; profit should be compared as a proportion of cost.
    b. i. 500%, 20.75%, 250%
    ii. Shoes, cookies, car
    iii. Fairer than in part a
    c. i. 83.3%, 17.2%, 71.4%
    ii. Shoes, cookies, car
    iii. Not fair; the profit should be calculated on the cost.

24. James paid $3240. The total percentage loss was 46%.

4.6 Goods and Services Tax (GST)

1. GST is a tax of 10% levied by the Australian federal government on goods and services.

2. a–e. Yes
   f. No

3. a. $0.45   b. $4.95

4. a. $0.21   b. $2.33

5. a. $0.39   b. $4.24

6. a. $3.50   b. $38

7. a. $5.91   b. $663.64

8. $2.75

9. $40.91

10. $60.21

11. $1020.80

12. a. $1.1(85t + 40)$ b. $371.25

13. a. $1.1(45x + 100)$ b. $1100

14. $98.18

15. a. $11.17   b. $55.83

16. NZ$92

17. *Inclusive* includes GST in the total price and *exclusive* excludes GST from the total price.

18. GST is equal to 10% of the pre-GST price, which is less than 10% of the GST-inclusive price. For example, if an item costs $100 pre-GST, the GST would be $10 and the GST-inclusive price would be $110. 10% of the GST-inclusive price would be $11, not $10.

19. 10% off the GST-inclusive price is equal to 99% of the pre-GST price. For example, if an item costs $100 pre-GST, the GST would be $10 and the GST-inclusive price would be $110. Taking 10% off the GST-inclusive price would be $99, not $100.

20. a. $1400   b. 11

21. a. $\$\dfrac{P}{10}$, $\$\dfrac{11P}{10}$ b. $\$\dfrac{10A}{11}$, $\$\dfrac{A}{11}$

22. a. 12 kopeks, 1233 plens
    b. $4.93 GST; total price $\approx$ $54.27

4. 13%

5. $77.03

6. 34%

7. 9.375%

8. $32.20

9. $144.65

10. a. $400   b. $640

11. $377.27

12. $216 364

13. $48.62

14. $83.68

15. 4.11 kg

16. $112.50

17. Loss of $463.03

18. $4171.93

19. a. $140.40   b. 44.4%

20. a. 488%   b. $415

21. a. $100.75   b. 40.5% saved

22. $50

Project

1. Metals used as alloying elements with gold are silver, copper, zinc, palladium, nickel, iron and aluminium.

2. 27 g gold, 8.01 g copper, 0.99 g silver

3. 6 g, $\dfrac{1}{6}$

4. From pink to rose to red gold, the percentage of silver decreases, causing the gold alloy to darken in colour. At the same time, the percentage of copper increases, also contributing to the darker colour.

5. The copper would colour the gold with its familiar reddish colour so that it would not be white.

6. Red gold and blue gold each have 75% gold and 25% of another metal. In the case of red gold, the contributing metal is copper; blue gold contains iron.

7. 19.2 carat

8. Sample responses can be found in the worked solutions in the online resources.

4.7 Review questions

1. a. $21.70   b. $178.35   c. $67.55

2. $5.95

3. a. 32%
   b. 24.24%
   c. The percentage profit is greater on the cost price.

5 Ratios and rates

LESSON
5.1 Overview

Why learn this?

Ratios are used to compare different values. They tell us how much there is of one thing compared to another. For example, what is the ratio of sultanas to bran flakes in a box of cereal? What is the best value for money from three different bags of dog food?

Ratios are commonly used in cooking and are particularly helpful if you need to adjust a recipe for more or fewer people. If a recipe for four people requires two cups of flour to one cup of water (a ratio of 2 : 1), then if we need to cook for only two people we can halve the ratio (1 : 0.5) and determine that we need one cup of flour to half a cup of water. Understanding ratios is important as they are used in money transactions, perspective drawings, enlarging or reducing measurements in building and construction, and dividing items equally within a group.

Rates are ratios that compare quantities in different units. Imagine you want to compare two different brands of chocolate to see which is the best value. One chocolate block weighs 250 grams and costs $3.50, and the other weighs 275 grams and costs $3.66. Since the cheaper chocolate block is also smaller, you do not know if it is better value than the larger block. Using rates, you will be able to determine which is the better buy. You will commonly see rates used for pricing petrol, displaying unit prices in supermarkets, paying hourly wages and determining problems based on speed, time and distance.

1. **MC** Select which one of the following options is the simplest form of the ratio $18 : 24$.

   A. $2 : 3$ **B.** $3 : 4$ **C.** $4 : 3$ **D.** $6 : 8$

2. A school has 3 boys to every 5 girls. A particular class where this ratio applies has 15 girls. Calculate how many boys are in this class.

3. **MC** Select all of the following that are the same as the ratio 5 cm to 1 m.

   A. $1 : 20$ **B.** $5 : 100$ **C.** $2 : 50$ **D.** $25 : 500$

4. **MC** In a hockey season, one team has a ratio for wins, losses and draws of $3 : 4 : 1$. The team had 3 draws in the season. What is the total number of games the team played this season?

   A. 9 **B.** 12 **C.** 24 **D.** 8

5. A recipe requires 240 g of flour, 4 eggs and 150 g of butter in order to make a dozen portions.
   a. Calculate how much flour is required to make 60 portions.
   b. Calculate how many eggs are required for 21 portions.
   c. Calculate how many portions can be made using 525 g of butter.

6. Calculate the value of m in each of the following proportion equations.

   a. $\dfrac{m}{6} = \dfrac{5}{12}$
   b. $\dfrac{10}{7} = \dfrac{6}{m}$

7. In a tennis tournament, Team A won 11 sets out of a possible 18, while Team B won 7 sets out of a possible 10. Compare these results to determine which team performed better.

8. Cecil, Darius and Eva combine funds to buy a ticket for a prize worth $2000. They contribute to the ticket in the ratio $3 : 1 : 4$ respectively, and decide that they will split the money in the same ratio if they win.
   Calculate how much Cecil would receive if they win the prize.

9. Express the following as rates in the units given.
   a. 464 words written in 8 minutes in the unit words/minute
   b. 550 litres of water flowing through a pipe in 12 minutes in the unit kL/h

10. Write the following ratios in their simplest forms.

    a. $3.5 : 0.45$
    b. $\dfrac{7}{10} : 1$
    c. $4m^2n : 2mn^2$

11. On signposts, the gradient of a hill can be written as a ratio. Write down the ratio (in its simplest form) of a hill with a vertical height of 200 m and a horizontal length of 1 km.

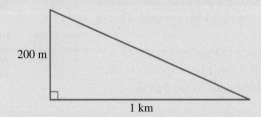

200 m

1 km

12. **MC** On a distance–time graph, the horizontal axis is measured in hours and the vertical axis is measured in kilometres. From the list select the correct representation for the units of speed.

A. km/h

B. kilometre/h

C. kms/hours

D. kilometres per hour

13. A boy rides a bicycle for 7 hours. Use the graph to choose the correct speeds he travels at and complete the following sentence.

He travels at _____ for 3 hours, stops for 2 hours at his friend's house and then travels home at a speed of _____.

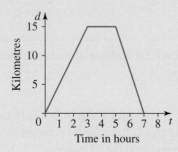

14. Three small towns have 7800 residents in total. Towns A and B have a ratio of 2 : 5 residents, and towns B and C have a ratio of 3 : 1. Calculate how many residents live in town B.

15. A piece of paper (P5) is 15 cm in height and 21 cm in length. A series of papers with the same height-to-length ratio exists as follows. The next piece of paper in the series is known as P4, and the height of P4 is equal to the length of P5. The next size, P3, has a height equal to the length of P4. Following this pattern, calculate the dimensions (height and length in cm) of a P2 piece of paper, giving your answer as an unrounded decimal.

LESSON
5.2 Ratios

LEARNING INTENTION

At the end of this lesson you should be able to:
• compare quantities using ratios.

▶ 5.2.1 Ratios

eles-3767

• **Ratios** are mathematical expressions that compare two or more quantities or numbers.
• In the diagram, the ratio of blue squares to pink squares is 2 : 4 (the quantities are separated by a colon). This is read as '2 to 4'.
• The order of the numbers in a ratio is important. If the ratio of flour to water is 1 : 4, 1 corresponds to the quantity of flour and 4 corresponds to the quantity of water.

$= 2 : 4 = 2$ to 4

- Ratios can also be written in fractional form: $1:4$ or $\frac{1}{4}$.

  Note: This does not mean that the quantity on the left of the ratio is one quarter of the total.
- Ratios can also be written as percentages. For example, if 20% of a class of Year 8 students walked to school, then 80% did not walk to school — a ratio of $20:80$, which simplifies to $1:4$.
- Ratios do not have names or units of measurement.
- Before ratios are written, the numbers must be expressed in the same unit of measurement. Once the units are the same, they can be omitted.
- Simplified ratios only contain whole numbers.

COMMUNICATING — COLLABORATIVE TASK: Ratio stations

Equipment: coloured blocks or counters (at least three different colours), paper, pen

1. Set up four stations around the room, each with at least 10 blocks of at least three different colours.
2. Divide the board into four sections, one for each station, so that people can write their ratios on the board.
3. In small groups, begin at any station and write at least *three* different ratios to describe the relationships between the blocks. At least one of these ratios must be a ratio of a part to a whole. Record these ratios in your book, simplifying where possible. Make sure that you write down what your ratio represents, such as red blocks to blue blocks.
4. Move around the stations and repeat step **3**, visiting all stations if time permits.
5. Share your findings by writing the ratios on the board.
6. What is the difference between comparing a part to a whole and comparing a part to another part?

WORKED EXAMPLE 1 Expressing quantities as ratios

Look at the completed game of noughts and crosses and write the ratios of:
a. **crosses to noughts**
b. **noughts to unmarked spaces.**

X	O	
X	X	X
O		O

THINK	WRITE
a. Count the number of crosses and the number of noughts. Write the two numbers as a ratio (the number of crosses must be written first).	a. $4:3$
b. Count the number of noughts and the number of unmarked spaces. Write the two numbers as a ratio, putting them in the order required (the number of noughts must be written first).	b. $3:2$

WORKED EXAMPLE 2 Writing a statement as a ratio

Rewrite the following statement as a ratio: 7 mm to 1 cm.

THINK	WRITE
1. Express both quantities in the same unit. To obtain whole numbers, convert 1 cm to mm (rather than 7 mm to cm).	7 mm to 1 cm 7 mm to 10 mm
2. Omit the units and write the two numbers as a ratio.	$7:10$

▶ 5.2.2 Expressing one part of a ratio as a fraction of the whole

eles-3768

- If you are given a ratio whose parts together represent the whole, you can express one part of the ratio as a fraction of the whole.
- To express one part of a ratio as a fraction of the whole, consider the total number of parts in the ratio.
- In the example shown, there are 2 blue squares and 4 pink squares, giving a ratio of blue : pink = 2 : 4. The total number of parts in the ratio is $2 + 4 = 6$. The blue squares make up 2 of the 6 total parts and can therefore be expressed as $\dfrac{2}{6}$, which can be simplified as $\dfrac{1}{3}$.

Expressing one part of a ratio as a fraction of the whole

In general, to express a part of a ratio as a fraction of the whole, write the number from the ratio as the numerator and the sum of both numbers in the ratio as the denominator.

For example, 3 in the ratio 3 : 4 can be expressed as $\dfrac{3}{3+4} = \dfrac{3}{7}$ of the whole.

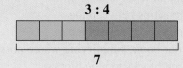

3 : 4

7

WORKED EXAMPLE 3 Expressing one part of a ratio as a fraction of the whole

The number of tries scored by two teams in a rugby match was in the ratio 1 : 3.
Express the number of tries the first team scored as a fraction of the whole number of tries scored.

THINK	WRITE
1. Determine the total number of parts in the ratio by summing the parts.	$1 + 3 = 4$
2. Express the first part (1) as a fraction of the whole (4).	$\dfrac{1}{4}$
3. Write the answer.	The first team scored $\dfrac{1}{4}$ of the total number of tries scored.

▶ 5.2.3 Ratios involving more than two numbers

- Ratios can be extended to compare more than two quantities of the same kind. Each quantity being compared should be split by the : symbol.
- For example, if a family has 3 dogs, 6 cats and 9 goldfish, the ratio of dogs to cats to goldfish is 3 : 6 : 9, which simplifies to 1 : 2 : 3.
- The simplified ratio can be read as 'For every 1 dog, there are 2 cats and 3 fish.'

| **3** | : | **6** | : | **9** |

Exercise 5.2 Ratios   **learn** on

| **5.2 Quick quiz** on | **5.2 Exercise** |

Individual pathways

■ PRACTISE	■ CONSOLIDATE	■ MASTER
1, 4, 8, 11, 15, 17, 20	2, 5, 7, 9, 13, 16, 18, 21	3, 6, 10, 12, 14, 19, 22

Fluency

1. **WE1** Examine the completed game of noughts and crosses and write the ratios of:

   a. noughts to crosses
   b. crosses to noughts
   c. crosses to total number of spaces
   d. total number of spaces to noughts
   e. noughts in the top row to crosses in the bottom row.

X	X	O
O	O	X
X	X	O

2. For the diagram shown, write the following ratios in simplest form, $a : b$.

   a. Shaded parts : unshaded parts
   b. Unshaded parts : shaded parts
   c. Shaded parts : total parts

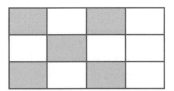

3. For the bag of numbers shown, write the ratios, in simplest form, of:

   a. even numbers to odd numbers
   b. prime numbers to composite numbers
   c. numbers greater than 3 to numbers less than 3
   d. multiples of 2 to multiples of 5
   e. numbers divisible by 3 to numbers not divisible by 3.

4. **WE2** Rewrite each of the following statements as a ratio.

   a. 3 mm to 5 mm   **b.** 6 s to 19 s   **c.** $4 to $11
   d. 7 teams to 9 teams   **e.** 1 goal to 5 goals

5. Rewrite each of the following statements as a ratio.

   a. 9 boys to 4 boys   **b.** 3 weeks to 1 month   **c.** 3 mm to 1 cm
   d. 17 seconds to 1 minute   **e.** 53 cents to $1

6. Rewrite each of the following statements as a ratio.

   a. 11 cm to 1 m   **b.** 1 g to 1 kg   **c.** 1 L to 2 kL
   d. 7 hours to 1 day   **e.** 5 months to 1 year

7. **WE3** Julia purchased 6 apples and 11 mandarins from the local
   farmers' market. Express the number of apples as a fraction of the
   whole number of pieces of fruit purchased.

Understanding

8. Out of 100 people selected for a school survey, 59 were junior students,
   3 were teachers and the rest were senior students. Write the following
   ratios.

   a. Teachers : juniors   **b.** Juniors : seniors
   c. Seniors : teachers   **d.** Teachers : students
   e. Juniors : other participants in the survey

9. Write each of the following as a mathematical ratio in simplest form.

   a. In their chess battles, Lynda has won 24 games and Karen has won 17.
   b. There are 21 first-division teams and 17 second-division teams.
   c. Nathan can long-jump twice as far as Rachel.
   d. At the school camp there were 4 teachers and 39 students.

10. Write each of the following as a mathematical ratio in simplest form.

    a. In the mixture there were 4 cups of flour and 1 cup of milk.
    b. Elena and Alex ran the 400 m in the same time.
    c. The radius and diameter of a circle were measured and their lengths recorded.
    d. The length of a rectangle is three times its width.

11. A pair of jeans originally priced at $215 was purchased for $179. State the
    ratio of:

    a. the original price compared to the selling price
    b. the original price compared to the discount.

12. Matthew received a score of 97% for his Maths test. Write the ratio of:

    a. the marks received compared to the marks lost
    b. the marks lost compared to the total possible marks.

13. Given the following fractions, express the first number in the ratio as a
    fraction of the whole.

    a. 3 : 7   **b.** 4 : 11   **c.** 13 : 12   **d.** 7 : 2

Communicating, reasoning and problem solving

14. For each comparison that follows, state whether a ratio could be written and give a reason for your answer. (*Remember:* Before ratios are written, numbers must be expressed in the same unit of measurement.)

    a. Anna's mass is 55 kg. Her cat has a mass of 7 kg.
    b. Brian can throw a cricket ball 40 metres, and John can throw the same ball 35 metres.
    c. The cost of painting the wall is $55; its area is 10 m².
    d. On one trip, the car's average speed was 85 km/h; the trip took 4 h.

15. For each comparison that follows, state whether a ratio could be written and give a reason for your answer. (*Remember:* Before ratios are written, numbers must be expressed in the same unit of measurement.)

    a. Brett's height is 2.1 m. Matt's height is 150 cm.
    b. Jonathan apples cost $2.40 per dozen; Delicious apples cost $3.20 per dozen.
    c. Mary is paid $108; she works 3 days a week.
    d. David kicked 5 goals and 3 behinds (33 points); his team scored 189 points.

16. a. If 17% of students in a class have sports training once a week, write as a ratio the number of students who have sports training once a week compared to the students who do not.
    b. A survey found that only 3% of workers take their lunch to work on a regular basis. Write as a ratio the number of workers who do not take their lunch to work compared to those who do.

17. A recipe to make enough spaghetti bolognaise to feed four people needs 500 grams of mincemeat. According to this recipe, how much mincemeat would be needed to make enough spaghetti bolognaise to feed a group of 9 people? Show all working.

18. Meredith wins two-thirds of her games of netball this season. Write down her win-to-loss ratio.

19. Why doesn't a simplified ratio have units?

20. Out of 100 people selected for a school survey, 68 were students, 5 were teachers and the rest were support staff.

    a. Write the ratio of teachers to students.
    b. Write the ratio of support staff to teachers.
    c. Write the ratio of students to people surveyed.
    d. Write the number of students as both a fraction and a percentage of the number of people surveyed.
    e. Write the number of students as both a fraction and a percentage of the number of teachers surveyed.

21. In class 8A, there are 4 boys and 5 girls. In class 8B, there are 12 boys and 15 girls. In class 8C, there are 8 boys and 6 girls.

    a. For each class, write the ratio of boys to girls.
    b. Compare the results from part a to identify the two classes in which the proportions are the same.
    c. Class 8A joins class 8B to watch a program for English. Determine the resulting boy-to-girl ratio.
    d. Class 8C joins class 8B for PE lessons. Determine the boy-to-girl ratio in the combined PE class.
    e. Is the ratio of boys to girls watching the English program equivalent to the ratio of boys to girls in the combined PE class?

22. The ratio of gold coins to silver coins in a purse is 2 : 5. If there are 10 gold coins in the purse, calculate the smallest number of coins that needs to be added to the purse so that the ratio of gold coins to silver coins changes to 3 : 4.

LESSON
5.3 Simplifying ratios

LEARNING INTENTION

At the end of this lesson you should be able to:
- convert between equivalent ratios
- simplify ratios.

▶ 5.3.1 Equivalent ratios

eles-3770

- When the numbers in a ratio are multiplied or divided by the same number to obtain another ratio, these two ratios are said to be equivalent. (This is a similar process to obtaining equivalent fractions.)
- Consider the diagram shown. A cordial mixture can be made by adding one part of cordial concentrate to four parts of water.

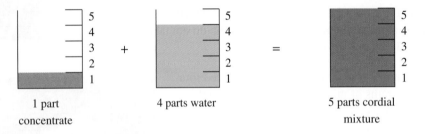

| 1 part | 4 parts water | 5 parts cordial |
| concentrate | | mixture |

- The ratio of concentrate to water is 1 to 4 and is written as **1 : 4**.
- By applying the knowledge of 'equivalent ratios', the amount of cordial can be doubled.
- This can be achieved by keeping the ratio of concentrate to water the same by doubling the amounts of both concentrate and water.

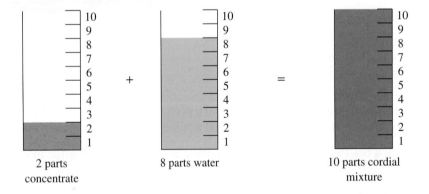

| 2 parts | 8 parts water | 10 parts cordial |
| concentrate | | mixture |

- The relationship between the amount of concentrate to water is now **2 : 8**, as shown in the diagram above.
- The ratios **2 : 8** and **1 : 4** are equivalent. In both ratios, there is 1 part of concentrate for every 4 parts of water.

Equivalent ratios

Ratios are equivalent if the numbers on either side have the same relationship. They work in the same way as equivalent fractions.

For example, 2 : 8 is equivalent to 1 : 4.

▶ 5.3.2 Simplifying ratios

eles-3771

- A ratio is simplified by dividing all numbers in the ratio by their highest common factor (HCF). For example, the ratio **4 : 8** can be simplified to **1 : 2**.

- Like fractions, ratios are usually written in simplest form; that is, reduced to their lowest terms.

WORKED EXAMPLE 4 Simplifying ratios

Express the ratio 16 : 24 in simplest form.

THINK	WRITE
1. Determine the largest number by which both 16 and 24 can be divided (i.e. the highest common factor).	The HCF is 8.
2. To go from the ratio with the bigger values to the ratio with the smaller values, divide both 16 and 24 by 8 to obtain an equivalent ratio in simplest form.	$\div 8 \underset{2\,:\,3}{\overset{16\,:\,24}{\huge(\ \ )}} \div 8$
3. Write the answer.	2 : 3

WORKED EXAMPLE 5 Simplifying ratios of quantities expressed in different units

Write the ratio of 45 cm to 1.5 m in simplest form.

THINK	WRITE
1. Write the question.	45 cm to 1.5 m
2. Express both quantities in the same unit by changing 1.5 m into cm. (1 m = 100 cm)	45 cm to 150 cm
3. Omit the unit and write the two numbers as a ratio.	45 : 150
4. Determine the HCF by which both 45 and 150 can be divided.	The HCF is 15.
5. Simplify the ratio by dividing both sides by the HCF.	$\div 15 \underset{3\,:\,10}{\overset{45\,:\,150}{\huge(\ \ )}} \div 15$
6. Write the answer.	3 : 10

Simplifying fractional and decimal ratios

- Sometimes you will come across ratios of quantities that are fractions or decimal numbers. In order to simplify them, you will first need to convert them into whole numbers.
- To convert ratios containing fractions into whole numbers, convert each fraction to the equivalent fraction with the lowest common denominator, then multiply both sides by the lowest common denominator.
- To convert ratios containing decimals into whole numbers, multiply both sides by the power of 10 that converts both sides into whole numbers.

WORKED EXAMPLE 6 Simplifying ratios involving fractions

Simplify the following ratios.

a. $\dfrac{2}{5} : \dfrac{7}{10}$

b. $\dfrac{5}{6} : \dfrac{5}{8}$

THINK	WRITE
a. 1. Write the fractions in ratio form.	**a.** $\dfrac{2}{5} : \dfrac{7}{10}$
2. Write equivalent fractions using the lowest common denominator (in this case, 10).	$= \dfrac{4}{10} : \dfrac{7}{10}$
3. Multiply both fractions by 10.	$\times 10 \left(\dfrac{4}{10} : \dfrac{7}{10} \right) \times 10$ $4 : 7$
4. Check if the remaining whole numbers that form the ratio can be simplified. In this case they cannot.	
5. Write the answer.	$4 : 7$
b. 1. Write the fractions in ratio form.	**b.** $\dfrac{5}{6} : \dfrac{5}{8}$
2. Write equivalent fractions using the lowest common denominator (in this case, 24).	$= \dfrac{20}{24} : \dfrac{15}{24}$
3. Multiply both fractions by 24.	$\times 24 \left(\dfrac{20}{24} : \dfrac{15}{24} \right) \times 24$ $20 : 15$
4. Check if the remaining whole numbers that form the ratio can be simplified. In this case, divide each number by their HCF, which is 5.	$\div 5 \left(20 : 15 \right) \div 5$ $4 : 3$
5. Write the answer.	$4 : 3$

WORKED EXAMPLE 7 Simplifying ratios involving decimals

Write the following ratios in simplest form.

a. 2.1 to 3.5

b. 1.4 : 0.75

THINK	WRITE
a. 1. Write the decimals in ratio form.	**a.** $2.1 : 3.5$
2. Both decimals have 1 decimal place, so multiplying each by 10 will produce whole numbers.	$\times 10 \left(2.1 : 3.5 \right) \times 10$ $21 : 35$
3. Determine the HCF of 21 and 35.	The HCF is 7.

4. Simplify the ratio by dividing both sides by the HCF.

$$\div 7 \quad\left(\begin{array}{c} 21 : 35 \\ 3 : 5 \end{array}\right)\quad \div 7$$

5. Write the answer.

$3 : 5$

b. 1. Write the decimals in ratio form.

b. $1.4 : 0.75$

2. Because 0.75 has 2 decimal places, multiply each decimal by 100 to produce whole numbers.

$$\times 100 \quad\left(\begin{array}{c} 1.4 : 0.75 \\ 140 : 75 \end{array}\right)\quad \times 100$$

3. Determine the HCF of 140 and 75.

The HCF is 5.

4. Simplify the ratio by dividing both sides by the HCF.

$$\div 5 \quad\left(\begin{array}{c} 140 : 75 \\ 28 : 15 \end{array}\right)\quad \div 5$$

5. Write the answer.

$28 : 15$

Digital technology

Since simplifying ratios follows a similar process to simplifying fractions, a calculator can be used to help check your answer. For example, in Worked example 7a the ratio 2.1 to 3.5 was simplified to $3 : 5$.

Using a calculator, input $2.1 \div 3.5$, then use the decimal-to-fraction button $S \Leftrightarrow D$ to display the answer as a fraction in simplest form.

$$2.1 \div 3.5$$

Math ▲

$$\frac{3}{5}$$

Simplifying ratios containing algebraic terms

- If the ratio contains algebraic terms, divide both parts of the ratio by the highest common factor including common algebraic terms.

WORKED EXAMPLE 8 Simplifying ratios containing algebraic terms

Simplify the following ratios.

a. $10a^2b : 15ab^2$

b. $3mn : 6mn$

THINK

a. 1. Write the ratios.

2. Determine the HCF of $10a^2b$ and $15ab^2$.

3. Simplify the ratio by dividing both sides by $5ab$.

4. Cancel common factors to obtain the ratio in simplest form and write the answer.

WRITE

a. $10a^2b : 15ab^2$

The HCF is $5ab$.

$$\frac{10a^2b}{5ab} : \frac{15ab^2}{5ab}$$

$2a : 3b$

b.	1.	Write the ratios.		b.	$3mn : 6mn$
	2.	Determine the HCF of $3mn$ and $6mn$.			The HCF is $3mn$.
	3.	Simplify the ratio by dividing both sides by $3mn$.			$\dfrac{3mn}{3mn} : \dfrac{6mn}{3mn}$
	4.	Cancel common factors to obtain the ratio in simplest form and write the answer.			$1 : 2$

 Resources

Interactivity Simplifying ratios (int-3734)

Exercise 5.3 Simplifying ratios

learnon

5.3 Quick quiz on	5.3 Exercise

Individual pathways

■ PRACTISE	■ CONSOLIDATE	■ MASTER
1, 3, 6, 8, 10, 12, 15, 19, 23, 24, 27	2, 4, 7, 9, 13, 16, 20, 21, 25, 28	5, 11, 14, 17, 18, 22, 26, 29

Fluency

1. **WE4** Express each ratio in simplest form.

   a. $5 : 10$
   d. $21 : 14$
   b. $6 : 18$
   e. $15 : 35$
   c. $24 : 16$

2. Express each ratio in simplest form.

   a. $27 : 36$
   d. $84 : 144$
   b. $45 : 54$
   e. $88 : 132$
   c. $50 : 15$

3. **WE5** Write the following ratios in simplest form.

   a. 8 cm to 12 cm
   d. 75 cents to $3
   b. $6 to $18
   e. 300 mL to 4 L
   c. 80 cm to 2 m

4. Write the following ratios in simplest form.

   a. 500 g to 2.5 kg
   d. 30 cents to $1.50
   b. $4 to $6.50
   e. 2 h 45 min to 30 min
   c. 2500 m to 2 km

5. Write the following ratios in simplest form.

   a. 0.8 km to 450 m
   d. $1.75 to $10.50
   b. $1\dfrac{1}{2}$ min to 300 s
   c. 3500 mg to 1.5 g

6. **WE6** Simplify the following ratios.

   a. $\dfrac{1}{3} : \dfrac{2}{3}$
   b. $\dfrac{5}{7} : \dfrac{6}{7}$
   c. $\dfrac{1}{4} : \dfrac{1}{2}$
   d. $\dfrac{3}{10} : 1$
   e. $1\dfrac{2}{3} : \dfrac{1}{3}$

7. **WE7** Write the following ratios in simplest form.

   a. $1\frac{1}{4} : 1\frac{1}{2}$ b. $3\frac{1}{3}$ to $2\frac{1}{2}$ c. $1 : 1\frac{3}{5}$ d. $3\frac{1}{4}$ to $2\frac{4}{5}$

8. Write the following ratios in simplest form.

   a. 0.7 to 0.9   b. 0.3 : 2.1   c. 0.25 : 1.5

9. Write the following ratios in simplest form.

   a. 0.375 to 0.8   b. 0.01 : 0.1   c. 1.2 : 0.875

10. **WE8** Simplify these ratios.

   a. $2a : 10b$ b. $6p : 3p$ c. $2x^2 : 3x$

11. Simplify these ratios.

   a. $36m^3n^2 : 48m^2n^2$ b. $ab : 4ab^2$ c. $10^3x : 10x^3$

Understanding

12. Complete the patterns of equivalent ratios.

   a.   1 : 3   b.   2 : 3
           2 : 6   4 : 6
        __ : 9   6 : __
        __ : 12   __ : 12
        5 : __   __ : 24

13. Complete the patterns of equivalent ratios.

   a.   2 : 1   b. 64 : 32
           4 : 2   __ : 16
        __ : 4   __ : 8
        __ : 8   8 : __
       20 : __   __ : 1

14. Complete the patterns of equivalent ratios.

       48 : 64
       24 : __
       12 : __
       __ : 8
       __ : __

15. Compare each of the following, using a mathematical ratio (in simplest form).

   a. The Magpies won 8 games and the Lions won 10 games.
   b. This jar of coffee costs \$4 but that one costs \$6.
   c. While Joanne made 12 hits, Holly made 8 hits.
   d. In the first innings, Ian scored 48 runs and Adam scored 12 runs.

16. Compare each of the following, using a mathematical ratio (in simplest form).

   a. During a car race, Rebecca's average speed was 200 km/h and Donna's average speed was 150 km/h.
   b. In a basketball game, the Tigers beat the Magic by 105 points to 84 points.
   c. The capacity of a plastic bottle is 250 mL and the capacity of a glass container is 2 L.
   d. Joseph ran 600 m in 2 minutes and Maya ran the same distance in 96 seconds.

17. One serving of a popular cereal contains:

- 3.6 g of protein
- 0.4 g of fat
- 20 g of carbohydrate
- 1 g of sugar
- 3.3 g of dietary fibre
- 84 mg of sodium.

Write the following ratios in simplest form.

a. Sugar to carbohydrate
b. Fat to protein
c. Protein to fibre
d. Sodium to protein

18. **MC** Wollongong's population is 232 000 and Sydney's population is 5.22 million. The ratio of Wollongong's population to that of Sydney is:

A. 2 : 45   **B.** 4 : 9   **C.** 1 : 1.8   **D.** 9 : 4

19. **MC** When Samuel was born, he was 30 cm long. Now, on his 20th birthday, he is 2.1 m tall. The ratio of his birth height to his present height is:

A. 3 : 7   **B.** 1 : 21   **C.** 7 : 10   **D.** 1 : 7

20. **MC** The cost of tickets to two different concerts is in the ratio 3 : 5. If the more expensive ticket is $110, the cheaper ticket is:

A. $180   **B.** $80   **C.** $50   **D.** $66

21. **MC** A coin was tossed 100 times and Tails appeared 60 times. The ratio of Heads to Tails was:

A. 2 : 3   **B.** 3 : 5   **C.** 3 : 2   **D.** 5 : 3

22. **MC** Out of a 1.25 -L bottle of soft drink, Fran has drunk 500 mL. The ratio of soft drink remaining to the original amount is:

A. 2 : 3   **B.** 3 : 5   **C.** 3 : 2   **D.** 5 : 3

Communicating, reasoning and problem solving

23. Explain how simplifying ratios is similar to simplifying fractions.

24. Simplify the following three-part ratios, explaining your method.

a. 50 : 20 : 15
b. 0.4 : 1.8 : 2.2

25. The total attendance at a large outdoor concert was 32 200 people. Of this total, 13 800 people were female.

a. Calculate how many males attended the concert.
b. Compare the number of females to males by writing the ratio in simplest form.
c. Write the number of females as a fraction of the number of males.
d. What percentage of the crowd was female? How does this compare with your answer to part **c**?

26. Simplify the following ratios.

a. $27wxy : 18wy$
b. $5t : t^2$
c. $6ab : 48ab$

27. In a primary school that has 910 students, 350 students are in the senior school and the remainder are in the junior school. Of the senior school students, 140 are females. There are as many junior males as there are junior females. Write the following ratios in simplest form.

a. Senior students to junior students
b. Senior females to senior males
c. Senior males to total senior students
d. Junior males to senior males
e. Junior females to the whole school population

28. Compare the following, using a mathematical ratio (in simplest form).

a. Of the 90 000 people who attended the test match, 23 112 were females. Compare the number of males to females.
b. A Concorde jet (no longer used) could cruise at 2170 km/h; a Cessna can cruise at 220 km/h. Compare their speeds.
c. A house and land package is sold for $750 000. If the land was valued at $270 000, compare the land and house values.
d. In a kilogram of fertiliser, there is 550 g of phosphorus. Compare the amount of phosphorus to other components of the fertiliser.
e. Sasha saves $120 out of his take-home pay of $700 each fortnight. Compare his savings with his expenses.

29. The table shown represents the selling price of a house over a period of time.

a. Compare the purchase price of the house in December 2006 with the purchase price in April 2010 as a ratio in simplest form.
b. Compare the purchase price of the house in December 2006 with the purchase price in December 2010 as a ratio in simplest form.

Date of sale	Selling price
March 2014	$1.275 million
December 2010	$1.207 million
April 2010	$1.03 million
December 2006	$850 000

c. Compare the purchase price of the house in December 2006 with the purchase price in March 2014 as a ratio in simplest form.
d. How much has the value of the house increased from December 2006 to March 2014?
e. Compare the increase obtained in part d to the purchase price of the house in December 2006 as a ratio in simplest form.

LESSON
5.4 Proportion

LEARNING INTENTION

At the end of this lesson you should be able to:
- apply the cross-multiplication method to check if a ratio is in proportion
- determine the value of pronumerals in a proportion.

▶ 5.4.1 Proportion

eles-3772

- A **proportion** is a statement of equality of two ratios.
  For example, $12 : 18 = 2 : 3$.
- When objects are in proportion, they have the same fractional relationship
  between the size of their parts and the size of the whole.
  For example, the rectangles below are in proportion to each other.

HEALTHY EATING TIPS

The following observations can be made:
- When you compare the coloured parts of each rectangle, there is

  1 blue square for every 2 pink squares; the number of blue squares is $\dfrac{1}{2}$
  of the number of pink squares.

- When you compare the parts with the whole rectangle, the blue squares form $\dfrac{1}{3}$ of the whole rectangle,

  and the pink squares form the other $\dfrac{2}{3}$.

- To determine whether a pair of ratios are in proportion, follow the steps given below.
  1. Write the ratios in fraction form.
  2. If the fractions you wish to compare do not have the same denominator:
     - use equivalent fractions to convert both fractions so that they have the same denominator
     - use the lowest common multiple of the two denominators.
  3. Once the fractions have the same denominator, determine whether the equivalent fractions are the
     same. If the equivalent fractions are same, then the pair of ratios are in proportion.
- We can also determine whether a pair of ratios are in proportion by using the **cross-multiplication** method.

Cross-multiplication

If $a : b = c : d$ or $\dfrac{a}{b} = \dfrac{c}{d}$, then using cross-multiplication, as shown below:

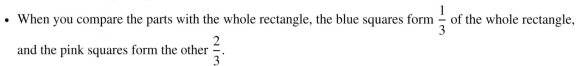

gives:

$$a \times d = c \times b$$

- If the cross products ($a \times d$ and $c \times b$) are equal, then the ratios form a proportion and therefore the pair of
  ratios are equivalent.

WORKED EXAMPLE 9 Determining whether a pair of ratios are in proportion

Determine whether the following pair of ratios are in proportion.

$$6 : 9; \quad 24 : 36$$

THINK	WRITE
Method 1: Equivalent fractions	
1. Write the ratios in fraction form.	$6 : 9 = \dfrac{6}{9}$ $24 : 36 = \dfrac{24}{36}$
2. As the fractions do not have the same denominator, determine the lowest common multiple of the denominators. First, list the multiples of 9 and 36. Identify the lowest number that is common to both lists.	Multiples of 9 are 9, 18, 27, ⟨36⟩, 45 ... Multiples of 36 are ⟨36⟩, 72, 108, 144, 180 ... The lowest common multiple is 36.
3. Write each fraction as an equivalent fraction using the lowest common multiple (36) as the denominator.	$\text{LHS} = \dfrac{6}{9} = \dfrac{6 \times 4}{9 \times 4} = \dfrac{24}{36}$ $\text{RHS} = \dfrac{24}{36} = \dfrac{24 \times 1}{36 \times 1} = \dfrac{24}{36}$
4. Decide whether the pair of ratios are in proportion. If the equivalent fractions are the same, then the pair of ratios are in proportion.	$\text{LHS} = \text{RHS}$ Therefore, $\dfrac{6}{9} = \dfrac{24}{36}$.
5. Write your answer in a sentence.	Therefore, the ratios are in proportion.
Method 2: Cross-multiplication	
1. Write the ratios in fraction form.	$\dfrac{6}{9} = \dfrac{24}{36}$
2. Perform cross-multiplication.	$\dfrac{6}{9} \bowtie \dfrac{24}{36}$ $6 \times 36 = 216; \ 24 \times 9 = 216$
3. Check whether the products are equal to determine whether the pair of ratios are in proportion.	$216 = 216$
4. Write your answer.	Therefore, the ratios are in proportion.

WORKED EXAMPLE 10 Determining the value of a pronumeral in a proportion

Determine the value of a in the proportion $\dfrac{a}{3} = \dfrac{6}{9}$.

THINK	WRITE
	$\dfrac{a}{3} = \dfrac{6}{9}$
1. Write the proportion statement.	
2. Cross-multiply and equate the products. (We could also use equivalent fractions to determine the value of a.)	$\dfrac{a}{3} \bowtie \dfrac{6}{9}$ $a \times 9 = 6 \times 3$

▶

3. Solve for a by dividing both sides of the equation by 9.

$$9a = 18$$
$$\frac{9a}{9} = \frac{18}{9}$$
$$a = 2$$

4. Write your answer.

The value of a is 2.

DISCUSSION

Provide an example to explain how proportion statements can be applied to other subject areas.

WORKED EXAMPLE 11 Determining the value of an unknown in a proportion

The ratio of girls to boys on the school bus was 4 : 3. If there were 28 girls, calculate how many boys there were.

THINK	WRITE
1. Let the number of boys be b and write a proportion statement. (Since the first number in the ratio represents girls, place the number of girls, 28, as the numerator.)	$\dfrac{4}{3} = \dfrac{28}{b}$
2. Cross-multiply and equate the products.	$\dfrac{4}{3} \ \diagdown\!\!\!\!\diagup \ \dfrac{28}{b}$ $4 \times b = 28 \times 3$
3. Solve for b by dividing both sides by 4.	$4b = 84$ $\dfrac{4b}{4} = \dfrac{84}{4}$ $b = 21$
4. Write the answer.	There are 21 boys.

 Resources

 Interactivity Proportion (int-3735)

5.4 Quick quiz on	5.4 Exercise

Individual pathways

■ PRACTISE	■ CONSOLIDATE	■ MASTER
1, 2, 5, 9, 11, 13, 16, 20, 23, 27, 30	3, 6, 7, 10, 14, 17, 18, 21, 25, 28, 31	4, 8, 12, 15, 19, 22, 24, 26, 29, 32

Fluency

1. **WE9** Determine whether the following pairs of ratios are in proportion.

$$2 : 3; \ 8 : 12$$

2. Determine whether the following pairs of ratios are in proportion.

$$4 : 7; \ 8 : 14$$

3. Determine whether the following pairs of ratios are in proportion.

$$5 : 7; \ 10 : 14$$

4. Determine whether the following pairs of ratios are in proportion.

$$5 : 8; \ 10 : 16$$

5. Determine whether the following pairs of ratios are in proportion.

$$\frac{7}{9}; \frac{21}{25}$$

6. Determine whether the following pairs of ratios are in proportion.

$$\frac{3}{8}; \frac{12}{32}$$

7. Determine whether the following pairs of ratios are in proportion.

$$\frac{14}{16}; \frac{5}{9}$$

8. Determine whether the following pairs of ratios are in proportion.

$$\frac{11}{12}; \frac{7}{8}$$

9. Determine whether the following pairs of ratios are in proportion.

$$\frac{13}{15}; \frac{6}{7}$$

10. Determine whether the following pairs of ratios are in proportion.

$$\frac{8}{9}; \frac{24}{27}$$

11. Determine whether the following pairs of ratios are in proportion.

$$\frac{3}{5}; \frac{6}{8}$$

12. Determine whether the following pairs of ratios are in proportion.

$$\frac{21}{18}; \frac{49}{42}$$

13. **WE10** Determine the value of a in each of the following proportions.

a. $\frac{a}{2} = \frac{4}{8}$

b. $\frac{a}{6} = \frac{8}{12}$

c. $\frac{a}{9} = \frac{2}{3}$

d. $\frac{3}{a} = \frac{9}{12}$

14. Determine the value of a in each of the following proportions.

a. $\frac{7}{a} = \frac{14}{48}$

b. $\frac{10}{a} = \frac{3}{15}$

c. $\frac{3}{7} = \frac{a}{28}$

d. $\frac{12}{10} = \frac{a}{5}$

15. Determine the value of a in each of the following proportions.

a. $\frac{8}{12} = \frac{a}{9}$

b. $\frac{35}{7} = \frac{5}{a}$

c. $\frac{24}{16} = \frac{6}{a}$

d. $\frac{30}{45} = \frac{2}{a}$

Understanding

16. **WE11** Solve each of the following, using a proportion statement and the cross-multiplication method.
   a. The ratio of boys to girls in a class is 3 : 4. If there are 12 girls, calculate how many boys there are in the class.
   b. In a room, the ratio of length to width is 5 : 4. If the width is 8 m, calculate the length.
   c. The team's win–loss ratio is 7 : 5. Calculate how many wins it has had if it has had 15 losses.
   d. A canteen made ham and chicken sandwiches in the ratio of 5 : 6. If 20 ham sandwiches were made, calculate how many chicken sandwiches were made.
   e. The ratio of concentrated cordial to water in a mixture is 1 : 5. Calculate how much concentrated cordial is needed for 25 litres of water.

17. **a.** The ratio of chairs to tables is 6 : 1. If there are 42 chairs, calculate how many tables there are.

    b. The ratio of flour to milk in a mixture is 7 : 2. If 14 cups of flour are used, calculate how much milk is required.

    c. The ratio of protein to fibre in a cereal is 12 : 11. If there are 36 grams of protein, calculate the mass of fibre.

    d. In a supermarket, the ratio of 600-mL cartons of milk to litre cartons is 4 : 5. If there are sixty 600-mL cartons, calculate how many litre cartons there are.

    e. In a crowd of mobile-phone users, the ratio of men to women is 7 : 8. Calculate the number of women if there are 2870 men.

18. Although we know that only whole numbers are used in ratios, sometimes in a proportion statement the answer can be a fraction or a mixed number. Consider the following proportion:

$$\frac{a}{6} = \frac{7}{4}$$
$$a \times 4 = 7 \times 6$$
$$4a = 42$$
$$a = 10.5$$

Calculate the value of a in each of the following proportion statements. Write your answers correct to 1 decimal place.

    a. $\dfrac{a}{7} = \dfrac{8}{5}$ **b.** $\dfrac{a}{6} = \dfrac{4}{5}$ **c.** $\dfrac{a}{3} = \dfrac{7}{10}$ **d.** $\dfrac{a}{9} = \dfrac{9}{10}$ **e.** $\dfrac{5}{a} = \dfrac{7}{10}$

19. Calculate the value of a in each of the following proportion statements. Write your answers correct to 1 decimal place.

    a. $\dfrac{8}{a} = \dfrac{6}{7}$ **b.** $\dfrac{9}{7} = \dfrac{a}{6}$ **c.** $\dfrac{13}{6} = \dfrac{a}{5}$ **d.** $\dfrac{9}{15} = \dfrac{7}{a}$ **e.** $\dfrac{7}{8} = \dfrac{9}{a}$

20. Write a proportion statement for each situation and then solve the problem. If necessary, write your answers correct to 1 decimal place.

    a. A rice recipe uses the ratio of 1 cup of rice to 3 cups of water. How many cups of rice can be cooked in 5 cups of water?

    b. Another recipe states that 2 cups of rice are required to serve 6 people. If you have invited 11 people, how many cups of rice will you need?

    c. In a chemical compound there should be 15 g of chemical A to every 4 g of chemical B. If my compound contains 50 g of chemical A, how many grams of chemical B should it contain?

    d. A saline solution contains 2 parts of salt to 17 parts of water. How much water should be added to 5 parts of salt?

    e. To mix concrete, 2 buckets of sand are needed for every 3 buckets of blue metal. For a big job, how much blue metal will be needed for 15 buckets of sand?

21. Decide whether a proportion statement could be made for each of the following ratios.

    a. Height : age

    b. Mass : age

    c. Intelligence : age

    d. Distance : time

    e. Cost : number

22. Decide whether a proportion statement could be made for each of the following ratios.

   a. Age : shoe size
   b. Sausages cooked : number of people
   c. Eggs : milk (in a recipe)
   d. Number of words : pages typed
   e. Length : area (of a square)

23. **MC** If $\dfrac{p}{q} = \dfrac{l}{m}$, then:

   A. $p \times q = l \times m$
   B. $p \times l = q \times m$
   C. $p \times m = l \times q$
   D. $\dfrac{p}{m} = \dfrac{l}{q}$

24. **MC** If $\dfrac{x}{3} = \dfrac{y}{6}$, then:

   A. $x = 2$ and $y = 4$
   B. $x = 1$ and $y = 2$
   C. $x = 3$ and $y = 6$
   D. all options are true.

25. **MC** If $\dfrac{23}{34} = \dfrac{x}{19}$, then, correct to the nearest whole number, x equals:

   A. 13
   B. 12
   C. 34
   D. 28

26. **MC** The directions on a cordial bottle suggest mixing 25 mL of cordial with 250 mL of water. How much cordial should be mixed with 5.5 L of water?

   A. 0.55 mL
   B. 5.5 mL
   C. 55 mL
   D. 550 mL

Communicating, reasoning and problem solving

27. In a family, 3 children receive their allowances in the ratio of their ages, which are 16 years, 14 years and 10 years. If the oldest child receives $32, determine how much the other two children receive.

28. Two classes each contain 8 boys. In one class, the ratio of boys to girls is 1 : 2; in the other, it is 2 : 1. If the two classes combine, determine the new ratio.

29. In jewellery, gold is often combined with other metals. 'Pink gold' is a mixture of pure gold, copper and silver in the ratio of 15 : 4 : 1.
   'White gold' is a mixture of pure gold and platinum in the ratio of 3 : 1.
   a. Calculate what fraction of both pink and white gold jewellery is pure gold.

   Pure gold is 24 carats and is not mixed with other metals. For most jewellery, however, 18-carat gold is used.

   b. Using your answer to part **a**, show why jewellery gold is labelled 18 carats.
   c. If the copper and silver in an 18-carat bracelet weigh a combined 2 grams, calculate the weight of gold in the bracelet.
   d. If the price of gold is $35 per gram, calculate the cost of the gold in the bracelet from part **c**.

30. The nutritional panel for a certain cereal is shown. Sugars are one form of carbohydrate. Calculate the ratio of sugars to total carbohydrates in the cereal.

	Quantity per 30-g serving	Percentage daily intake per 30-g serving	Quantity per 30-g serving with ½ cup skim milk
Energy	480 kJ	5.5%	670 kJ
Protein	6.6 g	13.1%	11.2 g
Fat			
— total	0.2 g	0.3%	0.3 g
— saturated	0.1 g	0.1%	0.2 g
Carbohydrate			
— total	20.8 g	6.7%	27.3 g
— sugars	9.6 g	10.7%	16.1 g

31. The solution to this question contains some gaps. Rewrite the solution, replacing the empty boxes with the appropriate numbers.

On a map for which the scale is 1 : 20 000, what distance in cm represents 2.4 km on the ground?

$$1 : 20\,000 = x\,\text{cm} : 2.4\,\text{km}$$
$$1 : 20\,000 = x\,\text{cm} : 2.4 \times 1000 \times 100\,\text{cm}$$
$$1 : 20\,000 = x : 240\,000$$
$$\frac{1}{20\,000} = \frac{x}{\Box}$$
$$\frac{1}{20\,000} \times \Box = \frac{x}{240\,000} \times \Box$$
$$x = \Box$$

So, $\Box$ cm on the map represents 2.4 km on the ground.

32. A concreter needs to make $1.0\,\text{m}^3$ of concrete for the base of a garden shed. How many cubic metres of each of the components should be used?

To make concrete mix, 1 part cement, 2 parts sand and 4 parts gravel are needed.

LESSON
5.5 Comparing ratios

LEARNING INTENTION

At the end of this lesson you should be able to:
- compare two ratios to determine which is larger.

▶ 5.5.1 Comparison of ratios

eles-3773

- Given two ratios, it is sometimes necessary to know which is larger. To determine which is larger, we must compare ratios.

Comparing ratios

To compare ratios, write them as fractions with a common denominator.

$$\frac{2}{3}, \frac{3}{5} = \frac{2}{3} \times \frac{5}{5}, \frac{3}{5} \times \frac{3}{3}$$
$$= \frac{10}{15}, \frac{9}{15}$$

$\dfrac{10}{15} > \dfrac{9}{15}$ therefore $\dfrac{2}{3}$ is the larger ratio

WORKED EXAMPLE 12 Comparing ratios

Determine which is the larger ratio in the following pair.

$$3 : 5 ; 2 : 3$$

THINK	WRITE
1. Write each ratio in fraction form.	$\dfrac{3}{5}$ $\dfrac{2}{3}$
2. Convert the fractions so that they have a common denominator.	$\dfrac{9}{15}$ $\dfrac{10}{15}$
3. Compare the fractions: since both fractions have a denominator of 15, the larger the numerator, the larger the fraction.	$\dfrac{9}{15} < \dfrac{10}{15}$
4. The second fraction is larger and corresponds to the second ratio in the pair. State your conclusion.	Therefore, 2 : 3 is the larger ratio.

 Resources

Interactivity Equivalent ratios (int-3736)

Exercise 5.5 Comparing ratios

learn on

5.5 Quick quiz   on

5.5 Exercise

Individual pathways

■ PRACTISE	■ CONSOLIDATE	■ MASTER
1, 4, 7, 9, 10, 13	2, 5, 8, 11, 14	3, 6, 12, 15

Fluency

1. **WE12** Determine which is the larger ratio in each of the following pairs.
   a. 1 : 4; 3 : 4
   b. 5 : 9; 7 : 9
   c. 6 : 5; 2 : 5
   d. 3 : 5; 7 : 10
   e. 7 : 9; 2 : 3

2. Determine which is the larger ratio in each of the following pairs.
   a. 2 : 5; 1 : 3
   b. 2 : 3; 3 : 4
   c. 5 : 6; 7 : 8
   d. 5 : 9; 7 : 12
   e. 9 : 8; 6 : 5

3. St Mary's won 2 football matches out of 5. Cessnock won 5 football matches out of 10. Who had the better record?

4. Newbridge won 13 football matches out of 18. Mitiamo won 7 football matches out of 12. Who had the better record?

5. Wedderburn won 12 football matches out of 20. Korong Vale won 14 football matches out of 25. Who had the better record?

6. Gresford won 6 soccer matches out of 13. Morpeth won 13 soccer matches out of 20. Who had the better record?

Understanding

7. In a cricket match, Jenny bowled 5 wides in her 7 overs and Lisa bowled 4 wides in her 6 overs. Compare the results to state which bowler had the higher wides-per-over ratio.

8. **MC** If $\frac{5}{6} > \frac{a}{5}$, then a could be:

   A. 4   B. 5   C. 6   D. 7

9. **MC** Which of the following must be true if $\frac{a}{b} < \frac{3}{5}$?

   A. $a < 3$ B. $b > 5$ C. $a < b$ D. $b > 5$

Communicating, reasoning and problem solving

10. Jamie likes his cordial strong in flavour and makes some in the ratio of 2 parts water to 5 parts concentrate. Monique also likes strong cordial and makes some in the ratio of 3 parts water to 6 parts concentrate. Giving reasons for your answer, determine who made the strongest cordial.

11. The steepness (or gradient) of a hill can be determined by finding the ratio $\frac{\text{vertical distance}}{\text{horizontal distance}}$. The two triangles shown represent different hills.

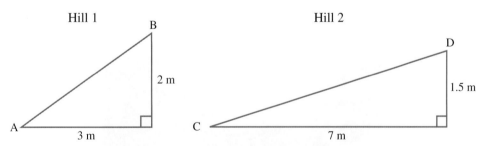

Explain which hill is steeper.

12. Draw a right-angled triangle on a piece of graph paper so that the two sides at right angles to each other are 6 cm and 8 cm. Measure the third side length of the triangle, which should be 10 cm.

   a. State the ratio of the three sides of this triangle.
   b. If you change the size of your triangle but keep the shape the same, explain what happens to the ratio of the three sides of the triangle.
   c. A piece of string is used to mark out a triangle with its sides in the same ratio as the one you have drawn. The smallest side of the triangle is 75 cm long. Calculate how long the other two sides are.

13. Janelle purchased a block of chocolate containing 24 small cube-like pieces. She is sharing the block with her mother Marina. Janelle suggests sharing the block in the ratio 2 : 1 in her favour, while Marina suggests sharing the block in the ratio 5 : 3.
    By which ratio should Janelle split the block of chocolate in order for her to maximise the number of pieces she gets to eat?

14. An internet search for a homemade lemonade recipe yielded the following results. All of the recipes had water added, but the sweetness of the lemonade is determined by the ratio of lemon juice to sugar.
    For each of the recipe ratios shown below, determine the mixtures that have the same taste. If they do not have the same taste, determine which website has the sweeter lemonade recipe.
    a. Website 1: 3 tablespoons of sugar for every 15 tablespoons of lemon juice
       Website 2: 4 tablespoons of sugar for every 20 tablespoons of lemon juice
    b. Website 1: 3 tablespoons of sugar for every 9 tablespoons of lemon juice
       Website 2: 5 tablespoons of sugar for every 16 tablespoons of lemon juice
    c. Website 1: 2 cups of sugar for every 3 cups of lemon juice
       Website 2: 5 tablespoons of sugar for every 8 tablespoons of lemon juice
    d. Website 1: 3 tablespoons of sugar for every 8 tablespoons of lemon juice
       Website 2: 7 tablespoons of sugar for every 12 tablespoons of lemon juice

15. The steepness of a hill can be written as a ratio of the vertical distance (height) to the horizontal distance. Hill A has a ratio of 1 : 5 and hill B has a ratio of 7 : 36.
    a. Use diagrams to determine which hill has the steeper slope.
    b. Is this an accurate method of determining the slope?
    c. Without using diagrams, investigate another method to determine which ratio produces the steeper slope.
    d. Explain whether your method from part c will always work when you want to compare ratios.

LESSON
5.6 Dividing in a given ratio

LEARNING INTENTION

At the end of this lesson you should be able to:
- divide a quantity in a given ratio
- use the unitary method to solve ratio problems.

▶ 5.6.1 Dividing in a given ratio

eles-3774

- When something is shared, we often use ratios to ensure that the sharing is fair.
- Consider the following situation.
  Two people buy a lottery ticket for $3. They win a prize of $60. How is the prize divided fairly?

Each person contributes $1.50.	**One person contributes $1 and the other $2.**
• The contribution for the ticket is in the ratio 1 : 1. • The prize is divided in the ratio 1 : 1. • The 1 : 1 ratio has $1 + 1 = 2$ total parts. Each person receives $\frac{1}{2}$ of the prize money ($30). $30 \quad\quad $30	• The contribution for the ticket is in the ratio 1 : 2. • The prize is divided in the ratio 1 : 2. • The 1 : 2 ratio has $1 + 2 = 3$ total parts. Person 1 paid for 1 part of the ticket and therefore receives $\frac{1}{3}$ of the prize money ($20); person 2 paid for 2 parts of the ticket and therefore receives $\frac{2}{3}$ of the prize money ($40). $20 \quad $20 \quad $20

COMMUNICATING — COLLABORATIVE TASK: Sharing and share size

1. As a pair, discuss the answers to the following questions.
   a. The lollies in the picture are to be shared between two people. How do you divide the lollies so that:
      i. each person receives the same amount
      ii. one person gets twice as many as the other person
      iii. for every one lolly that the first person receives, the other person receives three lollies (i.e. the lollies are divided in a ratio of 1 : 3)
      iv. they are divided in a ratio of 2 : 3?

   b. The lollies are now to be shared among three people. How do you divide the lollies so that they are divided in a ratio of 1 : 2 : 3?
   c. What fraction does each person receive in part **b** above?

2. As a class, discuss the relationship between the ratio in which the lollies were shared and the fraction of lollies that each person received as a result.

3. How can you determine the fraction of the whole amount each person will receive from the ratio by which the whole will be divided?

WORKED EXAMPLE 13 Sharing a quantity in a particular ratio

Share the amount of $500 000 in the ratio 3 : 7.

THINK	WRITE
1. Determine the total number of parts in the ratio.	Total number of parts $= 3 + 7$ $\qquad\qquad\qquad\qquad\quad = 10$
2. The first share represents 3 parts out of a total of 10, so calculate $\dfrac{3}{10}$ of the total amount.	First share $= \dfrac{3}{10} \times \$500\,000$ $\qquad\qquad\ = \$150\,000$
3. The second share could be calculated in two ways. **Method 1:** The second share is the remainder, so subtract the first share amount from the total amount.	Second share $= \$500\,000 - \$150\,000$ $\qquad\qquad\quad = \$350\,000$
Method 2: The second share represents 7 parts out of the total of 10, so calculate $\dfrac{7}{10}$ of the total amount.	Second share $= \dfrac{7}{10} \times \$500\,000$ $\qquad\qquad\quad = \$350\,000$
4. Write your answer.	First share $= \$150\,000$ Second share $= \$350\,000$

DISCUSSION

Discuss some other real-life examples where sharing things on the basis of ratios might be considered unfair.

WORKED EXAMPLE 14 Calculating quantities of components using ratios

Concrete mixture for a footpath was made up of 1 part cement, 2 parts sand and 4 parts blue metal. Calculate how much sand was used to make 4.2 m³ of concrete.

THINK	WRITE
1. Calculate the total number of parts.	Total number of parts $= 1 + 2 + 4$ $\qquad\qquad\qquad\qquad\quad = 7$
2. There are 2 parts of sand to be used in the mixture, so calculate $\dfrac{2}{7}$ of the total amount of concrete made.	Amount of sand $= \dfrac{2}{7} \times 4.2 \, \text{m}^3$ $\qquad\qquad\qquad = 1.2 \, \text{m}^3$
3. Write your answer.	$1.2 \, \text{m}^3$ sand was used to make $4.2 \, \text{m}^3$ of concrete.

⏵ 5.6.2 Using the unitary method to solve ratio problems

eles-3775

- The unitary method can also be used to solve ratio problems.
- It involves first calculating one part or one unit. This is why it is known as the unitary method.

Applying the unitary method

To use the unitary method, divide the quantity that you are sharing by the total amount of equal parts. Then multiply each share of the ratio by the quantity that one part represents.

WORKED EXAMPLE 15 Calculating contributions using the unitary method

Elena, Arjun and Megan contribute towards a lottery pool in the ratio 1 : 2 : 3.
If they win $1740 between them, use the unitary method to divide the winnings according to how much they contributed.

THINK	WRITE
1. Calculate the total number of parts.	Total number of parts $= 1 + 2 + 3$ $\qquad\qquad\qquad\qquad = 6$
2. Divide the quantity by the total amount of equal parts to obtain the amount for one unit or part.	$\div 6 \left(\begin{array}{c} 6 \text{ parts} = \$1740 \\ \\ 1 \text{ part} = \$290 \end{array} \right) \div 6$
3. Multiply each person's share by the quantity that one part represents.	$\boxed{\$290}\ \boxed{\$290}\ \boxed{\$290}\ \boxed{\$290}\ \boxed{\$290}\ \boxed{\$290}$ Elena: 1 part $290 \times 1 = \$290$ Arjun: 2 parts $290 \times 2 = \$580$ Megan: 3 parts $290 \times 3 = \$870$
4. Write the answer.	Elena wins $290, Arjun wins $580 and Megan wins $870.

DISCUSSION

Discuss some other real-life examples where you need to divide in a ratio other than 1 : 1.

---

on Resources

Interactivity Dividing in a given ratio (int-3737)

Exercise 5.6 Dividing in a given ratio

5.6 Quick quiz on	5.6 Exercise

Individual pathways

■ PRACTISE	■ CONSOLIDATE	■ MASTER
1, 4, 7, 9, 13, 14, 17, 20	2, 5, 8, 11, 15, 18, 21	3, 6, 10, 12, 16, 19, 22

Fluency

1. Write the total number of parts for each of the following ratios.
   a. $1:2$ b. $2:3$ c. $3:1$ d. $3:5$ e. $4:9$

2. Write the total number of parts for each of the following ratios.
   a. $5:8$ b. $6:7$ c. $9:10$ d. $1:2:3$ e. $3:4:5$

3. **WE13** Share the amount of $1000 in the following ratios.
   a. $2:3$ b. $1:4$ c. $1:1$

4. Share the amount of $1000 in the following ratios.
   a. $3:7$ b. $7:13$ c. $9:11$

5. If Nat and Sam decided to share their lottery winnings of $10 000 in the following ratios, how much would each receive?
   a. $1:1$ b. $2:3$ c. $3:2$

6. If Nat and Sam decided to share their lottery winnings of $10 000 in the following ratios, how much would each receive?
   a. $3:7$ b. $7:3$ c. $23:27$

Understanding

7. Rosa and Mila bought a lottery ticket costing $10. How should they share the first prize of $50 000 if their respective contributions were:
   a. $2 and $8   b. $3 and $7   c. $4 and $6
   d. $5 and $5   e. $2.50 and $7.50?

8. **WE14** Concrete mixture is made up of 1 part cement, 2 parts sand and 4 parts blue metal.
   a. Calculate how much sand is needed for $7\,m^3$ of concrete.
   b. Calculate how much cement is needed for $3.5\,m^3$ of concrete.
   c. Calculate how much blue metal is required for $2.8\,m^3$ of concrete.
   d. Calculate how much sand is used for $5.6\,m^3$ of concrete.
   e. Calculate how much cement is needed to make $8.4\,m^3$ of concrete.

9. Three of your teachers buy a lottery ticket costing $20. How should they share the first prize of $600 000 if they each contribute:
   a. $3, $7 and $10
   b. $6, $6 and $8
   c. $1, $8 and $11
   d. $5, $6 and $9
   e. $5, $7.50 and $7.50?

10. **WE15** In a family, 3 children receive their allowances in the ratio of their ages, which are 15 years, 12 years and 9 years. If the total of the allowances is $60, use the unitary method to determine how much each child receives.

11. In a school, the ratio of girls in Years 8, 9 and 10 is 6 : 7 : 11. If there is a total of 360 girls in the three year levels, calculate:
   a. how many Year 8 girls there are
   b. how many more Year 10 girls there are than Year 8 girls.

12. In a moneybox, there are 5-cent, 10-cent and 20-cent coins in the ratio 8 : 5 : 2. If there are 225 coins altogether, calculate:
   a. how many 5-cent coins there are
   b. how many more 10-cent coins than 20-cent coins there are
   c. the total value of the 5-cent coins
   d. the total value of the coins in the moneybox.

13. **MC** A square of side length 4 cm has its area divided into two sections in the ratio 3 : 5. The area of the larger section is:

   A. 3 cm²   B. 5 cm²   C. 8 cm²   D. 10 cm²

14. **MC** A block of cheese is cut in the ratio 2 : 3. If the smaller piece is 150 g, the mass of the original block was:

   A. 75 g   B. 200 g   C. 300 g   D. 375 g

15. **MC** Contributions of $1.75 and $1.25 were made to the cost of a lottery ticket. What fraction of the prize should the larger share be?

   A. $\dfrac{7}{12}$ B. $\dfrac{5}{7}$ C. $\dfrac{7}{5}$ D. $\dfrac{3}{5}$

16. **MC** A television channel that telecasts only news, movies and sport does so in the ratio 2 : 3 : 4 respectively. How many movies, averaging a length of $1\frac{1}{2}$ hours, would be shown during a 24-hour period?

A. 2   B. 3   C. 4   D. 5

Communicating, reasoning and problem solving

17. Three angles of a triangle are in the ratio 1 : 2 : 3. What is the magnitude of each angle? Justify your answer.

18. The angles of a quadrilateral are in the ratio 2 : 3 : 4 : 6. What is the difference in magnitude between the smallest and largest angles? Show your working.

19. The total amount of prize money in a photography competition is $20 000. It is shared between the first, second and third prizes in the ratio 12 : 5 : 3. The amounts the winners received were $6667, $4000 and $1667.
One of the winners claimed that these amounts were incorrect and asked for a fair share of the prize money. Explain what mistake was made when calculating the prizes and how much each winner should receive.

20. The ratio of boys to girls in Year 8 at a school is 3 : 2.

   a. If there are 75 boys in Year 8, calculate how many girls there are.

   Each class in Year 8 contains 25 students. One class contains boys only, but the remaining classes are a mixture of boys and girls.

   b. Determine how many classes in Year 8 contain a mixture of boys and girls.
   c. State the ratio of mixed classes to boys-only classes.

21. The wing of a model aeroplane is 4 cm long, and the scale of the model is 1 : 300.

   a. Calculate the ratio of the length of the model's wing to the length of the wing of the actual aeroplane.
   b. Calculate the wing length of the actual aeroplane.

22. a. There are apples, oranges and bananas in a fruit bowl. The ratio of apples to oranges is 3 : 4, and the number of apples is $\frac{3}{8}$ of the total number of pieces of fruit.
   Calculate the number of bananas.

   b. Next to the fruit bowl is a bowl of nuts. The ratio of cashews to peanuts is 15 : 4, the ratio of peanuts to walnuts is 4 : 7, and the ratio of hazelnuts to peanuts is 11 : 4.
   Calculate the minimum number of nuts there can be in the bowl.
   Somebody added some macadamias to the nut bowl in part b.
   The ratio of peanuts to macadamias is now 1 : 3. Calculate:

   c. the number of macadamia nuts added
   d. the total number of nuts now in the bowl.

LESSON
5.7 Rates

LEARNING INTENTION

At the end of this lesson you should be able to:
- understand the difference between ratios and rates
- calculate and express rates of two different quantities with correct units.

▶ 5.7.1 Rates

eles-3776

- A **rate** is a particular type of ratio that is used to compare two measurements of different kinds.
- Rates are often used to describe and compare how quantities change.
- A rate is different from a ratio. A ratio is used to compare the same types of quantities measured in the same unit, whereas a rate compares two different types of quantities measured in different units.
- Rates have units. An example of a rate is speed (measured in km/h or m/s).
- A forward slash, /, which is the mathematical symbol for 'per', is used to separate two different units. The forward slash works in the same way as a fraction line.
- A rate is in its simplest form if it is per one unit.

WORKED EXAMPLE 16 Expressing a rate in simplest form

Express the following statement using a rate in simplest form:
A 30-litre container was filled in 3 minutes.

THINK	WRITE
1. A suitable rate would be litres per minute (L/min). Put the capacity of the container in the numerator and the time in which it was filled in the denominator of the fraction.	$\text{Rate} = \dfrac{30\,\text{L}}{3\,\text{min}}$ $= \dfrac{10\,\text{L}}{1\,\text{min}}$
2. Simplify the fraction.	$= 10\,\text{L/min}$
3. Write the answer.	That is, the container was filled at the rate of 10 litres per minute.

WORKED EXAMPLE 17 Calculating a rate

Joseph is paid $8.50 per hour as a casual worker. At this rate, calculate how much he receives for 6 hours of work.

THINK	WRITE
1. The rate is given in $ per hour, so it tells us the amount of money earned in each hour (the hourly payment).	Payment per 1 hour $= \$8.50$
2. State the number of hours worked.	Hours worked $= 6$
3. To calculate the total payment, multiply the hourly payment by the total number of hours worked.	Total payment $= \$8.50 \times 6$ $= \$51$

▶ 5.7.2 Converting rates

eles-6176

- Rates can be converted from one unit to another; for example, speed in km/h can be expressed in m/s.

> ## Converting rates
>
> **To change the unit of a rate, follow these steps.**
> 1. **Convert the numerator of the rate to the new unit.**
> 2. **Convert the denominator of the rate to the new unit.**
> 3. **Divide the new numerator by the new denominator.**

WORKED EXAMPLE 18 Converting rates

Covert the following rates as shown to 2 decimal places.
a. **60 km/h to m/s**
b. **25 L/h to mL/min**

THINK	WRITE
a. 1. Convert the numerator from km to m by multiplying by 1000.	a. $60 \text{ km} = 60 \times 1000$ $= 60\,000 \text{ m}$
2. Convert the denominator from hours to seconds by multiplying first by 60 to change to minutes, and then by 60 again to change to seconds.	$1 \text{ h} = 1 \times 60 \times 60$ $= 3600 \text{ sec}$
3. Divide the numerator by the denominator.	$\text{Rate} = \dfrac{60\,000}{3600}$ $= 16.6667 \text{ m/s}$
4. Write the new rate to 2 decimal places.	$60 \text{ km/h} = 16.67 \text{ m/s}$
b. 1. Convert the numerator from L to mL by multiplying by 1000.	b. $25 \text{ L} = 25 \times 1000$ $= 25\,000 \text{ mL}$
2. Convert the denominator from hours to minutes by multiplying by 60.	$1 \text{ h} = 1 \times 60$ $= 60 \text{ min}$
3. Divide the numerator by the denominator.	$\text{Rate} = \dfrac{25\,000}{60}$ $= 416.6667 \text{ mL/min}$
4. Write the new rate to 2 decimal places.	$25 \text{ L/h} = 416.67 \text{ mL/min}$

on Resources

❖ **Interactivity** Rates (int-3738)

Exercise 5.7 Rates

5.7 Quick quiz on	5.7 Exercise

Individual pathways

■ PRACTISE	■ CONSOLIDATE	■ MASTER
1, 4, 7, 10, 11, 14, 17, 20, 23, 26, 31, 35	2, 5, 8, 12, 15, 18, 21, 24, 27, 30, 32, 33, 36, 37	3, 6, 9, 13, 16, 19, 22, 25, 28, 29, 34, 38

Fluency

1. **WE16** Express each of the following statements using a rate in simplest form.
   a. A lawn of $600\,\text{m}^2$ was mown in 60 min.
   b. A tank of capacity $350\,\text{kL}$ is filled in 70 min.
   c. A balloon of volume $4500\,\text{cm}^3$ was inflated in 15 s.
   d. The cost of 10 L of fuel was $13.80.
   e. A car used 16 litres of petrol in travelling 200 km.

2. Express each of the following statements using a rate in simplest form.
   a. A 12-m length of material cost $30.
   b. There were 20 cows grazing in a paddock that was $5000\,\text{m}^2$ in area.
   c. The gate receipts for a crowd of 20 000 people were $250 000.
   d. The cost of painting a 50-m^2 area was $160.
   e. The cost of a 12-minute phone call was $3.00.

3. Express each of the following statements using a rate in simplest form.
   a. The team scored 384 points in 24 games.
   b. Last year 75 kg of fertiliser cost $405.
   c. The winner ran the 100 m in 12 s.
   d. To win, Australia needs to make 260 runs in 50 overs.
   e. For 6 hours of work, Bill received $159.

4. Express each of the following statements using a rate in simplest form.
   a. The 5.5-kg parcel cost $19.25 to post.
   b. 780 words were typed in 15 minutes.
   c. From 6 am to noon, the temperature changed from $10\,°\text{C}$ to $22\,°\text{C}$.
   d. When Naoum was 10 years old he was 120 cm tall. When he was 18 years old he was 172 cm tall.
   e. A cyclist left home at 8:30 am and at 11:00 am had travelled 40 km.

5. **WE17** Sima is paid $15.50 per hour. At this rate, calculate how much she receives for 7 hours of work.

6. A basketball player scores, on average, 22 points per match. Calculate how many points he scores if he plays 18 matches.

7. A car's fuel consumption is 11 L/100 km. Calculate how much fuel it would use in travelling 550 km.

8. To make a solution of fertiliser, the directions recommend mixing 3 capfuls of fertiliser with 5 L of water. Calculate how many capfuls of fertiliser should be used to make 35 L of solution.

9. Anne can type 60 words per minute. Calculate how long she will take to type 4200 words.

10. Given the following information, express the rate in the units shown in the brackets.
    a. A runner travels 800 m in 4 minutes (km/min).
    b. Water runs into a water tank at 800 mL in 2 minutes (L/min).
    c. A phone call costs 64 cents for 2 minutes ($/min).
    d. Apples cost $13.50 for 3 kg (cents/gram).

11. `WE18` Convert the following rates as shown to 2 decimal places.
    a. 100 km/h to m/s
    b. 40 L/h to mL/min

12. Convert the following rates as shown to 2 decimal places.
    a. $5.50/m to cents/cm
    b. 10 m/s to km/h

13. Convert the following rates as shown to 2 decimal places.
    a. 80 km/h to m/s
    b. 25 m/s to km/h
    c. 50 g/ml to kg/L
    d. 360 mL/min to L/h

Understanding

14. Identify what quantities (such as distance, time, volume) are changing if the units of rate are:
    a. km/h
    b. cm^3/sec
    c. L/km
    d. $ per h
    e. $ per cm.

15. Identify what quantities (such as distance, time, volume) are changing if the units of rate are:
    a. kL/min
    b. cents/litre
    c. $ per dozen
    d. kg/year
    e. cattle/hectare.

16. State the units you would use to measure the changes taking place in each of the following situations.
    a. A rainwater tank being filled
    b. A girl running a sprint race
    c. A boy getting taller
    d. A snail moving across a path

17. State the units you would use to measure the changes taking place in each of the following situations.
    a. An ink blot getting larger
    b. A car consuming fuel
    c. A batsman scoring runs
    d. A typist typing a letter

18. Water flows from a hose at a rate of 3 L/min. Determine how much water flows in 2 h.

19. Tea bags in a supermarket can be bought for $1.45 per pack (pack of 10) or for $3.85 per pack (pack of 25). Determine which is the cheaper way of buying the tea bags.

20. Car A uses 41 L of petrol in travelling 500 km. Car B uses 34 L of petrol in travelling 400 km. Determine which car is more economical.

21. Coffee can be bought in 250-g jars for $9.50 or in 100-g jars for $4.10. Determine which is the cheaper way of buying the coffee and how large the saving is.

22. `MC` A case containing 720 apples was bought for $180. The cost could be written as:
    A. 30 cents each.
    B. 20 cents each.
    C. $3.00 per dozen.
    D. $2.00 per dozen.

23. `MC` Mark, a test cricketer, has a batting strike rate of 68, which means he has made 68 runs for every 100 balls he faced. What is Steve's strike rate if he has faced 65 overs and made 280 runs? (*Note:* Each over contains 6 balls.)
    A. 65
    B. 68.2
    C. 71.8
    D. 73.2

24. `MC` A carport measuring 8 m × 4 m is to be paved. The paving tiles cost $36 per m^2 and the tradesperson charges $12 per m^2 to lay the tiles. Select how much it will cost to pave the carport (to the nearest $50).
    A. $1400
    B. $1450
    C. $1500
    D. $1550

25. **MC** A tank of capacity 50 kL is to be filled by a hose whose flow rate is 150 L/min. If the tap is turned on at 8 am, identify when the tank will be full.

A. Between 1:00 pm and 1:30 pm
B. Between 1:30 pm and 2:00 pm
C. Between 2:00 pm and 2:30 pm
D. Between 2:30 pm and 3:00 pm

26. Mark is going on a trip with his school to NASA Space Camp. He did research and found that $1 Australian dollar is worth $0.70 US dollars. If Mark converted $500 Australian dollars to US dollars, calculate how much US money Mark would have.

27. While Mark was on his trip to the USA he realised that all the signs on the roads were in miles instead of kilometres. He wanted to understand how far he had to travel on the bus, so he needed to convert 78 miles into kilometres. Given there are 1.61 km for every mile, how far did Mark have to travel?

28. Layla is looking at a new phone plan. The plan she wants to go with charges a rate of $35 per month. Layla wants to organise her yearly budget and therefore needs to calculate the yearly cost of the phone plan. How much will Layla have to pay for one year on this phone plan?

29. For many years Aboriginal Peoples have cared for our country by using land management techniques that assisted our environment. One of these land management techniques is traditional burning, which created extensive grassland on good soil and reduced the likelihood of intense bushfires.
The grassland in turn encouraged kangaroos to come, which could then be hunted for food.
If during traditional burning a fire was burning at a rate of 550 m^2/minute, calculate how much land was burned in one hour.

30. Water has always been a very important resource for Aboriginal Peoples. In some areas, artwork and carvings on trees have pointed the way to water sources that were difficult to find. Due to the intense heat and low humidity in these areas, water would evaporate quickly.
If water was evaporating at a rate of 86 m^3/day, how much water would evaporate in the month of September?

Communicating, reasoning and problem solving

31. Explain the difference between a rate and a ratio.

32. A chiropractor sees 160 patients every week.
   a. Calculate how many patients he sees per hour if he works a 40-hour week.
   b. Calculate how long, on average, he spends with each patient.
   c. Using these rates, if the chiropractor wants to make at least $10 000 every week, determine the minimum charge for each patient.

33. If 4 monkeys eat 4 bananas in 4 minutes, calculate how long it takes 12 monkeys to eat 12 bananas.

34. If Bill takes 3 hours to paint a room and James takes 5 hours to paint a room, calculate how long it will take to paint a room if they work together.

35. The Stawell Gift is a 120-m handicap footrace. Runners who start from scratch run the full 120 m; for other runners, their handicap is how far in front of scratch they start. Joshua Ross has won the race twice. In 2003, with a handicap of 7 m, his time was 11.92 seconds. In 2005, from scratch, he won in 12.36 seconds.
Compare his results to identify which race he ran faster.

36. Jennifer's work is 40 km from her home. On the way to work one morning, her average speed was 80 km/h. Due to bad weather and roadworks, her average speed for the trip home was only 30 km/h.
Determine how much time Jennifer spent travelling to and from work on this day.

37. Beaches are sometimes unfit for swimming if heavy rain has washed pollution into the water. A beach is declared unsafe for swimming if the concentration of bacteria is more than 5000 organisms per litre. A sample of 20 millilitres was tested and found to contain 55 organisms.
Calculate the concentration in the sample (in organisms/litre) and state whether or not the beach should be closed.

38. **a.** Earth is a sphere with a mass of 6.0×10^{24} kg and a radius of 6.4×10^{6} m.

   i. Use the formula $V = \dfrac{4}{3} \pi r^3$ to calculate the volume of Earth.

   ii. Hence, calculate the density of Earth. (Remember density $= \dfrac{\text{mass}}{\text{volume}}$.)

   b. The planet Jupiter has a mass of 1.9×10^{27} kg and a radius of 7.2×10^{7} m.

   i. Calculate the volume of Jupiter.
   ii. Calculate the density of Jupiter.

   c. Different substances have their own individual densities. Does it seem likely that Earth and Jupiter are made of the same substance? Explain.

LESSON
5.8 Interpreting graphs

LEARNING INTENTION

At the end of this lesson you should be able to:
- interpret a variety of graphs
- calculate the gradient of a straight line
- determine speed from a distance–time graph
- determine the units of the gradient of a graph.

5.8.1 Interpretation of graphs

eles-3777

- A graph tells a story by describing data in everyday life.
- Graphs are frequently used in fields outside mathematics, including science, geography and economics.
- They compare two related quantities against each other. As one quantity changes, it affects the other.
- COVID-19 has seen mathematical modellers continuously track the spread of the virus and record important data such as the number of new infections each day, the number of deaths per day, the number of tests conducted and so on. The data are represented graphically for people to understand and interpret.

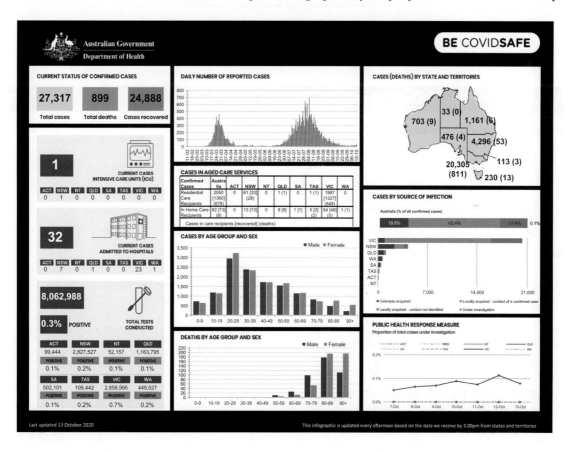

DISCUSSION

The infographic above displays important information regarding COVID-19 cases throughout Australia using a variety of graphs. Examine the infographic and discuss some of the points of interest as a class.

The gradient

- To measure the steepness of various slopes and hills, we need to calculate the **gradient**.
- Gradient is the ratio of vertical distance to horizontal distance between any two points. It is equal to $\dfrac{\text{vertical distance}}{\text{horizontal distance}}$, which is often called $\dfrac{\text{rise}}{\text{horizontal run}}$.
- The larger the gradient, the steeper the hill.

Calculating the gradient of a slope

The gradient of a hill (or a slope) is calculated by the formula:

$$\text{gradient} = \frac{\text{vertical distance}}{\text{horizontal distance}} = \frac{\text{rise}}{\text{run}}$$

rise = vertical distance

run = horizontal distance

WORKED EXAMPLE 19 Determining the gradient of a hill

Determine the gradient of the hill (AB) if AC = 2 m and BC = 10 m.

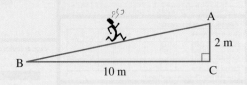

THINK	WRITE
1. Write the rule for calculating the gradient.	$\text{Gradient} = \dfrac{\text{vertical distance}}{\text{horizontal distance}}$
2. The vertical distance is 2 m and the horizontal distance is 10 m. Substitute these values into the formula for the gradient.	$= \dfrac{2}{10}$
3. Simplify by dividing both numerator and denominator by 2.	$= \dfrac{1}{5}$

Distance–time graphs

- Distance–time graphs are used to visually express journeys with respect to time.
- On a distance–time graph, time is represented on the horizontal axis, and the distance from a reference point (e.g. the starting point) is represented on the vertical axis.

> ### Speed
>
> - **In a distance–time graph, the gradient of the graph is the rate of change of distance over time, which represents the speed.**
> - **Speed is calculated as:**
>
> $$\text{speed } (s) = \frac{\text{distance } (d)}{\text{time } (t)}$$

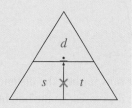

- The commonly used units for speed are m/s and km/h.
- The greater the gradient (slope), the greater the speed, and vice versa.
- A change in direction can be represented on a distance–time graph by a change in the sign of the gradient.
- In the graph shown, each of the sections A, C and D indicates a constant (steady) speed.
- A steeper line or gradient (e.g. section A) shows a faster speed than a less steep line (e.g. section C).
- A horizontal line or the zero gradient (e.g. section B) shows that the object is not moving or is at rest.
- The sign of the gradient does not matter when calculating speed, as speed can never be negative. In the example shown, section D shows the highest speed because it is the section with the steepest slope.

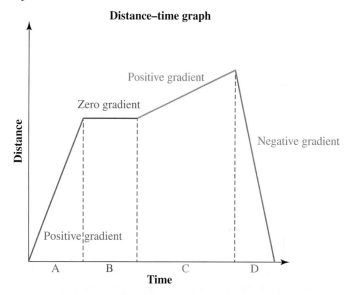

Distance–time graph

Positive gradient

Zero gradient

Negative gradient

Positive gradient

A   B   C   D

Time

Distance

WORKED EXAMPLE 20 Interpreting graphs

Three graphs (I, II and III) and three descriptions (a, b and c) are shown below. Match each graph to the description that best suits it and explain your choice.

I.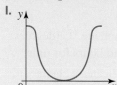

II.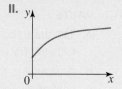

III.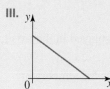

a. **The number of leaves on a deciduous tree over a calendar year**
b. **The distance run versus the distance from the finishing line for a runner in a 400-m race**
c. **The weight of a newborn baby in its first year of life**

THINK	WRITE
a. Deciduous trees lose all of their leaves in autumn and regrow them when the weather warms up in spring. 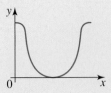	**a.** The number of leaves on a deciduous tree over a calendar year is best represented by graph I.

b. As the runner runs further, the distance to the finish line decreases, so the height of the curve should be decreasing. For example:

Distance run (m)	0	100	200	300	400
Distance to finish (m)	400	300	200	100	0

b. As the runner runs, the distance (x) from the starting line becomes greater and the distance (y) to the finish line becomes smaller. The graph should be nearing $y = 0$. Graph III is the only graph that does this. The distance run versus the distance left to run is therefore best represented by graph III.

THINK	WRITE
c. Babies grow rapidly at first, and then this growth rate begins to slow. 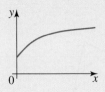	**c.** The growth rate of babies is quick at first, then slows but continues to increase. Graph II shows an increasing gradient that becomes less steep; therefore, graph II best represents the growth of a baby in the first year.

WORKED EXAMPLE 21 Analysing and interpreting graphs

Use the graph shown to answer the following questions.
a. Identify the units shown on the graph.
b. State what a change in the y-value represents.
c. State what a change in the x-value represents.
d. Calculate the gradient of the graph over the section in which the temperature was rising the fastest.
e. Describe what the gradient represents.

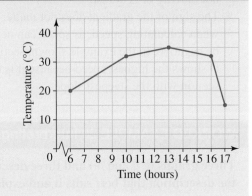

THINK	WRITE
a. The units are displayed in the axis labels.	**a.** The units shown on the graph are °C (degrees Celsius) for temperature and hours for time.
b. The y-axis shows the temperature.	**b.** A change in the y-value indicates a change in temperature.
c. The x-axis shows time.	**c.** A change in the x-value indicates a change in time.

d. The temperature is rising the fastest when the graph is increasing most steeply. This occurs between 6 am and 10 am, when the temperature increases from 20 °C to ~32 °C.

e. The gradient has units of °C/h, which are read as degrees Celsius per hour.

d. Gradient $= \dfrac{\text{rise}}{\text{run}}$

$= \dfrac{12\,°\text{C}}{4\,\text{h}}$

$= 3\,°\text{C/h}$

e. The gradient represents the rate at which the temperature is changing per hour.

COMMUNICATING — COLLABORATIVE TASK: The story of a car trip from Lakes Entrance to Jindabyne

1. As a pair, describe the trip from Lakes Entrance to Jindabyne shown in the graph. (Distances are approximate.)
2. What does the flat section of the graph represent?
3. Did the speed increase or decrease after passing through Cooma?
4. Determine the unit of the gradient of the graph.

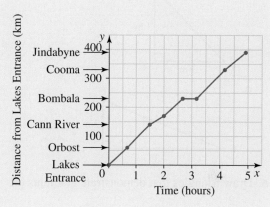

Exercise 5.8 Interpreting graphs

learn on

| 5.8 Quick quiz **on** | 5.8 Exercise |

Individual pathways

■ PRACTISE	■ CONSOLIDATE	■ MASTER
1, 4, 7, 10, 12, 13, 16	2, 5, 8, 11, 14, 17	3, 6, 9, 15, 18

Fluency

1. State which axis represents distance and which axis represents time on a distance–time graph.

2. If distance is measured in metres (vertical axis) and time is measured in seconds (horizontal axis), state how speed is measured.

3. An express train through country New South Wales is delayed for one hour because of a herd of cattle on the line. State how this event is graphed on a distance–time graph.

4. **WE19** Determine the gradient of each of the hills represented by the following triangles.

a.
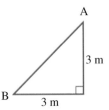

b.
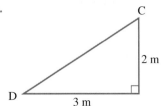

c.
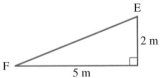

d.
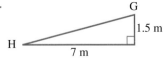

5. Draw triangles that demonstrate gradients of:

a. $\dfrac{2}{1}$ b. $\dfrac{3}{1}$ c. $\dfrac{4}{3}$ d. $\dfrac{3}{2}$ e. $\dfrac{2}{5}$

6. **MC** If the gradient of LN in the triangle shown is 1, then:

A. $a > b$ B. $a < b$

C. $a = b$ D. $a = 1$

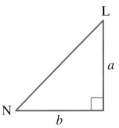

7. **MC** Select which of the following graphs best describes a student walking to school who dawdles at first, then meets up with some friends and drops in at the corner shop before stepping up the pace to get to school on time.

A.
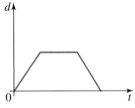

B.
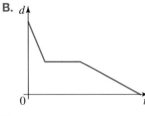

C.
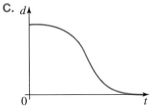

D.
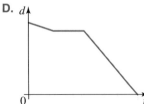

8. **MC** Select which of the following graphs best describes a student travelling to school on a bus.

A.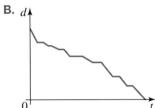

B.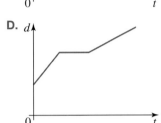

C.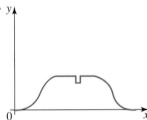

D.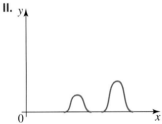

9. **WE20** Three descriptions (**a**, **b** and **c**) and three graphs (**I**, **II** and **III**) are shown below. Match each graph to the description that best suits it and explain your choice.

I.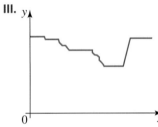

II.

III.

a. The number of students in the canteen line over a school day
b. The number of cans in a school's drink vending machine over a school day
c. The number of students on school grounds over a school day

Understanding

10. The graph shows the distance between Andrea's car and her house. Answer True or False for the following statements.

    a. Andrea is travelling faster at point C than at point A.
    b. Andrea is not moving at point B.
    c. Andrea is further away from home at point B compared to point E.
    d. At point D Andrea is travelling back towards home.
    e. Andrea is travelling slower at point E than at point C.

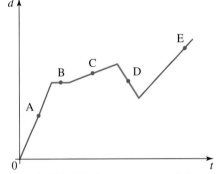

11. Draw sketches to represent each of the stories below and explain your sketches.

    a. The mass of a pig over its lifetime
    b. The altitude of a plane during a flight
    c. The height of water in a bathtub as it is emptied

Communicating, reasoning and problem solving

12. Describe what different gradients of straight-line segments of a distance–time graph represent.

13. Describe what a horizontal line segment on a distance–time graph represents.

14. Calculate the speed for segments A, B and C of the distance–time graph shown. If necessary, give your answers correct to 1 decimal place.

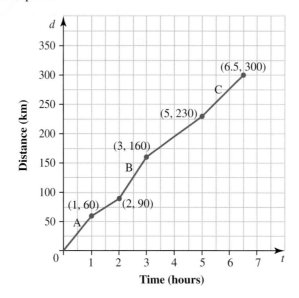

15. **WE21** The graph shows the cost of a mobile telephone call. Use the graph to answer the following questions.

   a. Identify the units shown on the graph.
   b. State what a change in the y-value represents.
   c. State what a change in the x-value represents.
   d. Calculate the gradient for each straight-line section of the graph using units.
   e. Describe what the gradient represents.

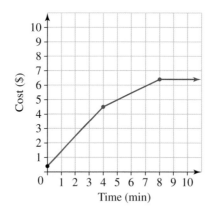

16. The price per kilogram for 3 different types of meat is illustrated in the graph shown.
    a. Calculate the gradient (using units) for each graph.
    b. State the cost of 1 kg of each type of meat.
    c. Calculate the cost of purchasing 0.5 kg of chicken.
    d. Calculate the cost of purchasing 2 kg of beef.

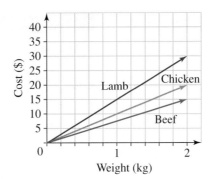

17. The graph shown represents the distance travelled by a vehicle versus time. Copy the graph onto some grid paper and, using red pen, show what the graph would look like if the vehicle was travelling twice as fast at any point in time.

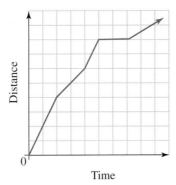

18. Draw a distance–time graph to represent the following story.
    Jordan decides to take his dog for a walk. He starts off at a steady, fast pace and continues at this pace for 2 minutes.
    After 2 minutes, he breaks out into a gentle jog for 3 minutes then takes a rest for 30 seconds.
    Following his rest, he walks back home in 10 minutes.

LESSON
5.9 Review

5.9.1 Topic summary

Equivalent ratios

- Ratios are equivalent if one can be converted into the other by multiplying or dividing by a factor (e.g. $1 : 5 = 2 : 10$).
- Ratios are generally written in simplest form.

Proportion

- If two ratios $a : b$ and $c : d$ are equivalent, then $a \times b = c \times d$.
- This relationship can be used to determine unknown values.

Ratios

- Ratios compare quantities in the same unit.
- $1 : 3$ is read as 'the ratio of 1 to 3'.
- Ratios contain only integer values without units.
- They can be converted into fractions.

  e.g. $2 : 5 \Rightarrow \dfrac{2}{7}$ and $\dfrac{5}{7}$

RATIOS AND RATES

Rates

- A rate compares two different types of quantities in different units (e.g. 12 km/h or 12 cm/day).
- A rate is in simplified form if it is per one unit.
- Rates can be converted from one unit to another (e.g. speed in km/h can be expressed in m/s).

- Graphs compare one quantity with another.
- The gradient or slope is a measure of how fast one quantity is changing with respect to another.
- A common example is the distance–time graph. In this graph, the slope is a measure of the speed. The greater the slope, the greater the speed.

Dividing an amount in a given ratio

Method 1:
1. Determine the total number of parts in the ratio.
2. Attribute each element of the ratio with a fraction.
3. Use the fractions to share the amount.

e.g. $1400 was shared in the ratio $2 : 5$.

The first amount is $\dfrac{2}{7} \times 1400 = \$400$.

The second amount is $\dfrac{5}{7} \times 1400 = \$1000$.

Method 2: The unitary method
1. Determine the total number of parts in the ratio.
2. Determine the amount that one part represents.
3. Multiply each person's share by the amount per one part.

e.g. Using the above example:

   Total parts $= 7$

   Each part $= \dfrac{1400}{7}$

            $= \$200$

   The first amount is $2 \times 200 = \$400$.

   The second amount is $5 \times 200 = \$1000$.

Comparing ratios

- To compare ratios, first write each ratio in fraction form, then convert the fractions to the lowest common denominator.
- Gradient (steepness of a slope) is a comparison of horizontal distance to vertical distance.

$$\text{Gradient} = \frac{\text{vertical distance}}{\text{horizontal distance}}$$

Speed

- Speed is an example of a rate.
- $\text{Speed} = \dfrac{\text{distance}}{\text{time}}$

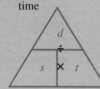

- $\text{Average speed} = \dfrac{\text{total distance travelled}}{\text{total time taken}}$

5.9.2 Project

The golden ratio

The Greeks believed that using a special ratio of numbers in building designs, paintings, sculptures and so on would automatically make them beautiful. This ratio is known as the golden ratio or golden number. The human body has many examples of the golden ratio.

$$\text{The golden ratio} = \frac{\sqrt{5}+1}{2}$$

Work out this number as a decimal correct to 3 decimal places.

Part A

The golden ratio is often represented by phi (φ). One of the interesting relationships of this ratio to the design of the human body is that there are:
- five appendages on the torso: arms, legs and head
- five appendages on each of these: fingers, toes and five openings on the face

The golden number is also based on the number 5 because the number phi can be written as:

$$5^{0.5} \times 0.5 + 0.5$$

Leonardo da Vinci's drawings of the human body emphasised its proportions. The ratios of the following distances equal the golden ratio:
- foot to navel : navel to head
- length of forearm : length of hand
- length of upper arm : length of hand and forearm.

Your task is to explore the golden ratio as it applies to your body.

Work in pairs to measure parts of your body. In the table below are some of the measurements you could take. Choose other parts of your body to measure and find as many golden ratios as possible.

Body measurement (cm)	Ratio	Decimal value
Foot to navel : navel to head		
Length of forearm : length of hand		
Length of upper arm : length of hand and forearm		

Part B

Another way to find the golden ratio is by using the Fibonacci sequence:

1, 1, 2, 3, 5, 8, 13, ...

1. Write the first 20 terms of the sequence in the table.
2. Take the terms two at a time and divide the larger number by the smaller (for example, divide 2 by 1, 3 by 2, 5 by 3), working as accurately as possible. Record your answers in the table.

Sequence	Ratio	Decimal value
	1	
1	$\frac{1}{1}$	
2		2
3	$\frac{3}{2}$	
5	$\frac{5}{3}$	1.6
8		
13	$\frac{13}{8}$	1.625

 Resources

 Interactivities Crossword (int-2624)
Sudoku puzzle (int-3185)

Exercise 5.9 Review questions

learnon

Fluency

1. On a farm there are 5 dogs, 3 cats, 17 cows and 1 horse. Write the following ratios.
   a. Cats : dogs
   b. Horses : cows
   c. Cows : cats
   d. Dogs : horses
   e. Dogs : other animals

2. Express each of the following ratios in simplest form.
   a. 8 : 16
   b. 24 : 36
   c. 35 mm : 10 cm
   d. $2 : 60 cents
   e. 20 s : $1\frac{1}{2}$ min

3. Express each of the following ratios in simplest form.
   a. $\frac{1}{12} : \frac{1}{3}$
   b. 4 : 10
   c. 56 : 80
   d. 2 hours : 40 min
   e. 1.5 km : 400 m

4. Find the value of n in each of the following proportions.

   a. $\dfrac{n}{3} = \dfrac{20}{5}$

   b. $\dfrac{n}{28} = \dfrac{5}{7}$

   c. $\dfrac{2}{3} = \dfrac{8}{n}$

5. Determine the value of n in each of the following proportions.

   a. $\dfrac{4}{5} = \dfrac{12}{n}$

   b. $\dfrac{6}{n} = \dfrac{5}{8}$

   c. $\dfrac{3}{10} = \dfrac{n}{4}$

Understanding

6. The directions for making lemon cordial require the mixing of 1 part cordial to 6 parts water.
   a. Express this as a ratio.
   b. How much cordial would you have to mix with 9 L of water?

7. Which is the larger ratio?

   a. $\dfrac{4}{5}, \dfrac{2}{3}$

   b. $\dfrac{7}{12}, \dfrac{5}{8}$

8. a. Divide \$25 in the ratio 2 : 3.
   b. Share \$720 in the ratio 7 : 5.

9. The horizontal and vertical distances between the top and bottom points of slide A are 3 m and 2 m respectively. For slide B, the horizontal distance between the top and bottom points is 10 m, and the vertical distance is 4 m.
   a. Calculate the gradients of slide A and slide B.
   b. Identify which slide is steeper. Justify your answer.

10. Three people share a lottery prize of \$6600 in the ratio of 4 : 5 : 6. Determine the difference between the smallest and largest shares.

11. A car travels 840 km on 72 litres of petrol. Calculate the fuel consumption of the car in L/100 km.

12. David's car has a fuel consumption rate of 12 km/L and Susan's car has a fuel consumption rate of 11 km/L.
    a. Identify which car is more economical.
    b. Calculate how far David's car can travel on 36 L of fuel.
    c. Determine how much fuel (to the nearest litre) Susan's car would use travelling 460 km.

13. A 1-kg packet of flour costs \$2.80 and a 750-g packet costs \$2.20. Compare the two and state which is the cheapest way to buy flour.

14. The sides of a triangle are in the ratio $3 : 4 : 5$. If the longest side of the triangle measures 40 cm, determine the perimeter of the triangle.

15. To make two $\frac{2}{3}$-cup servings of cooked rice, you add $\frac{3}{4}$ of a cup of rice, $\frac{1}{4}$ teaspoon of salt and 1 teaspoon of butter to $1\frac{1}{2}$ cups of water. Calculate how many $\frac{2}{3}$-cup servings of cooked rice you can make from a bag containing 12 cups of rice.

Communicating, reasoning and problem solving

16. Lachlan was driven from Newcastle to Branxton, a distance of 60 km, at an average speed of 80 km/h. He cycled back at an average speed of 20 km/h. Calculate his average speed for the whole journey. (*Hint:* It is not 50 km/h.)

17. The speed of the space shuttle *Discovery* in orbit was 17 400 miles per hour. Calculate this in km/h. (1 kilometre = 0.62 miles)

18. The rate of ascent of the space shuttle *Discovery* was 71 miles in 8.5 minutes.
   a. Calculate the speed in km/min.
   b. Calculate the speed in km/h.

19. You have a plastic bag that contains 80 tennis balls. The contents in the bag weigh 4 kg (the weight of the plastic bag is insignificant). You add 10 more balls to your bag. Calculate how much your bag weighs now.

20. A cyclist riding at 12 km/h completes a race in 3 h 45 min.
   a. Calculate the distance of the race.
   b. At what speed would he have to ride to complete the race in 3 h?

21. The steps of a staircase are to have a ratio of rise to run that is to be $\frac{2}{3}$. If the run is 30 cm, calculate the rise.

22. Travelling from Noort to Bastion takes Dexter 1 hour and 30 minutes by car at an average speed of 72 km per hour. Dexter stops for 15 minutes in Bastion before travelling to Smoop, which is 163 km away.
The trip from Bastion to Smoop takes him 2 hours and 12 minutes. Calculate his average speed for the whole trip.

23. It takes Deb 2 hours to mow her lawn. Her son takes 2.5 hours to mow the same lawn. If they work together using two lawnmowers, calculate how long it will take them to mow the lawn.
Give your answer in hours, minutes and seconds.

24. The graph shown represents the path of a ski lift.

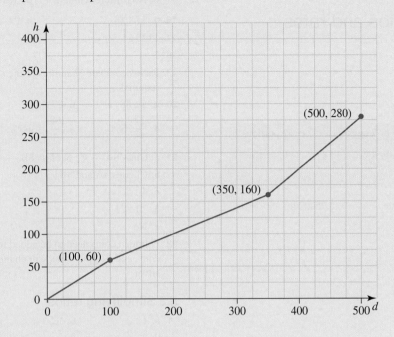

a. Calculate the gradients for each section.
b. Identify where the gradient is the steepest.
c. Calculate the average gradient of the ski lift.

on To test your understanding and knowledge of this topic, go to your learnON title at www.jacplus.com.au and complete the **post-test**.

Answers

Topic 5 Ratios and rates

5.1 Pre-test

1. B
2. 9
3. A, B, D
4. C
5. a. 1200 g   b. 7 eggs   c. 42 portions
6. a. 2.5   b. 4.2
7. Team B
8. $750
9. a. 58 words/minute   b. 2.75 kL/h
10. a. 70 : 9
    b. 7 : 10
    c. $2m : n$
11. 1 : 5
12. A
13. 5 km/h, 7.5 km/h
14. 4500
15. Height $= 41.16$ cm and length $= 57.624$ cm

5.2 Ratios

1. a. 4 : 5   b. 5 : 4   c. 5 : 9
   d. 9 : 4   e. 1 : 2
2. a. 5 : 7   b. 7 : 5   c. 5 : 12
3. a. 4 : 3   b. 3 : 4   c. 6 : 1
   d. 4 : 1   e. 2 : 5
4. a. 3 : 5   b. 6 : 19   c. 4 : 11
   d. 7 : 9   e. 1 : 5
5. a. 9 : 4   b. 3 : 4   c. 3 : 10
   d. 17 : 60   e. 53 : 100
6. a. 11 : 100   b. 1 : 1000   c. 1 : 2000
   d. 7 : 24   e. 5 : 12
7. $\dfrac{6}{17}$
8. a. 3 : 59   b. 59 : 38   c. 38 : 3
   d. 3 : 97   e. 59 : 41
9. a. 24 : 17   b. 21 : 17   c. 2 : 1
   d. 4 : 39
10. a. 4 : 1   b. 1 : 1   c. 1 : 2
    d. 3 : 1
11. a. 215 : 179   b. 215 : 36
12. a. 97 : 3   b. 3 : 100
13. a. $\dfrac{3}{10}$ b. $\dfrac{4}{15}$
    c. $\dfrac{13}{25}$ d. $\dfrac{7}{9}$
14. a. Yes (same unit)   b. Yes (same unit)
    c. No (different units)   d. No (different units)
15. a. Yes (same unit)   b. Yes (same unit)
    c. No (different units)   d. Yes (same unit)
16. a. 17 : 83   b. 97 : 3
17. 1125 g
18. 2 : 1
19. Ratios compare quantities of the same unit, so units are not required.
20. a. 5 : 68   b. 27 : 5   c. 17 : 25
    d. $\dfrac{17}{25}, 68\%$ e. $\dfrac{68}{5}, 1360\%$
21. a. Class 8A — 4 : 5; class 8B — 4 : 5; class 8C — 4 : 3
    b. Classes 8A and 8B
    c. 4 : 5
    d. 20 : 21
    e. No
22. Fourteen coins need to be added to the purse. Three silver coins and 11 gold coins will change the ratio of gold to silver.

5.3 Simplifying ratios

1. a. 1 : 2   b. 1 : 3   c. 3 : 2
   d. 3 : 2   e. 3 : 7
2. a. 3 : 4   b. 5 : 6   c. 10 : 3
   d. 7 : 12   e. 2 : 3
3. a. 2 : 3   b. 1 : 3   c. 2 : 5
   d. 1 : 4   e. 3 : 40
4. a. 1 : 5   b. 8 : 13   c. 5 : 4
   d. 1 : 5   e. 11 : 2
5. a. 16 : 9   b. 3 : 10   c. 7 : 3
   d. 1 : 6
6. a. 1 : 2   b. 5 : 6   c. 1 : 2
   d. 3 : 10   e. 5 : 1
7. a. 5 : 6   b. 4 : 3   c. 5 : 8
   d. 65 : 56
8. a. 7 : 9   b. 1 : 7   c. 1 : 6
9. a. 15 : 32   b. 1 : 10   c. 48 : 35
10. a. $a : 5b$ b. 2 : 1   c. $2x : 3$
11. a. $3m : 4$ b. $1 : 4b$ c. $10^2 : x^2$
12. a. 1 : 3
    2 : 6
    3 : 9
    4 : 12
    5 : 15
    b. 2 : 3
    4 : 6
    6 : 9
    8 : 12
    16 : 24
13. a. 2 : 1
    4 : 2
    8 : 4
    16 : 8
    20 : 10

b. 64 : 32
    32 : 16
    16 : 8
    8 : 4
    2 : 1

14. 48 : 64
    24 : 32
    12 : 16
    6 : 8
    3 : 4

15. a. 4 : 5   **b.** 2 : 3   **c.** 3 : 2
    d. 4 : 1

16. a. 4 : 3   **b.** 5 : 4   **c.** 1 : 8
    d. 5 : 4

17. a. 1 : 20   **b.** 1 : 9   **c.** 12 : 11
    d. 7 : 300

18. A

19. D

20. D

21. A

22. B

23. Both ratios and fractions are simplified by multiplying or dividing by the same number to obtain the lowest form.

24. a. 10 : 4 : 3
    b. 2 : 9 : 11

25. a. 18 400
    b. 3 : 4
    c. $\dfrac{3}{4}$
    d. 42.86%. In part **c** the number of females was compared with the number of males (part-to-part comparison), whereas in part **d** the number of females was compared with the total number in the crowd (part-to-whole comparison).

26. a. $3x : 2$ **b.** $5 : t$ **c.** 1 : 8

27. a. 5 : 8   **b.** 2 : 3   **c.** 3 : 5
    d. 4 : 3   **e.** 4 : 13

28. a. 929 : 321   **b.** 217 : 22   **c.** 9 : 16
    d. 11 : 9   **e.** 6 : 29

29. a. 85 : 103   **b.** 50 : 71   **c.** 2 : 3
    d. $425 000   **e.** 1 : 2

5.4 Proportion

1. Yes

2. Yes

3. Yes

4. Yes

5. No

6. Yes

7. No

8. No

9. No

10. Yes

11. No

12. Yes

13. a. $a = 1$ **b.** $a = 4$ **c.** $a = 6$
    d. $a = 4$

14. a. $a = 24$ **b.** $a = 50$ **c.** $a = 12$
    d. $a = 6$

15. a. $a = 6$ **b.** $a = 1$ **c.** $a = 4$
    d. $a = 3$

16. a. 9 boys
    b. 10 m
    c. 21 wins
    d. 24 chicken sandwiches
    e. 5 litres

17. a. 7 tables
    b. 4 cups
    c. 33 g
    d. 75 cartons
    e. 3280 women

18. a. 11.2   **b.** 4.8   **c.** 2.1
    d. 8.1   **e.** 7.1

19. a. 9.3   **b.** 7.7   **c.** 10.8
    d. 11.7   **e.** 10.3

20. a. $\dfrac{1}{3} = \dfrac{n}{5}; n = 1.7$

   b. $\dfrac{2}{6} = \dfrac{n}{11}; n = 3.7$

   c. $\dfrac{15}{4} = \dfrac{50}{n}; n = 13.3$

   d. $\dfrac{2}{17} = \dfrac{5}{n}; n = 42.5$

   e. $\dfrac{2}{3} = \dfrac{15}{n}; n = 22.5$

21. a. No   **b.** No   **c.** No
    d. Yes   **e.** Yes

22. a. No   **b.** Yes   **c.** Yes
    d. No   **e.** Yes

23. C

24. D

25. A

26. D

27. The 14-year-old receives $28 and the 10-year-old receives $20.

28. 4 : 5

29. a. White gold: $\dfrac{3}{4}$; pink gold: $\dfrac{3}{4}$

   b. Because $\dfrac{18}{24} = \dfrac{3}{4}$

   c. 6 g
    d. $210

30. 9.6 : 20.8 = 6 : 13

31.

$$1 : 20\,000 = x\,\text{cm} : 2.4\,\text{km}$$
$$1 : 20\,000 = x\,\text{cm} : 2.4 \times 1000 \times 100\,\text{cm}$$
$$1 : 20\,000 = x : 240\,000$$
$$\frac{1}{20\,000} = \frac{x}{240\,000}$$
$$\frac{1}{20\,000} \times 240\,000 = \frac{x}{240\,000} \times 240\,000$$
$$x = 12$$

So, 12 cm on the map represents 2.4 km on the ground.

32. $0.14\,\text{m}^3$ cement; $0.29\,\text{m}^3$ sand; $0.57\,\text{m}^3$ gravel

5.5 Comparing ratios

1. a. $3 : 4$ b. $7 : 9$ c. $6 : 5$ d. $7 : 10$ e. $7 : 9$

2. a. $2 : 5$ b. $3 : 4$ c. $7 : 8$ d. $7 : 12$ e. $6 : 5$

3. Cessnock had the better record.

4. Newbridge

5. Wedderburn

6. Morpeth

7. Jenny

8. A

9. C

10. Jamie made the strongest cordial.

11. Hill 1 is steeper.

12. a. $3 : 4 : 5$

   b. The ratio of the 3 sides stays the same.

   c. $1\,\text{m}$, $1.25\,\text{m}$

13. $2 : 1$

14. a. $\dfrac{3}{15} = \dfrac{4}{20}$; same taste

   b. $\dfrac{1}{3} \neq \dfrac{5}{16}$; website 1 has the sweeter recipe.

   c. $\dfrac{2}{3} \neq \dfrac{5}{8}$; website 1 has the sweeter recipe.

   d. $\dfrac{3}{8} \neq \dfrac{7}{12}$; website 2 has the sweeter recipe.

15. a. Hill A

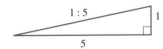

Hill B

Hill A is steeper. It is difficult to determine which hill has the steeper slope from these diagrams because the difference in slope is slight.

   b. Diagrams are accurate only if a scale drawing is used and there is significant difference in the scope.

c. Answers will vary. A sample answer is given here.

$$\frac{1}{5} : \frac{7}{36}$$
$$\frac{36}{180} : \frac{35}{180}$$
$$\frac{1}{5} > \frac{7}{36}$$

Therefore, Hill A is steeper.

d. Yes — the method will always work.

5.6 Dividing in a given ratio

1. a. 3 b. 5 c. 4 d. 8 e. 13

2. a. 13 b. 13 c. 19 d. 6 e. 12

3. a. $400, $600 b. $200, $800 c. $500, $500

4. a. $300, $700 b. $350, $650 c. $450, $550

5. a. $5000, $5000 b. $4000, $6000 c. $6000, $4000

6. a. $3000, $7000 b. $7000, $3000 c. $4600, $5400

7. a. $10\,000, $40\,000 b. $15\,000, $35\,000
   c. $20\,000, $30\,000 d. $25\,000, $25\,000
   e. $12\,500, $37\,500

8. a. $2\,\text{m}^3$ b. $0.5\,\text{m}^3$ c. $1.6\,\text{m}^3$
   d. $1.6\,\text{m}^3$ e. $1.2\,\text{m}^3$

9. a. $90\,000, $210\,000, $300\,000
   b. $180\,000, $180\,000, $240\,000
   c. $30\,000, $240\,000, $330\,000
   d. $150\,000, $180\,000, $270\,000
   e. $150\,000, $225\,000, $225\,000

10. $25, $20, $15

11. a. 90 b. 75

12. a. 120 b. 45 c. $6 d. $19.50

13. D

14. D

15. A

16. D

17. $30°$, $60°$, $90°$. Sum the ratio parts $(1 + 2 + 3 = 6)$, divide the angle sum of a triangle by the total number of ratio parts to calculate the value of 1 ratio part $(180 \div 6 = 30°)$, then multiply this by the number of ratio parts.

18. $96°$

19. The mistake made when determining the prize amounts was that the total prize money was simply divided by the 3 numbers in the ratio. The correct amounts should be:
First prize $= \$12\,000$
Second prize $= \$5000$
Third prize $= \$3000$

20. a. 50 b. 4 c. $4 : 1$

21. a. $4 : 1200$ b. $1200\,\text{cm}$

22. a. $\dfrac{1}{8}$ of the total number of fruit

   b. 37

   c. 12

   d. 49

5.7 Rates

1. a. $10\,\text{m}^2/\text{min}$ b. $5\,\text{kL}/\text{min}$
   c. $300\,\text{cm}^3/\text{s}$ d. $\$1.38/\text{L}$
   e. $8\,\text{L}/100\,\text{km}$ or $12.5\,\text{km}/\text{L}$

2. a. $\$2.50/\text{m}$
   b. 40 cows/hectare or $(250\,\text{m}^2/\text{cow})$
   c. $\$12.50/\text{person}$
   d. $\$3.20/\text{m}^2$
   e. 25 c/min

3. a. 16 points/game   b. $\$5.40/\text{kg}$
   c. $8\dfrac{1}{3}\,\text{m/s}$ d. 5.2 runs/over
   e. $\$26.50/\text{h}$

4. a. $\$3.50/\text{kg}$ b. 52 words/min   c. $2\,°\text{C/h}$
   d. 6.5 cm/year   e. 16 km/h

5. $\$108.50$

6. 396 points

7. $60.5\,\text{L}$

8. 21

9. 70 min

10. a. 0.2 km/min   b. 0.4 L/min
    c. $\$0.32/\text{min}$ d. 0.45 cents/g

11. a. 27.78 m/s   b. 666.67 mL/min

12. a. 5.50 cents/cm   b. 36.00 km/h

13. a. 22.22 m/s   b. 90.00 km/h
    c. 50.00 kg/L   d. 21.60 L/h

14. a. $\dfrac{\text{distance}}{\text{time}}$ b. $\dfrac{\text{volume}}{\text{time}}$ c. $\dfrac{\text{capacity}}{\text{distance}}$
    d. $\dfrac{\text{money}}{\text{time}}$ e. $\dfrac{\text{money}}{\text{length}}$

15. a. $\dfrac{\text{capacity}}{\text{time}}$ b. $\dfrac{\text{money}}{\text{capacity}}$ c. $\dfrac{\text{money}}{\text{number}}$
    d. $\dfrac{\text{mass}}{\text{time}}$ e. $\dfrac{\text{number}}{\text{area}}$

16. a. L/min   b. m/s   c. cm/year   d. cm/h

17. a. mm^2/sec b. L/km   c. Runs/ball
    d. Words/min

18. $360\,\text{L}$

19. Packs of 10

20. Car A

21. 250-g jar; 75c

22. C

23. C

24. D

25. B

26. USD $\$350$

27. 125.58 miles

28. $\$420$

29. $33\,000\,\text{m}^2$

30. $2580\,\text{m}^3$

31. A ratio compares two quantities measured in the same unit, whereas a rate compares two quantities measured in different units.

32. a. 4 patients per hour
    b. 15 min
    c. $\$62.50$ per patient

33. 4 min

34. $1\dfrac{7}{8}$ hours or 1 hour 52 minutes 30 seconds

35. The 2005 race

36. To work: 30 mins
    From work: $1\dfrac{1}{3}$ hours (1 hour and 20 minutes)
    Total travel time: $1\dfrac{5}{6}$ hours (1 hour and 50 minutes)

37. 2750 organisms/litre. The beach should not be closed.

38. a.i. $V = 1.098\,07 \times 10^{21}\,\text{m}^3$
    ii. $5464.15\,\text{kg/m}^3$
    b.i. $V = 1.563\,46 \times 10^{24}\,\text{m}^3$
    ii. $1215.26\,\text{kg/m}^3$
    c. Earth and Jupiter do not seem to be made of the same substance, because Jupiter's density is much lower than that of Earth.

5.8 Interpreting graphs

1. The vertical axis is distance and the horizontal axis is time.

2. m/s

3. Horizontal line stretching for 60 minutes (with time measured horizontally)

4. a. 1   b. $\dfrac{2}{3}$ c. $\dfrac{2}{5}$ d. $\dfrac{3}{14}$

5. a. (triangle, sides 1 and 2)   b. (triangle, sides 1 and 3)   c. (triangle, sides 3 and 4)
   d. (triangle, sides 2 and 3)   e. (triangle, sides 5 and 2)

6. C

7. D

8. B

9. a. II   b. III   c. I

10. a. False   b. True   c. False
    d. True   e. False

11. Personal response required. Example:
    a.

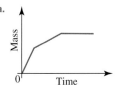

The mass of a pig will steadily increase over time.

b.

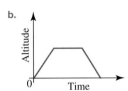

The altitude of a plane will steadily rise until cruising height, then stay at that height before descending back to ground.

c.

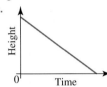

As a bathtub is emptied, the height of the water will steadily decrease.

12. A steeper gradient represents a faster speed, whereas a shallow gradient represents a slower speed.

13. Horizontal line segments represent when and where an object is not moving.

14. A: 60 km/h
    B: 70 km/h
    C: 46.7 km/h

15. a. Min (time) and $ (cost)
    b. Change in cost
    c. Change in time
    d. $m = \$1.00$ per min, $m = 50c$ per min, $m = 0c$ (i.e. free calls)
    e. The gradient represents the increase in cost for each increase in time.

16. a. Lamb: $m = \$15/kg$
       Chicken: $m = \$10/kg$
       Beef: $m = \$7.50/kg$
    b. Lamb: $15
       Chicken: $10
       Beef: $7.50
    c. $5
    d. $15

17. Sample responses can be found in the worked solutions in the online resources.
    Students should understand that the distance travelled at the end of each leg of the journey should not change, but the time taken to reach the end of each leg should be half the original time taken.
    The amount of time taken for a rest is not affected and, as such, the overall amount of time taken for the whole journey will not be exactly half the original time.

18.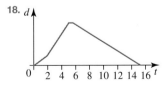

Project

$$\frac{\sqrt{5}+1}{2} = 1.618$$

Part A: Sample responses can be found in the worked solutions in the online resources.

Part B:

1. 1, 1, 2, 3, 5, 8, 13, 21, 34, 55, 89, 144, 233, 377, 610, 987, 1597, 2584, 4181, 6765

2.

Sequence	Ratio	Decimal value
1	–	–
1	$\frac{1}{1}$	1
2	$\frac{2}{1}$	2
3	$\frac{3}{2}$	1.5
5	$\frac{5}{3}$	1.67
8	$\frac{8}{5}$	1.6
13	$\frac{13}{8}$	1.625
21	$\frac{21}{13}$	1.615 38
34	$\frac{34}{21}$	1.619 05
55	$\frac{55}{34}$	1.617 65
89	$\frac{89}{55}$	1.618 18
144	$\frac{144}{89}$	1.617 98
233	$\frac{233}{144}$	1.618 06
377	$\frac{377}{233}$	1.618 03
610	$\frac{610}{377}$	1.618 04
987	$\frac{987}{610}$	1.618 03
1597	$\frac{1597}{987}$	1.618 03
2584	$\frac{2584}{1597}$	1.618 03
4181	$\frac{4181}{2584}$	1.618 03
6765	$\frac{6765}{4181}$	1.618 03

5.9 Review questions

1. a. $3 : 5$ b. $1 : 17$ c. $17 : 3$
   d. $5 : 1$ e. $5 : 21$

2. a. $1 : 2$ b. $2 : 3$ c. $7 : 20$
   d. $10 : 3$ e. $2 : 9$

3. a. $1 : 4$ b. $2 : 5$ c. $7 : 10$
   d. $3 : 1$ e. $15 : 4$

4. a. $n = 4$ b. $n = 20$ c. $n = 12$

5. a. $n = 15$ b. $n = 9.6$ c. $n = 1.2$

6. a. $1 : 6$ b. $1.5\,\text{L}$

7. a. $\dfrac{4}{5}$ b. $\dfrac{5}{8}$

8. a. $10, $15   b. $420, $300

9. a. Slide A: $\dfrac{2}{3}$, slide B: $\dfrac{2}{5}$
   b. Slide A

10. $880

11. $8.57\,\text{L}/100\,\text{km}$

12. a. David's   b. $432\,\text{km}$ c. $42\,\text{L}$

13. 1-kg packet

14. $96\,\text{cm}$

15. 32

16. $32\,\text{km/h}$

17. $28\,064.5\,\text{km/h}$

18. a. $13.5\,\text{km/min}$ b. $808.3\,\text{km/h}$

19. $4.5\,\text{kg}$

20. a. $45\,\text{km}$ b. $15\,\text{km/h}$

21. $20\,\text{cm}$

22. $68.61\,\text{km/h}$

23. $1\,\text{h}\,6\,\text{min}\,40\,\text{s}$

24. a. 0.6; 0.4; 0.8
   b. The third segment
   c. The average gradient is 0.56.

6 Algebraic techniques

LESSON
6.1 Overview

Why learn this?

Algebra is a fundamental building block of mathematics, used to create many of the things we use every day. Without algebra there would be no television, no smartphones and no internet — it would not be possible to have anything electrical at all. In its simplest form, algebra involves solving problems and working out unknown values. It is a systematic way of expressing and solving equations and is used to make problems easier. Studies of the Babylonians show that algebra has been used for over 4000 years.

Imagine you have $50 to spend at a bookstore and you select a book that costs $20. How much do you have left to spend on something else? Or you have a room that is 5 metres long in which you need to fit 10 chairs, all of which are 70 cm wide. Is it possible? You may not realise it, but solving problems such as these involves using algebra.

Algebra is also used in many fields, such as medicine, engineering, science, architecture and economics. If you wish to use geometry to build structures, or modelling to study financial markets or create new groundbreaking technology, algebra will be at the heart of your work.

1. **MC** Today Esther is y years old. Select all possible expressions that represent Esther's age in four years' time.

   A. $4y$ B. $4 + y$ C. $y - 4$ D. $y + 4$

2. Calculate the values of the following expressions if $t = 10$.

   a. $t + 7$ b. $-2t + 5$ c. t^2

3. Simplify the following expressions.

   a. $2a + 5a$ b. $4a - a$ c. $3ab + 2ba$

4. The area of a triangle is calculated using the formula $A = \dfrac{1}{2}bh$.

   a. Calculate the area of a triangle with $b = 20$ cm and $h = 6$ cm.
   b. Calculate the area of a triangle with $b = 7.6$ cm and $h = 2.4$ cm, correct to 2 decimal places.

5. State whether the following statement is True or False. If a number pattern has the formula $4n + 1$, where n is an integer, any value in that pattern will be odd.

6. If a has a value of 6 and b has a value of -3, calculate the values of the following expressions.

   a. $3a + 2b$ b. $a - b$

   c. $\dfrac{a}{b}$ d. $a + b$

7. **MC** Identify which one of the following operations follows the Commutative Law.

   A. Addition   B. Subtraction   C. Multiplication   D. Division

8. State whether the following statement is True or False.
   If $x = -5$, $y = -3$ and $z = -2$, then $x - y + z = x - (y - z)$.

9. Simplify the following expressions.

   a. $5a - 2b - 2a - 3b$ b. $7f + 3 - 2f - 1$ c. $5ab - 2ab + a$

10. Simplify the following expressions.

    a. $4x \times 2x \times 3y$ b. $\dfrac{45pq}{20p}$

11. Expand the following and simplify (if possible).

    a. $5(m + 10)$ b. $3m(2m - 5p)$ c. $3m(m + 5t) + 2t(3m - t)$

12. **MC** From the list of terms, select the highest common factor in all three of these expressions: $8mn$, $6m^2$ and $4mnp$.

    A. $24m^2np$ B. $2m$ C. $4m$ D. $2mn$

13. Factorise the following expressions.

    a. $4mn + 16$
    b. $3p^2q - 12pq^2$

14. Simplify the following as far as possible, but do not expand the brackets.

$$\frac{4t^2}{5(t-1)^3} \div \frac{2t}{3(t-1)}$$

15. If the first number in a series of consecutive odd numbers is written as n, write the simplest expression for the mean of the first five terms.

LESSON
6.2 Using pronumerals

LEARNING INTENTION

At the end of this lesson you should be able to:
- understand the meaning of and use the words *pronumeral*, *term*, *expression*, *coefficient* and *constant term*
- write simple algebraic expressions.

⏵ 6.2.1 The language of algebra

eles-3906

- Like English, French or a computer language, algebra is a type of language.
- In algebra there are important words that we need to be familiar with. They are *pronumeral, term, expression, coefficient* and *constant term*.

Pronumerals

- A **pronumeral** is a letter or symbol that is used in place of a number.
- Pronumerals are used to write general expressions or formulas and will allow us to determine their value when the values of the pronumerals become known.

  For example, the area of a triangle can be written as $A = \frac{1}{2}bh$ and contains the pronumerals b and h. Once we know the actual length of the base, b, and the height, h, of a triangle, we can determine its area using the formula.
- A pronumeral may also be described as a **variable**.

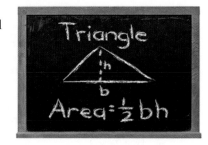

Terms, expressions, coefficients and constant terms

- A **term** is a group of letters and numbers that form an expression and are separated by an addition or subtraction sign.
- A **coefficient** is the number part of the term.
- A **constant term** is a number with no pronumeral attached to it.
- An **expression** is a mathematical statement made up of terms, operation symbols and/or brackets.

  For example, $4xy + 5x - 6$ is an expression made up of three terms:

  $4xy$, $5x$ and -6.

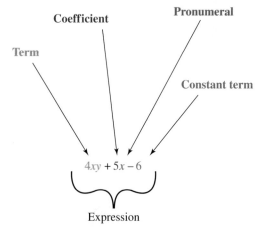

Algebra basics

- When we write expressions with pronumerals, the multiplication sign is omitted. For example, $8n$ means '$8 \times n$' and $\frac{1}{2}bh$ means '$\frac{1}{2} \times b \times h$'.

- The division sign is rarely used. For example, $y \div 6$ is usually written as $\frac{y}{6}$.

- When reading scenarios related to mathematics:
  - the words 'sum' or 'more than' refer to addition
  - the words 'difference' or 'less than' refer to subtraction
  - the words 'product' or 'times' refer to multiplication
  - the word 'quotient' refers to division
  - the word 'square' refers to the power of 2 (a number multiplied by itself).

WORKED EXAMPLE 1 Writing algebraic expressions from words

Suppose we use b to represent the number of ants in a nest.
a. **Write an expression for the number of ants in the nest if 25 ants died.**
b. **Write an expression for the number of ants in the nest if the original ant population doubled.**
c. **Write an expression for the number of ants in the nest if the original population increased by 50.**
d. **What would it mean if we said that a nearby nest contained $b + 100$ ants?**
e. **What would it mean if we said that another nest contained $b - 1000$ ants?**
f. **Another nest in very poor soil contains $\frac{b}{2}$ ants. How much smaller than the original is this nest?**

THINK	WRITE
a. The original number of ants (b) must be reduced by 25.	a. $b - 25$
b. The original number of ants (b) must be multiplied by 2. It is not necessary to show the $\times$ sign.	b. $2b$
c. 50 must be added to the original number of ants (b).	c. $b + 50$
d. This expression tells us that the nearby nest has 100 more ants.	d. The nearby nest has 100 more ants.
e. This expression tells us that the nest has 1000 fewer ants.	e. This nest has 1000 fewer ants.
f. The expression $\frac{b}{2}$ means $b \div 2$, so this nest is half the size of the original nest.	f. This nest is half the size of the original nest.

COMMUNICATING — COLLABORATIVE TASK: Algebraic symbols in other contexts

Use the internet to research where algebraic symbols are used in other contexts; for example, when representing cells in a spreadsheet. As a class, compile the responses into a list and analyse the different contexts in which variables appear.

  Resources

Interactivity Using variables (int-3762)

Exercise 6.2 Using pronumerals

6.2 Quick quiz **on**	6.2 Exercise

Individual pathways

■ PRACTISE	■ CONSOLIDATE	■ MASTER
1, 4, 7, 10, 12, 15	2, 5, 8, 11, 13, 16	3, 6, 9, 14, 17

Fluency

1. **WE1** Suppose we use x to represent the original number of ants in a nest.

   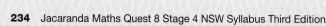

   a. Write an expression for the number of ants in the nest if 420 ants were born.
   b. Write an expression for the number of ants in the nest if the original ant population tripled.
   c. Write an expression for the number of ants in the nest if the original ant population decreased by 130.
   d. State what it would mean if we said that a nearby nest contained $x + 60$ ants.
   e. State what it would mean if we said that a nearby nest contained $x - 90$ ants.
   f. Another nest in very poor soil contains $\dfrac{x}{4}$ ants. Determine how much smaller this nest is than the original.

2. Suppose x people are in attendance at the start of an AFL match.

   a. If a further y people arrive during the first quarter, write an expression for the number of people at the ground.
   b. Write an expression for the number of people at the ground if a further 260 people arrive prior to the second quarter commencing.
   c. At half-time, 170 people leave. Write an expression for the number of people at the ground after they have left.
   d. In the final quarter, a further 350 people leave. Write an expression for the number of people at the ground after they have left.

3. Imagine that your cutlery drawer contains a knives, b forks and c spoons.

   a. Write an expression for the total number of knives and forks you have.
   b. Write an expression for the total number of items in the drawer.
   c. You put 4 more forks in the drawer. Write an expression for the number of forks that are in the drawer now.
   d. Write an expression for the number of knives in the drawer after 6 knives are removed.

4. If y represents a certain number, write expressions for the following numbers.

   a. A number 7 more than y b. A number 8 less than y
   c. A number that is equal to five times y

5. If y represents a certain number, write expressions for the following numbers.

   a. The number formed when y is subtracted from 14
   b. The number formed when y is divided by 3
   c. The number formed when y is multiplied by 8 and 3 is added to the result

6. Using a and b to represent numbers, write expressions for:

   a. the sum of a and b b. the difference between a and b
   c. three times a subtracted from two times b d. the product of a and b.

7. Using a and b to represent numbers, write expressions for:
   a. twice the product of a and b
   b. the sum of $3a$ and $7b$
   c. a multiplied by itself
   d. a multiplied by itself and the result divided by 5.

8. If tickets to a basketball match cost $27 for adults and $14 for children, write an expression for the cost of:

   a. y adult tickets
   b. d child tickets
   c. r adult and h child tickets.

Understanding

9. The canteen manager at Browning Industries orders m Danish pastries each day. Write a paragraph that could explain the table below.

Time	Number of Danish pastries available for sale
9:00 am	m
9:15 am	$m - 1$
10:45 am	$m - 12$
12:30 pm	$m - 12$
1:00 pm	$m - 30$
5:30 pm	$m - 30$

10. Naomi is now t years old.

    a. Write an expression for her age in 2 years' time.
    b. Write an expression for Steve's age if he is g years older than Naomi.
    c. Calculate Naomi's age 5 years ago.
    d. Naomi's father is twice her age. Write an expression for his age.

11. James is travelling by train into town one particular evening and observes that there are t passengers in his carriage. He continues to take note of the number of people in his carriage each time the train departs from a station, which occurs every 3 minutes.
    The table below shows the number of passengers.

Time (pm)	Number of passengers
7:10	t
7:13	$2t$
7:16	$2t + 12$
7:19	$4t + 12$
7:22	$4t + 7$
7:25	t
7:28	$t + 1$
7:31	$t - 8$
7:34	$t - 12$

a. Write a paragraph explaining what happened.
b. Determine when passengers first began to leave the train.
c. Determine the time at which the carriage had the most passengers.
d. Determine the time at which the carriage had the fewest passengers.

Communicating, reasoning and problem solving

12. List some reasons for using variables instead of numbers.

13. A microbiologist places *m* bacteria onto an agar plate.

She counts the number of bacteria at approximately 3-hour intervals from the starting time. The results are shown in the table below.

Time	Number of bacteria
9:00 am	m
Noon	$2m$
3:18 pm	$4m$
6:20 pm	$8m$
9:05 pm	$16m$
Midnight	$32m - 1240$

a. Explain what happens to the number of bacteria during the first 4 intervals.
b. Determine the possible cause of this bacterial increase.
c. Explain what is different about the last bacteria count.
d. Explain the cause of the difference in the bacteria count in part c.

14. *n* represents an even number.

a. Is the number $n + 1$ odd or even? Explain your answer.
b. Is $3n$ odd or even? Justify your answer.
c. Write an expression for the next three even numbers that are greater than *n*.
d. Write an expression for the even number that is 2 less than *n*.

15. Determine the 5 consecutive numbers that add to 120.

16. a. If the side of a square tile box is *x* cm long and the height is *h* cm, write expressions for the total surface area and the volume of the tile box.
    b. If a rectangular tile box has the same width and height as the square tile box in part a but is one and a half times as long, write expressions for the total surface area and the volume of the tile box.
    c. If the square tile box in part a has a side length of 20 cm and both boxes in parts a and b have a height of 15 cm, evaluate the surface area and volume of the square tile box and the surface area and volume of the rectangular tile box using your expressions.

17. Bill was describing his age to a group of people using two different algebraic expressions. Four times a certain number minus one and three times a certain number plus three both reveal Bill's age. Determine the number used and Bill's age.

LESSON
6.3 Substitution

▶ 6.3.1 Substitution of values

eles-3907

* If the value of a variable (or variables) is known, it is possible to **evaluate** (work out the value of) an expression by using **substitution**. The variable is replaced with its value.
* Substitution can also be used with a formula or rule.

WORKED EXAMPLE 2 Substituting values into expressions

Evaluate the following expressions if $a = 3$ and $b = 15$.

a. $6a$
b. $7a - \dfrac{2b}{3}$

THINK	WRITE
a. 1. Substitute the correct value for the variable (a) and insert the multiplication sign.	a. $6a = 6 \times 3$
2. Evaluate and write the answer.	$= 18$
b. 1. Substitute the correct values for each variable and insert the multiplication signs.	b. $7a - \dfrac{2b}{3} = 7 \times 3 - \dfrac{2 \times 15}{3}$
2. Perform the first multiplication.	$= 21 - \dfrac{2 \times 15}{3}$
3. Perform the second multiplication.	$= 21 - \dfrac{30}{3}$
4. Perform the division.	$= 21 - 10$
5. Perform the subtraction and write the answer.	$= 11$

WORKED EXAMPLE 3 Substituting values into an equation

The formula for calculating the area (A) of a rectangle of length l and width w is $A = l \times w$. Use this formula to determine the area of the following rectangle.

270 m
32 m

THINK	WRITE
1. Write the formula.	$A = l \times w$
2. Substitute the value for each variable.	$= 270 \times 32$
3. Perform the multiplication and state the correct units.	$= 8640 \, \text{m}^2$

▶ 6.3.2 Working with brackets

eles-3908

- Brackets are grouping symbols. The expression $3(a+5)$ can be thought of as 'three groups of $(a+5)$', or $(a+5)+(a+5)+(a+5)$.
- When substituting into an expression with brackets, remember to place a multiplication sign ($\times$) next to the brackets. For example, $3(a+5)$ is thought of as $3\times(a+5)$.

Evaluating expressions with brackets

Following the operation order, evaluate the brackets first and then multiply by the number outside the brackets.

WORKED EXAMPLE 4 Substituting into expressions containing brackets

a. **Substitute $r=4$ and $s=5$ into the expression $5(s+r)$ and evaluate.**
b. **Substitute $t=4$, $x=3$ and $y=5$ into the expression $2x(3t-y)$ and evaluate.**

THINK	WRITE
a. 1. Place the multiplication sign back into the expression.	a. $5(s+r)=5\times(s+r)$
2. Substitute the correct values for the variables.	$=5\times(5+4)$
3. Evaluate the expression in the pair of brackets first.	$=5\times9$
4. Perform the multiplication and write the answer.	$=45$
b. 1. Place the multiplication signs back into the expression.	b. $2x(3t-y)=2\times x\times(3\times t-y)$
2. Substitute the correct values for the variables.	$=2\times3\times(3\times4-5)$
3. Perform the multiplication inside the pair of brackets.	$=2\times3\times(12-5)$
4. Perform the subtraction inside the pair of brackets.	$=2\times3\times7$
5. Perform the multiplication and write the answer.	$=42$

---

on Resources

▶ **Video eLesson** Substitution (eles-1892)

Interactivities Substitution (int-3763)
Working with brackets (int-3764)

Exercise 6.3 Substitution

6.3 Quick quiz on	6.3 Exercise

Individual pathways

■ PRACTISE	■ CONSOLIDATE	■ MASTER
1, 3, 5, 7, 9, 15, 18, 21, 24	2, 6, 10, 11, 13, 16, 17, 22, 25	4, 8, 12, 14, 19, 20, 23, 26

Fluency

1. **WE2** Solve the following expressions, if $a = 2$ and $b = 5$.

   a. $6b$
   b. $\dfrac{a}{2}$
   c. $a + 7$
   d. $a + b$

2. Evaluate the following expressions, if $a = 1$ and $b = 3$.

   a. $b - a$
   b. $5 + \dfrac{b}{3}$
   c. $3a + 9$
   d. $2a + 3b$

3. Solve the following expressions, if $a = 4$ and $b = 10$.

   a. $\dfrac{8}{a}$
   b. $\dfrac{25}{b}$
   c. ab
   d. $5b - 30$

4. Evaluate the following expressions, if $a = 3$ and $b = 5$.

   a. $6b - 4a$
   b. $\dfrac{ab}{5}$
   c. $\dfrac{21}{a} + \dfrac{5}{b}$
   d. $\dfrac{9}{a} - \dfrac{5}{b}$

5. **WE4** Substitute $r = 5$ and $s = 7$ into the following expressions and solve.

   a. $3(r + s)$
   b. $2(s - r)$
   c. $7(r + s)$
   d. $9(s - r)$

6. Substitute $r = 2$ and $s = 3$ into the following expressions and evaluate.

   a. $s(r + 3)$
   b. $s(2r - 4)$
   c. $3r(r + 1)$
   d. $rs(3 + s)$

7. Substitute $r = 1$ and $s = 2$ into the following expressions and solve.

   a. $11r(s - 1)$
   b. $2r(s - r)$
   c. $s(4 + 3r)$
   d. $7s(4r - 2)$

8. Substitute $r = 5$ and $s = 7$ into the following expressions and evaluate.

   a. $s(3rs + 7)$
   b. $5r(24 - 2s)$
   c. $5sr(sr + 3s)$
   d. $8r(12 - s)$

9. Solve the following expressions, if $d = 5$ and $m = 2$.

   a. $-3d$
   b. $-2m$
   c. $6m + 5d$
   d. $\dfrac{3md}{2}$

10. Evaluate the following expressions, if $d = 10$ and $m = 4$.

    a. $25m - 2d$
    b. $\dfrac{7d}{15}$
    c. $4dm - 21$
    d. $\dfrac{15}{d} - \dfrac{m}{4}$

Understanding

11. Substitute $x = 6$ and $y = 3$ into the following expressions and solve.

    a. $3.2x + 1.7y$
    b. $11y - 2x$
    c. $\dfrac{13y}{3} - 2x$
    d. $\dfrac{4xy}{15}$

12. Substitute $x = 4$ and $y = 1$ into the following expressions and evaluate.

    a. $4.8x - 3.5y$
    b. $8.7y - x$
    c. $12.3x - 9.6x$
    d. $\dfrac{3x}{9} - \dfrac{y}{12}$

13. Solve each of the expressions below, if $x = 2$, $y = 4$ and $z = 7$.

a. $\left(7 - \dfrac{12}{x}\right)4y$ b. $\dfrac{6}{x}(xz + y - 3)$ c. $(y + 2)\dfrac{z}{x}$ d. $2x(xyz - 35)$

14. Evaluate each of the expressions below, if $x = 3$, $y = 5$ and $z = 9$.

a. $12(y - 1)(z + 3)$ b. $(3x - 7)\left(\dfrac{27}{x} + 7\right)$ c. $-2(4x + 1)\left(\dfrac{36}{z} - 3\right)$ d. $-3(2y - 11)\left(\dfrac{z}{x} + 8\right)$

15. **WE3** The area (A) of a rectangle of length l and width w can be found using the formula $A = lw$. Calculate the area of the following rectangles.

a. Length 12 cm, width 4 cm b. Length 200 m, width 42 m c. Length 4.3 m, width 104 cm

16. The formula $c = 0.1a + 42$ is used to calculate the cost in dollars (c) of renting a car for 1 day from Poole's Car Hire Ltd, where a is the number of kilometres travelled on that day.
Determine the cost of renting a car for 1 day if the distance travelled is 220 kilometres.

17. The formula for the perimeter (P) of a rectangle of length l and width w is $P = 2l + 2w$. This rule can also be written as $P = 2(l + w)$.
Use the rule to determine the perimeter of rectangular comic covers with the following measurements.

a. $l = 20$ cm, $w = 11$ cm b. $l = 27.5$ cm, $w = 21.4$ cm

18. The formula for the perimeter (P) of a square of side length l is $P = 4l$. Use this formula to calculate the perimeter of a square of side length 2.5 cm.

19. The formula $F = \dfrac{9}{5}C + 32$ is used to convert temperatures measured in degrees Celsius to an approximate Fahrenheit value.

F represents the temperature in degrees Fahrenheit (°F) and C the temperature in degrees Celsius (°C).

a. Calculate the value of F when $C = 100$ °C.
b. Convert 28 °C to Fahrenheit.
c. Water freezes at 0 °C. Calculate the freezing temperature of water in Fahrenheit.

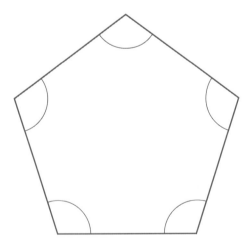

20. A rule for calculating the sum of the interior angles in a many-sided figure such as a pentagon is $S = 180(n - 2)°$, where S represents the sum of the angles inside the figure and n represents the number of sides.
The diagram shows the interior angles in a pentagon.
Use the rule to determine the sum of the interior angles for:
a. a hexagon (6 sides)
b. a pentagon
c. a triangle
d. a quadrilateral (4 sides)
e. a 20-sided figure.

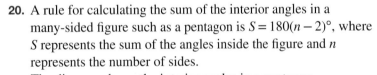

Communicating, reasoning and problem solving

21. Ben says that $\dfrac{4x^2}{2x} = 2x$. Emma says that that is not correct if $x = 0$. Explain Emma's reasoning.

22. It can be shown that $(x - a)(x - a) = x^2 - 2ax + a^2$. By substitution, show that this is true if:

    a. $x = 4$, $a = 1$ b. $x = 3p$, $a = 2p$ c. $x = 0$.

23. The width of a cuboid is x cm.
    a. If the length is 5 cm more than the width and the height is 2 cm less than the width, determine the volume, V cm^3, of the cuboid in terms of x.
    b. Evaluate V if x equals 10.
    c. Explain why x cannot equal 1.5.

24. The dimensions of the figure are given in terms of m and n. Write, in terms of m and n, an expression for:

    a. the length of CD
    b. the length of BC
    c. the perimeter of the figure.
    Show all of your workings.

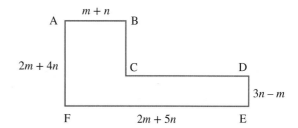

25. On the space battleship RAN *Fantasie*, there are p Pletons, each with 2 legs, $(p-50)$ Argors, each with 3 legs, and $(2p + 35)$ Kleptors, each with 4 legs.

    a. Determine the total number of legs, L, on board the *Fantasie*, in terms of p, in simplified form.
    b. If $p = 200$, evaluate L.

26. Answer the following question.

    a. Determine an expression for the area of a triangle whose base length is $(m + n)$ cm and whose height is $(m - n)$ cm.
    b. If $m = 15$ and $n = 6$, evaluate the area of the triangle.
    c. Show that $m > n$.
    d. Determine what happens to the triangle as m and n move closer in value.

LESSON
6.4 Substituting positive and negative numbers

LEARNING INTENTION

At the end of this lesson you should be able to:
- evaluate algebraic expressions that involve positive and negative numbers.

▶ 6.4.1 Substituting integer numbers

eles-3909

- If the variable you are substituting for has a negative value, simply remember the following rules for directed numbers.

Subtraction of positive and negative integers

- **To subtract a positive integer, move to the left. This is the same as adding a negative integer.**

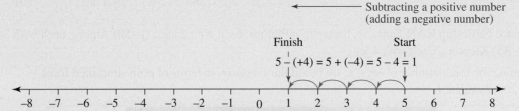

- **To subtract a negative integer, move to the right. This is the same as adding a positive integer.**

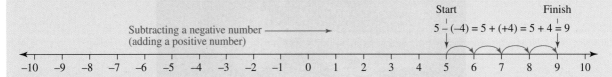

Multiplication and division of positive and negative integers

- **When multiplying or dividing two integers with the *same sign*, the answer is *positive*.**
  For example, $(-10) \times (-2) = +20$.

- **When multiplying or dividing two integers with *different signs*, the answer is *negative*.**
  For example, $\dfrac{+30}{-15} = -2$.

WORKED EXAMPLE 5 Substituting positive and negative numbers

a. Substitute $m = 5$ and $n = -3$ into the expression $m - n$ and evaluate.
b. Substitute $a = -2$ and $b = -1$ into the expression $2a + b$ and evaluate.
c. Substitute $x = -4$ and $y = -3$ into the expression $5y(6 + x)$ and evaluate.

THINK	WRITE
a. 1. Substitute the correct values for the variables.	a. $m - n = 5 - (-3)$
2. Combine the two **negative signs** and **add**.	$= 5 + 3$
3. Write the answer.	$= 8$

---

b. 1. Insert the multiplication sign between 2 and a.

b. $2a + b = 2 \times a + b$

2. Substitute the correct values for the variables and multiply.

$= 2 \times (-2) + (-1)$

3. Combine the different signs.

$= -4 + (-1)$

4. Perform the subtraction.

$= -4 - 1$

5. Write the answer.

$= -5$

c. 1. Insert multiplication signs between 5, y and $(6 + x)$.

c. $5y(6 + x) = 5 \times y \times (6 + x)$

2. Substitute the correct values for the variables.

$= 5 \times (-3) \times (6 + (-4))$

3. Perform the multiplication outside the brackets.

$= -15 \times (6 + (-4))$

4. Inside the set of brackets, combine the different signs and subtract.

$= -15 \times (6 - 4)$

5. Follow BIDMAS and evaluate the brackets first.

$= -15 \times 2$

6. Perform the multiplication and write the answer.

$= -30$

---

on Resources

💠 **Interactivity** Substituting positive and negative numbers (int-3765)

Exercise 6.4 Substituting positive and negative numbers **learn on**

6.4 Quick quiz **on**	6.4 Exercise

Individual pathways

■ PRACTISE	■ CONSOLIDATE	■ MASTER
1, 3, 5, 6, 10, 13, 16	2, 8, 11, 14, 17	4, 7, 9, 12, 15, 18

Fluency

1. **WE5a** Substitute $m = 6$ and $n = -3$ into the following expressions and solve.

a. $m + n$ **b.** $m - n$ **c.** $n - m$ **d.** $n + m$

2. Substitute $m = 3$ and $n = 2$ into the following expressions and evaluate.

a. $2n - m$ **b.** $n + 5$ **c.** $2m + n - 4$ **d.** $-5n - m$

3. Substitute $m = -4$ and $n = -2$ into the following expressions and solve.

a. $\dfrac{mn}{8}$ **b.** $\dfrac{4m}{n - 6}$ **c.** $\dfrac{4m}{n}$ **d.** $\dfrac{12}{2n}$

4. Substitute $m = 6$ and $n = -3$ into the following expressions and evaluate.

a. $\dfrac{9}{n} + \dfrac{m}{2}$ **b.** $6mn - 1$ **c.** $-\dfrac{3n}{2} + 1.5$ **d.** $14 - \dfrac{mn}{9}$

5. **WE5b** Substitute $a = -4$ and $b = -5$ into the following expressions and solve.

a. $a + b$ **b.** $a - b$ **c.** $b - 2a$ **d.** $2ab$

6. Substitute $a = 4$ and $b = 5$ into the following expressions and evaluate.

   a. $-2(b - a)$ b. $a - b - 4$ c. $3a(b + 4)$ d. $\dfrac{4}{b}$

7. Substitute $a = -4$ and $b = 5$ into the following expressions and evaluate.

   a. $\dfrac{16}{4a}$ b. $\dfrac{6b}{5}$ c. $(a - 5)(8 - b)$ d. $(9 - a)(b - 3)$

8. **WE5c** Substitute $x = -8$ and $y = -3$ into the following expressions and solve.

   a. $3(x - 2)$ b. $x(7 + y)$ c. $5y(x - 7)$ d. $xy(7 - x)$

9. Substitute $x = 8$ and $y = -3$ into the following expressions and evaluate.

   a. $(3 + x)(5 + y)$ b. $\dfrac{x}{2}(5 - y)$ c. $\left(\dfrac{x}{4} - 1\right)\left(\dfrac{2y}{6} + 4\right)$ d. $3(x - 1)\left(\dfrac{y}{3} + 2\right)$

Understanding

10. If $q = 4$ and $r = -5$, solve $\dfrac{2q(r - 3)}{2r + 2}$.

11. If $a = 17$ and $b = 13$, evaluate $\dfrac{3a}{b - 1}(b - a)$.

12. If $p = -2$ and $q = -3$, evaluate $\dfrac{3(-pq - p^2)}{q + 2p}$.

Communicating, reasoning and problem solving

13. a. Complete the following table by substituting the values of x into x^2.

x	-4	-3	-2	-1	0	1	2	3	4
x^2									

   b. Explain what you notice about the values of x^2 for values of x with the same magnitude but opposite sign.

14. Consider the expression $1 - 5x$. If x is a negative integer, explain why the expression will have a positive value.

15. Consider the equation $(a - b)(a + b) = a^2 - b^2$.

   a. By substituting $a = -3$ and $b = -2$, show that this is true.
   b. By substituting $a = -q$ and $b = -2q$, show that this is true.

16. Explain what can be said about the sign of x^2.

17. If $x = -2r$ is substituted into $\dfrac{(r - x)(x + r)}{(r - 2x)}$, determine whether the answer be positive or negative if:

   a. $r > 0$ b. $r < 0$.
   Show your working.

18. A circle is cut out of a square.

   a. If the side length of the square is x, the radius of the circle is $0.25x$ and the area of the circle is approximately $0.196x^2$, determine an expression for the remaining area.
   b. Evaluate the area when $x = 2$ by substituting into your expression. Give your answer to 3 decimal places.
   c. Determine the largest radius the circle can have.

LESSON
6.5 Number laws

LEARNING INTENTION

At the end of this lesson you should be able to:
- understand the Commutative, Associative, Identity and Inverse laws
- apply the Commutative, Associative, Identity and Inverse laws to algebraic expressions.

▶ 6.5.1 Commutative Law

eles-3910

- When dealing with any expression involving numbers, we must obey particular rules.
- The **Commutative Law** refers to the order in which two numbers are added or multiplied.
- The Commutative Law holds true for addition and multiplication but not for subtraction or division.

The Commutative Law for addition

- **The Commutative Law for addition states that:**
$$x + y = y + x$$
For example, $3 + 2 = 5$ and $2 + 3 = 5$.
- **This law does not hold true for subtraction.**
For example, $3 - 2 \neq 2 - 3$.

The Commutative Law for multiplication

- **The Commutative Law for multiplication states that:**
$$x \times y = y \times x$$
For example, $3 \times 2 = 6$ and $2 \times 3 = 6$.
- **This law does not hold true for division.**
For example, $3 \div 2 \neq 2 \div 3$.

WORKED EXAMPLE 6 Testing the Commutative Law

Evaluate the following expressions if $x = 4$ and $y = 7$. Comment on the results obtained.

a. i. $x + y$ b. i. $x - y$ c. i. $x \times y$ d. i. $x \div y$

  ii. $y + x$ ii. $y - x$ ii. $y \times x$ ii. $y \div x$

THINK	WRITE
a. i. 1. Substitute the correct value for each variable.	a. i. $x + y = 4 + 7$
2. Evaluate and write the answer.	$= 11$
ii. 1. Substitute the correct value for each variable.	ii. $y + x = 7 + 4$
2. Evaluate and write the answer.	$= 11$
3. Compare the result with the answer you obtained in part **a i**.	The same result is obtained; therefore, order is not important when adding two terms.
b. i. 1. Substitute the correct value for each variable.	b. i. $x - y = 4 - 7$
2. Evaluate and write the answer.	$= -3$

▶

ii. 1. Substitute the correct value for each variable. **ii.** $y - x = 7 - 4$

   2. Evaluate and write the answer. $= 3$

   3. Compare the result with the answer you obtained in part **b i**. Two different results are obtained; therefore, order is important when subtracting two terms.

c. i. 1. Substitute the correct value for each variable. **c. i.** $x \times y = 4 \times 7$

   2. Evaluate and write the answer. $= 28$

ii. 1. Substitute the correct value for each variable. **ii.** $y \times x = 7 \times 4$

   2. Evaluate and write the answer. $= 28$

   3. Compare the result with the answer you obtained in part **c i**. The same result is obtained; therefore, order is not important when multiplying two terms.

d. i. 1. Substitute the correct value for each variable. **d. i.** $x \div y = 4 \div 7$

   2. Evaluate and write the answer. $= \dfrac{4}{7} \ (\approx 0.57)$

ii. 1. Substitute the correct value for each variable. **ii.** $y \div x = 7 \div 4$

   2. Evaluate and write the answer. $= \dfrac{7}{4} \ (1.75)$

   3. Compare the result with the answer you obtained in part **d i**. Two different results are obtained; therefore, order is important when dividing two terms.

▶ 6.5.2 Associative Law

eles-3911

- The **Associative Law** refers to the grouping of three numbers that are added, subtracted, multiplied or divided.
- It holds true for addition and multiplication but not for subtraction or division.

The Associative Law for addition

- **The Associative Law for addition states that:**
$$(x + y) + z = x + (y + z) = x + y + z$$
For example, $(3 + 4) + 5 = 7 + 5 = 12$ and $3 + (4 + 5) = 3 + 9 = 12$.
- **This law does not hold true for subtraction.**
For example, $(3 - 4) - 5 \neq 3 - (4 - 5)$.

The Associative Law for multiplication

- **The Associative Law for multiplication states that:**
$$(x \times y) \times z = x \times (y \times z) = x \times y \times z$$
For example, $(2 \times 3) \times 4 = 6 \times 4 = 24$ and $2 \times (3 \times 4) = 2 \times 12 = 24$.
- **This law does not hold true for division.**
For example, $(2 \div 3) \div 4 \neq 2 \div (3 \div 4)$.

Evaluate the following expressions if $x = 12, y = 6$ and $z = 2$. Comment on the results obtained.

a. i. $x + (y + z)$ b. i. $x - (y - z)$ c. i. $x \times (y \times z)$ d. i. $x \div (y \div z)$

  ii. $(x + y) + z$ ii. $(x - y) - z$ ii. $(x \times y) \times z$ ii. $(x \div y) \div z$

THINK

a.   i.   1. Substitute the correct value for each variable.

          2. Evaluate the expression in the pair of brackets.

          3. Perform the addition and write the answer.

    ii.   1. Substitute the correct value for each variable.

          2. Evaluate the expression in the pair of brackets.

          3. Perform the addition and write the answer.

          4. Compare the result with the answer you obtained in part **a i.**

b.   i.   1. Substitute the correct value for each variable.

          2. Evaluate the expression in the pair of brackets.

          3. Perform the subtraction and write the answer.

    ii.   1. Substitute the correct value for each variable.

          2. Evaluate the expression in the pair of brackets.

          3. Perform the subtraction and write the answer.

          4. Compare the result with the answer you obtained in part **b i.**

c.   i.   1. Substitute the correct value for each variable.

          2. Evaluate the expression in the pair of brackets.

          3. Perform the multiplication and write the answer.

    ii.   1. Substitute the correct value for each variable.

          2. Evaluate the expression in the pair of brackets.

          3. Perform the multiplication and write the answer.

          4. Compare the result with the answer you obtained in part **c i.**

d.   i.   1. Substitute the correct value for each variable.

          2. Evaluate the expression in the pair of brackets.

          3. Perform the division and write the answer.

WRITE

a.   i. $x + (y + z) = 12 + (6 + 2)$

                  $= 12 + 8$

                  $= 20$

    ii. $(x + y) + z = (12 + 6) + 2$

                  $= 18 + 2$

                  $= 20$

          The same result is obtained; therefore, order is not important when adding 3 terms.

b.   i. $x - (y - z) = 12 - (6 - 2)$

                  $= 12 - 4$

                  $= 8$

    ii. $(x - y) - z = (12 - 6) - 2$

                  $= 6 - 2$

                  $= 4$

          Two different results are obtained; therefore, order is important when subtracting 3 terms.

c.   i. $x \times (y \times z) = 12 \times (6 \times 2)$

                  $= 12 \times 12$

                  $= 144$

    ii. $(x \times y) \times z = (12 \times 6) \times 2$

                  $= 72 \times 2$

                  $= 144$

          The same result is obtained; therefore, order is not important when multiplying 3 terms.

d.   i. $x \div (y \div z) = 12 \div (6 \div 2)$

                  $= 12 \div 3$

                  $= 4$

ii.	1.	Substitute the correct value for each variable.	ii. $(x \div y) \div z = (12 \div 6) \div 2$
	2.	Evaluate the expression in the pair of brackets.	$= 2 \div 2$
	3.	Perform the division and write the answer.	$= 1$
	4.	Compare the result with the answer you obtained in part **d i**.	Two different results are obtained; therefore, order is important when dividing 3 terms.

Identity Law

- The **Identity Law for addition** states that when zero is added to any number, the original number remains unchanged.
  For example, $5 + 0 = 0 + 5 = 5$.
- The **Identity Law for multiplication** states that when any number is multiplied by 1, the original number remains unchanged.
  For example, $3 \times 1 = 1 \times 3 = 3$.
- Since variables take the place of numbers, the Identity Law applies to all variables.

$$x + 0 = 0 + x = x$$
$$x \times 1 = 1 \times x = x$$

Inverse Law

- The **Inverse Law for addition** states that when a number is added to its additive inverse (opposite sign), the result is 0.
  For example, $5 + (-5) = 0$.
- The **Inverse Law for multiplication** states that when a number is multiplied by its multiplicative inverse (reciprocal), the result is 1.
  For example, $3 \times \dfrac{1}{3} = 1$.
- Since variables take the place of numbers, the Inverse Law applies to all variables.

$$x + (-x) = -x + x$$
$$= 0$$
$$x \times \dfrac{1}{x} = \dfrac{1}{x} \times x$$
$$= 1$$

DISCUSSION

Do the Identity and Inverse laws also apply to subtraction and division? If they do, how would they work?

 Resources

Interactivities Commutative Law (int-3766)
Associative Law (int-3767)
Identity Law (int-3768)
Inverse Law (int-3769)

6.5 Quick quiz on	6.5 Exercise

Individual pathways

■ PRACTISE	■ CONSOLIDATE	■ MASTER
1, 4, 7, 10, 13, 16, 19, 21, 23, 26	2, 5, 8, 11, 14, 17, 20, 22, 24, 27	3, 6, 9, 12, 15, 18, 25, 28

Fluency

1. **WE6a, b** Calculate the values of the following expressions if $x = 3$ and $y = 8$. Comment on the results obtained.

   a. i. $x + y$
      ii. $y + x$

   b. i. $5x + 2y$
      ii. $2y + 5x$

   c. i. $x - y$
      ii. $y - x$

   d. i. $4x - 5y$
      ii. $5y - 4x$

2. **WE6c, d** Determine the values of the following expressions if $x = -2$ and $y = 5$. Comment on the results obtained.

   a. i. $x \times y$
      ii. $y \times x$

   b. i. $4x \times y$
      ii. $y \times 4x$

   c. i. $x \div y$
      ii. $y \div x$

   d. i. $6x \div 3y$
      ii. $3y \div 6x$

For questions **3–14**, indicate whether each is true or false for all values of the variables.

3. $a + 5b = 5b + a$

4. $6x - 2y = 2y - 6x$

5. $7c + 3d = -3d + 7c$

6. $5 \times 2x \times x = 10x^2$

7. $4x \times -y = -y \times 4x$

8. $4 \times 3x \times x = 12x \times x$

9. $\dfrac{5p}{3r} = \dfrac{3r}{5p}$

10. $-7i - 2j = 2j + 7i$

11. $-3y \div 4x = 4x \div -3y$

12. $-2c + 3d = 3d - 2c$

13. $\dfrac{0}{3s} = \dfrac{3s}{0}$

14. $15 \times -\dfrac{2x}{3} = \dfrac{2x}{3} \times -15$

15. **WE7a, b** Calculate the values of the following expressions if $x = 3$, $y = 8$ and $z = 2$. Comment on the results obtained.

   a. i. $x + (y + z)$
      ii. $(x + y) + z$

   b. i. $2x + (y + 5z)$
      ii. $(2x + y) + 5z$

   c. i. $6x + (2y + 3z)$
      ii. $(6x + 2y) + 3z$

16. Determine the values of the following expressions if $x = 3$, $y = 8$ and $z = 2$. Comment on the results obtained.

   a. i. $x - (y - z)$
      ii. $(x - y) - z$

   b. i. $3x - (8y - 6z)$
      ii. $(3x - 8y) - 6z$

17. **WE7c, d** Calculate the values of the following expressions if $x = 8$, $y = 4$ and $z = -2$. Comment on the results obtained.

a. i. $x \times (y \times z)$
   ii. $(x \times y) \times z$

b. i. $x \times (-3y \times 4z)$
   ii. $(x \times -3y) \times 4z$

c. i. $2x \times (3y \times 4z)$
   ii. $(2x \times 3y) \times 4z$

18. Determine the values of the following expressions if $x = 8$, $y = 4$ and $z = -2$. Comment on the results obtained.

a. i. $x \div (y \div z)$
   ii. $(x \div y) \div z$

b. i. $x \div (2y \div 3z)$
   ii. $(x \div 2y) \div 3z$

c. i. $-x \div (5y \div 2z)$
   ii. $(-x \div 5y) \div 2z$

Understanding

19. Indicate whether each of the following is true or false for all values of the variables except 0.

a. $a - 0 = 0$

b. $a \times 1\,000\,000 = 0$

c. $15t \times -\dfrac{1}{15t} = 1$

20. Indicate whether each of the following is true or false for all values of the variables except 0.

a. $3d \times \dfrac{1}{3d} = 1$

b. $\dfrac{8x}{9y} \div \dfrac{8x}{9y} = 1$

c. $\dfrac{11t}{0} = 0$

21. **MC** The value of the expression $x \times (-3y \times 4z)$ when $x = 4$, $y = 3$ and $z = -3$ is:

A. 108   B. -432 C. 432   D. 112

22. **MC** The value of the expression $(x - 8y) - 10z$ when $x = 6$, $y = 5$ and $z = -4$ is:

A. -74 B. 74   C. -6 D. 6

Communicating, reasoning and problem solving

23. The Commutative Law does not hold for subtraction. Discuss the results of $x - a$ and $a - x$.

24. Evaluate each of the following expressions for $x = -3$, $y = 2$ and $z = -1$.
    a. $2x - (3y + 2z)$
    b. $x \times (y - 2z)$

25. a. If $x = -1$, $y = -2$ and $z = -3$, determine the value of:
    i. $(-x - y) - z$
    ii. $-x - (y + z)$
    b. Comment on the answers with special reference to the Associative Law.

26. Complete the following sentence.
    The Commutative Law holds true for _____ and _____. It does not hold true for _____ and _____.

27. Evaluate the following expressions if $a = 2$, $b = -3$ and $c = -1$. Comment on the results obtained.
    a. i. $a \times b + c$
       ii. $c + b \times a$
    b. i. $a - b + c$
       ii. $b - a + c$
    c. i. $a \div c$
       ii. $c \div a$

28. Answer the following questions.
    a. Determine the additive inverse of $(3p - 4q)$.
    b. Determine the multiplicative inverse of $(3p - 4q)$.
    c. Evaluate the answers to parts **a** and **b** when $p = -1$ and $q = 3$.
    d. Determine what happens if you multiply a term or pronumeral by its multiplicative inverse.

LESSON
6.6 Adding and subtracting terms

LEARNING INTENTION

At the end of this lesson you should be able to:
- simplify algebraic expressions by combining like terms.

▶ 6.6.1 Simplification of algebraic expressions

eles-3914

- Expressions can often be written in a simpler form by collecting (adding or subtracting) like terms.
- **Like terms** are terms that contain exactly the same pronumerals, raised to the same power (coefficients don't need to match).

Like terms	Unlike terms
$3x$ and $4x$ are like terms.	$3x$ and $3y$ are unlike terms.
$3ab$ and $7ab$ are like terms.	$7ab$ and $8a$ are unlike terms.
$2bc$ and $4cb$ are like terms.	$8a$ and $3a^2$ are unlike terms.
$3g^2$ and $45g^2$ are like terms.	

- To understand why $2a + 3a$ *can* be added but $2a + 3ab$ *cannot* be added, consider the following identical bags of lollies, each containing a lollies.

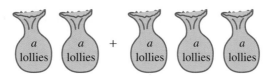

- As each bag contains the same number of lollies (a), we have 5 bags of a lollies. Therefore, $2a + 3a = 5a$.
- Then consider the following 2 bags containing a lollies and 3 bags containing $a \times b$ lollies.

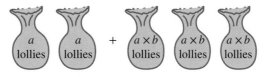

- Since we have bags containing different amounts of lollies, the expression cannot be simplified.
- Therefore, all we can say is that we have 2 bags containing a lollies and 3 bags containing ab lollies, $2a + 3ab = 2a + 3ab$.

Simplifying algebraic expressions

- **Like terms are terms that contain exactly the same pronumerals, raised to the same power (the coefficients don't need to match).**
- **Algebraic expressions can be simplified by combining like terms.**
  For example, $2a + 3a = 5a$.
- **Expressions can be rearranged to place like terms next to each other where necessary by ensuring that the signs of the terms remain the same.**
  For example, $4a + b - 3a = 4a - 3a + b = a + b$.
- **Unlike terms cannot be combined.**
  For example, $2a + 3ab$ cannot be simplified any further.

WORKED EXAMPLE 8 Simplifying algebraic expressions

Simplify the following expressions.
a. $3a + 5a$
b. $7ab - 3a - 4ab$
c. $2c - 6 + 4c + 15$

THINK	WRITE
a. 1. Write the expression and check that the two terms are like terms — that is, they contain exactly the same pronumerals.	a. $3a + 5a$
2. Add the like terms and write the answer.	$= 8a$
b. 1. Write the expression and check for like terms.	b. $7ab - 3a - 4ab$
2. Rearrange the terms so that the like terms are together. Remember to keep the correct sign in front of each term.	$= 7ab - 4ab - 3a$
3. Subtract the like terms and write the answer.	$= 3ab - 3a$
c. 1. Write the expression and check for like terms.	c. $2c - 6 + 4c + 15$
2. Rearrange the terms so that the like terms are together. Remember to keep the correct sign in front of each term.	$= 2c + 4c - 6 + 15$
3. Simplify by collecting like terms and write the answer.	$= 6c + 9$

DISCUSSION

How might you explain the important points to remember when showing someone else how to collect like terms in an algebraic expression?

 Resources

 Video eLesson Simplification of expressions (eles-1884)

 Interactivity Simplifying expressions (int-3771)

Exercise 6.6 Adding and subtracting terms

learn

6.6 Quick quiz on	6.6 Exercise

Individual pathways

■ PRACTISE	■ CONSOLIDATE	■ MASTER
1, 4, 7, 10, 13, 16, 19, 22	2, 5, 8, 11, 14, 17, 20, 23	3, 6, 9, 12, 15, 18, 21, 24

Fluency

1. **WE8** Simplify the following expressions.

   a. $4c + 2c$ b. $2c - 5c$ c. $3a + 5a - 4a$

2. Simplify the following expressions.

   a. $6q - 5q$ b. $-h - 2h$ c. $7x - 5x$

3. Simplify the following expressions.

   a. $3a - 7a - 2a$ b. $-3f + 7f$ c. $4p - 7p$

4. Simplify the following expressions.

   a. $-3h + 4h$ b. $11b + 2b + 5b$ c. $7t - 8t + 4t$

5. Simplify the following expressions.

   a. $5p + 3p + 2p$ b. $9g + 12g - 4g$ c. $18b - 4b - 11b$

6. Simplify the following expressions.

   a. $13t - 4t + 5t$ b. $-11j + 4j$ c. $-12l + 2l - 5l$

7. Simplify the following expressions.

   a. $3x + 7x - 2y$ b. $3x + 4x - 12$ c. $11 + 5f - 7f$

8. Simplify the following expressions.

   a. $3u - 4u + 6$ b. $2m + 3p + 5m$ c. $-3h + 4r - 2h$

9. Simplify the following expressions.

   a. $11a - 5b + 6a$ b. $9t - 7 + 5$ c. $12 - 3g + 5$

10. Simplify the following expressions.

   a. $2b - 6 - 4b + 18$ b. $11 - 12h + 9$ c. $12y - 3y - 7g + 5g - 6$

11. Simplify the following expressions.

   a. $8h - 6 + 3h - 2$ b. $11s - 6t + 4t - 7s$ c. $2m + 13l - 7m + l$

12. Simplify the following expressions.

   a. $3h + 4k - 16h - k + 7$ b. $13 + 5t - 9t - 8$ c. $2g + 5 + 5g - 7$

Understanding

13. Simplify the following expressions.

   a. $x^2 + 2x^2$ b. $3y^2 + 2y^2$ c. $a^3 + 3a^3$

14. Simplify the following expressions.

   a. $d^2 + 6d^2$ b. $7g^2 - 8g^2$ c. $3y^3 + 7y^3$

15. Simplify the following expressions.

   a. $a^2 + 4 + 3a^2 + 5$ b. $11x^2 - 6 + 12x^2 + 6$ c. $12s^2 - 3 + 7 - s^2$

16. Simplify the following expressions.

   a. $3a^2 + 2a + 5a^2 + 3a$ b. $11b - 3b^2 + 4b^2 + 12b$ c. $6t^2 - 6g - 5t^2 + 2g - 7$

17. Simplify the following expressions.

   a. $11g^3 + 17 - 4g^3 + 5 - g^2$ b. $12ab + 3 + 6ab$ c. $14xy + 3xy - xy - 5xy$

18. Simplify the following expressions.

   a. $4fg + 2s - fg + s$ b. $11ab + ab - 5$ c. $18ab^2 - 4ac + 2ab^2 - 10ac$

Communicating, reasoning and problem solving

19. Discuss what you need to remember when checking for like terms.

20. Three members of a fundraising committee are making books of tickets to sell for a raffle. Each book of tickets contains t tickets.

    If the first person has 14 books of tickets, the second person has 12 books of tickets and the third person has 13 books of tickets, write an expression for:

    a. the number of tickets that the first person has
    b. the number of tickets that the second person has
    c. the number of tickets that the third person has
    d. the total number of tickets for the raffle.

21. Explain, using mathematical reasoning and with diagrams if necessary, why the expression $2x + 2x^2$ cannot be simplified.

22. Write an expression for the total perimeter of the following shapes.

    a.
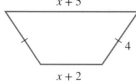

    b.
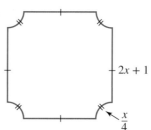

    c.
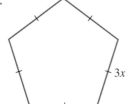

    d.
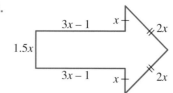

23. Rose owns an art gallery and sells items supplied to her by various artists. She receives a commission for all items sold.

Rose uses the following method to keep track of the money she owes the artists when their items are sold.

- Ask the artist how much they want for the item.
- Add 50% to that price, then mark the item for sale at this new price.
- When the item sells, take one-third of the sale price as commission, then return the balance to the artist.

Use algebra to show that this method does return the correct amount to the artist.

24. If 32 metres of rope are required to make the rectangular shape shown in the figure, evaluate x.

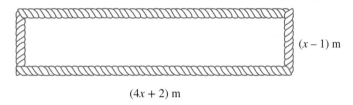

$(x - 1)$ m

$(4x + 2)$ m

LESSON
6.7 Multiplying and dividing terms

LEARNING INTENTION

At the end of this lesson you should be able to:
- simplify algebraic expressions by multiplying and dividing variables and constants.

▶ 6.7.1 Multiplying expressions with pronumerals

eles-3915

- When we multiply variables, the Commutative Law holds (as already stated), so order is not important.
  For example:

$$3 \times 6 = 6 \times 3$$
$$6 \times w = w \times 6$$
$$a \times b = b \times a$$

- The multiplication sign ($\times$) is usually omitted for reasons of convention.

$$3 \times g \times h = 3gh$$
$$2 \times x^2 \times y = 2x^2y$$

- Although order is not important, conventionally the pronumerals in each term are written in alphabetical order.
  For example:

$$2 \times b^2 \times a \times c = 2ab^2c$$

WORKED EXAMPLE 9 Multiplying algebraic expressions

Simplify the following.

a. $5 \times 4g$

b. $-3d \times 6ab \times 7$

THINK	WRITE
a. 1. Write the expression and insert the hidden multiplication signs.	a. $5 \times 4g = 5 \times 4 \times g$
2. Multiply the numbers.	$= 20 \times g$
3. Remove the multiplication sign.	$= 20g$
b. 1. Write the expression and insert the hidden multiplication signs.	b. $-3d \times 6ab \times 7$ $= -3 \times d \times 6 \times a \times b \times 7$
2. Place the numbers at the front.	$= -3 \times 6 \times 7 \times d \times a \times b$
3. Multiply the numbers.	$= -126 \times d \times a \times b$
4. Remove the multiplication signs and place the variables in alphabetical order to write the answer.	$= -126abd$

⊳ 6.7.2 Dividing expressions with pronumerals

eles-3916

- When dividing expressions with variables, rewrite the expression as a fraction and simplify by cancelling.
- Remember that when the same variable appears as a factor of both the numerator and denominator, it may be cancelled.

WORKED EXAMPLE 10 Dividing algebraic expressions

a. Simplify $\dfrac{16f}{4}$.

b. Simplify $15n \div (3n)$.

THINK	WRITE
a. 1. Write the expression.	a. $\dfrac{16f}{4}$
2. Simplify the fraction by cancelling 16 with 4 (divide both by 4).	$= \dfrac{\overset{4}{\cancel{16}}f}{\underset{1}{\cancel{4}}}$ $= \dfrac{4f}{1}$
3. Write the answer.	$= 4f$
b. 1. Write the expression and then rewrite it as a fraction.	b. $15n \div (3n) = \dfrac{15n}{3n}$
2. Simplify the fraction by cancelling 15 with 3 and n with n.	$= \dfrac{\overset{5}{\cancel{15}}\cancel{n}}{\underset{1}{\cancel{3}}\cancel{n}}$ $= \dfrac{5}{1}$
3. Write the answer.	$= 5$

WORKED EXAMPLE 11 Dividing algebraic expressions

Simplify $-12xy \div (27y)$.

THINK	WRITE
1. Write the expression and then rewrite it as a fraction.	$-12xy \div (27y) = -\dfrac{12xy}{27y}$
2. Simplify the fraction by cancelling 12 with 27 (divide both by 3) and y with y.	$= -\dfrac{\overset{4}{\cancel{12}}x\cancel{y}}{\underset{9}{\cancel{27}}\,\cancel{y}}$
3. Write the answer.	$= -\dfrac{4x}{9}$

▶ 6.7.3 Simplifying expressions with indices

eles-3917

- Write the expression in an expanded form.
- If the algebraic term is a fraction, cancel any common factors.
- Place the coefficients in front, as the orders of terms are not important when multiplying.
- Multiply and divide the coefficients and the pronumerals.

WORKED EXAMPLE 12 Simplifying algebraic expression containing indices

Simplify the following.

a. $3m^3 \times 2m$

b. $5p^{10} \times 3p^3$

c. $36x^7 \div (12x^4)$

d. $\dfrac{6y^3 4y^8}{12y^4}$

THINK	WRITE
a. 1. Write the expression.	a. $3m^3 \times 2m$
2. The order is not important when multiplying, so place the numbers first.	$= 3 \times 2 \times m^3 \times m$
3. Multiply the numbers.	$= 6 \times m^3 \times m$
4. Multiply the pronumerals.	$= 6 \times m^{3+1}$
5. Write the answer.	$= 6m^4$
b. 1. Write the expression.	b. $5p^{10} \times 3p^3$
2. The order is not important when multiplying, so place the numbers first.	$= 5 \times 3 \times p^{10} \times p^3$
3. Multiply the numbers.	$= 15 \times p^{10} \times p^3$
4. Multiply the pronumerals.	$= 15 \times p^{10+3}$
5. Write the answer.	$= 15p^{13}$

c. 1. Write the expression and show it as a fraction.

c. $36x^7 \div \left(12x^4\right)$

$$= \frac{36x^7}{12x^4}$$

2. Divide the numbers.

$$= \frac{3x^7}{x^4}$$

3. Divide the pronumerals and write the answer.

$$= 3x^{7-4}$$

$$= 3x^3$$

d. 1. Write the expression.

d. $\dfrac{6y^3 \times 4y^8}{12y^4}$

2. Perform the multiplication in the numerator.

$$= \frac{24y^{11}}{12y^4}$$

3. Divide the numbers.

$$= \frac{2y^{11}}{y^4}$$

4. Divide the pronumerals and write the answer.

$$= 2y^{11-4}$$

$$= 2y^7$$

▶ 6.7.4 The order of operations

eles-3918

- When working with algebra, the order of operations (BIDMAS) still applies.
- When dealing with fractions, remember that the numerator and denominator should be evaluated first before simplifying the fraction.

WORKED EXAMPLE 13 Simplifying algebraic expressions using BIDMAS

Simplify $\dfrac{16g \times (8g + 3g)}{5g - g}$.

THINK

WRITE

1. Evaluate the expression inside the brackets.

$$\frac{16g \times (8g + 3g)}{5g - g} = \frac{16g \times 11g}{5g - g}$$

2. Carry out the multiplication on the numerator.

$$= \frac{176g^2}{5g - g}$$

3. Carry out the subtraction on the denominator.

$$= \frac{176g^2}{4g}$$

4. Divide the numbers.

$$= \frac{44g^2}{g}$$

5. Divide the pronumerals and write the answer.

$$= 44g^{2-1}$$

$$= 44g$$

Resources

 Interactivities Dividing expressions with variables (int-3773)
Multiplying variables (int-3772)

Exercise 6.7 Multiplying and dividing terms

learn

6.7 Quick quiz on	6.7 Exercise

Individual pathways

■ PRACTISE	■ CONSOLIDATE	■ MASTER
1, 4, 7, 10, 13, 17, 19, 22, 25, 28	2, 5, 8, 11, 14, 16, 20, 23, 26, 29	3, 6, 9, 12, 15, 18, 21, 24, 27, 30

Fluency

1. **WE9** Simplify the following.
   - a. $4 \times 3g$
   - b. $7 \times 3h$
   - c. $4d \times 6$
   - d. $3z \times 5$

2. Simplify the following.
   - a. $5t \times 7$
   - b. $4 \times 3u$
   - c. $7 \times 6p$
   - d. $7gy \times 3$

3. Simplify the following.
   - a. $4x \times 6g$
   - b. $10a \times 7h$
   - c. $9m \times 4d$
   - d. $3c \times 5h$

4. Simplify the following.
   - a. $2 \times 8w \times 3x$
   - b. $11ab \times 3d \times 7$
   - c. $16xy \times 1.5$
   - d. $3.5x \times 3y$

5. **WE10** Simplify the following.
   - a. $\dfrac{8f}{2}$
   - b. $\dfrac{6h}{3}$
   - c. $\dfrac{15x}{3}$
   - d. $9g \div 3$

6. Simplify the following.

   a. $10r \div 5$ **b.** $4x \div (2x)$ **c.** $8r \div (4r)$ **d.** $\dfrac{16m}{8m}$

7. Simplify the following.

   a. $14q \div (21q)$ **b.** $\dfrac{3x}{6x}$ **c.** $\dfrac{12h}{14h}$ **d.** $50g \div (75g)$

8. Simplify the following.

   a. $27h \div (3h)$ **b.** $\dfrac{20d}{48d}$ **c.** $\dfrac{64q}{44q}$ **d.** $81l \div (27l)$

9. Simplify the following.

   a. $\dfrac{15fg}{3}$ **b.** $12cd \div 4$ **c.** $\dfrac{8xy}{12}$ **d.** $24cg \div 24$

10. Simplify the following.

    a. $\dfrac{132mnp}{60np}$ **b.** $\dfrac{11ad}{66ad}$ **c.** $18adg \div (45ag)$ **d.** $\dfrac{bh}{7h}$

11. Simplify the following.

    a. $3 \times -5f$ **b.** $-6 \times -2d$ **c.** $11a \times -3g$

12. Simplify the following.

    a. $5h \times 8j \times -k$ **b.** $75x \times 1.5y$ **c.** $12rt \times -3z \times 4p$

13. **WE11** Simplify the following.

    a. $\dfrac{-4a}{8}$ **b.** $\dfrac{-11ab}{33b}$ **c.** $60jk \div (-5k)$ **d.** $-3h \div (-6dh)$

14. Simplify the following.

    a. $\dfrac{-32g}{40gl}$ **b.** $-12xy \div (48y)$ **c.** $\dfrac{12ab}{-14ab}$ **d.** $\dfrac{6fgh}{30ghj}$

15. Simplify the following.

    a. $34ab \div (-17ab)$ **b.** $\dfrac{-2ab}{-3a}$ **c.** $\dfrac{-7dg}{35gh}$ **d.** $-60mn \div (55mnp)$

16. Simplify the following.

    a. $\dfrac{28def}{18d}$ **b.** $-72xyz \div (28yz)$ **c.** $\dfrac{54pq}{36pqr}$ **d.** $-\dfrac{121oc}{132oct}$

Understanding

17. **WE12** Simplify the following.

    a. $2a \times a$ **b.** $-5p \times -5p$ **c.** $-5 \times 3x \times 2x$ **d.** $ab \times 7a$

18. Simplify the following.

    a. $-5xy \times 4 \times 8x$ **b.** $7pq \times 3p \times 2q$ **c.** $5m \times n \times 6nt \times -t$ **d.** $-3 \times xyz \times -3z \times -2y$

19. Simplify the following.

    a. $2mn \times -3 \times 2n \times 0$ **b.** $w^2x \times -9z^2 \times 2xy^2$ **c.** $2a^4 \times 3a^7$ **d.** $2x^2y^3 \times x^3$

20. Simplify the following.

a. $20m^{12} \div (2m^3)$

b. $\dfrac{25p^{12} \times 4q^7}{15p^2 \times 8q^2}$

c. $\dfrac{8x^3 \times 7y^2 \times 2z^2}{6x \times 14y}$

d. $\dfrac{a \times ab \times 3b^2}{5a^2b^2}$

21. Simplify the following.

a. $\dfrac{3}{a} \times \dfrac{2}{a}$

b. $\dfrac{5b}{2} \times \dfrac{4b}{3}$

c. $w \times \dfrac{5}{w^2}$

d. $\dfrac{3rk}{2s} \times \dfrac{6st}{5rt}$

22. Simplify the following.

a. $-\dfrac{4ht}{3dk} \times \dfrac{-12hk}{9dt}$

b. $\dfrac{5t}{gn} \div \dfrac{1}{g}$

c. $\dfrac{-9th}{4g} \div \dfrac{tg}{6h}$

d. $\dfrac{4xy}{7wz} \div \dfrac{x}{14z}$

23. **WE13** Simplify the following.

a. $2x^2 + 3x \times 4x$

b. $4p^3 \times (5p^3 - 3p^3)$

c. $\dfrac{5s \times (11s - 5s)}{6s - 3s}$

24. Simplify the following.

a. $\dfrac{5x^3 \times (4y - 2y)}{x^2y}$

b. $\dfrac{(2s^2 + 6s^2) \times (6t - 3t)}{6}$

c. $\dfrac{(3a - 2a) \times (5bc + 2bc)}{abc}$

Communicating, reasoning and problem solving

25. Explain how multiplication and division of expressions with variables are similar to multiplication and division of numbers.

26. Explain why $x \div (yz)$ is equivalent to $\dfrac{x}{yz}$ and $x \div (y \times z)$, but is not equivalent to $x \div y \times z$ or $\dfrac{x}{y} \times z$.

27. The student's working here shows incorrect cancelling. Explain why the working is not correct.

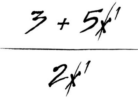

28. Evaluate $\dfrac{2x - (3y + 2z)}{-(-x - y) + z}$ if $x = 2$, $y = -1$ and $z = -4$.

29. Evaluate $\dfrac{x \times (y - 2z)}{3x - (y - z)}$ if $x = -2$, $y = 5$ and $z = -1$.

30. Two containers of lollies are placed one on top of the other.

a. Write an expression for the volume of each of the containers shown. (*Hint:* The volume of a container is found by multiplying the length by the width by the height.)

b. Determine how many times the contents of the smaller container will fit inside the larger container. (*Hint:* Divide the volumes.)

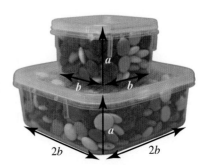

LESSON
6.8 Expanding brackets

LEARNING INTENTION

At the end of this lesson you should be able to:
- expand algebraic expressions containing brackets using the Distributive Law.

▶ 6.8.1 The Distributive Law

eles-3919

- The expression $3(a + b)$ means $3 \times (a + b)$ or $(a + b) + (a + b) + (a + b)$; when simplified, this becomes $3a + 3b$.
  That is, $3(a + b) = 3a + 3b$.
- Removing brackets from an expression is called **expanding** the expression.
- The rule that we have used to expand the expression above is called the **Distributive Law**.

The Distributive Law

The Distributive Law can be used to expand brackets and states that:

$$a(b + c) = a \times b + a \times c$$
$$= ab + ac$$

For example, $3(x + y) = 3 \times x + 3 \times y$
$$= 3x + 3y$$

- The Distributive Law can be demonstrated using the concept of rectangle areas.
- The expression $a(b + c)$ can be thought of as calculating the area of a rectangle with width a and length $b + c$, as shown below.

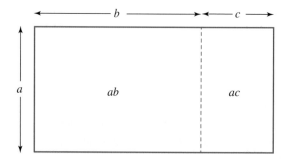

Therefore, area $= a(b + c) = a \times (b + c) = ab + ac$.
- An expression containing a bracket multiplied by a number (or pronumeral) can be written in expanded or factorised form.

Factorised form = expanded form
$$3(a + b) = 3a + 3b$$

- Expanding and factorising are the inverse of each other.

WORKED EXAMPLE 14 Expanding algebraic expressions

Use the Distributive Law to expand the following expressions.

a. $3(a + 2)$ b. $x(x - 5)$

THINK	WRITE
a. 1. Write the expression.	**a.** $3(a+2) = 3(a+2)$ 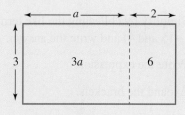
2. Use the Distributive Law to expand the brackets.	$= 3 \times a + 3 \times 2$
3. Perform the multiplication and write the answer.	$= 3a + 6$

ALTERNATIVE APPROACH

a. 1. Think of $3(a + 2)$ as calculating the area of a rectangle with width 3 and length $a + 2$. Draw a diagram to represent this situation and determine the areas of each smaller section.

2. Answer the problem by adding all the areas together.

$3(a + 2) = 3a + 6$

	WRITE
b. 1. Write the expression.	**b.** $x(x - 5) = x(x - 5)$
2. Use the Distributive Law to expand the brackets.	$= x \times x + x \times (-5)$
3. Perform the multiplication and write the answer.	$= x^2 - 5x$

ALTERNATIVE APPROACH

b. 1. Think of $x(x - 5)$ as calculating the area of a rectangle with length x and width $x - 5$. Draw a square with dimensions x and x and determine its area. To correctly represent $x - 5$ we need to take away 5 units from one of the sides of the square. This is shown by the shaded section outlined in pink. This is the piece we will need to *remove* from the original square.

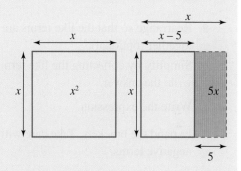

2. Determine the area of the original square and the area of the piece to be removed from the square.

Area of square $= x^2$
Area to be removed $= 5x$

3. Determine the area of the newly formed rectangle by subtracting the area of the shaded section from the areas of the larger square.

Area of new rectangle $= x^2 - 5x$

4. Write the answer.

$x(x - 5) = x^2 - 5x$

• Some expressions can be simplified further by collecting like terms after any brackets have been expanded.

Expand the expressions below and then simplify by collecting any like terms.
a. $3(x - 5) + 4$
b. $4(3x + 4) + 7x + 12$
c. $2x(3y + 3) + 3x(y + 1)$
d. $4x(2x - 1) - 3(2x - 1)$

THINK

WRITE

a. 1. Write the expression.

a. $3(x - 5) + 4$

    2. Expand the brackets.

$= 3 \times x + 3 \times (-5) + 4$
$= 3x - 15 + 4$

    3. Simplify by collecting the like terms (-15 and 4) and write the answer.

$= 3x - 11$

b. 1. Write the expression.

b. $4(3x + 4) + 7x + 12$

    2. Expand the brackets.

$= 4 \times 3x + 4 \times 4 + 7x + 12$
$= 12x + 16 + 7x + 12$

    3. Rearrange so that the like terms are together (optional).

$= 12x + 7x + 16 + 12$

    4. Simplify by collecting the like terms and write the answer.

$= 19x + 28$

c. 1. Write the expression.

c. $2x(3y + 3) + 3x(y + 1)$

    2. Expand the brackets.

$= 2x \times 3y + 2x \times 3 + 3x \times y + 3x \times 1$
$= 6xy + 6x + 3xy + 3x$

    3. Rearrange so that the like terms are together (optional).

$= 6xy + 3xy + 6x + 3x$

    4. Simplify by collecting the like terms and write the answer.

$= 9xy + 9x$

d. 1. Write the expression.

d. $4x(2x - 1) - 3(2x - 1)$

    2. Expand the brackets. Take care with negative terms.

$= 4x \times 2x + 4x \times (-1) - 3 \times 2x - 3 \times (-1)$
$= 8x^2 - 4x - 6x + 3$

    3. Simplify by collecting the like terms and write the answer.

$= 8x^2 - 10x + 3$

 Resources

▶ **Video eLesson** Expanding brackets (eles-1888)

🔧 **Interactivity** Expanding brackets: Distributive Law (int-3774)

Exercise 6.8 Expanding brackets

6.8 Quick quiz on	6.8 Exercise

Individual pathways

■ PRACTISE	■ CONSOLIDATE	■ MASTER
1, 4, 7, 9, 11, 14, 19, 22	2, 5, 8, 12, 15, 17, 20, 23	3, 6, 10, 13, 16, 18, 21, 24

Fluency

1. **WE14** Use the Distributive Law to expand the following expressions.
   a. $3(d+4)$
   b. $2(a+5)$
   c. $4(x+2)$

2. Use the Distributive Law to expand the following expressions.
   a. $5(r+7)$
   b. $6(g+6)$
   c. $2(t-3)$

3. Use the Distributive Law to expand the following expressions.
   a. $7(d+8)$
   b. $9(2x-6)$
   c. $12(4+c)$

4. Use the Distributive Law to expand the following expressions.
   a. $11(t-2)$
   b. $3(2t-6)$
   c. $t(t+3)$

5. Use the Distributive Law to expand the following expressions.
   a. $x(x+4)$
   b. $g(g+7)$
   c. $2g(g+5)$

6. Use the Distributive Law to expand the following expressions.
   a. $3(3x-2)$
   b. $3x(x-6y)$
   c. $5y(3x-9y)$

7. Use the Distributive Law to expand the following expressions.
   a. $50(2y-5)$
   b. $-3(c+3)$
   c. $-5(3x+4)$

8. Use the Distributive Law to expand the following expressions.
   a. $-4f(5-2f)$
   b. $9x(3y-2)$
   c. $-3h(2b-6h)$

9. Use the Distributive Law to expand the following expressions.
   a. $4a(5b+3c)$
   b. $-3a(2g-7a)$
   c. $5a(3b+6c)$

10. Use the Distributive Law to expand the following expressions.
    a. $-2w(9w-5z)$
    b. $12m(4m+10)$
    c. $-3k(-2k+5)$

Understanding

11. **WE15** Expand the expressions below and then simplify by collecting any like terms.
    a. $7(5x+4)+21$
    b. $3(c-2)+2$
    c. $2c(5-c)+12c$

12. Expand the expressions below and then simplify by collecting any like terms.
    a. $6(v+4)+6$
    b. $3d(d-4)+2d^2$
    c. $3y+4(2y+3)$

13. Expand the expressions below and then simplify by collecting any like terms.
    a. $24r+r(2+r)$
    b. $5-3g+6(2g-7)$
    c. $4(2f-3g)+3f-7$

14. Expand the expressions below and then simplify by collecting any like terms.
    a. $12+5(r-5)+3r$
    b. $12gh+3g(2h-9)+3g$
    c. $3(2t+8)+5t-23$

15. Expand the expressions below and then simplify by collecting any like terms.
    a. $4(d+7)-3(d+2)$
    b. $6(2h+1)+2(h-3)$
    c. $3(3m+2)+2(6m-5)$

16. Expand the expressions below and then simplify by collecting any like terms.

    a. $9(4f+3)-4(2f+7)$ **b.** $2a(a+2)-5(a^2+7)$ **c.** $3(2-t^2)+2t(t+1)$

17. Expand the expressions below and then simplify by collecting any like terms.

    a. $3h(2k+7)+4k(h+5)$ **b.** $6n(3y+7)-3n(8y+9)$ **c.** $4g(5m+6)-6(2gm+3)$

18. Expand the expressions below and then simplify by collecting any like terms.

    a. $7c(2f-3)+3c(8-f)$ **b.** $7x(4-y)+2xy-29$ **c.** $8m(7n-2)+3n(4+7m)$

Communicating, reasoning and problem solving

19. Using the concept of area as shown in this section, explain with diagrams and mathematical reasoning why $5(6+2)=5\times6+5\times2$.

20. Using the concept of area as shown in this section, explain with diagrams and mathematical reasoning why $4(x-y)=4\times x-4\times y$.

21. Discuss why the Distributive Law doesn't apply when there is a multiplication sign inside the brackets — that is, for $a(b\times c)$.

22. The price of a pair of jeans is $50. During a sale, the price of the jeans is discounted by d.

    a. Write an expression to represent the sale price of the jeans.
    b. If you buy three pairs of jeans during the sale, write an expression to represent the total purchase price:
       i. containing brackets
      ii. in expanded form (without brackets).
    c. Write an expression to represent the total change you would receive from $200 for the three pairs of jeans purchased during the sale.

23. A triptych is a piece of art that is divided into three sections or panels. The middle panel is usually the largest and it is flanked by two related panels.

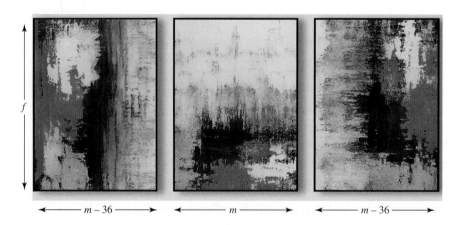

    a. Write a simplified expression for the area of each of the three paintings (excluding the frame).
    b. Write a simplified expression for the combined area of the triptych.
    c. The value of f is $m+102.5$. Substitute $(m+102.5)$ into your combined area formula and simplify the expression.
    d. The actual value of m is $122.5\,\text{cm}$. Sketch the shape of the three paintings in your workbook and show the actual measurements of each, including length, width and area.

24. Expressions of the form $(a + b)(c + d)$ can be expanded by using the Distributive Law twice. Distribute one of the factors over the other; for example, $\overbrace{(a + b)}(c + d)$.

The expression can then be fully expanded by applying the Distributive Law again. Fully expand the following expressions.

   a. $(x + 1)(x + 2)$ **b.** $(a + 3)(a + 4)$ **c.** $(c + 2)(c - 3)$

LESSON
6.9 Factorising

LEARNING INTENTION

At the end of this lesson you should be able to:
- factorise algebraic expressions.

▶ 6.9.1 Factorising algebraic expressions

eles-3920

- Factorising is the opposite process to expanding.
- To factorise a single algebraic term, the term must be broken up into all the factors of its individual parts. For example, $10xy = 2 \times 5 \times x \times y$ and $6abc = 2 \times 3 \times a \times b \times c$.
- Algebraic expressions could be split into several different factor combinations. For example, $10xy$ could also be written as $-2 \times -5 \times x \times y$ or $-10 \times -1 \times x \times y$.
- Generally, factorising more than one algebraic term involves identifying the highest common factors of the algebraic terms.
- To determine the highest common factor of the algebraic terms:
  1. determine the highest common factor of the number parts
  2. determine the highest common factor of the variable parts
  3. multiply these together.

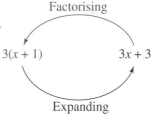

Factorising algebraic expressions

To factorise an expression, place the highest common factor of the terms outside the brackets and the remaining factors for each term inside the brackets.

For example, $3a + 6ab = 3a(1 + 2b)$.

WORKED EXAMPLE 16 Determining the HCF of algebraic terms

Determine the highest common factor of $6x$ and 10.

THINK	WRITE
1. Calculate the highest common factor of the number parts. Break 6 down into factors. Break 10 down into factors. The highest common factor is 2.	$6 = 3 \times ②$ $10 = 5 \times ②$ HCF = 2
2. Calculate the highest common factor of the variable parts. There isn't one, because only the first term has a variable part.	The HCF of $6x$ and 10 is 2.

WORKED EXAMPLE 17 Determining the HCF of algebraic terms

Determine the highest common factor of 14fg and 21gh.

THINK	WRITE
1. Determine the highest common factor of the number parts. Break 14 down into factors. Break 21 down into factors. The highest common factor is 7.	$14 = ⑦ \times 2$ $21 = ⑦ \times 3$ $HCF = 7$
2. Determine the highest common factor of the variable parts. Break fg down into factors. Break gh down into factors. Both contain a factor of g.	$fg = f \times ⑧$ $gh = ⑧ \times h$ $HCF = g$
3. Multiply the two factors together.	The HCF of 14fg and 21gh is 7g.

WORKED EXAMPLE 18 Factorising algebraic expressions

Factorise the expression $2x + 6$.

THINK	WRITE
1. Break down each term into its factors.	$2x + 6 = ② \times x + ② \times 3$
2. Write the highest common factor outside the brackets. Write the other factors inside the brackets.	$= 2 \times (x + 3)$
3. Remove the multiplication sign.	$= 2(x + 3)$
4. Check your answer by expanding.	CHECK $\overset{\times\ \times}{\overgroup{2(x + 3)}} = 2x + 6$

WORKED EXAMPLE 19 Factorising algebraic expressions

Factorise $12gh - 8g$.

THINK	WRITE
1. Break down each term into its factors.	$12gh - 8g$ $= ④ \times 3 \times ⑧ \times h - ④ \times 2 \times ⑧$
2. Write the highest common factor outside the brackets. Write the other factors inside the brackets.	$= 4 \times g \times (3 \times h - 2)$
3. Write the answer by removing the multiplication signs.	$= 4g(3h - 2)$
4. Check your answer by expanding.	CHECK $\overset{\times\ \times}{\overgroup{4g(3h - 2)}} = 12gh - 8g$

Factorise each of the following expressions.

a. $3a + 6b + 12c$

b. $4x + 6xy - 3xz$

c. $5m - 10m^2 - 15m^3$

THINK	WRITE
a. 1. Break down each term into its factors.	a. $3a + 6b + 12c = 3 \times a + 2 \times 3 \times b + 4 \times 3 \times c$
2. Write the highest common factor outside the brackets. Write the other factors inside the brackets.	$= 3 \times (a + 2 \times b + 4 \times c)$
3. Write the answer by removing the multiplication signs.	$= 3(a + 2b + 4c)$
4. Check your answer by expanding.	CHECK $3(a + 2b + 4c) = 3a + 6b + 12c$ $3(a + 2b + 4c) = 3a + 6b + 12c$
b. 1. Break down each term into its factors.	b. $4x + 6xy - 3xz = 4 \times x + 6 \times x \times y - 3 \times x \times z$
2. Write the highest common factor outside the brackets. Write the other factors inside the brackets.	$= x \times (4 + 6 \times y - 3 \times z)$
3. Write the answer by removing the multiplication signs.	$= x(4 + 6y - 3z)$
4. Check your answer by expanding.	CHECK $x(4 + 6y - 3z) = 4x + 6xy - 3xz$
c. 1. Break down each term into its factors.	c. $5m - 10m^2 - 15m^3 = 5 \times m - 5 \times 2 \times m \times m - 5 \times 3 \times m \times m \times m$
2. Write the highest common factor outside the brackets. Write the other factors inside the brackets.	$= 5 \times m \times (1 - 2 \times m - 3 \times m \times m)$
3. Write the answer by removing the multiplication signs.	$= 5m(1 - 2m - 3m^2)$
4. Check your answer by expanding.	CHECK $5m(1 - 2m - 3m^2) = 5m - 10m^2 - 15m^3$

on Resources

 Video eLesson Factorisation (eles-1887)

 Interactivity Factorising (int-3775)

Exercise 6.9 Factorising

6.9 Quick quiz	6.9 Exercise

Individual pathways

■ PRACTISE	■ CONSOLIDATE	■ MASTER
1, 4, 7, 12, 14, 16, 19, 22	2, 5, 8, 10, 13, 17, 20, 23	3, 6, 9, 11, 15, 18, 21, 24

Fluency

1. **WE16** Determine the highest common factor of the following.

   a. 4 and 6   b. 6 and 9   c. 12 and 18   d. 13 and 26

2. Determine the highest common factor of the following.

   a. 14 and 21   b. $2x$ and 4   c. $3x$ and 9   d. $12a$ and 16

3. **WE17** Determine the highest common factor of the following.

   a. $2gh$ and $6g$ b. $3mn$ and $6mp$ c. $11a$ and $22b$ d. $4ma$ and $6m$

4. Determine the highest common factor of the following.

   a. $12ab$ and $14ac$ b. $24fg$ and $36gh$ c. $20dg$ and $18ghq$ d. $11gl$ and $33lp$

5. Determine the highest common factor of the following.

   a. $16mnp$ and $20mn$ b. $28bc$ and $12c$ c. $4c$ and $12cd$ d. x and $3xz$

6. **WE18** Factorise the following expressions.

   a. $3x + 6$ b. $2y + 4$ c. $8x + 12$ d. $6f + 9$

7. Factorise the following expressions.

   a. $2d + 8$ b. $2x - 4$ c. $11h + 121$ d. $4s - 16$

8. Factorise the following expressions.

   a. $12g - 24$ b. $14 - 4b$ c. $48 - 12q$ d. $16 + 8f$

9. **WE19** Factorise the following. Check your answers by expanding the factorised expression.

   a. $3gh + 12$ b. $2xy + 6y$ c. $14g - 7gh$ d. $16jk - 2k$

10. Factorise the following. Check your answers by expanding the factorised expression.

    a. $7mn + 6m$ b. $5a - 15abc$ c. $8r + 14rt$ d. $4b - 6ab$

11. Factorise the following. Check your answers by expanding the factorised expression.

    a. $14x - 21xy$ b. $11jk + 3k$ c. $12ac - 4c + 3dc$ d. $4g + 8h - 16$

Understanding

12. Determine the highest common factor of $4ab$, $6a^2b^3$ and $12a^3b$.

13. Determine the highest common factor of $18e^2f$, $42efg$ and $30fg^2$.

14. **WE20a** Factorise each of the following expressions.

    a. $6p + 12pq + 18q$
    b. $32x + 8y + 16z$
    c. $16m - 4n + 24p$
    d. $72x - 8y + 64pq$

15. **WE20b** Factorise each of the following expressions.
    a. $xy + 9y - 3y^2$
    b. $5c + 3c^2d - cd$
    c. $3ab + a^2b + 4ab^2$
    d. $2x^2y + xy + 5xy^2$

16. **WE20c**
    a. $4x^2 + 2x + 8x^3$
    b. $8y^2 + 64y^4 - 32y^3$

17. Simplify $\dfrac{3x + 9}{12 - 15x}$.

18. Simplify $\dfrac{8}{x} \times \dfrac{15x^2 - 10x}{4 + 8x}$.

Communicating, reasoning and problem solving

19. Discuss some strategies that you can use to determine the highest common factor.

20. A farmer's paddock is a rectangle of length $(2x - 6)$ m and width $(3x + 6)$ m.
    a. Determine the area of the paddock in factorised form. Show your working.
    b. State the smallest possible value of x. Explain your reasoning.

21. Simplify $(5ax^2y - 6bxy + 2ax^2y - bxy) \div (ax^2 - bx)$. Show your working.

22. Factorise and hence simplify $\dfrac{4x - 4}{10x - 20} \times \dfrac{15x + 15}{3x - 3} \times \dfrac{6x - 12}{20x + 20}$.

23. Some expressions can be factorised by first grouping the terms, then factorising and removing the common factor. For example:

$$\begin{aligned} xy + 4y + 3x + 12 &= y(x + 4) + 3(x + 4) \\ &= (x + 4)(y + 3) \end{aligned}$$

Factorise the following by grouping first.
    a. $mn + 3m + 9n + 27$
    b. $-10xt + 30x - 2t + 6$

24. A cuboid measures $(3x - 6)$ cm by $(2x + 8)$ cm by $(ax - 5a)$ cm.
    a. Write an expression for its volume.
    b. If the cuboid weighs $(8x - 16)$ g, determine a factorised expression for its density in g/cm^3.

LESSON
6.10 Review

6.10.1 Topic summary

Algebraic expressions

- Algebraic expressions contain variables or pronumerals.
- A variable or pronumeral represents a value in an expression.
- When writing expressions, the multiplication and division signs are omitted.

  e.g. $5 \times a \times b = 5ab$

  $6 \div x = \dfrac{6}{x}$

Substitution

- Substitution is the process of replacing a variable with a value.
- The value of the expression can then be evaluated.

  e.g. $4(5 - x)$ when $x = -3$ becomes:

  $4(5 - (-3))$

  $= 4(5 + 3)$

  $= 4 \times 8$

  $= 32$

ALGEBRA

Like terms

- Like terms are terms with exactly the same pronumerals, raised to the same power.
- Only like terms can be added or subtracted.

  e.g. $2x$ and $-4x$ are like terms.

  $3ab$ and $5ba$ are like terms.

  $2x$ and $-x^2$ are **not** like terms.

Multiplying and dividing expressions

- When multiplying, multiply the numbers and write the pronumerals in alphabetical order. Remove the multiplication sign.

  e.g. $5ab \times 2a^2bc = 10a^3b^2c$

- When dividing, write the division as a fraction and simplify by cancelling.

  e.g. $12a^2b4c \div \left(4a^2b\right) = \dfrac{12a^2b^4c}{4a^2b} = 3b^3c$

- Don't forget BIDMAS.

Number laws

- The **Commutative Law** holds true for multiplication and addition and states that the order in which multiplication or addition happens does not matter.

  $$x \times y = y \times x \text{ and } x + y = y + x$$

- The **Associative Law** holds true for multiplication and addition and states that when adding or multiplying more than two terms, it does not matter how the terms are grouped together.

  $$(x \times y) \times z = x \times (y \times z) = (x \times z) \times y$$
  $$(x + y) + z = x + (y + z) = (x + z) + y$$

- The **Distributive Law** states that multiplying a number by a group of numbers added together is the same as doing each multiplication separately and then adding.

  $$a(b + c) = ab + ac$$

- The **Identity Law** states that adding 0 to a number or multiplying a number by 1 leaves the original number unchanged.

  $$x + 0 = 0 + x = x$$
  $$x \times 1 = 1 \times x = x$$

- The **Inverse Law** states that when a number is added to its additive inverse, the result is 0, and when a number is multiplied by its reciprocal, the result is 1.

  $$x + (-x) = x - x = 0$$
  $$x \times \dfrac{1}{x} = \dfrac{x}{x} = 1$$

Factorising

- Factorising is the opposite of expanding.
- First identify the highest common factor, then place this factor outside the brackets. The remaining factors of each term are inside the brackets.

  e.g. $20xy - 15x = 5x(4y - 3)$

6.10.2 Project

Readability index

Since you first learned how to read, you have probably read many books. These books would have ranged from picture books with simple words to books with short sentences. As you learned more words, you read short stories and more challenging books.

Have you ever picked up a book and put it down straight away because you thought there were too many 'difficult words' in it?

The reading difficulty of a text can be described by a readability index. There are several different methods used to calculate reading difficulty, and one of these methods is known as the Rix index.

The Rix index is obtained by dividing the number of long words by the number of sentences.

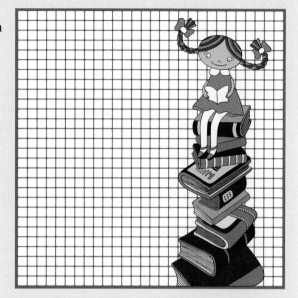

1. Use a variable to represent the number of long words and another to represent the number of sentences. Write a formula that can be used to calculate the Rix index.

When using the formula to determine the readability index, follow these guidelines:
- A long word is a word that contains seven or more letters.
- A sentence is a group of words that ends with a full stop, question mark, exclamation mark, colon or semicolon.
- Headings and numbers are not included and hyphenated words count as one word.

Consider this passage from a Science textbook.

> **A fatal fall . . . or was it murder?**
> In 1991, some German hikers found a body preserved in ice near the Italy–Austria border. Scientists used radiometric dating and found that the body was about 5300 years old! They thought that the person, known now as the Iceman, had died of hypothermia (extreme cold).
>
> Ten years later, another group of scientists using high-tech X-rays found the remains of an arrowhead lodged near his left lung. Specialists have not yet confirmed whether the Iceman fell back onto his arrow, or if he was murdered. And without any witnesses to question, the truth may never be known!

2. How many sentences and long words appear in this passage of text?
3. Use your formula to calculate the Rix index for this passage. Round your answer to 2 decimal places.

Once you have calculated the Rix index, the table shown can be used to work out the equivalent year level of the passage of text.

4. What year level is the passage of text equivalent to?

When testing the reading difficulty of a book, it is not necessary to consider the entire book.

Choose a section of text with at least 10 sentences and collect the required information for the formula.

5. Choose a passage of text from one of your school books, a magazine or newspaper. Calculate the Rix index and use the table to determine the equivalent year level.

6. Repeat question **5** using another section of the book, magazine or newspaper. Did the readability level change?

7. On a separate page, rewrite the passage from question **5** with minimal changes so that it is now suitable for a higher or lower year level. Explain the method you used to achieve this. Provide a Rix index calculation to prove that you changed the level of reading difficulty.

Rix index	Equivalent year level
Below 0.2	1
0.2–0.49	2
0.5–0.79	3
0.8–1.29	4
1.3–1.79	5
1.8–2.39	6
2.4–2.99	7
3.0–3.69	8
3.7–4.49	9
4.5–5.29	10
5.3–6.19	11
6.2–7.19	12
Above 7.2	University level

on Resources

 Interactivities Crossword (int-2630)
Sudoku puzzle (int-3188)

Exercise 6.10 Review questions

learn on

Fluency

1. Using x and y to represent numbers, write expressions for the following.
   a. The sum of x and y
   b. The difference between y and x
   c. Five times y subtracted from three times x
   d. The product of 5 and x

2. Using x and y to represent numbers, write expressions for the following.
   a. Twice the product of x and y
   b. The sum of $6x$ and $7y$
   c. y multiplied by itself
   d. $2x$ decreased by 7

3. If tickets to the school play cost $15 for adults and $9 for children, write expressions for the cost of:
   a. x adult tickets
   b. y child tickets
   c. k adult tickets and m child tickets.

4. Jake is now m years old.
   a. Write an expression for his age in 5 years' time.
   b. Write an expression for Jo's age if she is p years younger than Jake.
   c. Jake's mother is 5 times his age. Calculate her age.

5. Calculate the value of the following expressions if $a = 2$ and $b = 6$.

   a. $a + b$

   b. $b - a$

   c. $5 + \dfrac{b}{2}$

   d. $3a + 7$

6. Determine the value of the following expressions if $a = 2$ and $b = 6$.

   a. $2a + 3b$

   b. $\dfrac{20}{a}$

   c. $3b - 2a$

   d. $\dfrac{b}{a}$

Understanding

7. The formula $C = 2.2k + 4$ can be used to calculate the cost in dollars, C, of travelling by taxi for a distance of k kilometres. Calculate the cost of travelling 4.5 km by taxi.

8. The area (A) of a rectangle of length l and width w can be found using the formula $A = lw$.
   Calculate the width of a rectangle if $A = 65\ \text{cm}^2$ and $l = 13\ \text{cm}$.

9. Substitute $r = 3$ and $s = 5$ into the following expressions and evaluate.
   a. $5(r + s)$
   b. $8(s - r)$
   c. $s(2r - 3)$

10. Substitute $r = 3$ and $s = 5$ into the following expressions and evaluate.

    a. $rs(7 + s)$

    b. $r^2(5 - r)$

    c. $s^2(s + 15)$

11. Calculate the values of the following expressions if $a = 2$ and $b = -5$.
    a. $b(a - 4)$
    b. $12 - a(b - 3)$
    c. $5a + 6b$

12. State whether this equation is true or false for all values of the variables: $7x - 10y = 10y - 7x$.

13. State whether this equation is true or false for all values of the variables: $16 \times 2x \times x = 32x^2$.

14. State whether this equation is true or false for all values of the variables: $9x \times -y = -y \times 9x$.

15. State whether this equation is true or false for all values of the variables: $\dfrac{0}{5k} = \dfrac{5k}{0}$.

16. State whether this equation is true or false for all values of the variables: $21 \times \left(-\dfrac{7x}{3}\right) = \dfrac{7x}{3} \times (-21)$.

17. Simplify the following.
    a. $3 \times 7g$

    b. $6 \times 3y$

    c. $7d \times 6$

    d. $-3z \times 8$

18. Simplify the following.
    a. $-3gh \div (-6g)$

    b. $\dfrac{32t}{40stv}$

    c. $-36xy \div (-12y)$

    d. $\dfrac{5egh}{30ghj}$

▶

19. Simplify the following by collecting like terms.

a. $4x + 11 - 2x$ b. $2g + 5 - g - 6$ c. $2xy + 7xy$ d. $12t^2 + 3t + 3t^2 - t$

20. Expand the following and then simplify by collecting like terms.

a. $2(x + 5) + 5(x + 1)$ b. $2g(g - 6) + 3g(g - 7)$ c. $3(3t - 4) - 6(2t - 9)$

21. Factorise the following expressions.

a. $3g + 12$ b. $xy + 5y$ c. $5n - 20$

22. Factorise the following expressions.

a. $12mn + 4pn$ b. $12g - 6gh$ c. $24a - 6a^2 + 18ab$

Communicating, reasoning and problem solving

23. Using only $+, -, \times$ and $(\ )$, complete the following equations to demonstrate the Distributive Law.

a. 3   2   1 = 3   2   3   1

b. -10 8 -6 $-10 = -10$ 8 -6

c. 8   6   5 = 8   5   6   8

24. The base of a box has a length of $(x + 4)$ cm and a width of x cm.

a. Draw a diagram of the base, labelling the length and the width.

b. Write an expression for the area of the base of the box.

c. Expand part **b**.

d. If $x = 3$, determine the area of the base of the box.

e. If the height of the box is x cm, determine an expression for the volume of the box.

f. Determine the volume of the box if x is 3 cm.

25. Stephanie bought a skirt, a T-shirt and a pair of shorts during Target's annual sale. She spent $79.00. She paid $9 more for the T-shirt than for the shorts, and $7 more for the skirt than for the T-shirt. Calculate the cost of the skirt.

26. Aussie Rules Football is played in many Australian states. The scoring for the game is in goals (G) and behinds (B). Each goal (G) scores six points and each behind (B) scores one point.

To calculate the total number of points (P) scored by a team, use the rule $P = 6G + B$.

a. State the variables in the rule.

b. State the expression in the rule.

c. A team scored 11 goals and 10 behinds. Determine the total number of points the team scored.

d. A second team scored 9 goals and 18 behinds. Determine the total number of points the team scored.

e. Determine the number of goals and behinds a team might have scored if its total score was 87 points and it scored more than six goals.

27. Bobby the painter has two partially used 10-litre tins of paint, A and B.
There is more paint in Tin A than in Tin B.
He mixes the paint in the following fashion.
- He pours paint from Tin A into Tin B until the volume of paint in Tin B is doubled.
- He pours paint from Tin B into Tin A until the volume of paint in Tin A is doubled.
- He pours paint from Tin A into Tin B until the volume of paint in Tin B is doubled.

If Tin A originally contained x litres of paint and Tin B contained y litres of paint, determine an expression in terms of x and y for the volume of paint in Tin A after Bobby finished mixing.

28. If you add the first and last of any three consecutive integers together, determine a relationship to the middle number.

29. The Flesch–Kincaid Grade Level formula is used to determine the readability of a piece of text. It produces a score from 0 to 100 that can be used to determine the number of years of education generally required to understand a particular piece of text.

The formula is as follows:

$$0.39 \left(\frac{\text{total words}}{\text{total sentences}} \right) + 11.8 \left(\frac{\text{total syllables}}{\text{total words}} \right) - 15.59$$

Text suitable for a Year 8 student should have a value of roughly 8. A passage of text contains 30 sentences, with 500 words and 730 syllables.
Explain whether this would be suitable for a Year 8 student.

30. Consider the expression $x^{(x+1)^{(x+2)}}$.
This is called a *power tower*.
Evaluate the last digit of the resulting number when $x = 2$.
Note: You will have to look at patterns to determine the answer, as a calculator will not give you an exact answer to the power.

on To test your understanding and knowledge of this topic, go to your learnON title at www.jacplus.com.au and complete the **post-test**.

Answers

Topic 6 Algebraic techniques

6.1 Pre-test

1. B and D
2. a. 17   b. -15 c. 100
3. a. $7a$ b. $3a$ c. $5ab$
4. a. $60\,\text{cm}^2$ b. $9.12\,\text{cm}^2$
5. True
6. a. 12   b. 9   c. -2 d. 3
7. A and C
8. True
9. a. $3a - 5b$ b. $5f + 2$ c. $3ab + a$
10. a. $24x^2y$ b. $\dfrac{9q}{4}$
11. a. $5m + 50$ b. $6m^2 - 15mp$
    c. $3m^2 + 21mt - 2t^2$
12. B
13. a. $4(mn + 4)$ b. $3pq(p - 4q)$
14. $\dfrac{6t}{5(t-1)^2}$
15. $n + 4$

6.2 Using pronumerals

1. a. $x + 420$
   b. $3x$
   c. $x - 130$
   d. The nearby nest has 60 more ants.
   e. The nearby nest has 90 fewer ants.
   f. This nest is one-quarter of the size of the original nest.
2. a. $x + y$ b. $x + y + 260$
   c. $x + y + 90$ d. $x + y - 260$
3. a. $a + b$ b. $a + b + c$
   c. $b + 4$ d. $a - 6$
4. a. $y + 7$ b. $y - 8$ c. $5y$
5. a. $14 - y$ b. $\dfrac{y}{3}$ c. $8y + 3$
6. a. $a + b$ b. $a - b$
   c. $2b - 3a$ d. ab
7. a. $2ab$ b. $3a + 7b$
   c. a^2 d. $\dfrac{a^2}{5}$
8. a. $\$27y$ b. $\$14d$ c. $\$(27r + 14h)$
9. Between 9:00 am and 9:15 am, one Danish pastry was sold. In the next hour-and-a-half, a further 11 Danish pastries were sold. No more Danish pastries had been sold at 12:30 pm, but, in the next half-hour, 18 more were sold. No Danish pastries were sold after 1:00 pm.
10. a. $t + 2$ b. $t + g$
    c. $t - 5$ d. $2t$

11. a. Various answers are possible; an example is shown. The number of passengers doubled at the next stop and continued to increase, more than quadrupling in the first 9 minutes. At 7:22 pm 5 people left the train, and by 7:25 pm the same number of passengers were on the train as there were at the beginning. By 7:34 pm there were 12 fewer passengers than there were at the beginning.
    b. 7:22 pm
    c. 7:19 pm
    d. 7:34 pm
12. Variables are useful in situations where you do not know the value of a number, or where there are multiple possible values.
13. a. The number of bacteria in each of these intervals is double the number of bacteria in the previous interval.
    b. The bacteria could be dividing in two.
    c. It is lower than expected, based on the previous pattern of growth.
    d. Some of the bacteria may have died, or failed to divide and reproduce.
14. a. Odd   b. Even
    c. $n + 2$, $n + 4$ and $n + 6$ d. $n - 2$
15. 22, 23, 24, 25, 26
16. a. $\text{TSA} = 2x^2 + 4xh$; $V = x^2h$
    b. $\text{TSA} = 3x^2 + 5xh$; $V = \dfrac{3}{2}x^2h$
    c. $\text{TSA}_{\text{square box}} = 2000\,\text{cm}^2$; $V_{\text{square box}} = 6000\,\text{cm}^3$
    $\text{TSA}_{\text{rectangular box}} = 2700\,\text{cm}^2$; $V_{\text{rectangular box}} = 9000\,\text{cm}^3$
17. The number is 4; Bill's age is 15.

6.3 Substitution

1. a. 30   b. 1   c. 9   d. 7
2. a. 2   b. 6   c. 12   d. 11
3. a. 2   b. $\dfrac{5}{2}\left(2\dfrac{1}{2}\right)$ c. 40   d. 20
4. a. 18   b. 3   c. 8   d. 2
5. a. 36   b. 4   c. 84   d. 18
6. a. 15   b. 0   c. 18   d. 36
7. a. 11   b. 2   c. 14   d. 28
8. a. 784   b. 250   c. 9800   d. 200
9. a. -15 b. -4 c. 37   d. 15
10. a. 80   b. $\dfrac{14}{3}\left(4\dfrac{2}{3}\right)$ c. 139   d. $\dfrac{1}{2}$
11. a. 24.3   b. 21   c. 1   d. 4.8
12. a. 15.7   b. 4.7   c. 10.8   d. $\dfrac{5}{4}\left(1\dfrac{1}{4}\right)$
13. a. 16   b. 45   c. 21   d. 84
14. a. 576   b. 32   c. -26 d. 33
15. a. $48\,\text{cm}^2$ b. $8400\,\text{m}^2$
    c. $4.472\,\text{m}^2$ or $44\,720\,\text{cm}^2$

16. $c = \$64$

17. a. $62 \, \text{cm}$ b. $97.8 \, \text{cm}$

18. $10 \, \text{cm}$

19. a. $F = 212 \, °\text{F}$ b. $28 \, °\text{C} = 82.4 \, °\text{F}$
    c. $32 \, °\text{F}$

20. a. $720°$ b. $540°$ c. $180°$ d. $360°$ e. $3240°$

21. If $x = 0$, then the expression becomes $\dfrac{0}{0}$, which is indeterminate.

22. Sample responses can be found in the worked solutions in the online resources.

23. a. $V = x(x + 5)(x - 2)$ b. $1200 \, \text{cm}^3$
    c. Because $1.5 - 2 < 0$

24. a. $CD = m + 4n$ b. $BC = 3m + n$
    c. Perimeter $= 8m + 18n$

25. a. $13p - 10$ b. 2590

26. a. $A = \dfrac{1}{2}(m + n)(m - n)$

    b. $A = \dfrac{1}{2}(21 \times 9) = 94.5 \, \text{cm}^2$

    c. If $m < n$, then $(m - n) < 0$.
      Negative height is impossible. Also, $m \neq n$ as the height cannot equal 0.

    d. If m and n move closer in value, then the length of the base of the triangle gets closer to $2m$ (or $2n$) and the height gets closer to zero.

6.4 Substituting positive and negative numbers

1. a. 3 b. 9 c. -9 d. 3

2. a. 1 b. 7 c. 4 d. -13

3. a. 1 b. 2 c. 8 d. -3

4. a. 0 b. -109 c. 6 d. 16

5. a. -9 b. 1 c. 3 d. 40

6. a. -2 b. -5 c. 108 d. $\dfrac{4}{5}$

7. a. -1 b. 6 c. -27 d. 26

8. a. -30 b. -32 c. 225 d. 360

9. a. 22 b. 32 c. 3 d. 21

10. 8

11. -17

12. $\dfrac{30}{7}$ or $4\dfrac{2}{7}$

13. a.

x	-4	-3	-2	-1	0	1	2	3	4
y	16	9	4	1	0	1	4	9	16

    b. For two values of x with the same magnitude but different signs (e.g. -2 and 2), the values of x^2 are identical.

14. If x is negative, then $5x$ will also be a negative integer (less than or equal to -5). Subtracting this number is equivalent to adding a positive integer. The result will be positive.

15. Sample responses can be found in the worked solutions in the online resources.

16. Regardless of the sign of x, x^2 will be non-negative.

17. a. Negative   b. Positive

18. a. $\approx 0.804x^2$ b. 3.216 c. $\dfrac{x}{2}$

6.5 Number laws

1. a. i. 11 ii. 11, same
    b. i. 31 ii. 31, same
    c. i. -5 ii. 5, different
    d. i. -28 ii. 28, different

2. a. i. -10 ii. -10, same
    b. i. -40 ii. -40, same
    c. i. $-\dfrac{2}{5}$ ii. $-\dfrac{5}{2}$, different
    d. i. $-\dfrac{4}{5}$ ii. $-\dfrac{5}{4}$, different

3. True

4. False

5. False

6. True

7. True

8. True

9. False

10. False

11. False

12. True

13. False

14. True

15. a. i. 13 ii. 13, same
    b. i. 24 ii. 24, same
    c. i. 40 ii. 40, same

16. a. i. -3 ii. -7, different
    b. i. -43 ii. -67, different

17. a. i. -64 ii. -64, same
    b. i. 768 ii. 768, same
    c. i. -1536 ii. -1536, same

18. a. i. -4 ii. -1, different
    b. i. -6 ii. $-\dfrac{1}{6}$, different
    c. i. $\dfrac{8}{5}$ ii. $\dfrac{1}{10}$, different

19. a. False   b. False   c. False

20. a. True   b. True   c. False

21. C

22. D

23. These expressions will have different results unless $x = a$.

24. a. -10 b. -12

25. a. i. 6 ii. 6

    b. The answers are equal because of the use of the Associative Law.

26. The Commutative Law holds true for *addition* and *multiplication*. It does not hold true for *subtraction* and *division*.

27. a. i. -7 ii. -7

  Addition can be performed in any order and so can multiplication.

  b. i. 4 ii. -6

  Subtraction in a different order gives different results.

  c. i. -2 ii. $-\dfrac{1}{2}$

  Division in a different order gives different results.

28. a. $(-3p + 4q)$

  b. $\dfrac{1}{(3p - 4q)}$

  c. $-15, -\dfrac{1}{15}$

  d. The result is the identity, which is 1.

6.6 Adding and subtracting terms

1. a. $6c$ b. $-3c$ c. $4a$

2. a. q b. $-3h$ c. $2x$

3. a. $-6a$ b. $4f$ c. $-3p$

4. a. h b. $18b$ c. $3t$

5. a. $10p$ b. $17g$ c. $3b$

6. a. $14t$ b. $-7j$ c. $-15l$

7. a. $10x - 2y$ b. $7x - 12$ c. $11 - 2f$

8. a. $6 - u$ b. $7m + 3p$ c. $4r - 5h$

9. a. $17a - 5b$ b. $9t - 2$ c. $17 - 3g$

10. a. $12 - 2b$ b. $20 - 12h$ c. $9y - 2g - 6$

11. a. $11h - 8$ b. $4s - 2t$ c. $14l - 5m$

12. a. $3k - 13h + 7$ b. $5 - 4t$ c. $7g - 2$

13. a. $3x^2$ b. $5y^2$ c. $4a^3$

14. a. $7d^2$ b. $-g^2$ c. $10y^3$

15. a. $4a^2 + 9$ b. $23x^2$ c. $11s^2 + 4$

16. a. $8a^2 + 5a$ b. $b^2 + 23b$ c. $t^2 - 4g - 7$

17. a. $7g^3 - g^2 + 22$ b. $18ab + 3$ c. $11xy$

18. a. $3fg + 3s$ b. $12ab - 5$ c. $20ab^2 - 14ac$

19. The terms contain the exact same variables, raised to the same power.

20. a. $14t$ b. $12t$ c. $13t$ d. $39t$

21. They are not like terms.

22. a. $2x + 15$ b. $9x + 4$ c. $15x$ d. $13.5x - 2$

23. The correct amount is returned to the artist. Sample responses can be found in the worked solutions in the online resources.

24. $x = 3$

6.7 Multiplying and dividing terms

1. a. $12g$ b. $21h$ c. $24d$ d. $15z$

2. a. $35t$ b. $12u$ c. $42p$ d. $21gy$

3. a. $24gx$ b. $70ah$ c. $36dm$ d. $15ch$

4. a. $48wx$ b. $231abd$ c. $24xy$ d. $10.5xy$

5. a. $4f$ b. $2h$ c. $5x$ d. $3g$

6. a. $2r$ b. 2 c. 2 d. 2

7. a. $\dfrac{2}{3}$ b. $\dfrac{1}{2}$ c. $\dfrac{6}{7}$ d. $\dfrac{2}{3}$

8. a. 9 b. $\dfrac{5}{12}$

  c. $\dfrac{16}{11} = 1\dfrac{5}{11}$ d. 3

9. a. $5fg$ b. $3cd$ c. $\dfrac{2xy}{3}$ d. cg

10. a. $\dfrac{11m}{5}$ b. $\dfrac{1}{6}$ c. $\dfrac{2d}{5}$ d. $\dfrac{b}{7}$

11. a. $-15f$ b. $12d$ c. $-33ag$

12. a. $-40hjk$ b. $112.5xy$ c. $-144prtz$

13. a. $-\dfrac{a}{2}$ b. $-\dfrac{a}{3}$ c. $-12j$ d. $\dfrac{1}{2d}$

14. a. $-\dfrac{4}{5l}$ b. $-\dfrac{x}{4}$ c. $-\dfrac{6}{7}$ d. $\dfrac{f}{5j}$

15. a. -2 b. $\dfrac{2b}{3}$ c. $-\dfrac{d}{5h}$ d. $-\dfrac{12}{11p}$

16. a. $\dfrac{14ef}{9}$ b. $-\dfrac{18x}{7}$ c. $\dfrac{3}{2r}$ d. $-\dfrac{11}{12t}$

17. a. $2a^2$ b. $25p^2$ c. $-30x^2$ d. $7a^2b$

18. a. $-160x^2y$ b. $42p^2q^2$
  c. $-30mn^2t^2$ d. $-18xy^2z^2$

19. a. 0 b. $-18w^2x^2y^2z^2$
  c. $6a^{11}$ d. $2x^5y^3$

20. a. $10m^9$ b. $\dfrac{5p^{10}q^5}{6}$ c. $\dfrac{4x^2yz^2}{3}$ d. $\dfrac{3b}{5}$

21. a. $\dfrac{6}{a^2}$ b. $\dfrac{10b^2}{3}$ c. $\dfrac{5}{w}$ d. $\dfrac{9k}{5}$

22. a. $\dfrac{16h^2}{9d^2}$ b. $\dfrac{5t}{n}$ c. $-\dfrac{27h^2}{2g^2}$ d. $\dfrac{8y}{w}$

23. a. $14x^2$ b. $8p^6$ c. $10s$

24. a. $10x$ b. $4s^2t$ c. 7

25. They use the same principles. Multiplication of variables uses the Commutative Law, where order is not important. Division of variables is resolved the same way as numbers, where expressions are converted to fractions and then simplified and resolved.

26. The order of operations needs to be applied when dealing with algebraic expressions. Also, terms that contain more than one pronumeral should be dealt with as a single term. The first expressions are all equivalent to $\dfrac{x}{yz}$ and the final two terms are equivalent to $\dfrac{xz}{y}$.

27. The pronumeral must be a common factor of every term of the numerator and every term of the denominator before it can be cancelled.

28. -5

29. $\dfrac{7}{6}$

30. a. $V_{\text{small container}} = ab^2$, $V_{\text{large container}} = 4ab^2$

    b. 4 times

6.8 Expanding brackets

1. a. $3d + 12$ **b.** $2a + 10$ **c.** $4x + 8$

2. a. $5r + 35$ **b.** $6g + 36$ **c.** $2t - 6$

3. a. $7d + 56$ **b.** $18x - 54$ **c.** $48 + 12c$

4. a. $11t - 22$ **b.** $6t - 18$ **c.** $t^2 + 3t$

5. a. $x^2 + 4x$ **b.** $g^2 + 7g$ **c.** $2g^2 + 10g$

6. a. $9x - 6$ **b.** $3x^2 - 18xy$ **c.** $15xy - 45y^2$

7. a. $100y - 250$ **b.** $-3c - 9$ **c.** $-15x - 20$

8. a. $-20f + 8f^2$ **b.** $27xy - 18x$ **c.** $-6bh + 18h^2$

9. a. $20ab + 12ac$ **b.** $-6ag + 21a^2$ **c.** $15ab + 30ac$

10. a. $-18w^2 + 10wz$ **b.** $48m^2 + 120m$

    c. $6k^2 - 15k$

11. a. $35x + 49$ **b.** $3c - 4$ **c.** $22c - 2c^2$

12. a. $6v + 30$ **b.** $5d^2 - 12d$ **c.** $11y + 12$

13. a. $26r + r^2$ **b.** $9g - 37$ **c.** $11f - 12g - 7$

14. a. $8r - 13$ **b.** $18gh - 24g$ **c.** $11t + 1$

15. a. $d + 22$ **b.** $14h$ **c.** $21m - 4$

16. a. $28f - 1$ **b.** $4a - 3a^2 - 35$

    c. $6 - t^2 + 2t$

17. a. $10hk + 21h + 20k$ **b.** $15n - 6ny$

    c. $8gm + 24g - 18$

18. a. $11cf + 3c$ **b.** $28x - 5xy - 29$

    c. $77mn - 16m + 12n$

19.

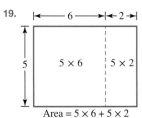

Area = $5 \times 6 + 5 \times 2$

20.

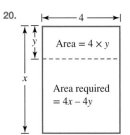

21. Multiplication of variables follows the Commutative Law, where order doesn't matter; therefore, brackets are not required.

22. a. $\$(50 - d)$

    b. i. Sale price $(P) = 3(50 - d)$

      ii. $P = 150 - 3d$

c. Amount of change $(C) = 200 - (150 - 3d) = 50 + 3d$

23. a. $A_{\text{left}} = fm - 36f$; $A_{\text{centre}} = fm$; $A_{\text{right}} = fm - 36f$

    b. $A = 3fm - 72f$

    c. $A = 3m^2 + 235.5\,m - 7380$

    d.

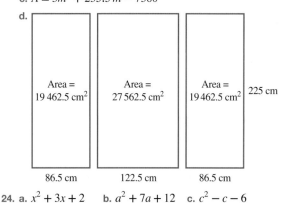

24. a. $x^2 + 3x + 2$ **b.** $a^2 + 7a + 12$ **c.** $c^2 - c - 6$

6.9 Factorising

1. a. 2   **b.** 3   **c.** 6   **d.** 13

2. a. 7   **b.** 2   **c.** 3   **d.** 4

3. a. $2g$ **b.** $3m$ **c.** 11   **d.** $2m$

4. a. $2a$ **b.** $12g$ **c.** $2g$ **d.** $11l$

5. a. $4mn$ **b.** $4c$ **c.** $4c$ **d.** x

6. a. $3(x + 2)$ **b.** $2(y + 2)$

    c. $4(2x + 3)$ **d.** $3(2f + 3)$

7. a. $2(d + 4)$ **b.** $2(x - 2)$

    c. $11(h + 11)$ **d.** $4(s - 4)$

8. a. $12(g - 2)$ **b.** $2(7 - 2b)$

    c. $12(4 - q)$ **d.** $8(2 + f)$

9. a. $3(gh + 4)$ **b.** $2y(x + 3)$

    c. $7g(2 - h)$ **d.** $2k(8j - 1)$

10. a. $m(7n + 6)$ **b.** $5a(1 - 3bc)$

    c. $2r(4 + 7t)$ **d.** $2b(2 - 3a)$

11. a. $7x(2 - 3y)$ **b.** $k(11j + 3)$

    c. $c(12a - 4 + 3d)$ **d.** $4(g + 2h - 4)$

12. $2ab$

13. $6f$

14. a. $6(p + 2pq + 3q)$ **b.** $8(4x + y + 2z)$

    c. $4(4m - n + 6p)$ **d.** $8(9x - y + 8pq)$

15. a. $y(x + 9 - 3y)$ **b.** $c(5 + 3cd - d)$

    c. $ab(3 + a + 4b)$ **d.** $xy(2x + 1 + 5y)$

16. a. $2x(2x + 1 + 4x^2)$ **b.** $8y^2(1 + 8y^2 - 4y)$

17. $\dfrac{x + 3}{4 - 5x}$

18. $\dfrac{10(3x - 2)}{1 + 2x}$

19. A sample response:

When determining the highest common factor, write down all the factors of the expression — that is, all the variables (making sure to repeat a variable by the number of the power it has been raised to) and all the numbers. Write each of the numbers as a product of its prime factors and then circle all the variables and numbers that each term has in common.

20. a. $6(x-3)(x+2)$

    b. $x > 3$. If x was smaller, the area would be negative, which is not possible.

21. $7y$

22. $\dfrac{3}{5}$

23. a. $(n+3)(m+9)$ b. $(-2t+6)(5x+1)$

24. a. $(3x-6)(2x+8)(ax-5a)$

    b. $\dfrac{4}{3a(x+4)(x-5)}$

Project

1. Let l represent the number of long words and s represent the number of sentences. Rix index $= \dfrac{l}{s}$.

2. Six sentences and 19 long words

3. 3.17

4. Grade 8

5. Students should include the numbers of sentences and long words that appeared in the selected passage.
   To determine the Rix index, they need to use the following formula: Rix index $= \dfrac{l}{s}$.
   Once students have obtained the Rix index, they need to use the table to determine the equivalent year level.

6. Students need to repeat the process stated in the answer for question 5.

7. Students could reduce the number of long words to lower the equivalent year level (as the Rix index will be lower). Students could increase the number of long words to raise the equivalent year level (as the Rix index will be higher).

6.10 Review questions

1. a. $x+y$ b. $y-x$ or $x-y$
   c. $3x-5y$ d. $5x$

2. a. $2xy$ b. $6x+7y$
   c. y^2 d. $2x-7$

3. a. $15x$ b. $9y$ c. $15k+9m$

4. a. $m+5$ b. $m-p$ c. $5m$

5. a. 8   b. 4   c. 8   d. 13

6. a. 22   b. 10   c. 14   d. 3

7. $13.90

8. 5 cm

9. a. 40   b. 16   c. 15

10. a. 180   b. 18   c. 500

11. a. 10   b. 28   c. -20

12. False

13. True

14. True

15. False

16. True

17. a. $21g$ b. $18y$ c. $42d$ d. $-24z$

18. a. $\dfrac{h}{2}$ b. $\dfrac{4}{5sv}$ c. $3x$ d. $\dfrac{e}{6j}$

19. a. $2x+11$ b. $g-1$ c. $9xy$ d. $15t^2+2t$

20. a. $7x+15$ b. $5g^2-33g$ c. $42-3t$

21. a. $3(g+4)$ b. $y(x+5)$ c. $5(n-4)$

22. a. $4n(3m+p)$ b. $6g(2-h)$ c. $6a(4-a+3b)$

23. a. $3(2+1)=3\times2+3\times1$ or $3(2-1)=3\times2-3\times1$

    b. $-10\times8+-10\times-6=-10(8+-6)$ or
       $-10\times8--10\times-6=-10(8--6)$

    c. $8(6+5)=8\times6+8\times5$ or $8(6-5)=8\times6-8\times5$

24. a.
$$\boxed{}\ x\text{ cm}$$
$(x+4)$ cm

    b. $x(x+4)\,\text{cm}^2$

    c. x^2+4x

    d. $21\,\text{cm}^2$

    e. Volume $= x^3+4x^2\,\text{cm}^3$

    f. $63\,\text{cm}^3$

25. $34

26. a. P (points), G (goals), B (behinds)

    b. $6G+B$

    c. 76

    d. 72

    e. (G, B): $(7, 45)$, $(8, 39)$, $(9, 33)$, $(10, 27)$, $(11, 21)$, $(12, 15)$, $(13, 9)$, $(14, 3)$

27. $(3x-5y)\,\text{L}$

28. The sum is twice the middle number.

29. Yes, the value is 8.138.

30. The final digit of 281 is 2.

Semester review 1

The learnON platform is a powerful tool that enables students to complete revision independently and allows teachers to set mixed and spaced practice with ease.

Student self-study

Review the **Course Content** to determine which topics and lessons you studied throughout the year. Notice the green bubbles showing which elements were covered.

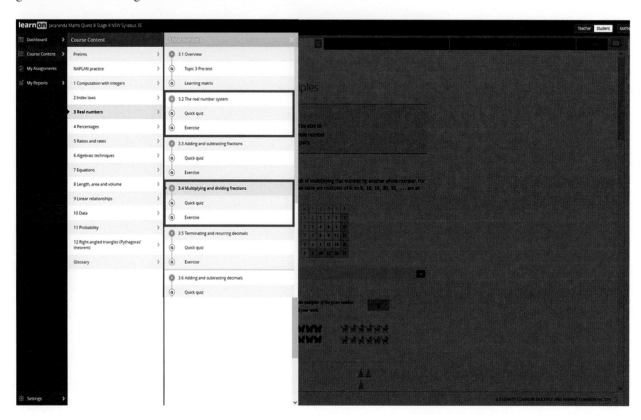

Review your results in **My Reports** and highlight the areas where you may need additional practice.

Use these and other tools to help identify areas of strengths and weakness and target those areas for improvement.

Teachers

It is possible to set questions that span multiple topics. These assignments can be given to individual students, to groups or to the whole class in a few easy steps.

Go to **Menu** and select **Assignments** and then **Create Assignment**. You can select questions from one or many topics simply by ticking the boxes as shown below.

Once your selections are made, you can assign to your whole class or subsets of your class, with individualised start and finish times. You can also share with other teachers.

More instructions and helpful hints are available at www.jacplus.com.au.

7 Equations

LESSON SEQUENCE

LESSON
7.1 Overview

Why learn this?

Equations are mathematical statements that tell us that two expressions are equal. All equations have an equals sign, and the value on the left-hand side of the equals sign needs to equal that of the right-hand side. The expressions in an equation can be either numerical (e.g. $10 + 3 = 13$) or algebraic (e.g. $2x + 3 = x - 1$). Statements such as $y + 2 = 8$ and $x - 5 = 10$ are both examples of an equation.

Linear equations form part of algebra and are used to describe everyday situations using mathematics. Let us say that a plumber charges an $80 call-out fee and $50 per hour for each job. An equation can be created to determine the cost: $\text{cost} = 80 + 50 \times \text{hours}$. Using this equation will make it easier for the plumber to calculate the bill for each customer. Equations are used widely in many aspects of life, including science, engineering, business and economics.

Meteorologists use equations with many variables to predict the weather for days into the future. Equations are also used in mathematical modelling that helps us determine and predict trends such as investments, house prices or the spread of COVID-19. One of the most useful things you will learn in algebra is how to solve equations. Solving equations will allow you to find the answers to problems that contain unknown values. The algebraic techniques you learn will be used throughout your future schooling and beyond.

1. **MC** State which of the following is the solution to the equation $21y = 7$.

   A. $y = 2$ **B.** $y = 3$ **C.** $y = 7$ **D.** $y = 21$

2. Write the new equation when both sides of the equation $2x = 14$ are divided by 2.

3. Determine:
   a. the output number for the following flowchart

   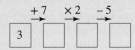

   b. the input number for the following flowchart.

   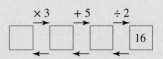

4. **MC** If $p = 4$, state which one of the following equations is false.

   A. $\dfrac{p}{4} = 1$ **B.** $6p = 24$ **C.** $p - 5 = 9$ **D.** $p - 4 = 0$

5. Solve the following one-step equations.

   a. $p - 3 = -9$ **b.** $\dfrac{q}{4} = -5$

6. Solve the following equations for x.

   a. $\dfrac{x}{4} + 1 = 3$ **b.** $3(x - 2) = -6$

   c. $6 = 1 - \dfrac{x}{9}$ **d.** $\dfrac{6 + 2x}{3} + 1 = -5$

7. Solve the following. Where appropriate, give your answers as decimal approximations, to 2 decimal places.

   a. $x^2 = 25$ **b.** $4y^2 = 256$ **c.** $t^2 = 53$

8. A square paving stone has an area of 144 cm^2. By solving a quadratic equation, determine the side length of the paving stone.

9. Solve $7x^2 + 44 = 5x^2 + 286$.

10. Tyson is y years old and his brother, Ted, is 3 years younger. In 3 years' time, the total of their ages will be 33. Determine how old Tyson is now.

11. Solve the following equation for p.

$$\frac{5(p + 1)}{3} - 2 = 8$$

12. An isosceles triangle has two sides of length $2x - 1$ and a third side of length $x - 4$. If the perimeter of the triangle is 24 cm, determine the length of the shortest side.

13. Solve the following equations for x.
    a. $5x - 2 = 3x + 12$ b. $7 - x = -3 - 6x$ c. $5x - 2(x - 3) = x + 9$

14. A plumber charges a \$75 call-out fee and \$60 per hour. An electrician charges a \$150 call-out fee and \$50 per hour. Both are called out on the same day for the same length of time and charge the same amount. Evaluate the earnings of each of them for the day's work.

15. Sarah wants to buy her first car, which costs \$10 695. She already has \$2505 in the bank. She decides on a savings plan to put money in the bank each month. In the first month, she puts x dollars in the bank. Each month she deposits double the previous month's amount into the bank. If Sarah saved enough money for the car in one year, evaluate her first payment to the bank.

LESSON
7.2 Backtracking and inverse operations

LEARNING INTENTION

At the end of this lesson you should be able to:
- understand the need for strategies to solve equations
- use a flowchart to determine the output number
- backtrack through a flowchart to determine the input number
- draw a flowchart to represent a series of operations.

▶ 7.2.1 Solving equations

eles-6210

- An equation is a mathematical statement indicating that two mathematical expressions are equal.
- If an equation is always a true statement (e.g. $6 - 2 = 4$), then the equation is true. If an equation is always a false statement (e.g. $x + 1 = x + 2$), then the equation is false.
- In an equation with an unknown, the unknown may have a value that makes the equation true. For example, the equation $2x + 3 = 13$ is true only if $x = 5$. The value that makes the equation true is called the solution to the equation.
- Equations can have no solutions, one or two solutions, or infinitely many solutions.
- There are many strategies to solve equations, such as guess–check–improve, backtracking and inverse operations. Generally, one strategy to solve an equation will be more efficient than others, depending on the equation.

DISCUSSION

What process would you use to solve the logic puzzle shown?

$$\text{▦} + \text{▦} + \heartsuit = \text{❋}$$
$$\text{▦} + \heartsuit = 8$$
$$\text{❋} - \heartsuit = 6$$
$$\text{❋} = \, ?$$

▶ 7.2.2 Using flowcharts

eles-4348

- A flowchart can be used to represent a series of operations.
- In a flowchart, the starting number is called the input number and the final number is called the output number.
- Flowcharts are useful for building up an expression and they provide a simple visual method for solving equations.
- **Backtracking** is a method used to work backwards through a flowchart. It involves moving from the output number to the input number.
- When working backwards through a flowchart, use inverse (opposite) operations.

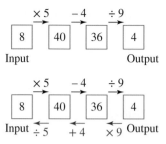

Inverse operations

+ and − are inverse operations of each other.

× and ÷ are inverse operations of each other.

WORKED EXAMPLE 1 Using backtracking to determine the input number

Determine the input number for this flowchart.

THINK	WRITE/DRAW
1. Copy the flowchart.	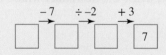
2. Backtrack to determine the input number. The inverse operation of +3 is −3 $(7 - 3 = 4)$. The inverse operation of ÷ − 2 is × − 2 $(4 \times -2 = -8)$. The inverse operation of −7 is +7 $(-8 + 7 = -1)$. Fill in the missing numbers.	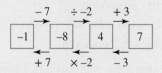
3. Write the input number.	The input number is −1.

WORKED EXAMPLE 2 Using a flowchart to determine the output

Determine the output expression for this flowchart.

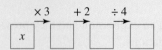

THINK	WRITE/DRAW
1. Copy the flowchart and look at the operations that have been performed.	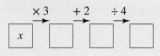
2. Multiplying x by 3 gives $3x$.	
3. Adding 2 gives $3x + 2$.	
4. Now place a line beneath all of $3x + 2$ and divide by 4.	
5. Write the output expression.	The output expression is $\dfrac{3x + 2}{4}$.

WORKED EXAMPLE 3 Drawing a flowchart to build up an expression

Starting with x, draw the flowchart whose output number is given by the expression:
a. $6 - 2x$ **b. $-2(x + 6)$.**

THINK	WRITE/DRAW
a. 1. Rearrange the expression. *Note:* $6 - 2x$ is the same as $-2x + 6$.	a. 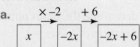
2. Multiply x by -2, then add 6.	
b. 1. The expression $x + 6$ is grouped in a pair of brackets, so we must obtain this part first. Therefore, add 6 to x.	b. 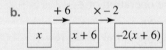
2. Multiply the whole expression by -2.	

 Resources

🧩 **Interactivity** Backtracking and inverse operations (int-3803)

7.2 Quick quiz on	7.2 Exercise

Individual pathways

■ PRACTISE	■ CONSOLIDATE	■ MASTER
1, 4, 7, 10, 13	2, 5, 8, 11, 14	3, 6, 9, 12, 15

Fluency

1. **WE1** Calculate the input number for each of the following flowcharts.

   a.    b. 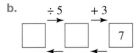   c. 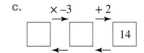   d.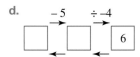

2. Calculate the input number for each of the following flowcharts.

   a. 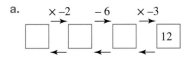   b.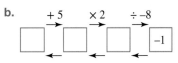

   c. 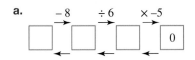   d.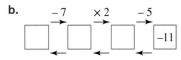

3. Calculate the input number for each of the following flowcharts.

   a. 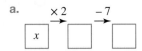   b.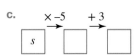

   c. 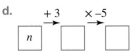   d.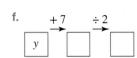

4. **WE2** Determine the output expression for each of the following flowcharts.

   a.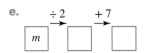

   b.

   c.

   d.

   e.

   f.

5. Determine the output expression for each of the following flowcharts.

a.
$$z \xrightarrow{\times 6} \square \xrightarrow{-3} \square \xrightarrow{\div 2} \square$$

b.
$$d \xrightarrow{+5} \square \xrightarrow{\times -3} \square \xrightarrow{\div 4} \square$$

c.
$$e \xrightarrow{\times 2} \square \xrightarrow{\div 5} \square \xrightarrow{+1} \square$$

d.
$$x \xrightarrow{\times -1} \square \xrightarrow{+3} \square \xrightarrow{\times 4} \square$$

e.
$$w \xrightarrow{-5} \square \xrightarrow{\times -2} \square \xrightarrow{\div 7} \square$$

f.
$$z \xrightarrow{+6} \square \xrightarrow{\times -3} \square \xrightarrow{-11} \square$$

6. Determine the output expression for each of the following flowcharts.

a.
$$v \xrightarrow{-3} \square \xrightarrow{\div 6} \square \xrightarrow{-8} \square$$

b.
$$m \xrightarrow{\times 8} \square \xrightarrow{-4} \square \xrightarrow{\times -7} \square$$

c.
$$k \xrightarrow{\div 6} \square \xrightarrow{\times -5} \square \xrightarrow{+2} \square$$

d.
$$p \xrightarrow{\times -5} \square \xrightarrow{-7} \square \xrightarrow{\div 3} \square$$

Understanding

7. **WE3** Starting with x, draw the flowchart whose output expression is:

a. $2(x+7)$ **b.** $-2(x-8)$ **c.** $3m-6$ **d.** $-3m-6$ **e.** $\dfrac{x-5}{8}$ **f.** $\dfrac{x}{8}-5$

8. Starting with x, draw the flowchart whose output expression is:

a. $-5x+11$ **b.** $-x+11$ **c.** $-x-13$ **d.** $5-2x$ **e.** $\dfrac{3x-7}{4}$ **f.** $\dfrac{-3(x-2)}{4}$

9. Starting with x, draw the flowchart whose output expression is:

a. $\dfrac{x+5}{8}-3$ **b.** $-7\left(\dfrac{x}{5}-2\right)$ **c.** $3\left(\dfrac{2x}{7}+4\right)$ **d.** $\dfrac{1}{4}\left(\dfrac{6x}{11}-3\right)$

Communicating, reasoning and problem solving

10. a. Draw a flowchart to convert from degrees Fahrenheit (F) to degrees Celsius (C) using the formula $C=\dfrac{5}{9}(F-32)$.

b. Use the flowchart to convert 50 °F to degrees Celsius by substituting into the flowchart.

c. Backtrack through the flowchart to evaluate 35 °C in degrees Fahrenheit.

11. The rectangle shown has an area of 255 cm².

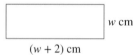

w cm

$(w+2)$ cm

a. Discuss the strategies you might use to determine the value of the pronumeral w. Explain whether you can use backtracking.

b. Determine the value of the pronumeral w and explain the method you used.

12. Linda and Amy are discussing their answers to the problem below. Determine the output for the following flowchart.

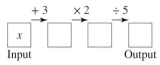

Linda's answer is $\dfrac{2x+3}{5}$ and Amy's answer is $\dfrac{2x}{5} - 3$.

Is one of them correct? Explain your reasoning.

13. A third of a certain number is added to four and the result is then doubled. Write this statement as an expression and represent the expression as a flowchart.

14. Consider the following puzzle.

> Think of a number.
>
> Double it.
>
> Add 10.
>
> Divide by 2.
>
> Subtract the number you first thought of.

   a. Represent this puzzle as a flowchart, with *n* representing the unknown number.
   b. Discuss what you notice about the final number.
   c. Repeat this with different starting numbers.

15. Tan, Bart and Matthew decided to share 20 chocolates. Tan took 8 chocolates and Bart took 3 times as many as Matthew.

   a. Let *x* be the number of chocolates that Matthew took. Develop an expression to represent how many chocolates Tan, Bart and Matthew took altogether.
   b. Use any problem-solving strategy to determine the number of chocolates Matthew and Bart took.

LESSON
7.3 Keeping equations balanced

LEARNING INTENTION

At the end of this lesson you should be able to:
- manipulate equations without unbalancing them.

▶ 7.3.1 Balancing a set of scales

eles-4349

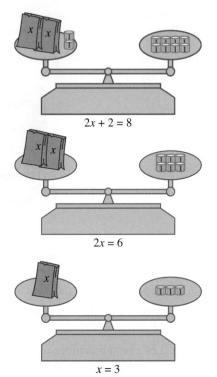

- An equation contains two expressions with an equals sign between them, which means the left-hand side (LHS) is equal to the right-hand side (RHS).
- Equations can be thought of as a balanced scale. The diagram shown represents the equation $x = 3$.
- If the amount on the LHS is doubled, the scale will stay balanced provided that the amount on the RHS is also doubled.
- Similarly, the scale will stay balanced if we remove the same quantity from both sides.
- As long as we do the same thing to both sides, equations will remain balanced.
- Different versions of balanced equations are equivalent to each other. For example, $2x = 6$ is equivalent to $2x + 2 = 8$, as the second equation is simply the first equation with 2 added to both sides.
- An equation is true if the LHS is equal to the RHS; for example, $5 + 5 = 10$ is a true statement.
- An equation is false if the LHS is not equal to the RHS; for example, $4 + 5 = 10$ is a false statement.

$2x + 2 = 8$

$2x = 6$

$x = 3$

Keeping equations balanced

As long as the same operation is done to both sides, equations will remain balanced.

This includes:
- **adding the same number to both sides**
- **subtracting the same number from both sides**
- **multiplying both sides by the same number**
- **dividing both sides by the same number.**

WORKED EXAMPLE 4 Keeping equations balanced

Starting with the equation $x = 4$, write the new equation when we:

a. multiply both sides by 4

b. subtract 6 from both sides

c. divide both sides by $\dfrac{2}{5}$.

THINK	WRITE
a. 1. Write the equation.	**a.** $x = 4$
2. Multiply both sides by 4.	$x \times 4 = 4 \times 4$
3. Simplify by removing the multiplication signs. Write numbers before variables.	$4x = 16$
b. 1. Write the equation.	**b.** $x = 4$
2. Subtract 6 from both sides.	$x - 6 = 4 - 6$
3. Simplify and write the answer.	$x - 6 = -2$
c. 1. Write the equation.	**c.** $x = 4$
2. Dividing by a fraction is the same as multiplying by its reciprocal. Multiply both sides by $\dfrac{5}{2}$.	$x \div \dfrac{2}{5} = 4 \div \dfrac{2}{5}$ $x \times \dfrac{5}{2} = 4 \times \dfrac{5}{2}$ $\dfrac{5x}{2} = \dfrac{20}{2}$
3. Simplify and write the answer.	$\dfrac{5x}{2} = 10$

COMMUNICATING — COLLABORATIVE TASK: Keeping it balanced

Equipment: Pan balance scale, small paper bags, blocks of equal mass

1. In pairs, put a number of blocks inside a paper bag.
2. Put your paper bag and a number of blocks on one side of the scale.
3. Swap scales with another pair and:
   a. use blocks to balance their bag and blocks
   b. work out how many blocks they have put in their bag.
4. Compare your answer with the other pair — do you both get the right answer?
5. What strategies would you recommend to a classmate to solve this kind of problem?

 Resources

♦ **Interactivity** Keeping equations balanced (int-3804)

Exercise 7.3 Keeping equations balanced

7.3 Quick quiz **on**	7.3 Exercise

Individual pathways

■ PRACTISE	■ CONSOLIDATE	■ MASTER
1, 4, 6, 8, 12, 15	2, 5, 9, 10, 13, 16	3, 7, 11, 14, 17

Fluency

1. **WE4** Starting with the equation $x = 6$, write the new equation when we:
   a. add 5 to both sides
   b. multiply both sides by 7
   c. subtract 4 from both sides
   d. divide both sides by 3.

2. Starting with the equation $x = 6$, write the new equation when we:
   a. multiply both sides by -4
   b. multiply both sides by -1
   c. divide both sides by -1
   d. subtract 9 from both sides.

3. Starting with the equation $x = 6$, write the new equation when we:
   a. multiply both sides by $\dfrac{2}{3}$
   b. divide both sides by $\dfrac{2}{3}$
   c. subtract $\dfrac{2}{3}$ from both sides
   d. add $\dfrac{5}{6}$ to both sides.

4. a. Write the equation that is represented by the diagram shown.
   b. Show what happens when you halve the amount on both sides. Write the new equation.

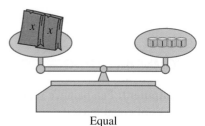

Equal

5. a. Write the equation that is represented by the diagram shown.
   b. Show what happens when you take 3 from both sides. Write the new equation.

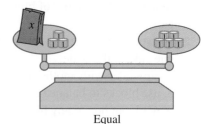

Equal

6. a. Write the equation that is represented by the diagram shown.
   b. Show what happens when you take 1 from both sides. Write the new equation.
   c. Show what happens when you then divide the amount on each side by 3. Write the new equation.

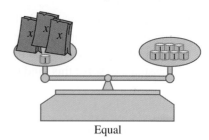

Equal

7. **a.** Write the equation that is represented by the diagram shown.
   b. State which steps enable you to determine x.

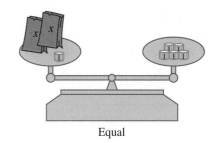

Equal

Understanding

8. **MC** If $x = 5$, determine which of these equations is false.

   A. $x + 2 = 7$ **B.** $3x = 8$ **C.** $-2x = -10$ **D.** $\dfrac{x}{5} = 1$

9. **MC** If $x = 3$, determine which of these equations is false.

   A. $\dfrac{2x}{3} = 2$ **B.** $-2x = -6$ **C.** $2x - 6 = 0$ **D.** $x - 5 = 2$

10. **MC** If $x = -6$, determine which of these equations is false.

    A. $-x = 6$ **B.** $2x = -12$ **C.** $x - 6 = 0$ **D.** $x + 4 = -2$

11. **MC** If $2x = 12$, determine which of these equations is false.

    A. $\dfrac{2x}{3} = 4$ **B.** $-2x = -12$ **C.** $2x - 6 = 2$ **D.** $4x = 24$

Communicating, reasoning and problem solving

12. Starting with $4(x + 1) = 26$, determine the calculation that has been performed on both sides of the equation to give the new equation $2(x + 1) = 13$.

13. Write an equation to represent each of the following situations.

    a. Five bags of sugar weigh three kilos.
    b. Four chocolate bars and an ice cream cost $9.90. The ice cream costs $2.70.
    c. A shopkeeper weighed three apples at 920 g. He thought this was heavy, and realised somebody had left a 500-g weight on his balance scale.

14. Two friends, Liam and Trent, are discussing how to write an equation.
    Liam has written the equation $2x + 4 = 12$ and Trent has written $2(x + 4) = 12$.
    Explain whether these two equations are the same.

15. Develop an equivalent equation for each of the equations listed below by performing the operation given in brackets on both sides.

    a. $m + 8 = 9$ (-8)
    b. $n - 3 = 5$ $(+3)$
    c. $2p + 4 = 8$ $(\div 2)$

16. Some errors have been made in keeping this equation balanced. Start with $2x = 4$.
    a. Add 4 to both sides. $6x = 8$
    b. Subtract 1 from both sides. $2x - 1 = 5$
    c. Multiply both sides by 5. $10x = -1$

    Identify the errors and rewrite the equations to correct them.

17. You have eight $1 coins, one of which is heavier than the rest. Using a set of balance scales, discuss a method to determine the heavy coin in the fewest weighings.

LESSON
7.4 Solving linear equations and simple quadratic equations

LEARNING INTENTION

At the end of this lesson you should be able to:
- solve linear equations algebraically
- solve simple quadratic equations
- verify solutions to equations using substitution.

▶ 7.4.1 Solving one-step equations

eles-4350

- Solving equations is the process of finding pronumeral values that make the equation true. For example, the equation $5x - 2 = 18$ is true only when $x = 4$, so the solution of the equation is $x = 4$.
- Equations can be solved using many different techniques, including inspection, guess and check, balancing with inverse operations and backtracking.
- As previously shown, by performing the same operation on both sides of an equation, it remains balanced.
- The method for solving equations we will investigate here will be using inverse operations with balancing; that is, applying an inverse operation to both sides of the equation.
- We do this because sometimes the equations are not simple enough to use guess and check, and we need to use a more algebraic way of solving these more complex equations.

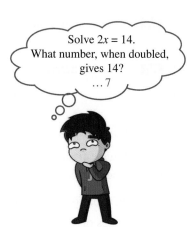

Solve $2x = 14$.
What number, when doubled, gives 14?
... 7

Solving equations

- **To solve an equation, perform inverse operations on both sides until the pronumeral (unknown) is left by itself (isolated on one side of the equation).**
- **Recall that:**
  - **addition and subtraction are inverse operations of each other**
  - **multiplication and division are inverse operations of each other.**

- Consider the following example: How would we isolate the x in $x + 4 = 10$?
  - We need to remove the $+4$ from the LHS so that only x remains.
  - To remove $+4$ we will need to subtract 4.
  - Subtracting 4 from both sides gives us $x = 6$.

$$-4 \left(\begin{array}{c} x + 4 = 10 \\ x = 6 \end{array} \right) -4 \quad \text{(or)} \quad \begin{array}{c} x + 4 = 10 \\ x + 4 - 4 = 10 - 4 \\ x = 6 \end{array}$$

- The following examples demonstrate how the unknown can be isolated in other one-step equations.

$$+3 \left(\begin{array}{c} x-3=8 \\ x=11 \end{array} \right) +3 \quad \text{(or)} \quad \begin{array}{c} x-3=8 \\ x-3+3=8+3 \\ x=11 \end{array}$$

$$\div 4 \left(\begin{array}{c} 4x=20 \\ x=5 \end{array} \right) \div 4 \quad \text{(or)} \quad \begin{array}{c} 4x=20 \\ \dfrac{4x}{4}=\dfrac{20}{4} \\ x=5 \end{array}$$

$$\times 2 \left(\begin{array}{c} \dfrac{x}{2}=7 \\ x=14 \end{array} \right) \times 2 \quad \text{(or)} \quad \begin{array}{c} \dfrac{x}{2}=7 \\ \dfrac{x}{2} \times 2 = 7 \times 2 \\ x=14 \end{array}$$

WORKED EXAMPLE 5 Solving one-step equations

Solve the following one-step equations.

a. $p - 5 = 11$

b. $\dfrac{x}{16} = -2$

THINK

a. 1. Write the equation.

2. To isolate the unknown, add 5 to both sides.

3. Simplify.

b. 1. Write the equation.

2. To isolate the unknown, multiply both sides by 16.

3. Simplify and write the answer.

WRITE

a. $\quad p - 5 = 11$

$p - 5 + 5 = 11 + 5$

$p = 16$

b. $\quad \dfrac{x}{16} = -2$

$\dfrac{x}{16} \times 16 = -2 \times 16$

$x = -32$

DISCUSSION

$c^2 = a^2 + b^2$ is a very popular equation from Pythagoras. This allows us to determine a missing side of a right-angled triangle.

What other popular equations do you know, and how do you or people you know use them?

7.4.2 Solving linear equations

eles-4351

- Linear equations are equations where the pronumeral has an index (power) of 1.
  For example, $2x - 6 = 10$, $m - 6 = 11$ and $\dfrac{5(y-4)}{6} = 15$ are all linear equations.
- Linear equations never contain terms such as x^2 or $\sqrt{x}$.
- Many linear equations require more than one operation to isolate the pronumeral.
- One of the most effective ways to solve multi-step equations is to use backtracking. Backtracking involves developing the given equation, beginning with the pronumeral, and then working backwards to determine the solution.
- When working backwards, use inverse operations.

- If the equation contains brackets, it is often best (but not necessary) to expand the brackets first before developing the equation.
- Backtracking helps understand which operations should be undone first.

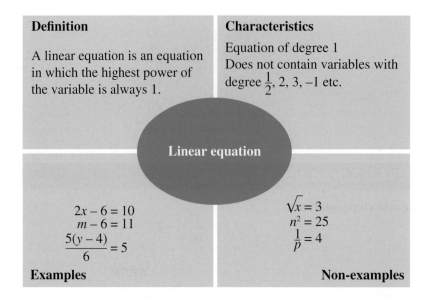

Definition

A linear equation is an equation in which the highest power of the variable is always 1.

Characteristics

Equation of degree 1
Does not contain variables with degree $\frac{1}{2}$, 2, 3, –1 etc.

Linear equation

$2x - 6 = 10$
$m - 6 = 11$
$\dfrac{5(y - 4)}{6} = 5$

Examples

$\sqrt{x} = 3$
$n^2 = 25$
$\dfrac{1}{p} = 4$

Non-examples

WORKED EXAMPLE 6 Solving linear equations using backtracking

Solve the following linear equations using backtracking.

a. $2x - 6 = 10$

b. $\dfrac{3(y + 1)}{2} = 12$

THINK

a. 1. Develop the equation starting with the pronumeral.
   Use words or a flowchart.

2. Starting with the right-hand side (10), backtrack through the flowchart using inverse operations.
   The inverse of subtraction is addition.
   The inverse of multiplication is division.

3. State the answer.

WRITE

a. The pronumeral x has been multiplied by 2 and then 6 has been subtracted. This equals 10.

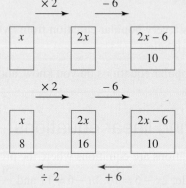

The solution to the equation is $x = 8$.

b. 1. Develop the equation starting with the pronumeral.
Use words or a flowchart.

b. 1 has been added to the pronumeral y. This result has then been multiplied by 3.
This result was then divided by 2.
This equals 12.

$$\begin{array}{cccc} \xrightarrow{+1} & \xrightarrow{\times 3} & \xrightarrow{\div 2} & \\ \boxed{y} & \boxed{y+1} & \boxed{3(y+1)} & \boxed{\dfrac{3(y+1)}{2}} \\ & & & \boxed{12} \end{array}$$

2. Starting with the right-hand side (12), backtrack through the flowchart using inverse operations.
The inverse of division is multiplication.
The inverse of multiplication is division.
The inverse of addition is subtraction.

$$\begin{array}{cccc} \xrightarrow{+1} & \xrightarrow{\times 3} & \xrightarrow{\div 2} & \\ \boxed{y} & \boxed{y+1} & \boxed{3(y+1)} & \boxed{\dfrac{3(y+1)}{2}} \\ \boxed{7} & \boxed{8} & \boxed{24} & \boxed{12} \\ \xleftarrow{-1} & \xleftarrow{\div 3} & \xleftarrow{\times 2} & \end{array}$$

3. State the answer.

The solution to the equation is $y = 7$.

- Backtracking can be used to solve linear equations.
- After solving an equation, it is highly recommended that you check your answer using substitution.

WORKED EXAMPLE 7 Solving linear equations

Solve the following linear equations.

a. $\dfrac{x}{3} + 1 = 7$

b. $2(x + 5) = 18$

THINK

a. 1. Write the equation.

2. Subtract 1 from both sides.

3. Simplify.

4. Multiply both sides by 3.

5. Simplify.

6. Check the solution by substituting $x = 18$ into the left-hand side of the equation.

WRITE

a. $\dfrac{x}{3} + 1 = 7$

$\dfrac{x}{3} + 1 - 1 = 7 - 1$

$\dfrac{x}{3} = 6$

$\dfrac{x}{3} \times 3 = 6 \times 3$

$x = 18$

If $x = 18$,

$\text{LHS} = \dfrac{18}{3} + 1$

$= 6 + 1$

$= 7$

$\text{RHS} = 7$

7. Comment on the answers obtained.	Since the LHS and RHS are equal, the solution is $x = 18$.	

b. 1. Write the equation.

b. $2(x + 5) = 18$

2. Divide both sides by the number in front of the brackets, 2.

$$\frac{2(x + 5)}{2} = \frac{18}{2}$$

3. Simplify.

$$x + 5 = 9$$

4. Subtract 5 from both sides.

$$x + 5 - 5 = 9 - 5$$

5. Simplify.

$$x = 4$$

6. Check the solution by substituting $x = 4$ into the left-hand side of the equation.

If $x = 4$,

$$LHS = 2(4 + 5)$$
$$= 2(9)$$
$$= 18$$
$$RHS = 18$$

7. Comment on the answers obtained.

Since the LHS and RHS are equal, the solution is $x = 4$.

WORKED EXAMPLE 8 Solving complex linear equations

Solve the following linear equations. They will require more than two steps.

a. $3(m - 4) + 8 = 5$

b. $6\left(\dfrac{x}{2} + 5\right) = -18$

THINK	WRITE
a. 1. Write the equation.	**a.** $3(m - 4) + 8 = 5$
2. Subtract 8 from both sides.	$3(m - 4) + 8 - 8 = 5 - 8$
3. Simplify.	$3(m - 4) = -3$
4. Divide both sides by 3.	$\dfrac{3(m - 4)}{3} = \dfrac{-3}{3}$
5. Simplify.	$m - 4 = -1$
6. Add 4 to both sides.	$m - 4 + 4 = -1 + 4$
7. Simplify.	$m = 3$
8. Check the solution by substituting $m = 3$ into the left-hand side of the equation.	If $m = 3$, $LHS = 3(3 - 4) + 8$ $= 3(-1) + 8$ $= -3 + 8$ $= 5$ $RHS = 5$
9. Comment on the answers obtained.	Since the LHS and RHS are equal, the solution is $m = 3$.

b. 1. Write the equation.

b. $6\left(\dfrac{x}{2}+5\right)=-18$

2. Divide both sides by 6.

$$\dfrac{6\left(\dfrac{x}{2}+5\right)}{6}=\dfrac{-18}{6}$$

3. Simplify.

$$\dfrac{x}{2}+5=-3$$

4. Subtract 5 from both sides.

$$\dfrac{x}{2}+5-5=-3-5$$

5. Simplify.

$$\dfrac{x}{2}=-8$$

6. Multiply both sides by 2.

$$\dfrac{x}{2}\times 2=-8\times 2$$

7. Simplify.

$$x=-16$$

8. Check the solution by substituting $x=-16$ into the left-hand side of the equation.

If $x=-16$,

$$\text{LHS}=6\left(\dfrac{-16}{2}+5\right)$$
$$=6(-8+5)$$
$$=6(-3)$$
$$=-18$$
$$\text{RHS}=-18$$

9. Comment on the answers obtained.

Since the LHS and RHS are equal, the solution is $x=-16$.

COMMUNICATING — COLLABORATIVE TASK: Modelling and solving word problems

Equipment: paper, pen

1. In groups, select one of the five following word problems:
   - Uyen has to finish reading a 230-page book for her English class in 4 days. She has already read 90 pages. How many pages a day should she read to finish in time?
   - Lucas and Yannis have 13 lollies. They eat 6 of them, buy 10 more, and give 5 of them to one of their friends. How many lollies do Lucas and Yannis have?
   - Toni is making pancakes for breakfast for her family. It takes her 4 minutes to prepare enough batter for 12 pancakes, and 3 minutes to cook each pancake. If Toni wants to serve the 12 pancakes at 9 am, at what time must she start preparing the batter?
   - Amelia, Naomi and Luke are doing a survey for their mathematics class. They have to survey at least 25 people to complete their assignment, and it takes 3 minutes to go through their questionnaire with each person. If they evenly share the work, how long will it take them to survey 30 people?
   - Alexei runs 6 km every Mondays and Thursdays, and 10 km every Saturdays to prepare for a competition. How many kilometres does he run in 8 weeks?
2. Write an equation to model your word problem.
3. Solve the equation.
4. In turns, explain to the other groups how you solved your word problem.
5. Discuss the methods used and work together to determine an efficient problem-solving strategy for this type of word problems.

▶ 7.4.3 Solving simple quadratic equations

eles-6211

- A **quadratic equation** is an equation in which the pronumeral is squared.
- Simple quadratic equations are quadratic equations in the form $x^2 = c$.
- The inverse operation of squaring a number is taking the square root.

DISCUSSION

Consider the following pattern.

$1 \times 1 = 1$	$(-1) \times (-1) = 1$
$2 \times 2 = 4$	$(-2) \times (-2) = 4$
$3 \times 3 = 9$	$(-3) \times (-3) = 9$
$4 \times 4 = 16$	$(-4) \times (-4) = 16$
$5 \times 5 = 25$	$(-5) \times (-5) = 25$

Can you think of a number that, when multiplied by itself, gives a negative number?

- When taking the square root of a positive number, there are two solutions: a positive value and a negative value. For example, $3^2 = 9$ and $(-3)^2 = 9$, so $\sqrt{9} = +3$ and $\sqrt{9} = -3$. This is written as $\sqrt{9} = \pm 3$, where $\pm$ is the plus–minus sign. It is used to indicate both the positive and the negative values. For example, ± 3 is read as positive or negative 3.
- A simple quadratic equation in the form $x^2 = c$, where $c > 0$, has exactly two solutions, $-\sqrt{c}$ and $+\sqrt{c}$, which are written as $\pm \sqrt{c}$.
- The square of a positive number or of a negative number is always positive. For example, $-5 \times -5 = 25$. Therefore, $x^2 = -c$, where $c > 0$, does not have any solution.
- When a quadratic equation represents a real-life situation, it is possible that one of the values of the pronumeral is not valid. For example, say x is the width of a room and $2x$ is its length. When determining the values of x, in metres, for which the area of the room is equal to $72 \, \text{m}^2$ that is solving $x \times 2x = 72$ or $x^2 = 36$, only $x = 6$ is a possible solution in real life, as a length is always a positive quantity.
- Solutions to simple quadratic equations can be given in exact form and as decimal approximations where appropriate. For example, if $x^2 = 5$, then, in exact form, $x = \pm \sqrt{5}$, and as a decimal approximation to 2 decimal places, $x \approx \pm 2.24$.

WORKED EXAMPLE 9 Solving quadratic equations

Solve the following quadratic equations.
a. $x^2 = 16$
b. $6x^2 = 150$

THINK	WRITE
a. 1. Write the equation.	a. $x^2 = 16$
2. Take the square root of both sides.	$x = \pm\sqrt{16}$
3. Write the answer.	$x = \pm 4$
b. 1. Write the equation in the form $x^2 = c$.	b. $6x^2 = 150$ $x^2 = \dfrac{150}{6}$ $x^2 = 25$
2. Take the square root of both sides.	$x = \pm\sqrt{25}$
3. Write the answer.	$x = \pm 5$

WORKED EXAMPLE 10 Solutions in exact form and in decimal approximations

Solve the quadratic equation $s^2 = 27$, giving your answer in exact form and then as a decimal approximation, to 2 decimal places.

THINK	WRITE
1. Write the equation.	$s^2 = 27$
2. Take the square root of both sides.	$s = \pm\sqrt{27}$
3. Give a decimal approximation to 2 decimal places.	$s = \pm 5.20$

 Resources

 Video eLesson Solving linear equations (eles-1895)

 Interactivity Using algebra to solve problems (int-3805)

Exercise 7.4 Solving linear equations and simple quadratic equations

learn on

7.4 Quick quiz on	7.4 Exercise

Individual pathways

■ PRACTISE	■ CONSOLIDATE	■ MASTER
1, 2, 4, 5, 7, 12, 17, 18, 20, 24	3, 6, 8, 10, 13, 15, 19, 21, 25	9, 11, 14, 16, 22, 23, 26

Fluency

1. **WE5** Solve the following one-step equations.
   a. $x + 8 = 7$
   b. $12 + r = 7$
   c. $31 = t + 7$
   d. $w + 4.2 = 6.9$

2. Solve the following one-step equations.
   a. $q - 8 = 11$
   b. $-16 + r = -7$
   c. $21 = t - 11$
   d. $y - 5.7 = 8.8$

3. Solve the following one-step equations.
   a. $v - 21 = -26$
   b. $-3 + n = 8$
   c. $14 = 3 + k$
   d. $142 = z + 151$

4. Solve the following one-step equations.
   a. $5e = 20$
   b. $\dfrac{d}{6} = -7$
   c. $9k = 54$
   d. $-30 = \dfrac{r}{2}$

5. Solve the following one-step equations.
   a. $11d = 88$
   b. $\dfrac{t}{8} = 3$
   c. $7p = -98$
   d. $2.5g = 12.5$

6. Solve the following one-step equations.
   a. $-4m = 28$
   b. $90 = \dfrac{g}{3}$
   c. $16f = 8$
   d. $-4 = \dfrac{L}{18}$

7. **WE6** Solve the following equations using backtracking.

   a. $3m + 5 = 14$ **b.** $-2w + 6 = 16$ **c.** $-5k - 12 = 8$ **d.** $4t - 3 = -15$

8. Solve the following linear equations.

   a. $2(m - 4) = -6$ **b.** $-3(n + 12) = 18$ **c.** $5(k + 6) = -15$ **d.** $-6(s + 11) = -24$

9. Solve the following linear equations.

   a. $2m + 3 = 10$ **b.** $40 = -5(p + 6)$ **c.** $5 - 3g = 14$ **d.** $11 - 4f = -9$

10. **WE7** Solve the following linear equations.

    a. $\dfrac{x}{3} + 2 = 9$ **b.** $\dfrac{x - 5}{4} = 1$ **c.** $\dfrac{m + 3}{2} = -7$ **d.** $\dfrac{h}{-3} + 1 = 5$

11. Solve the following linear equations.

    a. $\dfrac{-m}{5} - 3 = 1$ **b.** $\dfrac{2w}{5} = -4$ **c.** $\dfrac{-3m}{7} = -1$ **d.** $\dfrac{c - 7}{3} = -2$

Understanding

12. **WE8** Solve the following linear equations by doing the same to both sides. This will require more than two steps.

    a. $2(m + 3) + 7 = 3$ **b.** $\dfrac{-2(x + 5)}{5} = 6$ **c.** $\dfrac{5m + 6}{3} = 4$ **d.** $\dfrac{4 - 2x}{3} = 6$

13. Solve the following linear equations by doing the same to both sides. This will require more than two steps.

    a. $\dfrac{3x}{7} - 2 = 1$ **b.** $\dfrac{7f}{9} + 2 = -5$ **c.** $8 - \dfrac{6m}{5} = 2$ **d.** $-9 - \dfrac{5u}{11} = -4$

14. Below is Alex's working to solve the equation $2x + 3 = 14$.

$$2x + 3 = 14$$
$$\frac{2x}{2} + 3 = \frac{14}{2}$$
$$x + 3 = 7$$
$$x + 3 - 3 = 7 - 3$$
$$x = 4$$

   a. Determine whether the solution is correct.
   b. If not, identify the error and show the correct working.

15. Simplify the LHS of the following equations by collecting like terms, and then solve the equations.

    a. $3x + 5 + 2x + 4 = 19$ **b.** $-3y + 7 + 4y - 2 = 9$
    c. $-3m + 6 - 5m + 1 = 15$ **d.** $13v - 4v + 2v = -22$

16. Simplify the LHS of the following equations by collecting like terms, and then solve the equations.

    a. $5w + 3w - 7 + w = 13$ **b.** $w + 7 + w - 15 + w + 1 = -5$
    c. $7 - 3u + 4 + 2u = 15$ **d.** $7c - 4 - 11 + 3c - 7c + 5 = 8$

17. **WE9** Solve the following quadratic equations without using a calculator.

    a. $a^2 = 36$ **b.** $b^2 = 169$ **c.** $c^2 = 64$ **d.** $d^2 = 81$ **e.** $e^2 = 121$

18. **WE10** Solve the following equations, giving your answers as decimal approximations, to 2 decimal places.

    a. $a^2 = 26$ **b.** $b^2 = 48$ **c.** $c^2 = 70$ **d.** $d^2 = 35$ **e.** $e^2 = 99$

19. Solve the following equations, giving your answers correct to 2 decimal places. (*Hint:* First divide both sides by the coefficient of x^2.)

   a. $3x^2 = 45$
   b. $7y^2 = 21$
   c. $8p^2 = 80$
   d. $9s^2 = 85$
   e. $4t^2 = 110$

Communicating, reasoning and problem solving

20. Lyn and Peta together raised $517 from their cake stalls at the school fete. If Lyn raised l dollars and Peta raised $286, write an equation that represents the situation and determine the amount Lyn raised.

21. If four times a certain number equals nine minus a half of the number, determine the number. Show your working.

22. Tom is 5 years old and his dad is 10 times his age, being 50 years old. Explain whether it is possible, at any stage, for Tom's dad to be twice the age of his son.

23. A traffic sign in the shape of a square has an area of 5329 cm^2. What is the length of one side of the sign?

24. A repair person calculates his service fee using the equation $F = 40t + 55$, where F is the service fee in dollars and t is the number of hours spent on the job.

   a. Determine how long a particular job took if the service fee was $155.
   b. Discuss what costs the numbers 40 and 55 could represent in the service fee equation.

25. a. Write an equation that represents the perimeter of the triangle and solve for x.

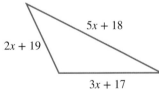

   5x + 18
   2x + 19
   3x + 17
   Perimeter = 184 cm

   b. Write an equation that represents the perimeter of the quadrilateral and solve for x.

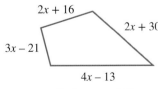

   2x + 16
   2x + 30
   3x − 21
   4x − 13
   Perimeter = 287 cm

26. Lauren earns the same amount for mowing four of the neighbours' lawns every month. Each month she saves all her pay except $30, which she spends on her mobile phone. If she has $600 at the end of the year, calculate how much she earned each month. Show your working in two different ways.

LESSON
7.5 Equations with an unknown on both sides

LEARNING INTENTION

At the end of this lesson you should be able to:
- solve equations that have an unknown on both sides.

▶ 7.5.1 Solving equations and checking solutions

eles-4352

- Some equations have unknowns on both sides of the equation.
- If an equation has unknowns on both sides, eliminate the unknowns from one side and then solve as usual.
- Consider the equation $4x + 1 = 2x + 5$.
- Drawing the equation on a pair of scales looks like the diagram shown.

  - The scales remain balanced if $2x$ is eliminated from both sides.
  - Writing this algebraically, we have:

$$4x + 1 = 2x + 5$$
$$4x + 1 - 2x = 2x + 5 - 2x$$
$$2x + 1 = 5$$

  - We can then solve as usual:

$$2x + 1 - 1 = 5 - 1$$
$$2x = 4$$
$$\frac{2x}{2} = \frac{4}{2}$$
$$x = 2$$

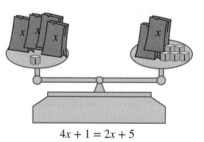

$4x + 1 = 2x + 5$

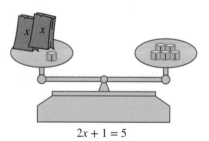

$2x + 1 = 5$

- When solving equations with unknowns on both sides, it is best to remove the unknown with the lowest coefficient from its relevant side.

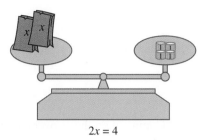

$2x = 4$

- Substitution can be used to check that the answer to an equation you have solved is correct. Once substituted, if the LHS equals the RHS, then your answer is correct. If the LHS does not equal the RHS, then your answer is incorrect and you should have another try.

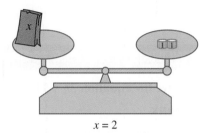

$x = 2$

Solving equations with an unknown on both sides

To solve an equation with pronumerals (unknowns) on both sides, follow the steps below.
1. **Bring all terms with the pronumeral to one side. Do this by removing the pronumeral with the lowest coefficient from its relevant side by using inverse operations.**
2. **Collect like terms with the pronumerals.**
3. **Solve the equation using balancing and inverse operations to determine the solution.**
4. **Check your answer using substitution.**

WORKED EXAMPLE 11 Solving equations with unknowns on both sides

Solve the equation $5t - 8 = 3t + 12$ and check your solution by substitution.

THINK	WRITE
1. Write the equation.	$5t - 8 = 3t + 12$
2. Remove the pronumeral term with the lowest coefficient ($3t$) from the RHS by subtracting it from both sides and simplifying.	$5t - 8 - 3t = 3t + 12 - 3t$ $2t - 8 = 12$
3. Add 8 to both sides and simplify.	$2t - 8 + 8 = 12 + 8$ $2t = 20$
4. Divide both sides by 2 and simplify.	$\dfrac{2t}{2} = \dfrac{20}{2}$ $t = 10$
5. Check the solution by substituting $t = 10$ into the left-hand side and then the right-hand side of the equation.	If $t = 10$, $\text{LHS} = 5t - 8$ $= 5 \times 10 - 8$ $= 50 - 8$ $= 42$

$$If\ t = 10,$$
$$RHS = 3t + 12$$
$$= 3 \times 10 + 12$$
$$= 30 + 12$$
$$= 42$$

6. Comment on the answers obtained.

Since the LHS and RHS are equal, the solution is $t = 10$.

WORKED EXAMPLE 12 Solving equations with unknowns on both sides

Solve the equation $3n + 11 = 6 - 2n$ and check your solution by substitution.

THINK	WRITE
1. Write the equation.	$3n + 11 = 6 - 2n$
2. The inverse of $-2n$ is $+2n$. Therefore, add $2n$ to both sides and simplify.	$3n + 11 + 2n = 6 - 2n + 2n$ $5n + 11 = 6$
3. Subtract 11 from both sides and simplify.	$5n + 11 - 11 = 6 - 11$ $5n = -5$
4. Divide both sides by 5 and simplify.	$\dfrac{5n}{5} = -\dfrac{5}{5}$ $n = -1$
5. Check the solution by substituting $n = -1$ into the left-hand side and then the right-hand side of the equation.	If $n = -1$, $\quad LHS = 3n + 11$ $\qquad = 3 \times (-1) + 11$ $\qquad = -3 + 11$ $\qquad = 8$ If $n = -1$, $\quad RHS = 6 - 2n$ $\qquad = 6 - 2 \times (-1)$ $\qquad = 6 + 2$ $\qquad = 8$
6. Comment on the answers obtained.	Since the LHS and RHS are equal, the solution is $n = -1$.

WORKED EXAMPLE 13 Solving equations with unknowns on both sides

Expand the brackets and then solve the following equations, checking your solution by substitution.
a. $3(s + 2) = 2(s + 7) + 4$ **b.** $4(d + 3) - 2(d + 7) + 4 = 5(d + 2) + 7$

THINK	WRITE
a. 1. Write the equation.	**a.** $\quad 3(s + 2) = 2(s + 7) + 4$
2. Expand the brackets on each side of the equation first and then simplify.	$3s + 6 = 2s + 14 + 4$ $3s + 6 = 2s + 18$

3. Subtract the smaller unknown term $(2s)$ from both sides and simplify.

$$3s + 6 - 2s = 2s + 18 - 2s$$
$$s + 6 = 18$$

4. Subtract 6 from both sides and simplify.

$$s + 6 - 6 = 18 - 6$$
$$s = 12$$

5. Check the solution by substituting $s = 12$ into the left-hand side and then the right-hand side of the equation.

If $s = 12$,
$$\text{LHS} = 3(s + 2)$$
$$= 3(12 + 2)$$
$$= 3(14)$$
$$= 42$$

If $s = 12$,
$$\text{RHS} = 2(s + 7) + 4$$
$$= 2(12 + 7) + 4$$
$$= 2(19) + 4$$
$$= 38 + 4$$
$$= 42$$

6. Comment on the answers obtained.

Since the LHS and RHS are equal, the solution is $s = 12$.

b. 1. Write the equation.

b. $4(d + 3) - 2(d + 7) + 4 = 5(d + 2) + 7$

2. Expand the brackets on each side of the equation first, then simplify.

$$4d + 12 - 2d - 14 + 4 = 5d + 10 + 7$$
$$2d + 2 = 5d + 17$$

3. Subtract the smaller unknown term $(2d)$ from both sides and simplify.

$$2d + 2 - 2d = 5d + 17 - 2d$$
$$2 = 3d + 17$$

4. Rearrange the equation so that the unknown is on the left-hand side of the equation.

$$3d + 17 = 2$$

5. Subtract 17 from both sides and simplify.

$$3d + 17 - 17 = 2 - 17$$
$$3d = -15$$

6. Divide both sides by 3 and simplify.

$$\frac{3d}{3} = -\frac{15}{3}$$
$$d = -5$$

7. Check the solution by substituting $d = -5$ into the left-hand side and then the right-hand side of the equation.

If $d = -5$,
$$\text{LHS} = 4(d + 3) - 2(d + 7) + 4$$
$$= 4(-5 + 3) - 2(-5 + 7) + 4$$
$$= 4(-2) - 2(2) + 4$$
$$= -8$$

If $d = -5$,
$$\text{RHS} = 5(-5 + 2) + 7$$
$$= 5(-3) + 7$$
$$= -15 + 7$$
$$= -8$$

8. Comment on the answers obtained.

Since the LHS and RHS are equal, the solution is $d = -5$.

Equipment: paper, pen, small container or hat

1. In groups, write two equations similar to $4(w + 1) = 3w - 2$ and $3(a + 9) - 2(4a + 7) = 5a + 1$, and place them in the hat.
2. When the class is ready, each group randomly selects an equation to solve.
   - First, use a trial-and-error process to solve the equation. The group might like to nominate a scribe to record the process in a table.
   - Next, solve the equation using inverse operations. Record your steps in solving the problem and check against your answer from the trial-and-error process.
3. When you have recorded the solution to the equation for *both* processes, return the equation to the hat and select another.
4. Repeat the activity and see if you can reduce the number of steps you take in the trial-and-error process.
5. You might like to make the task more challenging by setting a time limit or perhaps having a race — the first group to solve five equations wins.
6. Discuss your results as a class. Which process did you find easier? Which process was quicker?

DISCUSSION

Does it really matter which side of the equation you eliminate the variable from?

 Resources

 Video eLesson Solving linear equations with the pronumeral on both sides (eles-1901)

 Interactivity Equations with the unknown on both sides (int-3806)

Exercise 7.5 Equations with an unknown on both sides **learn**on

| 7.5 Quick quiz on | 7.5 Exercise |

Individual pathways

■ PRACTISE	■ CONSOLIDATE	■ MASTER
1, 4, 6, 9, 13, 17	2, 5, 7, 10, 14, 15, 18	3, 8, 11, 12, 16, 19, 20

Fluency

1. **WE11** Solve the following equations and check your solution by substitution.
   a. $8x + 5 = 6x + 11$
   b. $5y - 5 = 2y + 7$
   c. $11n - 1 = 6n + 19$
   d. $6t + 5 = 3t + 17$

2. Solve the following equations and check your solution by substitution.
   a. $2w + 6 = w + 11$
   b. $4y - 2 = y + 9$
   c. $3z - 15 = 2z - 11$
   d. $5a + 2 = 2a - 10$

3. Solve the following equations, checking your solution by substitution.

   a. $2s + 9 = 5s + 3$
   b. $k + 5 = 7k - 19$
   c. $4w + 9 = 2w + 3$
   d. $7v + 5 = 3v - 11$

4. **WE12** Solve the following equations and check your solution by substitution.

   a. $3w + 1 = 11 - 2w$
   b. $2b + 7 = 13 - b$
   c. $4n - 3 = 17 - 6n$
   d. $7m + 2 = -3m + 22$

5. Solve the following equations, checking your solution by substitution.

   a. $p + 7 = -p + 15$
   b. $5 + m = 5 - m$
   c. $3t - 7 = -17 - 2t$
   d. $16 - 2x = x + 4$

6. **WE13** Expand the brackets and solve the equations, checking your solution by substitution.

   a. $3(2x + 1) + 3x = 30$
   b. $2(4m - 7) + m = 76$
   c. $3(2n - 1) = 4(n + 5) + 1$
   d. $t + 4 = 3(2t - 7)$

7. Expand the brackets and solve the equations, checking your solution by substitution.

   a. $3d - 5 = 3(4 - d)$
   b. $4(3 - w) = 5w + 1$
   c. $2(k + 5) - 3(k - 1) = k - 7$
   d. $4(2 - s) = -2(3s - 1)$

8. Expand the brackets and solve the equations, checking your solution by substitution.

   a. $2m + 3(2m - 7) = 4 + 5(m + 2)$
   b. $3d + 2(d + 1) = 5(3d - 7)$
   c. $4(d + 3) - 2(d + 7) + 5 = 5(d + 12)$
   d. $5(k + 11) + 2(k - 3) - 7 = 2(k - 4)$

Understanding

9. Solve the equation $\dfrac{(x - 2)}{3} + 5 = 2x$.

10. Solve the equation $\dfrac{-3(x + 4)}{7} - \dfrac{5(1 - 3x)}{3} = 3x - 4$.

11. Determine the value of x if the length of the rectangle shown is equal to four times the width.

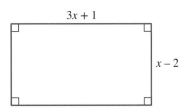

3x + 1

x − 2

12. The two shapes shown have the same area.

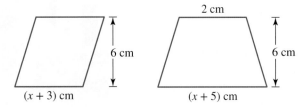

6 cm

(x + 3) cm

2 cm

6 cm

(x + 5) cm

   a. Write an equation to show that the parallelogram and the trapezium have the same area.
   b. Solve the equation for x.
   c. State the dimensions of the shapes.

Reasoning

13. Jasmin is thinking of a number. First she doubles it and adds 2. She realises that if she multiplies it by 3 and subtracts 1, she gets the same result. Determine the value of the number.

14. An animal park advertises two options for pony rides (as shown).

a. Evaluate the number of rides that you need so that option 1 and option 2 cost the same.
b. If you were planning on having only two rides, explain which option you would choose and why.

15. Given the following number line for a line segment PR, determine the length of PR.

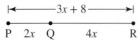

16. A balanced scale contains boxes of Smarties and loose Smarties in the pans. There are three full boxes of Smarties and another box with four Smarties missing in one pan. The other pan contains four empty boxes and 48 loose Smarties. Evaluate the number of Smarties in a full box.

Problem solving

17. In 8 years' time, Tess will be 5 times as old as her age 8 years ago. Evaluate Tess's age at present.

18. The length of the rectangle shown is 7 centimetres less than three times its width.
If the perimeter of the rectangle is the same as the perimeter of the triangle, determine the side lengths of the rectangle and the triangle.

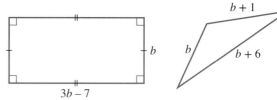

19. You have 12 more than three times the number of marker pens in your pencil case than your friend has in his pencil case. The teacher has 5 more than four times the number of marker pens in your friend's pencil case.

a. Write an expression for the number of marker pens in:
   i. your pencil case
   ii. your teacher's pencil case.
b. You have the same number of marker pens as the teacher. Write an equation to show this.
c. Evaluate the number of marker pens in your friend's pencil case by solving the equation from part b.

20. On her birthday today, a mother is three times as old as her daughter will be in eight years' time. It just so happens that the mother's age today is the same as their house number, which is four times the value of 13 minus the daughter's age.

a. Write each sentence above as an expression.
   (*Hint:* Make the daughter's age the unknown variable.)
b. Use these two expressions to develop an equation and solve it.
c. Determine the ages of the mother and the daughter.
d. Evaluate the number of their house.

LESSON
7.6 Review

7.6.1 Topic summary

EQUATIONS

Balancing equations

- An equation contains a left-hand side, a right-hand side and an equals sign.
- Equations must be kept balanced; that is, whatever is done to one side must also be done to the other side.
  e.g. If 2 is added to the left-hand side, then 2 must be added to the right-hand side.

$$x = 5$$
$$x + 2 = 5 + 2$$
$$x + 2 = 7$$

Linear equations

- A linear equation is an equation in which the power of the unknown variable is 1.
  e.g.

$$2x + 3 = 5$$
$$\frac{x - 3}{4} = 5$$
$$-3(x + 3) = 10$$

Simple quadratic equations

- A simple quadratic equation is an equation in the form $x^2 = c$.
  e.g.

$$x^2 = 45$$
$$m^2 = 36$$

- The inverse operation of squaring a number is taking the square root.

Solving linear equations

- Inverse operations are used to isolate the unknown.
- Addition and subtraction are inverses of each other.
- Multiplication and division are inverses of each other.

Equations with pronumerals on both sides (Extending)

- When an equation has pronumerals on both sides, a more algebraic method is used. Generally, it is best to remove the unknown with the lowest coefficient from its corresponding side.
  e.g.

$$4x - 5 = 3x + 4$$
$$4x - 5 - 3x = 3x + 4 - 3x$$
$$x - 5 = 4$$
$$x - 5 + 5 = 4 + 5$$
$$x = 9$$

Solving simple quadratic equations

- $x^2 = c$ has no solutions when $c < 0$.
- $x^2 = c$ has two solutions when $c > 0$, $x = \pm\sqrt{c}$.
  e.g.

$$3x^2 = 75$$
$$\frac{3x^2}{3} = \frac{75}{3}$$
$$x^2 = 25$$
$$x = \pm 5$$

Checking solutions

- Substitution can be used to check your solution.
  e.g.

$$3x - 5 = 10$$
$$3x - 5 + 5 = 10 + 5$$
$$3x = 15$$
$$\frac{3x}{3} = \frac{15}{3}$$
$$x = 5$$

7.6.2 Project

The Olympic freestyle final

Swimming is one of the most popular sports of the Olympic Games. Over the years the Olympics have been held, competitors have been swimming more quickly and their times have correspondingly reduced.

The table below displays the winning times for the men's and women's 100-metre freestyle final for the Olympic Games from 1960 to 2012.

Year	Men's time (seconds)	Women's time (seconds)
1960	55.2	61.2
1964	53.4	59.5
1968	52.2	60.0
1972	51.2	58.6
1976	50.0	55.7
1980	50.4	54.8
1984	49.8	55.9
1988	48.6	54.9
1992	49.0	54.7
1996	48.7	54.5
2000	48.3	53.8
2004	48.2	53.8
2008	47.2	53.1
2012	47.5	53.0

Note: All times have been rounded to 1 decimal place.

The women's times have been graphed as shown. There is no straight line that passes through all the points, so a line of best fit has been selected to approximate the swimming times. The equation of this line is given by $t = 59.9 - 0.1504x$, where t represents the time taken and x represents the number of years after 1956.

1. What year is represented by $x = 16$?
2. Substitute $x = 16$ into the equation to find an approximation for the time taken. How close is this approximation to the actual time given in the table?
3. What x-value would you use for the 2016 Olympics? Use the equation to predict the women's 100-metre freestyle final time for the 2016 Olympics.
4. The women's time in 2016 was 52.7 seconds. How does your prediction compare with the actual time?
5. Use a calculator to solve the equation to the nearest whole number when $t = 55.7$ seconds. What year does your solution represent?

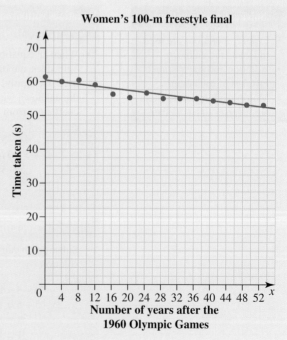

Women's 100-m freestyle final

Time taken (s)

Number of years after the 1960 Olympic Games

An equation to approximate the men's times is given by $t = 53.4 - 0.1299x$. Use this equation to answer the following questions.

6. The men's time in 1984 was 49.8 seconds. How does this compare with the time obtained from the equation?

7. Use the equation to predict the men's time for the 2016 Olympic Games.

8. Plot the men's times on the set of axes provided. Between the points, draw in a line of best fit.

9. Extend your line of best fit so that it passes the x-value that represents the year 2016. How does this value compare with the prediction you obtained in question 7 above?

10. The men's time in 2016 was 47.6 seconds. How does your prediction compare with the actual time?

11. Use your graphs to compare both the men's and women's times with the actual results obtained during the 2004 Athens Olympics.

12. How long will it be before the men and women are swimming identical times?

Investigate this by plotting the times given on the same set of axes and drawing a line of best fit. Extend both lines until they intersect. The point of intersection represents the time when the swimming times are identical. Present your findings on graph paper.

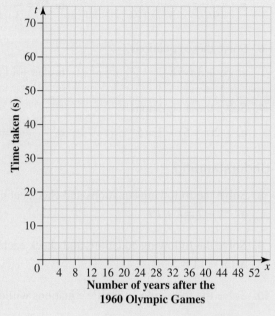

Number of years after the 1960 Olympic Games

---

on Resources

 Interactivities Crossword (int-2634)
              Sudoku puzzle (int-3190)

Exercise 7.6 Review questions

learn on

Fluency

1. **a.** Write an equation that is represented by the diagram shown.
   b. Show what happens when you take 2 from both sides, and write the new equation.

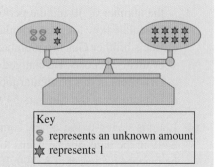

Key
 represents an unknown amount
 represents 1

2. **MC** If $x = 5$, state which of the following equations is false.

   A. $x + 2 = 7$ **B.** $3x = 12$

   C. $-2x = -10$ **D.** $\dfrac{x}{5} = 1$

3. **MC** If $x = 4$, state which of the following equations is false.

   A. $\dfrac{2x}{3} = \dfrac{8}{3}$ **B.** $-2x = -8$ **C.** $2x - 8 = 0$ **D.** $x - 5 = 1$

4. Solve these equations.

a. $z + 7 = 18$

b. $-25 + b = -18$

c. $-8.7 = \dfrac{l}{5}$

5. Solve these equations.

a. $-\dfrac{8}{9} = z - \dfrac{4}{3}$

b. $9t = \dfrac{1}{3}$

c. $-\dfrac{6}{13} = \dfrac{h}{8}$

6. Solve these equations.

a. $5v + 3 = 18$

b. $5(s + 11) = 35$

c. $\dfrac{d - 7}{4} = 10$

7. Solve these equations.

a. $-2(r + 5) - 3 = 5$

b. $\dfrac{2y - 3}{7} = 9$

c. $\dfrac{x}{5} - 3 = 2$

Understanding

8. Solve the following equations and check each solution.

a. $5k + 7 = k + 19$

b. $4s - 8 = 2s - 12$

9. Solve the following equations and check each solution.

a. $3t - 11 = 5 - t$

b. $5x + 2 = -2x + 16$

10. Solve the following quadratic equations without using a calculator.

a. $n^2 = 64$

b. $\dfrac{1}{4}y^2 = 36$

c. $5p^2 = 20$

11. Solve the following equations, giving your answers as decimal approximations to 2 decimal places.

a. $n^2 = 7$

b. $\dfrac{1}{5}x^2 = 3$

c. $\pi r^2 = 12$

12. Expand the brackets first and then solve the following equations.

a. $5(2v + 3) - 7v = 21$

b. $3(m - 4) + 2m = m + 8$

Communicating, reasoning and problem solving

13. Rae the electrician charges $80 for a call-out visit and then $65 per
half-hour.
a. Develop an equation for her fees, where C is her call-out cost and t is
the number of 30-minute periods she spent on the job.
b. Calculate how long a particular job took if she charged $275.
c. Rae's brother Gus is a plumber and uses the equation $C = 54t + 86$ to
evaluate his costs. He charged $275 for one job. Calculate how long
Gus spent at this particular job.
d. Explain what the numbers 54 and 86 could mean.

14. Liam is saving to buy a new computer, which costs $3299. So far he has
$449 in the bank and he wants to make regular deposits each month until
he reaches his target of $3299. If he wants to buy the computer in
8 months' time, calculate how much he needs to save as a monthly
deposit.

15. Three children were each born 2 years apart. Their combined ages add to 63 years. Determine the age
of the eldest child.

16. A rectangular vegetable patch is $(3x + 4)$ metres long and $(2x - 5)$ metres wide. Its perimeter is 58 metres. Determine the dimensions of the vegetable patch.

17. You lend three friends a total of $45. You lend the first friend x dollars. To the second friend you lend $5 more than you lent to the first friend. To the third friend you lend three times as much as you lent the second friend. Evaluate how much each person receives.

18. At the end of the year, Katie's teacher gave the class their average scores. They had done four tests for the year. Katie's average was 76%. She had a mark of 83% for Probability, 72% for Geometry and 91% for Measurement but had forgotten what she got for Algebra. Write an equation to show how Katie would work out her Algebra test score and then solve this equation.

19. A truck carrying 50 bags of cement weighs 7.43 tonnes. After delivering 15 bags of cement, the truck weighs 6.755 tonnes. Determine how much an empty truck would weigh.

20. While on holiday, Amy hired a bicycle for $9 an hour and paid $3 for the use of a helmet. Her brother, Ben, found a cheaper hire place, which charged $6 per hour, but the hire of the helmet was $5 and he had to pay $5 for insurance. Each hire place measures the time and charges in 20-minute blocks. They both ended up paying the same amount for the same number of hours. Construct a table of values to evaluate how long they were gone and how much it cost.

21. You are 4 times as old as your sister. In 8 years' time you will be twice as old as your sister. Determine your ages now.

22. Michael checked his bank balance before going shopping. He had $450. While shopping, he paid with his debit card. He bought two suits, which each cost the same, and three pairs of shoes, each of which cost half the price of a suit. He also had lunch for $12. When he checked his balance again, he was $33 overdrawn. Evaluate the cost of one suit.

23. Evaluate the greatest possible perimeter of a triangle with sides $5x + 20$, $3x + 76$ and $x + 196$, given that the triangle is isosceles. All sides are in mm.

on To test your understanding and knowledge of this topic, go to your learnON title at www.jacplus.com.au and complete the **post-test**.

Answers

Topic 7 Equations

7.1 Pre-test

1. B
2. $x = 7$
3. a. 15 b. 9
4. C
5. a. $p = -6$ b. $q = -20$
6. a. $x = 8$ b. $x = 0$ c. $x = -45$ d. $x = -12$
7. a. $x = -5$ or $x = +5$
   b. $y = -8$ or $y = +8$
   c. $t = \pm 7.28$
8. The side length is 12 cm.
9. $x = \pm 11$
10. 15
11. $p = 5$
12. 10
13. a. $x = 7$ b. $x = -2$ c. $x = 1\frac{1}{2}$
14. Each earned \$525 for working 7.5 hours.
15. \$2

7.2 Backtracking and inverse operations

1. a. 8 b. 20 c. -4 d. -19
2. a. -1 b. -1 c. -2 d. -44
3. a. 8 b. 4 c. 1.075 d. 9
4. a. $2x - 7$ b. $2(w - 7)$ c. $-5s + 3$
   d. $-5(n + 3)$ e. $\frac{m}{2} + 7$ f. $\frac{y + 7}{2}$
5. a. $\frac{6z - 3}{2}$ b. $\frac{-3(d + 5)}{4}$ c. $\frac{2e}{5} + 1$
   d. $4(3 - x)$ e. $\frac{-2(w - 5)}{7}$ f. $-3(z + 6) - 11$
6. a. $\frac{v - 3}{6} - 8$ b. $-7(8m - 4)$ c. $\frac{-5k}{6} + 2$
   d. $\frac{-5p - 7}{3}$

7. a.
$$x \xrightarrow{+7} x + 7 \xrightarrow{\times 2} 2(x + 7)$$
b.
$$x \xrightarrow{-8} x - 8 \xrightarrow{\times -2} -2(x - 8)$$
c.
$$m \xrightarrow{\times 3} 3m \xrightarrow{-6} 3m - 6$$
d.
$$m \xrightarrow{\times -3} -3m \xrightarrow{-6} -3m - 6$$
e.
$$x \xrightarrow{-5} x - 5 \xrightarrow{\div 8} \frac{x - 5}{8}$$
f.
$$x \xrightarrow{\div 8} \frac{x}{8} \xrightarrow{-5} \frac{x}{8} - 5$$

8. a.
$$x \xrightarrow{\times -5} -5x \xrightarrow{+11} -5x + 11$$
b.
$$x \xrightarrow{\times -1} -x \xrightarrow{+11} -x + 11$$
c.
$$x \xrightarrow{\times -1} -x \xrightarrow{-13} -x - 13$$
d.
$$x \xrightarrow{\times -2} -2x \xrightarrow{+5} -2x + 5$$
e.
$$x \xrightarrow{\times 3} 3x \xrightarrow{-7} 3x - 7 \xrightarrow{\div 4} \frac{3x - 7}{4}$$
f.
$$x \xrightarrow{-2} x - 2 \xrightarrow{\times -3} -3(x - 2) \xrightarrow{\div 4} \frac{-3(x - 2)}{4}$$

9. a.
$$x \xrightarrow{+5} x + 5 \xrightarrow{\div 8} \frac{x + 5}{8} \xrightarrow{-3} \frac{x + 5}{8} - 3$$
b.
$$x \xrightarrow{\div 5} \frac{x}{5} \xrightarrow{-2} \frac{x}{5} - 2 \xrightarrow{\times -7} -7\left(\frac{x}{5} - 2\right)$$
c.
$$x \xrightarrow{\times 2} 2x \xrightarrow{\div 7} \frac{2x}{7} \xrightarrow{+4} \frac{2x}{7} + 4 \xrightarrow{\times 3} 3\left(\frac{2x}{7} + 4\right)$$
d.
$$x \xrightarrow{\times 6} 6x \xrightarrow{\div 11} \frac{6x}{11} \xrightarrow{-3} \frac{6x}{11} - 3 \xrightarrow{\times \frac{1}{4}} \frac{1}{4}\left(\frac{6x}{11} - 3\right)$$

10. a.
$$F \xrightarrow{-32} F - 32 \xrightarrow{\times 5} 5(F - 32) \xrightarrow{\div 9} \frac{5(F - 32)}{9}$$
   b. 10
   c. 95
11. a. Strategies include 'guess and check'. Backtracking will not work as the pronumeral w appears more than once in the equation and cannot be simplified.
    b. $w = 15$ cm
12. Both answers are incorrect. The correct expression is $\frac{2(x + 3)}{5}$.
13. $2\left(\frac{x}{3} + 4\right)$
$$x \xrightarrow{\div 3} \frac{x}{3} \xrightarrow{+4} \frac{x}{3} + 4 \xrightarrow{\times 2} 2\left(\frac{x}{3} + 4\right)$$

14. a. See the flowchart at the bottom of the page.*

   b. It is 5.

   c. It is always 5.

15. a. $4x + 8$

   b. Matthew $= 3$ chocolates

      Bart $= 9$ chocolates

7.3 Keeping equations balanced

1. a. $x + 5 = 11$ **b.** $7x = 42$

  c. $x - 4 = 2$ **d.** $\dfrac{x}{3} = 2$

2. a. $-4x = -24$ **b.** $-x = -6$

  c. $-x = -6$ **d.** $x - 9 = -3$

3. a. $\dfrac{2x}{3} = 4$ **b.** $\dfrac{3x}{2} = 9$

  c. $x - \dfrac{2}{3} = 5\dfrac{1}{3}$ **d.** $x + \dfrac{5}{6} = 6\dfrac{5}{6}$

4. a. $2x = 4$ **b.** $x = 2$

5. a. $x + 3 = 5$ **b.** $x = 2$

6. a. $3x + 1 = 7$ **b.** $3x = 6$ **c.** $x = 2$

7. a. $2x + 1 = 5$

   b. Take 1 from both sides $(2x = 4)$, then halve the amount on each side $(x = 2)$.

8. B

9. D

10. C

11. C

12. Both sides of the equation have been divided by 2 $(\div 2)$.

13. a. $5x = 3$ **b.** $4x + 2.70 = 9.90$

   c. $3x + 500 = 920$

14. No, the equations are different.

$2x + 4 = 12$: x is multiplied by 2 and then 4 is added to get the answer of 12.

$2(x + 4) = 12$: 4 is added to x and the result is multiplied by 2 to get the answer of 12.

15. a. $m = 1$ **b.** $n = 8$ **c.** $p + 2 = 4$

16. Errors are shown in pink.

	Error	Correct working
a.	$6x = 8$	$2x + 4 = 8$
b.	$2x - 1 = 5$	$2x - 1 = 3$
c.	$10x = -1$	$10x = 20$

17. The least number of weighings is two. Take 6 unweighed coins and place 1 coin on each side of the balance to determine which is the heaviest.

If the coins balance, the heavy coin is the coin not on the balance. If the coins do not balance, the heavy coin is on the heavy side of the balance.

7.4 Solving linear equations and simple quadratic equations

1. a. $x = -1$ **b.** $r = -5$ **c.** $t = 24$ **d.** $w = 2.7$

2. a. $q = 19$ **b.** $r = 9$ **c.** $t = 32$ **d.** $y = 14.5$

3. a. $v = -5$ **b.** $n = 11$ **c.** $k = 11$ **d.** $z = -9$

4. a. $e = 4$ **b.** $d = -42$ **c.** $k = 6$ **d.** $r = -60$

5. a. $d = 8$ **b.** $t = 24$ **c.** $p = -14$ **d.** $g = 5$

6. a. $m = -7$ **b.** $g = 270$

   c. $f = \dfrac{1}{2}$ **d.** $l = -72$

7. a. $m = 3$ **b.** $w = -5$

   c. $k = -4$ **d.** $t = -3$

8. a. $m = 1$ **b.** $n = -18$

   c. $k = -9$ **d.** $s = -7$

9. a. $m = 3.5$ **b.** $p = -14$

   c. $g = -3$ **d.** $f = 5$

10. a. $x = 21$ **b.** $x = 9$

   c. $m = -17$ **d.** $h = -12$

11. a. $m = -20$ **b.** $w = -10$

   c. $m = 2\dfrac{1}{3}$ **d.** $c = 1$

12. a. $m = -5$ **b.** $x = -20$

   c. $m = 1\dfrac{1}{5}$ **d.** $x = -7$

13. a. $x = 7$ **b.** $f = -9$

   c. $m = 5$ **d.** $u = -11$

14. a. The solution is not correct.

   b. Alex should have subtracted 3 first.

15. a. $x = 2$ **b.** $y = 4$

   c. $m = -1$ **d.** $v = -2$

16. a. $w = 2\dfrac{2}{9}$ **b.** $w = \dfrac{2}{3}$

   c. $u = -4$ **d.** $c = 6$

17. a. $a = \pm 6$ **b.** $b = \pm 13$

   c. $c = \pm 8$ **d.** $d = \pm 9$

   e. $d = \pm 11$

18. a. $a \approx \pm 5.10$ **b.** $b \approx \pm 6.93$

   c. $c \approx \pm 8.37$ **d.** $d \approx \pm 5.92$

   e. $d \approx \pm 9.95$

19. a. $x = \pm 3.97$ **b.** $y = \pm 1.73$

   c. $p = \pm 3.16$ **d.** $s = \pm 3.07$

   e. $t = \pm 5.24$

20. $l + 286 = 517$, $l = \$231$

21. 2

22. Yes, when Tom is 45.

23. The side length is 73 cm.

---

*14. a.

$$\boxed{n} \xrightarrow{\times 2} \boxed{2n} \xrightarrow{+\,10} \boxed{2n + 10} \xrightarrow{\div\,2} \boxed{\dfrac{2n + 10}{2}} \xrightarrow{-\,n} \boxed{\dfrac{2n + 10}{2} - n}$$

24. a. $2\dfrac{1}{2}$ hours

b. 40 represents the hourly rate ($40 per hour). The 55 could be a call-out fee covering travel costs and other expenses ($55 call-out or flat fee).

25. a. $10x + 54 = 184, x = 13$ cm

b. $11x + 12 = 287, x = 25$ cm

26. $80 per month

7.5 Equations with an unknown on both sides

1. a. $x = 3$ **b.** $y = 4$ **c.** $n = 4$ **d.** $t = 4$

2. a. $w = 5$ **b.** $y = 3\dfrac{2}{3}$ **c.** $z = 4$ **d.** $a = -4$

3. a. $s = 2$ **b.** $k = 4$ **c.** $w = -3$ **d.** $v = -4$

4. a. $w = 2$ **b.** $b = 2$ **c.** $n = 2$ **d.** $m = 2$

5. a. $p = 4$ **b.** $m = 0$ **c.** $t = -2$ **d.** $x = 4$

6. a. $x = 3$ **b.** $m = 10$ **c.** $n = 12$ **d.** $t = 5$

7. a. $d = 2\dfrac{5}{6}$ **b.** $w = 1\dfrac{2}{9}$ **c.** $k = 10$ **d.** $s = -3$

8. a. $m = 11\dfrac{2}{3}$ **b.** $d = 3\dfrac{7}{10}$

c. $d = -19$ **d.** $k = -10$

9. $x = \dfrac{13}{5}$

10. $x = -\dfrac{13}{33}$

11. $x = 9$

12. a. $6(x + 3) = \dfrac{1}{2} \times 6 \times (x + 5 + 2)$

b. $x = 1$ cm

c. Parallelogram: base $= 4$ cm
Trapezium: base $= 6$ cm

13. 3

14. a. $2.5x + 10 = 3.5x + 5$
$x = 5$
Both options cost the same for 5 rides.

b. Option 1: $2.5 \times 2 + 10 = \$15$ rides
Option 2: $3.5 \times 2 + 5 = \$12$
Therefore, you would choose option 2.

15. 16 units

16. 13 Smarties

17. 12

18. $2(3b - 7) + 2b = 3b + 7$
$6b - 14 + 2b = 3b + 7$
$5b = 21$
$b = 4.2$
Rectangle: 4.2 and 5.6. Triangle: 4.2, 5.2 and 10.2

19. a. i. $3x + 12$ **ii.** $4x + 5$

b. $3x + 12 = 4x + 5$

c. $x = 7$

20. a. Let daughter's age $= x$.
$3(x + 8)$ and $4(13 - x)$

b. $3x + 24 = 52 - 4x$
$x = 4$

c. The daughter is 4 and the mother is 36.

d. The number of their house is 36.

Project

1. 1976

2. $t = 57.5$ s, 1.1 s away from actual time.

3. $x = 56$, $t = 51.5$ s

4. Sample responses can be found in the worked solutions in the online resources.

5. $x = 28$; the year is 1988.

6. $t = 50.3$ s, 0.5 s difference

7. $t = 46.1$ s

8–12. Sample responses can be found in the worked solutions in the online resources.

7.6 Review questions

1. a. $2x + 2 = 8$

b. $2x = 6$

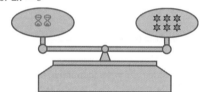

2. B

3. D

4. a. 11 **b.** 7 **c.** -43.5

5. a. $\dfrac{4}{9}$ **b.** $\dfrac{1}{27}$ **c.** $-\dfrac{48}{13}$

6. a. $v = 3$ **b.** $s = -4$ **c.** $d = 47$

7. a. $r = -9$ **b.** $y = 33$ **c.** $x = 25$

8. a. $k = 3$ **b.** $s = -2$

9. a. $t = 4$ **b.** $x = 2$

10. a. $n = \pm 8$ **b.** $y = \pm 12$ **c.** $p = \pm 2$

11. a. $n = \pm 2.65$ **b.** $x = \pm 3.87$ **c.** $r = \pm 1.95$

12. a. $v = 2$ **b.** $m = 5$

13. a. $C = 65t + 80$

b. $1\dfrac{1}{2}$ hours

c. $1\dfrac{3}{4}$ hours

d. 54 represents the fee charged per half-hour ($54 per 30 min). 86 represents a call-out fee ($86 call-out or flat fee).

14. $356.25

15. The eldest is 23 years old.

16. $22\,\text{m} \times 7\,\text{m}$

17. $5, $10, $30

18. $\dfrac{83 + 72 + 91 + x}{4} = 76$
Katie scored 58% for her Algebra test.

19. The empty truck would weigh 5.18 tonnes.

20. 2 hours and 24 minutes; $24

Time (mins)	Amy	Ben
0	$3	$10
20	$6	$12
40	$9	$14
60	$12	$16
80	$15	$18
100	$18	$20
120	$21	$22
140	$24	$24

21. You are 16 and your sister is 4.

22. $134.57

23. 832 mm

8 Length, area and volume

LESSON SEQUENCE

LESSON
8.1 Overview

Why learn this?

Measurement is used in many aspects of our everyday life. You would measure your feet when you need new shoes, the area of your backyard to lay new grass on, or the correct amount of flour and sugar to bake a cake. Being able to measure and having a good understanding of length, area, volume and time is particularly important and helpful. How would you know how much paint to buy if you were unable to calculate the area of the room that needs painting? How would you know the time in London if you were unable to calculate the time difference? Many professions rely on measurement — imagine being a dressmaker, designer, architect or builder without a good understanding of measurement. It would be very difficult for you to complete your work! Nurses and doctors use measurement to administer the correct amount of medication and take our body temperature and blood pressure.

Without measurement it would be difficult for scientists to conduct experiments and draw conclusions. Professional athletes use measurement to estimate the distance needed to make a pass or goal, or to determine which club to use to land the golf ball in the perfect spot. Measurements are used so often that you may not even realise when you are measuring something. A good understanding of measurement and being able to calculate length, area, volume and time is crucial for everyday life.

Hey students! Bring these pages to life online

Watch videos

Engage with interactivities

Answer questions and check solutions

Find all this and MORE in jacPLUS

Reading content and rich media, including interactivities and videos for every concept

Extra learning resources

Differentiated question sets

Questions with immediate feedback, and fully worked solutions to help students get unstuck

1. **MC** Convert 0.23 km into metres.

   A. 2.3 m   **B.** 23 m   **C.** 230 m   **D.** 203 m

2. **MC** Calculate the area of a triangle with a base of 14 cm and a perpendicular height of 10 cm.

   A. 140 m   **B.** 70 m   **C.** 140 cm^2 **D.** 70 cm^2

3. A square of material with a side length of 11 cm has a tiny square (with a side length of 1 cm) cut out of it. Calculate the area of the remaining material.

4. **MC** A cylinder with a face area of 95 cm^2 and a height of 10 cm has a volume of:

   A. 95 cm^3 **B.** 9.5 cm^3 **C.** 950 cm^3 **D.** 0.95 cm^3

5. **MC** Convert 0.62 cm^2 into mm^2.

   A. 620   **B.** 62   **C.** 6.2   **D.** 6200

6. **a.** Calculate the circumference of a circle with a radius of 5.5 cm, correct to 1 decimal place.
   b. Calculate the perimeter of a semicircle with a diameter of 10 cm, correct to 1 decimal place.

7. Calculate the area of a kite with diagonal lengths of 10 cm and 13 cm.

8. An upright flowerpot viewed from the side appears to be a trapezium. The diameter of the top is 26 cm and the diameter of the base is 18 cm. The height of the pot is 25 cm. Determine the area of the trapezium.

9. A piece of wire is in the shape of a circle with a diameter of 11 cm. That wire is then bent to form a square with the same area as the circle.
   Determine the side length of the square, correct to 1 decimal place.

10. A shed is in the shape of a rectangular box with a triangular prism on top. The shed is 12 m long. The base of the rectangular box is 4 m wide and the walls are 2.5 m high. The full height, including the walls and the height of the triangular prism, is 5 m.
    Calculate the volume of the shed to the nearest cubic metre.

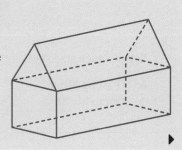

11. Calculate the area of the following shape (correct to 2 decimal places).

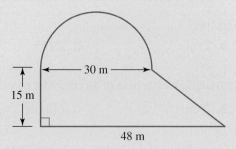

12. Calculate the area of the following shape.

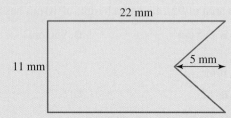

13. A fenced area in a garden is a 15 m by 12 m rectangle. Inside the fenced area is a 12 m by 8 m rectangular pool. The area from the edge of the pool to the fence will be filled with concrete to a depth of 150 mm. The cost of the concrete is \$125 per m^3. Evaluate the cost of the concrete around the pool.

14. **MC** A cylinder-shaped silo has a volume of 75.40 m^3 (correct to 2 decimal places). If the silo has a radius of 2 meters, the height of the silo must be closest to:
A. 4 m
B. 6 m
C. 8 m
D. 12 m

15. **MC** Astor wants to make a rectangular fish tank and needs to know how much glass is required. The tank will be 80 cm long, 40 cm wide and 30 cm tall. Each piece of the tank must be cut from a single piece of glass. The lid will be made of plastic and does not need to be included in the calculation.

Select which size sheets of glass would be best to use to ensure minimum wastage.

A. 80 cm by 80 cm
B. 120 cm by 80 cm
C. 200 cm by 70 cm
D. 170 cm by 80 cm

LESSON
8.2 Length and perimeter

LEARNING INTENTION

At the end of this lesson you should be able to:
- convert one unit of length to another unit of length
- calculate the perimeter of a closed shape.

▶ 8.2.1 Units of length

eles-4041

- Metric units of length include millimetres (mm), centimetres (cm), metres (m) and kilometres (km).
- The figure below shows how to convert between different units of length. Later in this topic, you'll see how it can be adapted to convert units of area and volume.

Converting units of length

To convert between the units of length, use the following conversion chart:

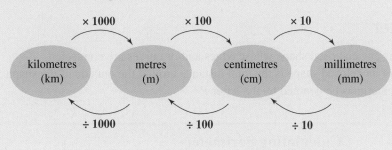

- When converting from a larger unit to a smaller unit, multiply by the conversion factor.
- When converting from a smaller unit to a larger unit, divide by the conversion factor.

WORKED EXAMPLE 1 Converting length

Complete the following metric length conversions.
a. $1.027\text{ m} = \underline{\hspace{1.5cm}}\text{cm}$
b. $0.0034\text{ km} = \underline{\hspace{1.5cm}}\text{m}$
c. $76\,500\text{ m} = \underline{\hspace{1.5cm}}\text{km}$
d. $3.069\text{ m} = \underline{\hspace{1.5cm}}\text{mm}$

THINK

a. Look at the conversion chart above. To convert metres to centimetres, multiply by 100. Multiplying by 100 is the same as moving the decimal point 2 places to the right.

WRITE

a.
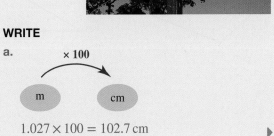

$1.027 \times 100 = 102.7\text{ cm}$

▶

---

b. To convert kilometres to metres, we need to multiply by 1000.
Multiplying by 1000 is the same as moving the decimal point 3 places to the right.

b.

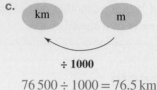

$$0.0034 \times 1000 = 3.4 \, \text{m}$$

c. To convert metres to kilometres, divide by 1000.
Dividing by 1000 is the same as moving the decimal point 3 places to the left.

c.

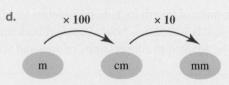

$$76\,500 \div 1000 = 76.5 \, \text{km}$$

d. To convert metres to millimetres, multiply by 100 and then by 10.
This is the same as multiplying by 1000, so the decimal point can be moved 3 places to the right.

d.

$$3.069 \times 100 \times 10 = 3069 \, \text{mm}$$

▶ 8.2.2 Perimeter

eles-4042

- A closed shape is any enclosed shape whose edges and/or curves are connected.
- The **perimeter** of any closed shape is the total distance around the outside of the shape.
- Perimeter can sometimes be denoted by the letter P.

> **Calculating perimeter**
>
> **To calculate the perimeter of a shape:**
> 1. **identify the length of each side**
> 2. **change all lengths to the same unit if needed**
> 3. **add all side lengths together.**

WORKED EXAMPLE 2 Calculating the perimeter of a given shape

Calculate the perimeter of each of the shapes below.

a. A kite

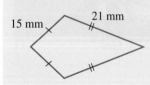

15 mm
21 mm

b. A trapezium

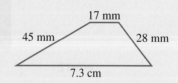

17 mm
45 mm
28 mm
7.3 cm

c. An irregular shape

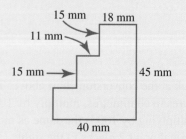

15 mm
18 mm
11 mm
15 mm
45 mm
40 mm

THINK	WRITE/DRAW
a. 1. Lines with the same marking are equal in length. All measurements are in the same units. Calculate the perimeter by adding the lengths of all four sides.	**a.** $P = 15 + 15 + 21 + 21$ $= 72$
2. State the answer and include the units.	The perimeter of the kite shown is 72 mm.
b. 1. Notice that the measurements are not all in the same metric unit. Convert to the smaller unit (in this case convert 7.3 cm to mm).	**b.** $7.3\,\text{cm} = 7.3 \times 10\,\text{mm}$ $= 73\,\text{mm}$
2. Calculate the perimeter by adding the lengths of all four sides.	$P = 45 + 17 + 28 + 73$ $= 163$
3. State the answer and include the units.	The perimeter of the trapezium shown is 163 mm.
c. 1. Ensure all measurements are in the same unit.	**c.** 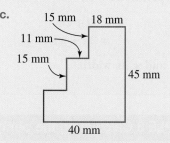 Pink line $= 40 - (18 + 11) = 11$ mm Purple line $= 45 - (15 + 15) = 15$ mm
2. Determine the lengths of the unknown sides.	
3. Calculate the perimeter by adding the lengths of all sides.	$P = 45 + 18 + 15 + 11 + 15 + 11 + 15 + 40$ $= 170$
4. State the answer and include the units.	The perimeter of the irregular shape shown is 170 mm.

COMMUNICATING — COLLABORATIVE TASK: Different ways to calculate perimeter

There are several approaches to calculating perimeter, particularly with composite shapes. Analyse the shape shown.

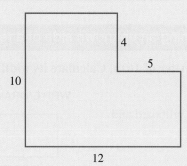

In pairs, discuss different ways to calculate the perimeter of the shape. Share with the rest of the class. How many different ways can you come up with?

8.2.3 Calculating the perimeter of a square and a rectangle

eles-4043

- The perimeter of some common shapes can be determined through the use of a formula.
- The perimeter (*P*) of a square and a rectangle can be calculated using the following formulas.

Perimeter of a square

For a square, the perimeter *P* is:

$$P = 4l$$

where *l* is its side length.

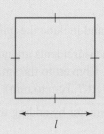

Perimeter of a rectangle

For a rectangle, the perimeter *P* is:

$$P = 2l + 2w$$
$$= 2(l + w)$$

where *l* is its length and *w* its width.

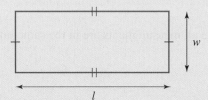

WORKED EXAMPLE 3 Calculating the perimeter of a rectangle

Calculate the perimeter of a rectangular block of land that is 20.5 m long and 9.8 m wide.

THINK	WRITE/DRAW
1. Draw a diagram of the block of land and include its dimensions.	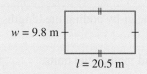
2. Write the formula for the perimeter of a rectangle.	$P = 2(l + w)$
3. Substitute the values of *l* and *w* into the formula and evaluate.	$P = 2 \times (20.5 + 9.8)$ $= 2 \times 30.3$ $= 60.6$
4. State the answer and include the units.	The perimeter of the block of land is 60.6 m.

WORKED EXAMPLE 4 Determining an unknown side of a rectangle given its perimeter

A rectangular billboard has a perimeter of 16 m. Calculate its width if the length is 4.5 m.

THINK	WRITE/DRAW
1. Draw a diagram of the rectangular billboard and include its known measurements.	 $P = 16 \text{ m}$
2. Write the formula for the perimeter of a rectangle.	$P = 2(l + w)$ $= 2l + 2w$

3. Substitute the values of P and l into the formula and solve the equation:

$$16 = 2 \times 4.5 + 2w$$
$$16 = 9 + 2w$$

- Subtract 9 from both sides.

$$16 - 9 = 9 - 9 + 2w$$
$$7 = 2w$$

- Divide both sides by 2.

$$\frac{7}{2} = \frac{2w}{2}$$

$$3.5 = w$$

- Simplify if appropriate.

$$w = 3.5$$

4. State the answer and include the units.

The width of the rectangular billboard is 3.5 m.

 Resources

▶ **Video eLesson** Perimeter (eles-1874)

✦ **Interactivities** Units of length (int-3779)
Perimeter (int-3780)
Perimeter of squares and rectangles (int-3781)

Exercise 8.2 Length and perimeter

learn

8.2 Quick quiz on	8.2 Exercise

Individual pathways

■ PRACTISE	■ CONSOLIDATE	■ MASTER
1, 4, 7, 10, 13, 16, 19, 21, 24	2, 5, 8, 11, 14, 17, 20, 22, 25	3, 6, 9, 12, 15, 18, 23, 26

Fluency

1. **WE1** Complete the following metric length conversions.

   a. 20 mm = _____ cm
   b. 13 mm = _____ cm
   c. 130 mm = _____ cm
   d. 1.5 cm = _____ mm
   e. 0.03 cm = _____ mm
   f. 1.005 cm = _____ mm

2. Fill in the gaps for each of the following metric length conversions.

   a. 2.8 km = _____ m
   b. 0.034 m = _____ cm
   c. 2400 mm = _____ m
   d. 1375 mm = _____ m
   e. 2.7 m = _____ mm
   f. 30.05 cm = _____ mm

3. Fill in the gaps for each of the following metric length conversions.

   a. 0.08 m = _____ mm
   b. 6.071 km = _____ m
   c. 670 cm = _____ m
   d. 0.0051 km = _____ m
   e. 0.0005 cm = _____ mm
   f. 1.75 m = _____ km

4. **WE2** Calculate the perimeters of the shapes below.

a.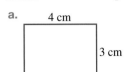
4 cm
3 cm

b.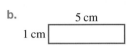
5 cm
1 cm

c.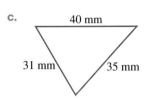
40 mm
31 mm
35 mm

d.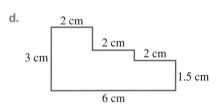
2 cm
2 cm
2 cm
3 cm
1.5 cm
6 cm

5. Calculate the perimeters of the shapes below.

a.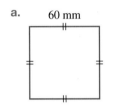
60 mm

b.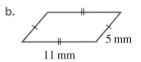
5 mm
11 mm

c.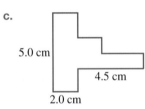
5.0 cm
4.5 cm
2.0 cm

d.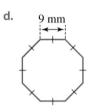
9 mm

6. Determine the perimeters of the shapes below.

a.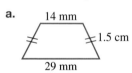
14 mm
1.5 cm
29 mm

b.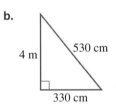
530 cm
4 m
330 cm

c.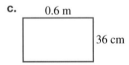
0.6 m
36 cm

d.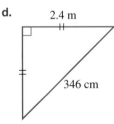
2.4 m
346 cm

Understanding

7. Chipboard sheets are sold in three sizes. Convert each of the measurements below into centimetres and then into metres:
   a. 1800 mm × 900 mm
   b. 2400 mm × 900 mm
   c. 2700 mm × 1200 mm

8. A particular type of chain is sold for $2.25 per metre. Calculate the cost of 240 cm of this chain.

9. Fabric is sold for $7.95 per metre. Calculate the cost of 480 cm of this fabric.

10. The standard marathon distance is 42.2 km. If a marathon race starts and finishes with one lap of a stadium that is 400 m in length, calculate the distance run on the road outside the stadium.

11. Maria needs 3 pieces of timber of lengths 2100 mm, 65 cm and 4250 mm to construct a clothes rack.

    a. Calculate the total length of timber required in metres.
    b. Determine how much the timber will cost if one metre costs $3.80.

12. **WE3** Calculate the perimeter of a basketball court that is 28 m long and 15 m wide.

13. A piece of modern art is in the shape of a rhombus with a side length of 65 cm. Calculate the perimeter of the piece of art in metres.

14. A woven rectangular rug is 175 cm wide and 315 cm long. Determine the perimeter of the rug.

15. A line is drawn to form a border 2 cm from each edge of a piece of A4 paper. If the paper is 30 cm long and 21 cm wide, calculate the length of the border line.

16. A rectangular paddock 144 m long and 111 m wide requires a new single-strand wire fence.

    a. Calculate the length of fencing wire required to complete the fence.
    b. Calculate how much it will cost to rewire the fence if the wire cost $1.47 per metre.

17. A computer desk with the dimensions shown in the diagram needs to have table edging.

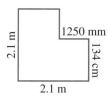

    If the edging cost $1.89 per metre, calculate the cost of the table edging required for the desk.

18. Calculate the unknown side lengths in each of the given shapes.

    a.

    Perimeter = 30 cm

    b.

    Perimeter = 176 cm

    c.
    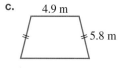
    Perimeter = 23.4 m

19. **WE4** A rectangular billboard has a perimeter of 25 m. Calculate its width if the length is 7 m.

20. The ticket shown has a perimeter of 42 cm.

    a. Calculate the unknown side length.
    b. Olivia wishes to decorate the ticket by placing a gold line along the slanted sides. Calculate the length of the line on each ticket.
    c. A bottle of gold ink will supply enough ink to draw 20 m of line. Calculate the number of bottles of ink needed for 200 tickets to be decorated.

Communicating, reasoning and problem solving

21. A rectangle has a width of 5 cm and a length that is 4 cm longer than the width. The rectangle needs to be enlarged so that the perimeter is 36 cm, and the length needs to remain 4 cm longer than the width. Determine the values of the new width and length (all values are whole numbers).

22. One lap around the block is 613 metres. If you wish to walk at least 3 km, determine the minimum number of laps you will need to complete around the full block.

23. A square and an equilateral triangle have the same perimeter. The side of the triangle is 3 cm longer than the side of the square. Determine the side length of the square. Show your working.

24. As a warm-up activity for PE class, you are required to run laps of the basketball court. The dimensions of a full-size basketball court are shown.

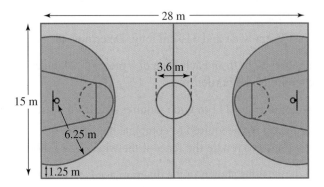

a. Evaluate how far you will run if you run three full laps of the court.
b. Determine how far you will run if you run three laps of half the court.

25. a. A pool fence is to be placed around a rectangular pool that is 6 m by 8 m. If the fence is to be 1 m away from the edge of the pool and also rectangular in shape, evaluate the length of fencing required. (Include the width of the gate in the length of the fence.)
b. Write a rule that relates the length of the fence to the length and width of any rectangular pool. Let l represent the length of the pool and w the width of the pool.

26. A present is to be wrapped using a box, as shown here.

a. If the box is 30 cm long, 25 cm wide and 18 cm high, determine the total length of ribbon that would be needed if the bow used 35 cm of ribbon. (Assume no overlap at the start or finish of the ribbon.)
b. Write a rule that will allow you to evaluate the length of ribbon required for any rectangular box. Let l represent the length of the box, w the width of the box, h the height of the box and b the length of the ribbon.

LESSON
8.3 Circumference

▶ 8.3.1 Circumference

eles-4044

- The distance around a circle is called the **circumference** (C).
- The **diameter** (d) of a circle is the straight-line distance across a circle through its centre.
- The **radius** (r) of a circle is the distance from the centre to the circumference.
- The diameter and radius of a circle are related by the formula $d = 2r$. That is, the diameter is twice as long as the radius.

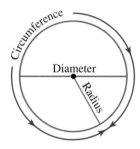

COMMUNICATING — COLLABORATIVE TASK: The diameter and circumference of a circle

Measure the diameters and circumferences of a variety of different circles and cylinders (for example, coins and drink bottles), and record your results in a table.

What do you notice about the value of $\dfrac{C}{d}$ for each circle you measured?

Pi

- The ratio of a circle's circumference to its diameter, $\dfrac{C}{d}$, gives the same value for any circle, no matter how large or small the circle is.
- This special number is known as pi (π). That is, $\pi = \dfrac{C}{d}$.
- Pi (π) is an irrational number and cannot be written as a fraction.
- When pi is written as a decimal number, the decimal places continue forever with no repeated pattern.
- Pi written to 8 decimal places is $3.141\,592\,65$.

Calculating circumference

The circumference, C, of a circle can be determined using one of the following formulas.

$$C = 2\pi r, \text{ where } r \text{ is the radius of the circle}$$

or

$$C = \pi d, \text{ where } d \text{ is the diameter of the circle}$$

- All scientific calculators have a π button and this feature can be used when completing calculations involving pi.

Digital technology

Scientific calculators can be used to assist with calculations involving pi. Locate the π button on your calculator and become familiar with accessing this feature.

$$\pi \qquad \text{DEG} \quad 3.141592654$$

WORKED EXAMPLE 5 Calculating the circumference of a circle

Calculate the circumference of each of the following circles, giving answers:
i. in terms of π
ii. correct to 2 decimal places.

a.

24 cm

b.

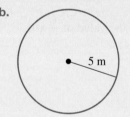
5 m

THINK				WRITE		
a.	i.	1.	Write the formula for the circumference of a circle. *Note:* Since the diameter of the circle is given, use the formula that relates the circumference to the diameter.	a.	i.	$C = \pi d$
		2.	Substitute the value $d = 24$ cm into the formula.			$= \pi \times 24$
		3.	Write the answer and include the units.			$= 24\pi$ cm
	ii.	1.	Write the formula for the circumference of a circle.		ii.	$C = \pi d$
		2.	Substitute the value $d = 24$ cm into the formula.			$= \pi \times 24$
		3.	Evaluate the multiplication using a calculator and the π button.			$= 75.398\ldots$
		4.	Round correct to 2 decimal places and include the units.			$= 75.40$ cm

b.	**i.**	**1.**	Write the formula for the circumference of a circle. *Note:* Since the radius of the circle is given, use the formula that relates the circumference to the radius.	**b. i.** $C = 2\pi r$
		2.	Substitute the value $r = 5$ m into the formula.	$= 2 \times \pi \times 5$
		3.	Write the answer and include the units.	$= 10\pi$ m
	ii.	**1.**	Write the formula for the circumference of a circle.	**ii.** $C = 2\pi r$
		2.	Substitute the value $r = 5$ m into the formula.	$= 2 \times \pi \times 5$
		3.	Evaluate the multiplication using a calculator and the π button.	$= 31.415 \ldots$
		4.	Round the answer correct to 2 decimal places and include the units.	$= 31.42$ m

WORKED EXAMPLE 6 Calculating the perimeter of a semicircle

Calculate the perimeter of the following shape, correct to 2 decimal places.

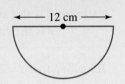

THINK	**WRITE**
1. Identify the parts that constitute the perimeter of the given shape.	$P = \dfrac{1}{2}$ circumference + straight-line section
2. Write the formula for the circumference of a semicircle. *Note:* If the circle were complete, the straight-line segment shown would be its diameter. So the formula that relates the circumference to the diameter is used.	$P = \dfrac{1}{2}\pi d$ + straight-line section
3. Substitute the value $d = 12$ cm into the formula.	$= \dfrac{1}{2} \times \pi \times 12 + 12$
4. Evaluate the multiplication using a calculator and the π button.	$= 18.849 \ldots + 12$ $= 30.849 \ldots$
5. Round the answer correct to 2 decimal places and include the units.	$= 30.85$ cm

⊙ 8.3.2 Parts of a circle

eles-4045

- A **sector** is the region of a circle between two radii. It looks like a slice of pizza.
- An **arc** is a section of the circumference of a circle.
- A **chord** is a straight line joining any two points on the circumference of a circle. The diameter is a type of chord.
- A **segment** of a circle is a section bounded by a chord and an arc.
- A **tangent** to a circle can be described as a straight line that passes through a point on a circle and is perpendicular to the radius.

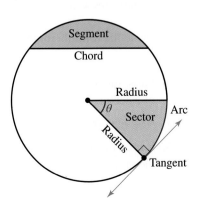

Calculating arc length

- An arc is a portion of the circumference of a circle.
- The length, l, of an arc can be determined using one of the following formulas:

$$l = \frac{\theta}{360°} \times 2\pi r$$

where: θ is the angle (in degrees) at the centre of the circle
r is the radius of the circle.

Calculating the perimeter of a sector

- A sector consists of an arc and 2 radii.
- The perimeter, P, of a sector can therefore be calculated using the following formula:

$$P = l + 2r = \frac{\theta}{360°} \times 2\pi r + 2r$$

where: θ is the angle (in degrees) at the centre of the circle
r is the radius of the circle.

- Recalling that $d = 2r$, the perimeter of a sector could also be determined using the following formula:

$$P = l + d$$
$$= \frac{\theta}{360°} \times \pi d + d$$

WORKED EXAMPLE 7 Calculating the perimeter a sector

Calculate the perimeter of the following sector, correct to 2 decimal places.

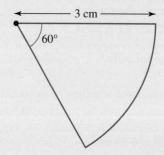

THINK

1. Identify the values of θ and r.

2. Write the formula for the perimeter of a sector.

3. Substitute these values into the formula for the perimeter of a sector.

4. Evaluate the formula.

5. Round the answer correct to 2 decimal places and include the units.

WRITE

$\theta = 60°, r = 3$ cm

$P = \dfrac{\theta}{360} \times 2\pi r + 2r$

$P = \dfrac{60}{360} \times 2\pi \times 3 + 2 \times 3$

$= 9.141 \ldots$

$= 9.14$ cm

- If the circumference of a circle is known, it is possible to determine its radius or diameter using one of the formulas below.

If $C = 2\pi r$, then $r = \dfrac{C}{2\pi}$.

If $C = \pi d$, then $d = \dfrac{C}{\pi}$.

WORKED EXAMPLE 8 Determining the radius of a circle from its circumference

Calculate the radius of a cylindrical water tank with a circumference of 807 cm, correct to 2 decimal places.

THINK	WRITE
1. Write the rule relating circumference and radius.	If $C = 2\pi r$ then $r = \dfrac{C}{2\pi}$.
2. Substitute the value of the circumference into the formula.	$r = \dfrac{807}{2\pi}$ cm
3. Evaluate the expression using a calculator.	$r = 128.438\ldots$ cm
4. Round the answer correct to 2 decimal places.	$r = 128.44$ cm

COMMUNICATING — COLLABORATIVE TASK: Calculating distance along the equator

Objective: To estimate the approximate distance between two locations on the equator using an assumed value for the Earth's radius.

Instructions:
1. Divide students into pairs or small groups of 3–4.
2. Provide each group with the following information:
   - assumed radius of the Earth: 6371 kilometers
   - latitude of the locations: 0° (since they are on the equator)
   - longitude of two different locations (e.g., Location A: 30°, Location B: 60°).
3. Instruct the students to calculate the approximate distance between the two locations using the formula:

$$\text{approximate distance} = \text{radius of the earth} \times \text{difference in longitude}$$

   where radius of the Earth is in kilometres and difference in longitude is in radians

4. Tell students to convert degrees to radians by multiplying the difference in longitude by $\dfrac{\pi}{180°}$.

   For example, for the given locations, the difference in longitude $= 60° - 30°$
   $$= 30°$$

   Convert degrees to radians:

   $$30° = 30° \times \dfrac{\pi}{180°}$$
   $$= 0.5233 \text{ radians}$$

   Therefore, the approximate distance on the equator could be calculated as:

   $$\text{approximate distance} = 6371 \times 0.5233$$
   $$\approx 3334 \text{ km}$$

5. Encourage the groups to discuss their calculations, share ideas, and collaborate to solve the task.
6. Ask each group to record their calculations and present their results to the class.
7. Facilitate a class discussion to compare the different results obtained by each group.
   - Discuss any variations in the calculations and possible sources of error.
   - Encourage students to explain their reasoning and share their approaches.
   - Ask students to reflect on the limitations of using an assumed value for the Earth's radius and discuss how this might affect the accuracy of their results.

Note: You may consider providing additional resources or guidance to help students understand the concepts of radians, latitude, longitude, and the Earth's radius before conducting this collaborative task.

on Resources

⬩ **Interactivity** Circumference (int-3782)

Exercise 8.3 Circumference

learnon

8.3 Quick quiz **on**	8.3 Exercise

Individual pathways

■ PRACTISE	■ CONSOLIDATE	■ MASTER
1, 4, 7, 10, 12, 14, 19, 22	2, 5, 8, 11, 15, 17, 20, 23	3, 6, 9, 13, 16, 18, 21, 24

Fluency

1. **WE5a** Calculate the circumference of each of these circles, giving answers in terms of π.

   a.
   2 cm

   b.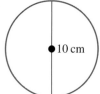
   10 cm

   c.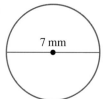
   7 mm

2. Calculate the circumference of each of these circles, giving answers correct to 2 decimal places.

   a.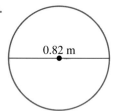
   0.82 m

   b.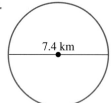
   7.4 km

   c. 34 m

3. **WE5b** Determine the circumference of each of the following circles, giving answers in terms of π.

a.
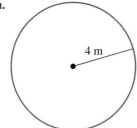
4 m

b.

17 mm

c.
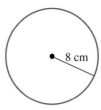
8 cm

4. Calculate the circumference of each of the following circles, giving answers correct to 2 decimal places.

a.

1.43 km

b.

0.4 m

c.

10.6 m

5. **WE6** Determine the perimeter of each of the shapes below. Give your answers to 2 decimal places. (Remember to add the lengths of the straight sections.)

a.
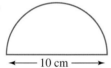
10 cm

b.

16 mm

c.
24 m
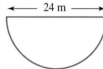

d.

11 mm

e.

20 cm

6. Calculate the perimeter of each of the shapes below. Give your answers to 2 decimal places. (Remember to add the lengths of the straight sections.)

a.
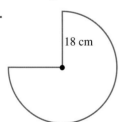
18 cm

b.
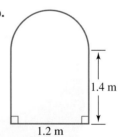
1.4 m
1.2 m

c.
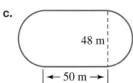
48 m
50 m

d.
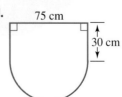
75 cm
30 cm

7. **MC** The circumference of a circle with a radius of 12 cm is:

A. $\pi \times 12$ cm

B. $2 \times \pi \times 12$ cm

C. $2 \times \pi \times 24$ cm

D. $\pi \times 6$ cm

8. **MC** The circumference of a circle with a diameter of 55 m is:

A. $2 \times \pi \times 55$ m

B. $\pi \times \dfrac{55}{2}$ m

C. $\pi \times 55$ m

D. $\pi \times 110 \times 2$ m

9. **WE7** Calculate the perimeter of the following sectors, correct to 2 decimal places.

a.
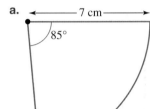

b.
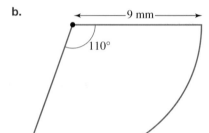

c.
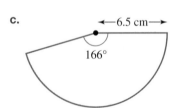

d.
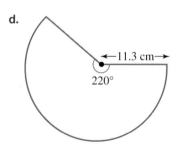

Understanding

10. Calculate the circumference of the seaweed around the outside of this sushi roll, correct to 2 decimal places.

11. A scooter tyre has a diameter of 32 cm. Determine the circumference of the tyre. Give your answer to 2 decimal places.

12. Calculate the circumference of the Ferris wheel shown, correct to 2 decimal places.

13. In a Physics experiment, students spin a metal weight around on the end of a nylon thread. Calculate how far the metal weight travels if it completes 10 revolutions on the end of a 0.88 m thread. Give your answer to 2 decimal places.

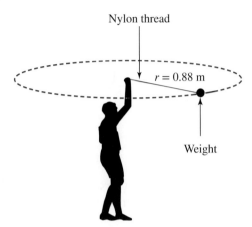

14. **WE8** Calculate the radius of a tyre with a circumference of 135.56 cm. Give your answer to 2 decimal places.

15. Determine the diameter of a circle (correct to 2 decimal places where appropriate) with a circumference of:
   a. 18.84 m
   b. 64.81 cm
   c. 74.62 mm.

16. Calculate the radius of a circle (correct to 2 decimal places where appropriate) with a circumference of:
   a. 12.62 cm
   b. 47.35 m
   c. 157 mm.

17. Determine the total length of metal pipe needed to assemble the wading pool frame shown. Give your answer in metres to 2 decimal places.

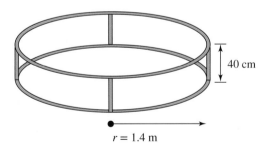

18. Nathan runs around the inside lane of a circular track that has a radius of 29 m. Rachel runs in the outer lane, which is 2.5 m further from the centre of the track.
   Calculate the distance Rachel covers in each lap. Give your answer to 2 decimal places.

Communicating, reasoning and problem solving

19. To cover a total distance of 1.5 km, a student needs to run around a circular track three times. Calculate the radius of the track correct to the nearest metre.

20. A shop sells circular trampolines of four different sizes. Safety nets that go around the trampoline are optional and can be purchased separately. The entrance to the trampoline is via a zip in the net. The following table shows the diameters of all available trampolines and their net lengths.

Diameter of trampoline		Length of safety net	
a.	1.75 m	i.	6.03 m
b.	1.92 m	ii.	9.86 m
c.	2.46 m	iii.	5.50 m
d.	3.14 m	iv.	7.73 m

Determine which safety net matches which trampoline.

21. In *Around the world in eighty days* by Jules Verne, Phileas Fogg boasts that he can travel around the world in 80 days or fewer. This was in the 1800s, so he couldn't take a plane. Determine the average speed required to go around Earth at the equator in 80 days.
    Assume that you travel for 12 hours each day and that the radius of Earth is approximately 6390 km. Give your answer in km/h to 2 decimal places.

22. Liesel's bicycle covers 19 m in 10 revolutions of the wheel while Jared's bicycle covers 20 m in 8 revolutions of the wheel. Determine the difference between the radii of the two bicycle wheels. Give your answer in cm to 2 decimal places.

23. Evaluate the perimeter of each of the following shapes. Give your answers correct to 2 decimal places.

a.
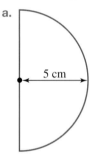
5 cm

b.
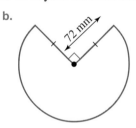
72 mm

c.
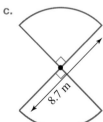
8.7 m

d.
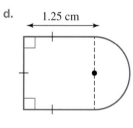
1.25 cm

24. Determine the perimeter of the segment shown. Give your answer correct to 2 decimal places.

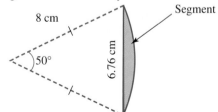
8 cm   Segment   50°   6.76 cm

LESSON
8.4 Areas of rectangles, triangles, parallelograms, rhombuses and kites

---

LEARNING INTENTION

At the end of this lesson you should be able to:
- convert one unit of area to another unit of area
- calculate the area of rectangles, squares, triangles, parallelograms, rhombuses and kites using the formulas.

▶ 8.4.1 Area

eles-4046

- The **area** of a shape is the amount of flat surface enclosed by the shape.
- Area is measured in square units, such as square millimetres (mm^2), square centimetres (cm^2), square metres (m^2) and square kilometres (km^2).
- The figure below shows how to convert between different units of area. This conversion table is simply the square of the length conversion table in section 8.2.1.

Converting units of area

To convert between units of area, use the following conversion chart:

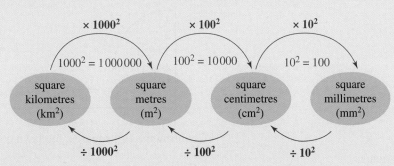

Large areas of land can be measured in hectares (ha); $1\,ha = 10\,000\,m^2$.

WORKED EXAMPLE 9 Converting units of area

Complete the following metric conversions.
a. $0.081\,km^2 = \underline{\hspace{1.5cm}}\,m^2$ b. $19\,645\,mm^2 = \underline{\hspace{1.5cm}}\,m^2$

THINK

a. Look at the metric conversion chart. To convert square kilometres to square metres, multiply by 1000^2 (or $1\,000\,000$); that is, move the decimal point 6 places to the right.

WRITE

a. $0.081\,km^2 = 0.081 \times 1000^2\,m^2$
$\qquad\quad = 0.081 \times 1\,000\,000\,m^2$
$\qquad\quad = 81\,000\,m^2$

$\times 1\,000\,000$

$km^2 \qquad\qquad m^2$

---

b. Look at the metric conversion chart. To convert square millimetres to square metres, divide by 10^2(100) first, then divide by 100^2(10 000). This is the same as dividing by 1 000 000. Move the decimal point 6 places to the left.

b. $19\,645\,\text{mm}^2 = 19\,645 \div 1\,000\,000\,\text{m}^2$
$= 0.019\,645\,\text{m}^2$

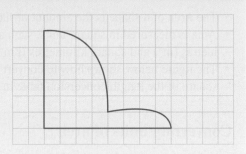

⊙ 8.4.2 Approximate area

eles-6191

Not all shapes have a formula to calculate their area. One technique to approximate the area of these shapes is to fit squares or rectangles into the irregular shape and add them to determine the approximate area of the shape.

WORKED EXAMPLE 10 Estimating area

The following shape represents and area of lawn that needs to be fertilised. Given each square represents 1 m².
a. Approximate the area of the shape by counting squares
b. If the fertiliser is spread at 0.5 kg per square metre, approximately how much fertiliser is required?

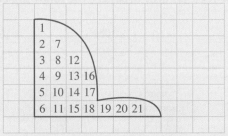

THINK

a. Count all the full squares.

Count the part squares.

Add the full squares and part squares together to calculate the total approximate area.

Write the answer.

b. Multiply the amount of fertiliser per square metre by the area.

Write the answer.

WRITE

a.

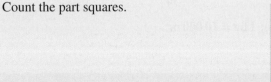

Counting the full squares, the lawn covers 21 m².

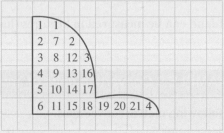

Counting the part squares, the lawn covers $4 \times \dfrac{1}{2} = 2\,\text{m}^2$.

Total approximate area $= 21 + 2 = 23\,\text{m}^2$.

Approximate area is 23 m².

b. Fertiliser $= 0.5 \times 23$
$= 11.5\,\text{kg}$

Requires 11.5 kg of fertiliser.

⏵ 8.4.3 Area of a rectangle

eles-4047

- The area of a rectangle and a square can be calculated by using a formula.

Area of a rectangle

The area, A, of a rectangle is given by the rule:

$$A = l \times w$$
$$= lw$$

where l is its length and w its width.

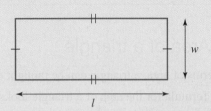

Area of a square

The area, A, of a square is given by the rule:

$$A = l \times l$$
$$= l^2$$

where l is the length of its side.

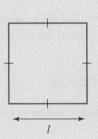

WORKED EXAMPLE 11 Calculating the area of a rectangle

a. **Calculate the area of the rectangle given each square is $1\,\text{cm}^2$.**

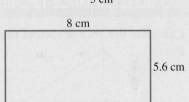

3 cm

5 cm

b. **Calculate the area of a rectangle with the dimensions shown.**

8 cm

5.6 cm

THINK	WRITE
a. **1.** Each square represents $1\,\text{cm}^2$. Count how many squares in length and width.	**a.** Length $= 5$ squares Width $= 3$ squares
2. Calculate the area by counting all the squares.	Total squares $= 15$ squares $= 15\,\text{cm}^2$
3. Compare to using the area of a rectangle formula.	$A = l \times w$ $= 5 \times 3$ $= 15\,\text{cm}^2$
4. Write the answer.	$A = 15\,\text{cm}^2$
b. **1.** Write the formula for the area of a rectangle.	**b.** $A = l \times w$
2. Identify the values of l and w.	$l = 8\,\text{cm}$ and $w = 5.6\,\text{cm}$

⏵

3.	Substitute the values of l and w into the formula.	$A = 8 \times 5.6$
4.	Evaluate the multiplication.	$= 44.8$
5.	Write the answer and include the units.	The area of the rectangle is $44.8\,\text{cm}^2$.

▶ 8.4.4 Area of a triangle

eles-4048

- In terms of area, a triangle can be thought of as half a rectangle.
- The formula for the area of a triangle looks like that of a rectangle with an added factor of $\frac{1}{2}$.
- The dimensions used to calculate the area of a triangle are the length of the base (b) and the perpendicular height (h).

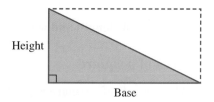

Area of a triangle

The area, A, of a triangle is given by the rule:

$$A = \frac{1}{2} \times b \times h$$
$$= \frac{1}{2}bh$$

where b is the base and h the perpendicular height.

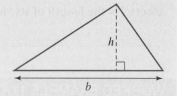

WORKED EXAMPLE 12 Calculating the area of a triangle

Calculate the area of each of these triangles in the smaller unit of measurement.

a.

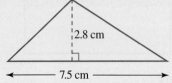

2.8 cm
7.5 cm

b.

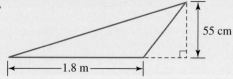

55 cm
1.8 m

THINK	WRITE
a. 1. Write the formula for the area of a triangle.	a. $A = \frac{1}{2}bh$
2. Identify the values of b and h.	$b = 7.5,\ h = 2.8$
3. Substitute the values of b and h into the formula.	$A = \frac{1}{2} \times 7.5 \times 2.8$
4. Evaluate the multiplication.	$= 3.75 \times 2.8$ $= 10.5$
5. Write the answer and include the units.	The area of the given triangle is $10.5\,\text{cm}^2$.

b. 1. Write the formula for the area of a triangle.	**b.** $A = \dfrac{1}{2}bh$	
2. Convert measurements to cm.	$1.8\,\text{m} = 1.8 \times 100\,\text{cm}$ $\qquad\quad = 180\,\text{cm}$	
3. Identify the values of b and h.	$b = 180,\ h = 55$	
4. Substitute the values of b and h into the formula.	$A = \dfrac{1}{2} \times 180 \times 55$	
5. Evaluate the multiplication.	$= 90 \times 55$ $= 4950$	
6. Write the answer and include the units.	The area of the given triangle is $4950\,\text{cm}^2$.	

⊳ 8.4.5 Area of a parallelogram

eles-4049

- A **parallelogram** is a quadrilateral with two pairs of parallel sides. Each parallel pair of opposite sides is of equal length.

Area of a parallelogram

The area, A, of a parallelogram is given by the rule:

$$A = b \times h$$
$$= bh$$

where b is the base and h the perpendicular height.

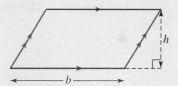

WORKED EXAMPLE 13 Calculating the area of a parallelogram

Calculate the area of the parallelogram shown.

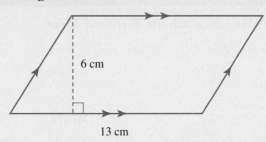

6 cm

13 cm

THINK	**WRITE**
1. Write the formula for the area of a parallelogram.	$A = bh$
2. Identify the values of b and h.	$b = 13\,\text{cm},\ h = 6\,\text{cm}$
3. Substitute the values of b and h into the formula.	$A = 13 \times 6$
4. Evaluate the multiplication.	$= 78$
5. Write the answer and include the units.	The area of the parallelogram is $78\,\text{cm}^2$.

▶ 8.4.6 Area of a rhombus and a kite

eles-4050

- A **rhombus** is a parallelogram with all four sides of equal length and each pair of opposite sides parallel.

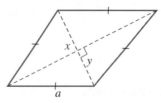

- A **kite** is a quadrilateral with two pairs of equal, adjacent sides and one pair of equal angles.

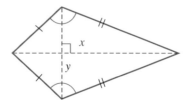

- The diagonals of both the rhombus and the kite divide the shapes into triangles.
- The areas of the rhombus and kite can be found using the lengths of their diagonals, x and y.

Area of a rhombus or a kite

The area, A, of a rhombus or a kite is given by the rule:

$$A = \frac{1}{2} \times x \times y$$
$$= \frac{1}{2}xy$$

where x and y are the lengths of the diagonals.

WORKED EXAMPLE 14 Calculating the area of a kite

Calculate the area of the kite whose diagonals are 30 cm and 20 cm.

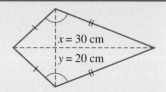

THINK	WRITE
1. Write the formula for the area of a kite.	$A = \frac{1}{2}xy$
2. Identify the values of x and y.	$x = 30\,\text{cm},\ y = 20\,\text{cm}$
3. Substitute the values of x and y into the formula.	$A = \frac{1}{2} \times 30 \times 20$
4. Evaluate the multiplication.	$= 300$
5. Write the answer and include the units.	The area of the kite is 300 cm².

▶ 8.4.7 Areas of composite shapes

eles-4051

- Composite shapes are a combination of shapes placed together.
- To determine the area of a composite shape:
  1. identify smaller known shapes within the composite shape
  2. calculate the area of these smaller shapes
  3. add the areas together.

Composite shape:

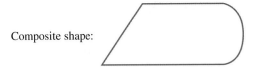

Split into smaller shapes:

Area of composite shape $= A_1 + A_2 + A_3$

WORKED EXAMPLE 15 Calculating the area of a composite shape

Calculate the area of the following shape.

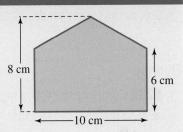

THINK	WRITE
1. The shape can be divided into a rectangle and a triangle. The height of the triangle is $8 - 6 = 2$ cm.	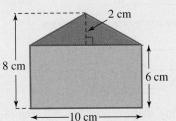
2. Write the formula for the area of a rectangle.	$A_{\text{rectangle}} = l \times w$
• Identify the values of the pronumerals.	$l = 10$ cm and $w = 6$ cm
• Substitute these values into the formula and evaluate.	$A_{\text{rectangle}} = 10 \times 6$ $= 60 \text{ cm}^2$
• Write the formula for the area of a triangle.	$A_{\text{triangle}} = \dfrac{1}{2} \times b \times h$
• Identify the values of the pronumerals.	$b = 10$ cm and $h = 2$ cm
• Substitute these values into the formula and evaluate.	$A_{\text{triangle}} = \dfrac{1}{2} \times 10 \times 2$ $= 10 \text{ cm}^2$
3. Add the area of the rectangle and the area of the triangle to calculate the total area of the composite shape.	Total area $= A_{\text{rectangle}} + A_{\text{triangle}}$ $= 60 + 10$ $= 70$
4. Write the answer and include the units.	The area of the composite shape is 70 cm^2.

Exercise 8.4 Areas of rectangles, triangles, parallelograms, rhombuses and kites

learn on

8.4 Quick quiz	8.4 Exercise

Individual pathways

■ PRACTISE	■ CONSOLIDATE	■ MASTER
1, 3, 6, 8, 11, 14, 15, 19, 23, 26, 27	2, 4, 7, 9, 12, 16, 18, 21, 24, 28, 29	5, 10, 13, 17, 20, 22, 25, 30, 31, 32

Fluency

1. **WE9** Complete the following metric conversions.

   a. $0.53 \, \text{km}^2 = $ _____ m^2

   b. $235 \, \text{mm}^2 = $ _____ cm^2

   c. $2540 \, \text{cm}^2 = $ _____ mm^2

   d. $542\,000 \, \text{cm}^2 = $ _____ m^2

   e. $74\,000 \, \text{mm}^2 = $ _____ m^2

2. Complete the following metric conversions.

   a. $3\,000\,000 \, \text{m}^2 = $ _____ km^2

   b. $98\,563 \, \text{m}^2 = $ _____ ha

   c. $1.78 \, \text{ha} = $ _____ m^2

   d. $0.987 \, \text{m}^2 = $ _____ mm^2

   e. $0.000\,127\,5 \, \text{km}^2 = $ _____ cm^2

3. **WE10** The following shape represents and area of lawn that needs to be watered. Given each square represents $1 \, \text{m}^2$.

   a. Approximate the area of the shape by counting squares
   b. If the water is spread at 5 L per square metre, approximately how much water is required?

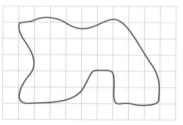

4. The following shape represents a vegetable patch that requires sugar cane mulch. Given each square represents $4 \, \text{m}^2$.

   a. Approximate the area of the shape by counting squares
   b. If the sugar cane mulch is spread at 0.25 kg per square metre, approximately how much water is required?

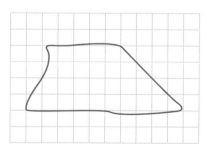

5. **WE11** Calculate the area of each of the rectangles below.

a.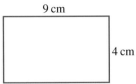
9 cm
4 cm

b.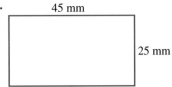
45 mm
25 mm

c.
1.5 m
3 m

6. Determine the area of each of the rectangles below.

a.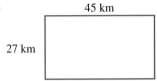
45 km
27 km

b.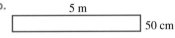
5 m
50 cm

c.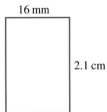
16 mm
2.1 cm

7. Calculate the area of each of the squares below.

a.
5 mm

b.
16 cm

c.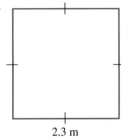
2.3 m

Questions 8 and 9 relate to the diagram shown.

8. **MC** The height and base respectively of the triangle are:

   A. 32 mm and 62 mm
   B. 32 mm and 134 mm
   C. 32 mm and 187 mm
   D. 62 mm and 187 mm

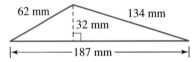
62 mm   134 mm
32 mm
187 mm

9. **MC** The area of the triangle is:

   A. 2992 mm
   B. 2992 mm^2
   C. 5984 mm
   D. 5984 mm^2

10. **WE12** Calculate the areas of the following triangles.

a.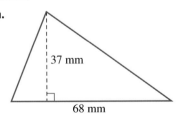
37 mm
68 mm

b.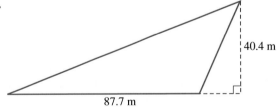
40.4 m
87.7 m

c.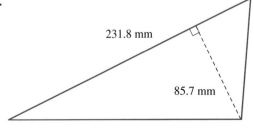
231.8 mm
85.7 mm

11. Determine the areas of the following triangles. Where there are two units given, answer using the smaller unit.

a.
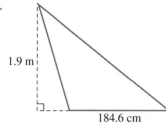
1.9 m
184.6 cm

b.
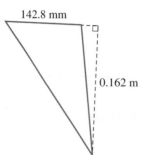
142.8 mm
0.162 m

c.
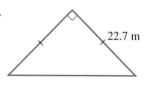
22.7 m

12. **WE13** Calculate the areas of the following parallelograms.

a.
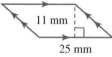
11 mm
25 mm

b.
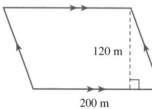
120 m
200 m

c.
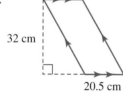
32 cm
20.5 cm

d.
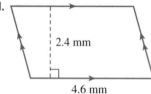
2.4 mm
4.6 mm

e.
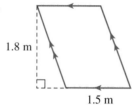
1.8 m
1.5 m

13. Calculate the areas of the following parallelograms.

a.
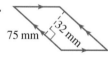
75 mm
32 mm

b.
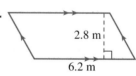
2.8 m
6.2 m

c.
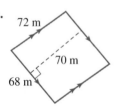
72 m
70 m
68 m

d.
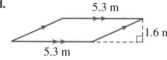
5.3 m
1.6 m
5.3 m

14. **WE14** Determine the area of:

a. a rhombus whose diagonals are 10 cm and 6 cm
b. a rhombus whose diagonals are 8 cm and 6 cm
c. a kite whose diagonals are 20 cm and 9 cm.

Understanding

15. Calculate the area of the block of land in the figure.

76 m
27 m

16. **WE15** Calculate the areas of the following composite shapes.

a.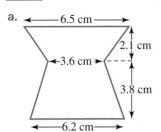

b.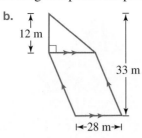

c.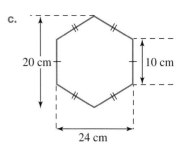

17. A triangular pyramid can be constructed from the net shown. Calculate the total area of the net.

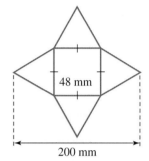

18. Calculate the area of gold braid needed to make the four military stripes shown.

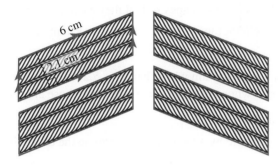

19. **MC** Select which of the following statements about parallelograms is false.

   A. The opposite sides of a parallelogram are parallel.
   B. The height of the parallelogram is perpendicular to its base.
   C. The area of a parallelogram is equal to the area of the rectangle whose length is the same as the base and whose breadth is the same as the height of the parallelogram.
   D. The perimeter of the parallelogram is given by the formula $P = 2(b + h)$.

20. Zorko has divided his vegetable patch, which is in the shape of a regular (all sides equal) pentagon, into 3 sections as shown in the diagram.

   a. Calculate the area of each individual section, correct to 2 decimal places.
   b. Calculate the area of the vegetable patch, correct to 2 decimal places.

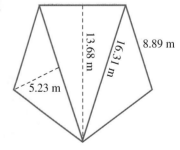

21. Determine the approximate area of the triangle used to rack up the pool balls in the image.

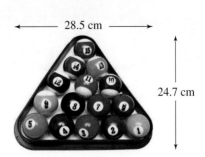

22. The pyramid has 4 identical triangular faces with the dimensions shown.

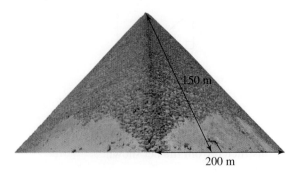

150 m

200 m

Calculate:

a. the area of one of the triangular faces
b. the total area of the four faces.

Communicating, reasoning and problem solving

23. Show possible dimensions for each of the following shapes so that they each have an area of $36 \, \text{cm}^2$.

    a. Rectangle   **b.** Parallelogram   **c.** Rhombus

24. Answer the following questions, showing full working.

    a. Determine the length of the base of a parallelogram whose height is $5.2 \, \text{cm}$ and whose area is $18.72 \, \text{cm}^2$.
    b. Determine the height of a parallelogram whose base is $7.5 \, \text{cm}$ long and whose area is $69 \, \text{cm}^2$.

25. Answer the following questions, showing full working.

    a. If you increase the side lengths of a rectangle by a factor of 2, determine the effect this has on the area of the rectangle.
    b. If you decrease the side lengths of a rectangle by a factor of 2, determine the effect this has on the area of the rectangle.
    c. If you square the side lengths of a rectangle, determine the effect this has on the area of the rectangle.

26. Georgia is planning to create a feature wall in her lounge room by painting it a different colour. The wall is $4.6 \, \text{m}$ wide and $3.4 \, \text{m}$ high.

    a. Calculate the area of the wall to be painted.
    b. Georgia knows that a 4-litre can of paint is sufficient to cover 12 square metres of wall. Determine the number of cans she must purchase if she needs to apply two coats of paint.

27. Calculate the base length of the give-way sign shown.

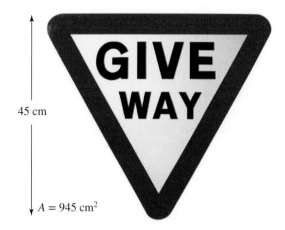

45 cm

$A = 945 \text{ cm}^2$

28. A designer vase has a square base of side length 12 cm and four identical sides, each of which is a parallelogram. If the vertical height of the vase is 30 cm, determine the total area of the glass used to make this vase. (Assume no waste and do not forget to include the base.)

29. The base of a parallelogram is 3 times as long as its height. Calculate the area of the parallelogram, given that its height is 2.4 cm long.

30. The length of the base of a parallelogram is equal to its height. If the area of the parallelogram is 90.25 cm^2, evaluate its dimensions.

31. **a.** Calculate the width of a rectangular sportsground if it has an area of 30 ha and a length of 750 m.
b. The watering system at the sportsground covers 8000 square metres in 10 minutes.
Determine how long it takes to water the sportsground.

32. A rectangular flowerbed measures 20 m by 16 m. A gravel path 2 m wide surrounds it.

a. Draw a diagram representing the flower bed and path.
b. Determine the area of the flower bed.
c. Determine the area of the gravel path.
d. If gravel costs $5 per square metre, evaluate the cost of covering the path.

LESSON
8.5 Areas of circles

LEARNING INTENTION

At the end of this lesson you should be able to:
- calculate the area of a circle using the formula
- calculate the area of parts of a circle, including semicircles, quadrants and sectors.

▶ 8.5.1 The area of a circle

eles-4052

- The formula for the area of a circle can be discovered in the following activity.

COMMUNICATING — COLLABORATIVE TASK: Finding a formula to calculate the area of a circle

Equipment: paper, pair of compasses, protractor, scissors, pencil

1. Use your compasses to draw a circle with a radius of 10 cm.
2. Use a protractor to mark off 10° angles along the circumference. Join the markings to the centre of the circle with straight lines. Your circle will be separated into 36 equal sectors.
3. Cut out the sectors and arrange them in a pattern as shown below. Half of the sectors will be pointing up and half will be pointing down.
4. The resultant shape resembles a _____ .
5. Express the values of the pronumerals a and b shown on the diagram in terms of r, the radius of the original uncut circle. Hence, calculate the area.

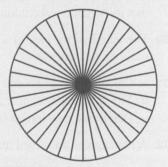

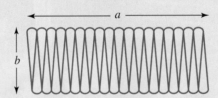

6. Explain why the area of the shape is the same as the area of the original circle. Hence, state the formula for the area of the circle.

Area of a circle

The area, A, of a circle is given by the rule:

$$A = \pi r^2$$

where r is the radius of the circle.

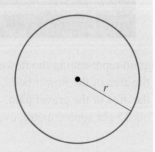

- *Note:* When calculating the area of a circle where the diameter is given, recall that the diameter, d, is twice the length of the radius, r. That is, $d = 2r$ or $r = \dfrac{d}{2}$.

WORKED EXAMPLE 16 Calculating the area of a circle

Calculate the area of each of the following circles correct to 2 decimal places.

a.

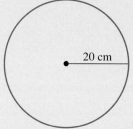

20 cm

b.

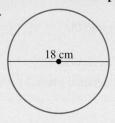

18 cm

THINK	WRITE
a. 1. Write the formula for the area of a circle.	**a.** $A = \pi r^2$
2. Identify the value of r from the diagram.	$r = 20\,\text{cm}$
3. Substitute the value for r into the formula.	$A = \pi \times 20^2$
4. Evaluate the multiplication using a calculator and the π button.	$= \pi \times 400$
	$= 1256.637\ldots$
Round the answer correct to 2 decimal places.	$= 1256.64$
5. Write the answer and include the unit.	The area of the circle is $1256.64\,\text{cm}^2$.
b. 1. Write the formula for the area of a circle.	**b.** $A = \pi r^2$
2. Since the diameter is given, state the relation between the radius and the diameter.	$d = 18\,\text{cm};\ r = \dfrac{d}{2}$
3. Determine the value of r.	$r = \dfrac{18}{2}$
	$= 9\,\text{cm}$
4. Substitute the value of r into the formula.	$A = \pi \times 9^2$
5. Evaluate the multiplication using a calculator and the π button.	$= \pi \times 81$
	$= 254.469\ldots$
Round the answer correct to 2 decimal places.	$= 254.47$
6. Write the answer and include the units.	The area of the circle is $254.47\,\text{cm}^2$.

- If the area of a circle is known, it is possible to determine its radius using the following formula.

$$\text{If } A = \pi r^2 \text{ then } r = \sqrt{\frac{A}{\pi}}, \text{ since } r > 0.$$

Calculate the radius of a circle with area 106 cm², correct to 2 decimal places.

THINK	WRITE
1. Write the rule connecting area and radius.	If $A = \pi r^2$ then $r = \sqrt{\dfrac{A}{\pi}}$, since $r > 0$.
2. Substitute the value of the area into the formula.	$r = \sqrt{\dfrac{106}{\pi}}$ cm
3. Evaluate the expression using a calculator.	$r = 5.808\ldots$ cm
4. Round the answer correct to 2 decimal places.	$r = 5.81$ cm

▶ 8.5.2 The areas of quadrants, semicircles and sectors

eles-4053

- By adjusting the formula for the area of a circle, we can develop rules for calculating the area of common circle portions.

Area of a semicircle

The area, A, of a semicircle is given by the rule:

$$A = \frac{1}{2} \times \pi r^2$$

where r is the radius of the circle.

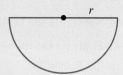

Area of a quadrant

The area, A, of a quadrant is given by the rule:

$$A = \frac{1}{4} \times \pi r^2$$

where r is the radius of the circle.

Area of a sector

The area, A, of a sector is given by the rule:

$$A = \frac{\theta}{360°} \times \pi r^2$$

where r is the radius of the circle and θ is the angle (in degrees) at the centre.

Calculate the area of each of the following shapes correct to 2 decimal places.

a.

13 cm

b.

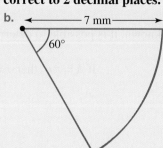

7 mm

60°

THINK	WRITE
a. 1. Write the formula for the area of a quadrant.	**a.** $A = \dfrac{1}{4} \times \pi r^2$
2. Identify and substitute the value of r into the formula.	$A = \dfrac{1}{4} \times \pi \times 13^2$
3. Evaluate and round the answer correct to 2 decimal places.	$= 132.732\dots$
	$= 132.73$
4. Write the answer and include the units.	The area of the quadrant is 132.73 cm^2.
b. 1. Write the formula for the area of a sector.	**b.** $A = \dfrac{\theta}{360°} \times \pi r^2$
2. Identify and substitute the values of θ and r into the formula.	$A = \dfrac{60°}{360°} \times \pi \times 7^2$
3. Evaluate and round the answer correct to 2 decimal places.	$= 25.6563\dots$
	$= 25.66$
4. Write the answer and include the units.	The area of the sector is 25.66 mm^2.

 Resources

 Interactivity Area of circles (int-3788)

Exercise 8.5 Areas of circles

learn on

8.5 Quick quiz on	**8.5 Exercise**

Individual pathways

■ **PRACTISE**	■ **CONSOLIDATE**	■ **MASTER**
1, 5, 7, 10, 11, 14	2, 4, 8, 12, 15	3, 6, 9, 13, 16

Fluency

1. **WE16** Calculate the area of each of the following circles correct to 2 decimal places.

a.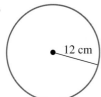

12 cm

b.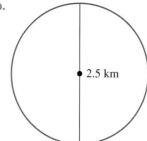

2.5 km

c.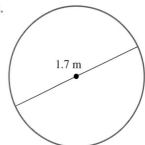

1.7 m

2. Calculate the area of each of the following circles correct to 2 decimal places.

a.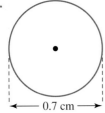

0.7 cm

b.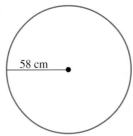

58 cm

c.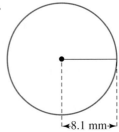

8.1 mm

3. Calculate the area, correct to 2 decimal places, of:

a. a circle of radius 5 cm

b. a circle of radius 12.4 mm

c. a circle of diameter 28 m

d. a circle of diameter 18 cm.

4. **WE17** Calculate the radius, correct to 2 decimal places, of:

a. a circle of area 50 cm^2

b. a circle of area 75 mm^2

c. a circle of area 333 cm^2.

5. **WE18** Calculate the area of each of the following shapes correct to 2 decimal places.

a.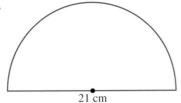

21 cm

b.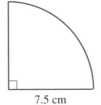

7.5 cm

c.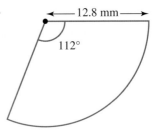

12.8 mm

112°

Understanding

6. The word *annulus* means *ring* in Latin. An annulus is the shape formed between two circles with a common centre (called concentric circles). To calculate the area of an annulus, calculate the area of the smaller circle and subtract it from the area of the larger circle.

Calculate the area of the annulus, correct to 2 decimal places, for each of the following sets of concentric circles.

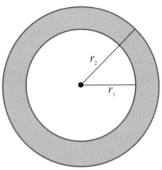

An annulus is the shaded area between concentric circles.

$r_1 = $ radius of smaller circle

$r_2 = $ radius of large circle

$\text{Area}_{\text{annulus}} = \pi r_2^2 - \pi r_1^2$

a.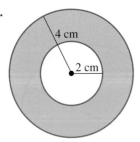

4 cm

2 cm

b.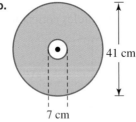

41 cm

7 cm

c.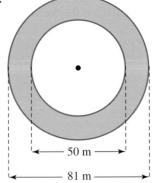

50 m

81 m

7. Determine the area of each of the following shapes. Give your answers correct to 2 decimal places.

a.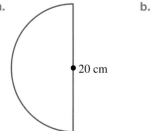
20 cm

b.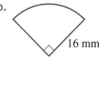
16 mm

c.
4.2 m

d.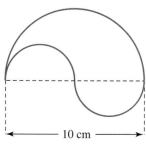
10 cm

8. Calculate the area of each of the following shapes. Give your answers correct to 2 decimal places.

a.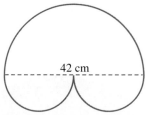
42 cm

b.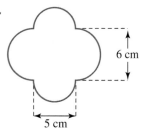
6 cm
5 cm

c.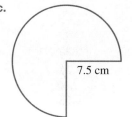
7.5 cm

d.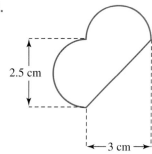
2.5 cm
3 cm

9. Determine the minimum area of aluminium foil, correct to 2 decimal places, that could be used to cover the top of the circular tray with diameter 38 cm.

10. Calculate the area of material in a circular mat of diameter 2.4 m. Give your answer correct to 2 decimal places.

Communicating, reasoning and problem solving

11. Determine the number of packets of lawn seed Joanne should buy to sow a circular bed of diameter 27 m, if each packet of seed covers 23 m². Show your full working.

12. Investigate what happens to the diameter of a circle if its area is:

   a. doubled
   b. quadrupled
   c. halved.

13. The two small blue circles in the diagram have a diameter that is equal to the radius of the medium-sized orange circle. The diameter of the medium-sized circle is equal to the radius of the large yellow circle. If the large circle has a radius of 8 cm, determine the area of the orange-shaded section. Show your full working.

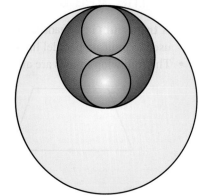

14. A large circular mural is going to be painted at a local school. The diameter of the mural will be 2.2 metres. The smallest paint tin available for purchase is 500 mL and it contains enough paint to cover 3.2 m² of wall. Determine whether one tin of paint will be enough to paint the mural.
Justify your response.

15. Calculate the shaded area in each of the following shapes.

a.

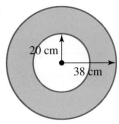

b.

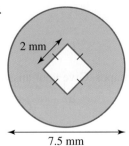

c.

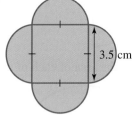

16. The total area of wood used to make a set of six identical round coasters is 425 cm².

a. Evaluate the area of the wood used in each coaster.
b. Determine the radius of each coaster correct to 2 decimal places.
c. An 8.5 cm wide cylindrical coffee mug is placed on one of the coasters. Use your answer to part b to decide whether it will fit entirely within the coaster's surface.

LESSON
8.6 Areas of trapeziums

LEARNING INTENTION

At the end of this lesson you should be able to:
- calculate the area of a trapezium using the formula.

eles-4054

▶ 8.6.1 The area of a trapezium

- A **trapezium** is a quadrilateral with one pair of parallel unequal sides. (Remember that the arrow symbol is used to indicate parallel lines.)
- The following figures are all trapeziums.

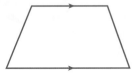  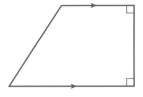

- The height of a trapezium is perpendicular to each of its parallel bases.

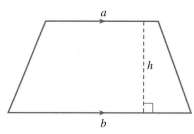

- The area of trapezium can be developed by bisecting the non-parallel sides, dissecting the trapezium and rotating triangles to form a rectangle with the same perpendicular height.

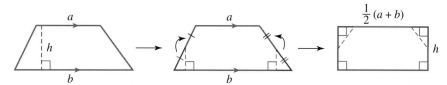

- The length can now be determined as half of the sum of the two parallel sides a and b.
- Multiplying the new length with the height (the same as for a rectangle) gives the formula for the area of trapezium.

Area of a trapezium

The area, A, of a trapezium is given by the rule:

$$A = \frac{h}{2}(a+b)$$

where a and b are the lengths of the parallel sides, and h is the perpendicular height.

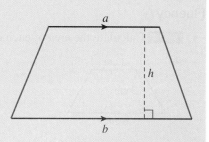

WORKED EXAMPLE 19 Calculating the area of a trapezium

Calculate the area of the trapezium shown.

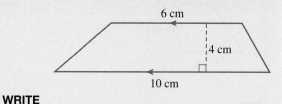

THINK	WRITE
1. Write the formula for the area of the trapezium.	$A = \dfrac{h}{2}(a+b)$
2. Identify the values of a, b and h. *Note:* It does not matter which of the parallel sides is a and which one is b.	$a = 6\,\text{cm}$, $b = 10\,\text{cm}$ and $h = 4\,\text{cm}$
3. Substitute the values of a, b and h into the formula.	$A = \dfrac{4}{2}(6+10)$

▶

4. Calculate the multiplication inside the brackets and simplify the fraction out the front. $= 2 \times 16$

5. Multiply together. $= 32$

6. Write the answer and include the units.   The area of the trapezium is 32 cm^2.

 Resources

 Interactivity Area of trapeziums (int-3789)

Exercise 8.6 Areas of trapeziums

learn on

| 8.6 Quick quiz **on** | 8.6 Exercise |

Individual pathways

■ **PRACTISE**	■ **CONSOLIDATE**	■ **MASTER**
1, 4, 7, 9, 13	2, 5, 8, 11, 14	3, 6, 10, 12, 15

Fluency

1. **WE19** Calculate the area of each of the following trapeziums.

   a.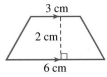

   b.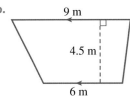

   c.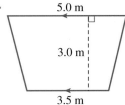

2. Calculate the area of each of the following trapeziums.

   a.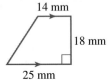

   b.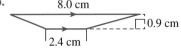

   c.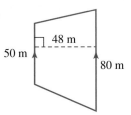

3. **MC** Select the correct way to calculate the area of the trapezium shown.

   A. $\dfrac{1}{2} \times (3 + 5) \times 11$

   B. $\dfrac{1}{2} \times (11 - 3) \times 5$

   C. $\dfrac{1}{2} \times (11 + 5) \times 3$

   D. $\dfrac{1}{2} \times (3 + 11) \times 5$

   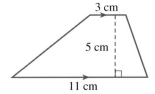

4. A dress pattern contains these two pieces. Calculate the total area of material needed to make both pieces.

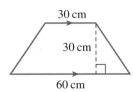

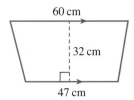

Understanding

5. A science laboratory has four benches with dimensions as shown.
   Calculate the cost, to the nearest 5 cents, of covering all four benches with a protective coating that costs $38.50 per square metre.

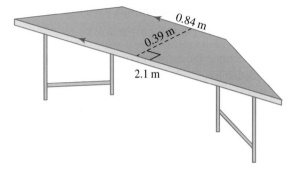

6. Stavros has accepted a contract to concrete and edge a yard, the dimensions of which are shown in the figure below.

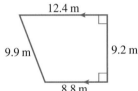

   a. Calculate the cost, to the nearest 5 cents, of concreting the yard if concrete costs $28.00 per square metre.
   b. The yard must be surrounded by edging strips, which cost $8.25 per metre. Determine the cost of the edging strips to the nearest 5 cents.
   c. Calculate the total cost of materials for the job.

7. The side wall of this shed is in the shape of a trapezium and has an area of $4.6\,\text{m}^2$. Calculate the perpendicular distance between the parallel sides if one side of the wall is 2.6 m high and the other 2 m high.

8. **MC** Two trapeziums have corresponding parallel sides of equal length. The height of the first trapezium is twice as large as the height of the second. The area of the second trapezium is:

   A. twice the area of the first trapezium.
   B. half the area of the first trapezium.
   C. quarter of the area of the first trapezium.
   D. four times the area of the first trapezium.

9. Select the side (A, B, or C) that represents the height of the following trapeziums.

   a.
   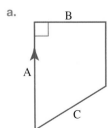

   b.
   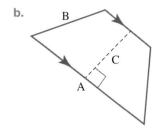

   c.
   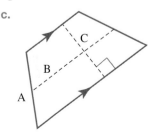

Communicating, reasoning and problem solving

10. Calculate the areas of the following composite shapes (correct to 2 decimal places where necessary).

a.

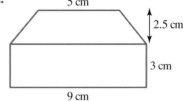

b.

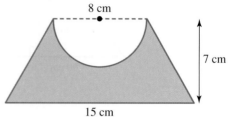

11. The following working was used to calculate the area of the figure shown.

$$\text{Area} = \frac{1}{2}(12+11)\times 6 + \pi \times 5.5^2$$

Determine the error or errors in the working and show the correct working needed to calculate the area.

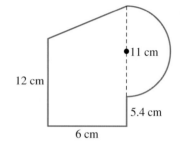

12. The formula for the area of a trapezium can be proved by dividing it into two triangles, as shown.

a. Calculate the area of triangle 1 in terms of the pronumerals shown in the diagram.

b. Calculate the area of triangle 2 in terms of the pronumerals shown in the diagram.

c. Calculate the area of the trapezium by adding the areas of the two triangles together.

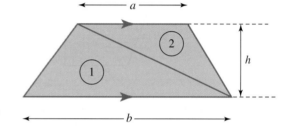

13. A shade sail being installed at a kindergarten is shown in the diagram.

Determine the cost of the shade sail if the material is $98 per metre squared, plus $2300 for installation.

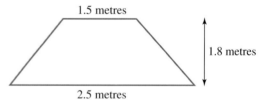

14. A section of a garden is in the shape of a trapezium, as shown in the diagram.

a. Calculate the area of this section of the garden.

b. This section of the garden is to be covered with mulch. According to the information on the pack, each bag contains enough mulch to cover $5\,\text{m}^2$ of surface. Determine the number of bags of mulch needed to complete the job.

c. If each bag costs $8.95, evaluate the total cost of the mulch.

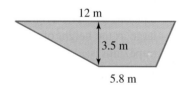

15. A grassed area is in the shape of a trapezium with parallel sides that measure 128 m and 92 m, and a height that measures 40 m. A 4 m wide walkway is constructed, running perpendicular between the two parallel sides. Evaluate the area of the grassed region after the addition of the walkway.

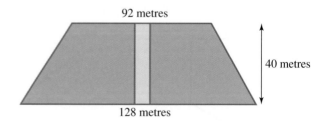

LESSON
8.7 Volumes of prisms and other solids

> **LEARNING INTENTION**
>
> At the end of this lesson you should be able to:
> • calculate the volume of prisms and solids with uniform cross-sections.

▶ 8.7.1 Converting units of volume

eles-4055

- **Volume** is the amount of space inside a three-dimensional object.
- Volume is measured in cubic units such as mm^3, cm^3, m^3 or km^3.
- The figure below shows how to convert between different units of volume. This conversion table is simply the cube of the length conversion table in section 8.2.1.

Converting units of volume

To convert between the units of volume, use the following conversion chart:

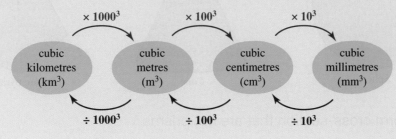

▶ 8.7.2 Volume of prisms and cylinders

eles-6192

Prisms

- A **prism** is a solid object with identical ends, flat faces and the same cross-section along its length.
- Prisms are named according to the shape of their cross-section. The objects below are all prisms.

Triangular prism

Rectangular prism

Hexagonal prism

- A cross-section is the shape made by cutting straight across an object.
- The cross-section of the loaf of bread shown is a rectangle and it is the same all along its length. Hence, it is a rectangular prism.
- The base of a prism is identical to the area of its uniform cross-section and is not simply the 'bottom' of the prism.
- The volume of a prism can be determined by multiplying the area of the base by the length of the prism.

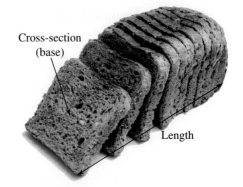

Cross-section (base)

Length

Volume of a prism

The volume, V, of a prism is given by the rule:

$$V = \text{area of cross-section} \times \text{height}$$
$$= A \times h$$
$$= Ah$$

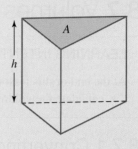

- Solids with identical ends that have curved sides (edges) are *not* prisms.
- Objects that do not have a uniform cross-section cannot be classified as prisms. For example, the objects below are *not* prisms.

Sphere

Cone

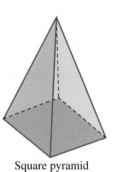

Square pyramid

Solids with uniform cross-section that are not prisms

- Objects with a uniform cross-section whose ends are not polygons (i.e. those whose ends are not joined by straight edges) cannot be classified as prisms.
  For example, the shapes shown are not prisms, even though they have uniform cross-sections.
- The formula $V = Ah$, where A is the cross-sectional area and h is the dimension perpendicular to it, will give the volume of any solid with a uniform cross-section, even if it is not a prism.
- The cross-sectional area of a cylinder is a circle.

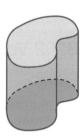

Volume of a cylinder

The volume, V, of a cylinder is given by the rule:

$$V = \pi r^2 h$$

where r is the radius of the circular cross-section, and h is the height of the cylinder.

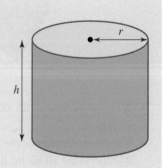

Calculate the volume of each of the following solids. Give answers correct to 2 decimal places where appropriate.

a.

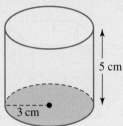

b.

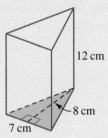

c.

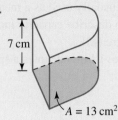

THINK

WRITE

a. 1. Write the formula for the volume of a cylinder.

a. $V = \pi r^2 h$

2. Identify the values of r and h.

$r = 3\,\text{cm}, \ h = 5\,\text{cm}$

3. Substitute the values of r and h into the formula.

$V = \pi \times 3^2 \times 5$

$= 141.371\ldots$

4. Calculate the multiplication and round the answer correct to 2 decimal places.

$= 141.37$

5. Write the answer and include the units.

The volume of the cylinder is $141.37\,\text{cm}^3$.

b. 1. Write the formula for the volume of a prism.

b. $V = A \times h$

2. Identify the shape of the cross-section and hence write the formula to calculate its area.

The base area is a triangle.

$A_{\text{triangle}} = \dfrac{1}{2}bh$

3. Identify the values of b and h.
(*Note: h* is the height of the triangle, *not* the height of the prism.)

$b = 7\,\text{cm}, \ h = 8\,\text{cm}$

4. Substitute the values of b and h into the formula and evaluate the area of the triangle.

$A = \dfrac{1}{2} \times 7 \times 8$

$= 28\,\text{cm}^2$

5. Identify the height of the prism.

Height of prism: $h = 12\,\text{cm}$

6. Write the formula for the volume of the prism, then substitute values of A and h.

$V = A \times h$

$= 28 \times 12$

7. Calculate the multiplication.

$= 336$

8. Write the answer and include the units.

The volume of the prism is $336\,\text{cm}^3$.

c. 1. Write the formula for the volume of the given shape.

c. $V = Ah$

2. Identify the values of the cross-sectional area and the height of the shape.

$A = 13\,\text{cm}^2, \ h = 7\,\text{cm}$

3. Substitute the values of A and h into the formula.

$V = Ah$

$= 13 \times 7$

4. Calculate the multiplication.

$= 91$

5. Write the answer and include the units.

The volume of the given space is $91\,\text{cm}^3$.

▶ 8.7.3 Capacity

eles-6193

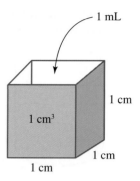

- If a 3-dimensional object is hollow, it can hold another substance. In this case the volume is also referred to as **capacity**.
- Capacity is usually used as a measure of the volume of liquid a container can hold.
- The units to describe capacity include the millilitre (mL), litre (L), kilolitre (kL) and megalitre (ML).
  For example, 1 litre occupies a space of $1000\,\text{cm}^3$.

Units of capacity

Units of capacity are related as shown below.

$$1\,L = 1\,000\,mL$$
$$1\,kL = 1\,000\,L$$
$$1\,ML = 1\,000\,000\,L$$

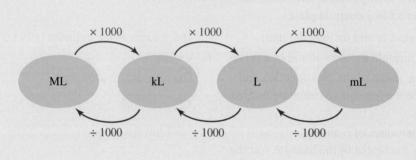

- When calculating the capacity of an object, it is sometimes useful to calculate the volume of the object first, then convert to units of capacity.
- The metric unit, $1\,\text{cm}^3$, is defined as having a capacity of 1 mL.

Relation between volume and capacity units

Units of volume and capacity are related as shown.

$$1\,\text{cm}^3 = 1\,\text{mL}$$
$$1000\,\text{cm}^3 = 1000\,\text{mL} = 1\,\text{L}$$
$$1\,\text{m}^3 = 1000\,\text{L} = 1\,\text{kL}$$

Complete the following unit conversions.

a. $70 \text{ mL} = \underline{\hspace{1cm}} \text{ cm}^3$

b. $530 \text{ mL} = \underline{\hspace{1cm}} \text{ L}$

c. $0.382 \text{ L} = \underline{\hspace{1cm}} \text{ cm}^3$

THINK	WRITE
a. There is 1 cm^3 in each 1 mL.	a. $1 \text{ mL} = 1 \text{ cm}^3$ Therefore, $70 \text{ mL} = 70 \text{ cm}^3$.
b. There are 1000 mL in 1 L, so to convert millilitres to litres divide by 1000.	b. $530 \text{ mL} = (530 \div 1000) \text{ L}$ $= 0.53 \text{ L}$
c. 1. There are 1000 mL in 1 L, so to convert litres to millilitres multiply by 1000.	c. $0.382 \text{ L} = (0.382 \times 1000) \text{ mL}$ $= 382 \text{ mL}$
2. Convert millilitres to cubic centimetres: $1 \text{ mL} = 1 \text{ cm}^3$.	$382 \text{ mL} = 382 \text{ cm}^3$

Exercise 8.7 Volumes of prisms and other solids

learn on

8.7 Quick quiz on	8.7 Exercise

Individual pathways

■ PRACTISE	■ CONSOLIDATE	■ MASTER
1, 4, 7, 9, 11, 14, 18	2, 5, 8, 12, 15, 19	3, 6, 10, 13, 16, 17, 20

Fluency

1. Select which of the three-dimensional shapes below are prisms.

a.

b.

c.

d.

e.

2. `WE20` Calculate the volume of each of the following solids.

a.

6 cm

$A = 14 \text{ cm}^2$

b.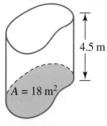

4.5 m

$A = 18 \text{ m}^2$

c.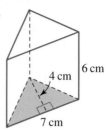

6 cm

4 cm

7 cm

3. Calculate the volume of each of the following solids.

a.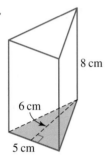

8 cm

6 cm

5 cm

b.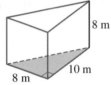

8 m

10 m

8 m

c.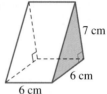

7 cm

6 cm

6 cm

4. Calculate the volume of each of the following solids. Give your answers to 2 decimal places.

a.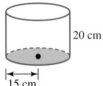

20 cm

15 cm

b.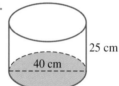

25 cm

40 cm

c.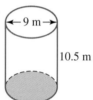

9 m

10.5 m

5. Calculate the volume of each of the following solids. Give your answers to 2 decimal places where appropriate.

a.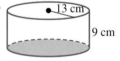

13 cm

9 cm

b.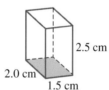

2.5 cm

2.0 cm

1.5 cm

c.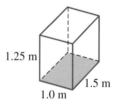

1.25 m

1.5 m

1.0 m

6. State whether the following measures are a measure of volume or capacity.
   a. 35 mL
   b. 120 m^3
   c. 1.2 cm^3
   d. 1750 kL
   e. 432 mm^3
   f. 97.37 L

7. Complete the following unit conversions.
   a. 12 L = _____ mL
   b. 3125 L = _____ mL
   c. 397 mL = _____ L
   d. 0.0078 L = _____ mL
   e. 4893 mL = _____ L
   f. 36.97 L = _____ mL

8. `WE21` Complete the following unit conversions.
   a. 372 cm^3 = _____ mL
   b. 1630 L = _____ cm^3
   c. 3.4 L = _____ cm^3
   d. 0.38 mL = _____ cm^3
   e. 163 L = _____ cm^3
   f. 49.28 cm^3 = _____ mL

9. Complete the following unit conversions.
   a. 578 mL = _____ L
   b. 750 mL = _____ L
   c. 0.429 L = _____ mL

Understanding

10. Determine the volume of water that a rectangular swimming pool with dimensions shown in the photograph will hold if it is completely filled. The pool has no shallow or deep end; it has the same depth everywhere.

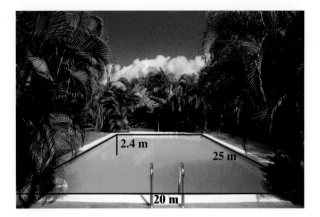

11. Determine how many cubic metres of cement will be needed to make the cylindrical foundation shown in the figure. Give your answer to 2 decimal places.

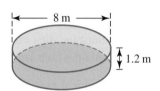

12. Calculate the volumes of these pieces of cheese.

a.

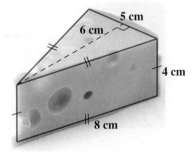

b.

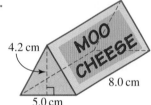

13. Calculate the volume of the bread bin shown in the diagram.

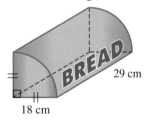

Give your answer to the nearest whole number.

14. Determine how much water this pig trough, with dimensions shown in the diagram, will hold if it is completely filled.

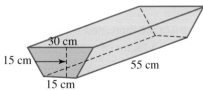

Give your answer in litres correct to 1 decimal place. (*Hint:* 1 litre = 1000 cm³)

Communicating, reasoning and problem solving

15. A cylindrical water bottle has a radius of 4 cm and a perpendicular height of 20 cm. Determine the capacity of the water bottle in mL correct to the nearest whole number. (*Hint:* 1 mL = 1 cm^3)

16. A rectangular prism has a volume of 96 cm^3. Determine its length, width and height.

17. A rectangular metal sheet with length 2.5 m and width 1.9 m is used to make a cylinder. The metal sheet can be rolled on its length or on its width. The two cylinders that can be formed are shown in the diagram.

    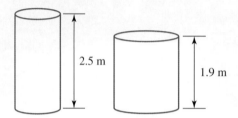

    a. Evaluate the radius to 1 decimal place for:
       i. the taller cylinder   ii. the shorter cylinder.
    b. Determine the two volumes.
    c. Explain which volume is larger.

18. A cylindrical vase has a diameter of 20 cm and a height of 85 cm. The volume of water in the vase is 17 278.76 cm^3. Determine the height of the water in the vase. Round your answer to the nearest whole number.

19. A set of cabin tents is being made for a historical documentary. The dimensions of each tent are shown.

    a. Calculate the area of material required to make each tent, including the floor.
    b. Determine the amount of space inside the tent.

    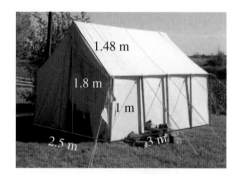

20. Consider a set of three food containers of different sizes, each of which is in the shape of a rectangular prism.
    The smallest container is 6 cm long, 4 cm wide and 3 cm high. All the dimensions of the medium container are double those of the smallest one; all the dimensions of the largest container are triple those of the smallest one.

    a. Calculate the volume of the smallest container.
    b. Determine the dimensions and hence calculate the volume of:
       i. the medium container
       ii. the largest container.
    c. Determine the ratio of the volumes of:
       i. the smallest container to the medium container
       ii. the smallest container to the largest container.
    d. Consider your answers to part c and use them to copy and complete the following:
       If all dimensions of a rectangular prism are increased by a factor of n, the volume of the prism is increased by a factor of _____.

LESSON
8.8 Review

8.8.1 Topic summary

Perimeter
• The perimeter is the distance around a shape.
• Units are km, m, cm and mm.
• $P_{\text{square}} = 4l$
• $P_{\text{rectangle}} = 2(l + w)$

Circumference
• The circumference is the distance around a circle.
• $C = 2\pi r$ or $C = \pi d$
• Arc length $l = \dfrac{\theta^\circ}{360^\circ} \times 2\pi r$

Length, area and volume

Area
• Area is the amount of space inside a flat object or shape.
• Units are km², m², cm² and mm².

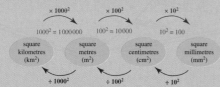

• $A_{\text{square}} = l^2$

• $A_{\text{rectangle}} = lw$

• $A_{\text{triangle}} = \dfrac{1}{2}bh$

• $A_{\text{parallelogram}} = bh$

• $A_{\text{rhombus/kite}} = \dfrac{1}{2}xy$

• $A_{\text{trapezium}} = \dfrac{h}{2}(a+b)$

• To calculate the area of a composite shape, break the shape into two or more known shapes, then calculate the individual areas.

Volume and capacity
Volume
• Volume is the space inside a three-dimensional object.
• Units are km³, m³, cm³ and mm³.

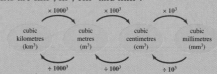

• $V_{\text{prism}} = Ah$
• $V_{\text{rectangular prism}} = lwh$
• $V_{\text{cylinder}} = \pi r^2 h$
• Capacity is the volume of liquid a container can hold.
• Common units are mL, L, kL and ML.

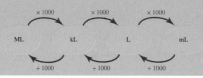

Area of circle and area of circle portions
• $A_{\text{circle}} = \pi r^2$
• $A_{\text{semicircle}} = \dfrac{1}{2}\pi r^2$
• $A_{\text{quadrant}} = \dfrac{1}{4}\pi r^2$
• $A_{\text{sector}} = \dfrac{\theta^\circ}{360^\circ} \times \pi r^2$

8.8.2 Project

Designing a skate park

Skateboarding is an extremely popular activity. Skateboard parks are constructed in many areas for enthusiasts to demonstrate their skills and practise new moves. Two popular pieces of equipment in skate parks are ramps known as the half-pipe and the quarter-pipe. Your local council has decided to build a skate park in your area. The park will have, among the collection of permanent equipment, a half-pipe ramp and a quarter-pipe ramp. You have been asked to provide precise diagrams and measurements to assist in the building of these two structures.

Two different views of your quarter-pipe design are shown with measurements. The ramp section is made from metal and its length is equal to the arc length of a quarter circle with a radius of 2 m.

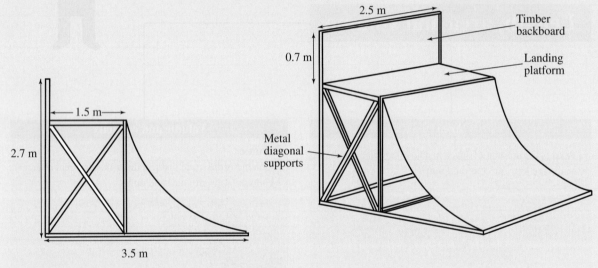

1. Calculate the length of the curved part of the ramp.
2. Use a scale diagram to estimate the length of the diagonal metal supports within the frame.
3. Calculate the combined area of the timber backboard, landing platform and the curved section of the ramp.
4. The frame is made from strong square metal piping. If there are 10 horizontal supports in the frame, determine the total length of metal piping used in the construction.
5. How many times will a wheel of a skateboard turn on the curved section of the ramp, given that the diameter of the skateboard wheel is 55 mm?

Your design for the half-pipe includes two quarter-pipes separated by a 2-metre flat section. The quarter-pipes have the same measurements shown earlier.

6. Draw a side-on view of your half-pipe, showing all relevant measurements.
7. Determine the total area of metal required to construct the skating section of the half-pipe and landing platforms.

**Resources**

 Interactivities Crossword (int-2757)
Sudoku puzzle (int-3189)

Exercise 8.8 Review questions

learnon

Fluency

1. Convert each of the following to the units shown in brackets.
   a. 5.3 mm (cm)   **b.** 7.6 cm (mm)   **c.** 15 cm (m)   **d.** 4.6 m (cm)

2. Convert each of the following to the units shown in brackets.
   a. 250 m (km)   **b.** 6.5 km (m)   **c.** 1.5 m (mm)   **d.** 12 500 cm (km)

3. Determine the perimeter of each of the shapes below. Where necessary, change to the smaller unit.

   a. 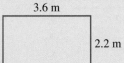   **b.** 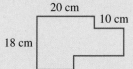   **c.**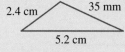

4. Calculate the circumference of each of these circles correct to 2 decimal places.

   a. 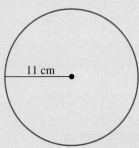   **b.** 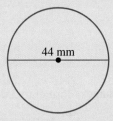   **c.**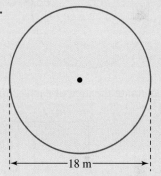

5. Calculate the perimeter of each of these shapes correct to 2 decimal places.

   a. 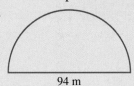   **b.** 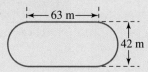   **c.**

6. Determine the area of each of the following triangles.

   a. 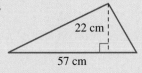   **b.** 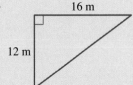   **c.**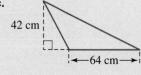

7. Calculate the area of each of the following shapes.

a.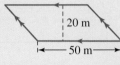
20 m
50 m

b.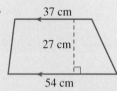
37 cm
27 cm
54 cm

c.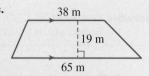
38 m
19 m
65 m

Understanding

8. Calculate the area of each of the following shapes (correct to 2 decimal places where necessary).

a.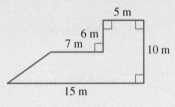
5 m
6 m
7 m
10 m
15 m

b.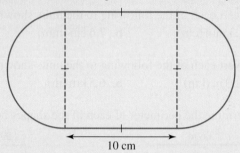
10 cm

c.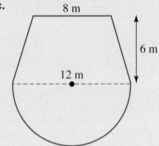
8 m
6 m
12 m

9. Calculate the area of each of the following shapes (correct to 2 decimal places where necessary).

a.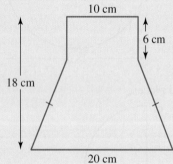
10 cm
6 cm
18 cm
20 cm

b.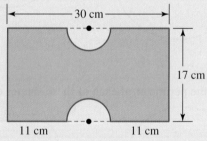
30 cm
17 cm
11 cm   11 cm

c.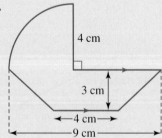
4 cm
3 cm
4 cm
9 cm

10. Calculate the volume of each of the following solids. Where necessary give your answers to 1 decimal place.

a.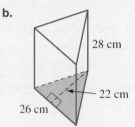
64 cm
35 cm

b.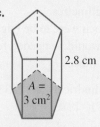
28 cm
22 cm
26 cm

c.
2.8 cm
A = 3 cm²

Communicating, reasoning and problem solving

11. A give-way sign is in the shape of a triangle with a base of 0.5 m. If the sign is 58 cm high, calculate the amount (in m²) of aluminium needed to make 20 such signs. Assume no waste.

12. Restaurant owners want a dome such as the one shown over their new kitchen. Pink glass is more expensive and they want to estimate how much of it is needed for the dome.

The shortest sides of the pink triangles are 40 cm, the longer sides are 54 cm and their heights are 50 cm. The trapeziums around the central light are also 50 cm high. The lengths of their parallel sides are 30 cm and 20 cm.

Evaluate:
a. the area of the pink triangles
b. the area of the pink trapeziums
c. the total area of pink glass in m².

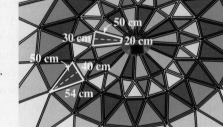

13. Determine the area of cardboard that would be required to make the poster shown below.

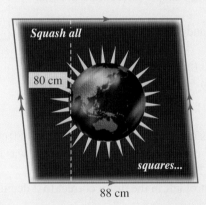

Squash all
80 cm
squares...
88 cm

14. Evaluate the area that the 12-mm-long minute hand of a watch sweeps out in one revolution. Give your answer correct to 2 decimal places.

15. Of the two parallel sides of a trapezium, one is 5 cm longer than the other. Determine the height of the trapezium if the longer side is 12 cm and its area is 57 cm².

16. The diagram shows the design for a brooch.
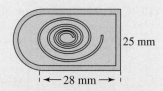

    a. Determine the total area, correct to 2 decimal places, of the brooch in square millimetres.

    b. If the brooch were to be edged with gold, evaluate the length of gold strip that would be needed for the edge. Give your answers correct to 2 decimal places.

17. A narrow cylindrical vase is 33 cm tall and has a volume of 2592 cm³. Determine (to the nearest cm) the radius of its base.

18. Nathan and Rachel ride around a circular track that has an inner radius of 30 m. The track is 2 m wide and Rachel rides along the outer lane.

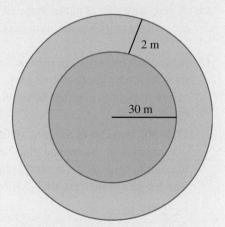

    a. Evaluate how much further Rachel rides than Nathan in one lap of the track. Give your answer correct to 2 decimal places.

    b. The track area needs to be repaved before the next big race. Determine how much bitumen will need to be laid (in m²). Give your answer correct to 2 decimal places.

    c. The centre of the circular field will be grass except for a rectangular area in the centre, which will be a large shed used to store extra bikes and equipment. If the shed sits on a slab of concrete that is 18 m by 10 m, evaluate how much area will be grass. Give your answer correct to 2 decimal places.

    d. Determine the number of packets of lawn seed that will be required if each packet covers 25 m².

19. A sandpit is to be built in the grounds of a kindergarten. As the garden already has play equipment in it, the sandpit is shaped as shown in the diagram.

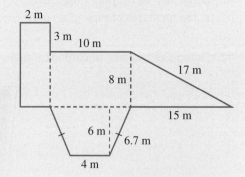

    a. Evaluate the perimeter of the sandpit correct to 2 decimal places.

    b. Calculate the area of the sandpit correct to 2 decimal places.

    c. Determine how much sand (in m³) will be needed if the sandpit is to be 30 cm deep.

20. A cylindrical petrol drum has a base diameter of 40 cm and is 80 cm high. The drum was full on Friday morning but then 30 L of petrol was used for the mower and 5 L was used for the whipper snipper.

Evaluate the height of the petrol left in the drum correct to 2 decimal places. (*Hint:* 1 cm³ = 1 mL)

21. Refer to the diagram squares. The area of square F is 16 square units. The area of square B is 25 square units. The area of square H is 25 square units.
    a. Evaluate the areas of all the other squares and explain how you got your answers.
    b. Calculate the area of the total shape.
    c. Determine the perimeter of the shape.

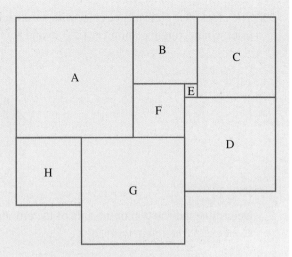

22. A cube with edges 36 cm long is cut into smaller cubes.
    a. Determine how many small cubes there will be if the smaller cubes have an edge of 12 cm.
    b. Evaluate the length of each side of the smaller cubes if there are 64 cubes.

23. Cameron was mowing the backyard when he stubbed his toe and had to go inside. If he had mowed a 3 m strip in the middle before he stopped, determine the area his sister needed to mow to finish the job for him.

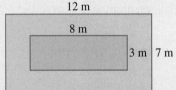

24. Polly divided her rectangular vegetable garden into four plots, as shown. Three of the plots were rectangular and one was square. Evaluate the area of the shaded plot.

15 m²	9 m²
	12 m²

25. The width of a rectangle is 6 cm and its perimeter is 26 cm. Evaluate the area of the rectangle.

26. While practising his karate skills, Alex accidentally made a large round hole in the wall of his office. The area of the hole is about 154 cm². To cover up the hole, Alex plans to use a square photo frame (with a photo of himself in full karate uniform, including his black belt). Determine the smallest side length, correct to 2 decimal places, that the photo frame can have.

27. The Great Wall of China stretches from the east to the west of China. It is the largest of a number of walls that were built bit by bit over thousands of years to serve as protection against military invasions. Some estimates are that the Great Wall itself is about 6300 km long.
    Assume your pace length is 70 cm. At 10 000 paces a day, determine how many days it would take you to walk the length of the Great Wall of China.

28. The clock face shown is circular with a radius of 10 cm. It has a decorative mother-of-pearl square inset, connecting the numbers 12, 3, 6 and 9.

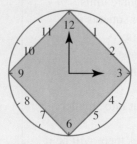

Determine the fraction of the area of the circular face that the square inset represents. Give your answer as an exact fraction in terms of π.

29. The two smaller circles in this diagram have a diameter that is equal to the radius of the medium-sized circle.

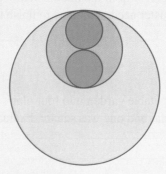

The diameter of the medium-sized circle is equal to the radius of the large circle. Evaluate the fraction of the large circle that the two small circles represent.

30. A square field is enclosed by a square fence, using 48 posts. The posts are 5 m apart, with one at each corner of the field. Evaluate the area bounded by the fence.

 To test your understanding and knowledge of this topic, go to your learnON title at www.jacplus.com.au and complete the **post-test**.

Answers

Topic 8 Length, area and volume

8.1 Pre-test

1. C
2. D
3. $120\,cm^2$
4. C
5. B
6. a. $34.6\,cm$ b. $25.7\,cm$
7. $65\,cm^2$
8. $550\,cm^2$
9. $9.7\,cm$
10. $180\,m^3$
11. $938.43\,m^2$
12. $214.5\,mm^2$
13. $1575
14. B
15. D

8.2 Length and perimeter

1. a. $20\,mm = 2\,cm$ b. $13\,mm = 1.3\,cm$
   c. $130\,mm = 13\,cm$ d. $1.5\,cm = 15\,mm$
   e. $0.03\,cm = 0.3\,mm$ f. $10.05\,mm$
2. a. $2.8\,km = 2800\,m$ b. $0.034\,m = 3.4\,cm$
   c. $2400\,mm = 2.4\,m$ d. $1375\,mm = 1.375\,m$
   e. $2.7\,m = 2700\,mm$ f. $300.5\,mm$
3. a. $0.08\,m = 80\,mm$ b. $6.071\,km = 6071\,m$
   c. $670\,cm = 6.7\,m$ d. $0.0051\,km = 5.1\,m$
   e. $0.005\,mm$ f. $0.001\,75\,km$
4. a. $14\,cm$ b. $12\,cm$ c. $106\,mm$ d. $18\,cm$
5. a. $240\,mm$ b. $32\,mm$ c. $23\,cm$ d. $72\,mm$
6. a. $73\,mm$ b. $1260\,cm$ $(12.6\,m)$
   c. $192\,cm$ $(1.92\,m)$ d. $826\,cm$
7. a. $1800\,mm \times 900\,mm = 180\,cm \times 90\,cm$
   $$= 1.8\,m \times 0.9\,m$$
   b. $2400\,mm \times 900\,mm = 240\,cm \times 90\,cm$
   $$= 2.4\,m \times 0.9\,m$$
   c. $2700\,mm \times 1200\,mm = 270\,cm \times 120\,cm$
   $$= 2.7\,m \times 1.2\,m$$
8. $5.40
9. $38.16
10. $41\,400\,m$ or $41.4\,km$
11. a. $7\,m$ b. $26.60
12. $86\,m$
13. $2.6\,m$
14. $980\,cm$
15. $86\,cm$
16. a. $510\,m$ b. $749.70
17. $15.88
18. a. $11\,cm$ b. $22\,cm$ c. $6.9\,m$
19. $5.5\,m$

20. a. $5\,cm$ b. $20\,cm$ c. 2 bottles
21. Width $= 7\,cm$; length $= 11\,cm$
22. 5
23. $9\,cm$
24. a. $258\,m$ b. $174\,m$
25. a. Fence length $(l) = 10\,m$ and width $(w) = 8\,m$, so perimeter $= 36\,m$.
    b. Fence length $= 2\,(l + w + 4)$
26. a. $217\,cm$
    b. Length $= 2w + 2l + 4h + b$

8.3 Circumference

1. a. $2\pi\,cm$ b. $10\pi\,cm$ c. $7\pi\,mm$
2. a. $2.58\,m$ b. $23.25\,km$ c. $106.81\,m$
3. a. $8\pi\,m$ b. $34\pi\,mm$ c. $16\pi\,cm$
4. a. $8.98\,km$ b. $2.51\,m$ c. $66.60\,m$
5. a. $25.71\,cm$ b. $82.27\,mm$ c. $61.70\,m$
   d. $39.28\,mm$ e. $71.42\,cm$
6. a. $120.82\,cm$ b. $5.88\,m$
   c. $250.80\,m$ d. $252.81\,cm$
7. B
8. C
9. a. $24.38\,cm$ b. $35.28\,mm$
   c. $31.83\,cm$ d. $65.99\,cm$
10. $119.38\,mm$
11. $100.53\,cm$
12. $25.13\,m$
13. $55.29\,m$
14. $21.58\,cm$
15. a. $6.00\,m$ b. $20.63\,cm$ c. $23.75\,mm$
16. a. $2.01\,cm$ b. $7.54\,m$ c. $24.99\,mm$
17. $19.19\,m$
18. $15.71\,m$
19. $80\,m$
20. a. iii   b. i   c. iv   d. ii
21. $41.82\,km/h$
22. $9.55\,cm$
23. a. $25.71\,cm$ b. $483.29\,mm$
    c. $31.07\,m$ d. $5.71\,cm$
24. Length of arc $= 6.98\,cm$; perimeter $= 13.74\,cm$

8.4 Areas of rectangles, triangles, parallelograms, rhombuses and kites

1. a. $530\,000\,m^2$ b. $2.35\,cm^2$ c. $254\,000\,mm^2$
   d. $54.2\,m^2$ e. $0.074\,m^2$
2. a. $3\,km^2$ b. $9.8563\,ha$ c. $17\,800\,m^2$
   d. $987\,000\,mm^2$ e. $1\,275\,000\,cm^2$
3. a. Approx. $34\,m^2$ b. $170\,L$
4. a. $118\,m^2$ b. $29.5\,kg$
5. a. $36\,cm^2$ b. $1125\,mm^2$ c. $4.5\,m^2$
6. a. $1215\,km^2$ b. $2.5\,m^2$ c. $336\,mm^2$

7. a. $25\,\text{mm}^2$ b. $256\,\text{cm}^2$ c. $5.29\,\text{m}^2$

8. C

9. B

10. a. $1258\,\text{mm}^2$ b. $1771.54\,\text{m}^2$ c. $9932.63\,\text{mm}^2$

11. a. $17\,537\,\text{cm}^2$ b. $11\,566.8\,\text{mm}^2$ c. $257.645\,\text{m}^2$

12. a. $275\,\text{mm}^2$ b. $24\,000\,\text{m}^2$ c. $656\,\text{cm}^2$
    d. $11.04\,\text{mm}^2$ e. $2.7\,\text{m}^2$

13. a. $2400\,\text{mm}^2$ b. $17.36\,\text{m}^2$
    c. $4760\,\text{m}^2$ d. $8.48\,\text{m}^2$

14. a. $30\,\text{cm}^2$ b. $24\,\text{cm}^2$ c. $90\,\text{cm}^2$

15. $2052\,\text{m}^2$

16. a. $29.225\,\text{cm}^2$ b. $756\,\text{m}^2$ c. $360\,\text{cm}^2$

17. $9600\,\text{mm}^2$

18. $50.4\,\text{cm}^2$

19. D

20. a. $42.65\,\text{m}^2$, $60.81\,\text{m}^2$, $42.65\,\text{m}^2$
    b. $146.11\,\text{m}^2$

21. $351.98\,\text{cm}^2$

22. a. $15\,000\,\text{m}^2$ b. $60\,000\,\text{m}^2$

23. Examples of possible answers are given. More sample responses can be found in the worked solutions in the online resources.
    a. Length $=12\,\text{cm}$; width $=3\,\text{cm}$
    b. Base $=4\,\text{cm}$; height $=9\,\text{cm}$
    c. Diagonals $12\,\text{cm}$ and $6\,\text{cm}$

24. a. $3.6\,\text{cm}$ b. $9.2\,\text{cm}$

25. a. The area will increase by a factor of 4.
    b. The area will decrease by a factor of 4.
    c. The area will be squared.

26. a. $15.64\,\text{m}^2$ b. 3

27. $42\,\text{cm}$

28. $1584\,\text{cm}^2$

29. $17.28\,\text{cm}^2$

30. $b = h = 9.5\,\text{cm}$

31. a. $400\,\text{m}$ b. $375\,\text{min}$ or $6\frac{1}{4}\,\text{h}$

32. a.

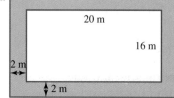

    b. $320\,\text{m}^2$
    c. $160\,\text{m}^2$
    d. \$800

8.5 Areas of circles

1. a. $452.39\,\text{cm}^2$ b. $4.91\,\text{km}^2$ c. $2.27\,\text{m}^2$

2. a. $0.38\,\text{cm}^2$ b. $10\,568.32\,\text{cm}^2$ c. $206.12\,\text{mm}^2$

3. a. $78.54\,\text{cm}^2$ b. $483.05\,\text{mm}^2$
   c. $615.75\,\text{m}^2$ d. $254.47\,\text{cm}^2$

4. a. $3.99\,\text{cm}$ b. $4.89\,\text{mm}$ c. $10.30\,\text{cm}$

5. a. $173.18\,\text{cm}^2$ b. $44.18\,\text{cm}^2$ c. $160.13\,\text{mm}^2$

6. a. $37.70\,\text{cm}^2$ b. $1281.77\,\text{cm}^2$ c. $3189.50\,\text{m}^2$

7. a. $157.08\,\text{cm}^2$ b. $201.06\,\text{mm}^2$
   c. $13.85\,\text{m}^2$ d. $39.27\,\text{cm}^2$

8. a. $1039.08\,\text{cm}^2$ b. $77.91\,\text{cm}^2$
   c. $132.54\,\text{cm}^2$ d. $9.74\,\text{cm}^2$

9. $1134.11\,\text{cm}^2$

10. $4.52\,\text{m}^2$

11. 25 packets

12. a. Diameter changes by a factor of $1.414\left(=\sqrt{2}\right)$.
    b. Diameter changes by a factor of 2.
    c. Diameter changes by a factor of $0.707\left(=\sqrt{\dfrac{1}{2}}\right)$.

13. $8\pi\,\text{cm}^2$

14. No, the area to paint is $3.8\,\text{m}^2$.

15. a. $3279.82\,\text{cm}^2$ b. $40.18\,\text{mm}^2$ c. $31.49\,\text{cm}^2$

16. a. $A = 70.83\,\text{cm}^2$
    b. $r = 4.75\,\text{cm}$
    c. The diameter of the coaster is larger than that of the mug, so the coffee mug will fit within the coaster's surface.

8.6 Areas of trapeziums

1. a. $9\,\text{cm}^2$ b. $33.75\,\text{m}^2$ c. $12.75\,\text{m}^2$

2. a. $351\,\text{mm}^2$ $(3.51\,\text{cm}^2)$ b. $4.68\,\text{cm}^2$
   c. $3120\,\text{m}^2$

3. D

4. $3062\,\text{cm}^2$

5. \$88.30

6. a. \$2730.55   b. \$332.50   c. \$3063.05

7. $2\,\text{m}$

8. B

9. a. B   b. C   c. C

10. a. $44.5\,\text{cm}^2$ b. $55.37\,\text{cm}^2$

11. Errors in pink: Area $= \dfrac{1}{2}(12 + 11) \times 6 + \pi \times 5.5^2$
    Correct working: Area $= \dfrac{1}{2}(12 + 16.4) \times 6 + \dfrac{\pi \times 5.5^2}{2}$

12. a. $A_1 = \dfrac{1}{2}bh$
    b. $A_2 = \dfrac{1}{2}ah$
    c. $A = A_1 + A_2 = \dfrac{1}{2}ah + \dfrac{1}{2}bh$

13. \$2652.80

14. a. $31.15\,\text{m}^2$ b. 7   c. \$62.65

15. $4240\,\text{m}^2$

8.7 Volumes of prisms and other solids

1. b and e

2. a. $84\,\text{cm}^3$ b. $81\,\text{m}^3$ c. $84\,\text{cm}^3$

3. a. $120\,cm^3$ **b.** $320\,m^3$ **c.** $126\,cm^3$

4. a. $14\,137.17\,cm^3$ **b.** $31\,415.93\,cm^3$ **c.** $667.98\,m^3$

5. a. $4778.36\,cm^3$ **b.** $7.5\,cm^3$ **c.** $1.88\,m^3$

6. a. Capacity **b.** Volume
   c. Volume **d.** Capacity
   e. Volume **f.** Capacity

7. a. $12\,000\,mL$ **b.** $0.003125\,mL$
   c. $0.397\,L$ **d.** $7.8\,mL$
   e. $4.893\,L$ **f.** $0.00\,003\,697\,mL$

8. a. $372\,mL$ **b.** $1\,630\,000\,cm^3$
   c. $3400\,cm^3$ **d.** $0.38\,cm^3$
   e. $163\,000\,cm^3$ **f.** $49.28\,mL$

9. a. $0.578\,L$ **b.** $0.75\,L$ **c.** $429\,mL$

10. $1200\,m^3$

11. $60.32\,m^3$

12. a. $60\,cm^3$ **b.** $84\,cm^3$

13. $7380\,cm^3$

14. $18.6\,L$

15. $1005\,mL$

16. Answers will vary. Some sample responses are:
   $2\times6\times8$ $4\times3\times8$ $1\times12\times8$ $6\times4\times4$ $48\times2\times1$
   $16\times1\times6$ $24\times1\times4$ $16\times2\times3$ $12\times4\times2$

17. a. i. $0.3\,m$ **ii.** $0.4\,m$
   b. $0.72\,m^3, 0.94\,m^3$
   c. The cylinder with the bigger radius is larger. The area of the base is πr^2. The radius has a bigger influence on the volume than the height because the radius is squared.

18. $55\,cm$

19. a. $29.38\,m^2$ **b.** $10.5\,m^3$

20. a. $72\,cm^3$
   b. i. $12\,cm$ long, $8\,cm$ wide, $6\,cm$ high; $V = 576\,cm^3$
   ii. $18\,cm$ long, $12\,cm$ wide, $9\,cm$ high; $V = 1944\,cm^3$
   c. i. $1 : 8$ **ii.** $1 : 27$
   d. If all dimensions of a rectangular prism are increased by a factor of n, the volume of the prism is increased by a factor of n^3.

Project

1. $3.14\,m$

2. $2.5\,m$

3. $13.35\,m^2$

4. $46.0\,m$

5. 18.18 times

6.

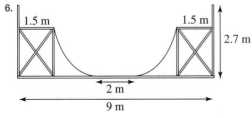

7. $28.2\,m^2$

1. a. $0.53\,cm$ **b.** $76\,mm$
   c. $0.15\,m$ **d.** $460\,cm$

2. a. $0.25\,km$ **b.** $6500\,m$
   c. $1500\,mm$ **d.** $0.125\,km$

3. a. $11.6\,m$ **b.** $96\,cm$ **c.** $111\,mm$

4. a. $69.12\,cm$ **b.** $138.23\,mm$ **c.** $56.55\,m$

5. a. $241.65\,m$ **b.** $257.95\,m$ **c.** $28.57\,cm$

6. a. $627\,cm^2$ **b.** $96\,m^2$ **c.** $1344\,cm^2$

7. a. $1000\,m^2$ **b.** $1228.5\,cm^2$ **c.** $978.5\,m^2$

8. a. $84\,m^2$ **b.** $178.54\,cm^2$ **c.** $116.55\,m^2$

9. a. $240\,cm^2$ **b.** $459.73\,cm^2$ **c.** $32.07\,cm^2$

10. a. $112\,594.7\,cm^3$ **b.** $8008\,cm^3$
   c. $8.4\,cm^3$

11. $2.9\,m^2$

12. a. $16\,000\,cm^2$ **b.** $10\,000\,cm^2$ **c.** $2.6\,m^2$

13. $7040\,cm^2$

14. $452.39\,mm^2$

15. $6\,cm$

16. a. $945.44\,mm^2$ **b.** $120.27\,mm$

17. $5\,cm$

18. a. $12.57\,m$ further **b.** $389.56\,m^2$
   c. $2647.43\,m^2$ **d.** 106 packets

19. a. $77.40\,m$ **b.** $204\,m^2$ **c.** $61.2\,m^3$

20. $52.15\,cm$

21. a. If B is 25 square units, then each side is 5 units long. If F is 16 square units, then each side is 4 units long. Then E is $1 \times 1 = 1$ square unit, C is $6 \times 6 = 36$ square units and D is $7 \times 7 = 49$ square units. A is made up of B and F, which means it is $9 \times 9 = 81$ square units, and then G is $8 \times 8 = 64$ square units.
   b. Total area: A = 81, B = 25, C = 36, D = 49, E = 1, F = 16, G = 64, H = 25.
   Sum of all areas = 297 square units.
   c. 74 units

22. a. 27 cubes
   b. $9\,cm$

23. $60\,m^2$

24. The shaded plot has dimensions $5\,m \times 4\,m$ and an area of $20\,m^2$.

25. $42\,cm^2$

26. $14.00\,cm$

27. 900 days

28. $\dfrac{2}{\pi}$

29. $\dfrac{1}{8}$

30. $3600\,m^2$

9 Linear relationships

LESSON SEQUENCE

LESSON
9.1 Overview

Why learn this?

Coordinates are a set of numbers used to locate a point on a map or a graph. They can be shown on a grid known as the Cartesian plane, which is described by two symbols (usually letters or numbers) that indicate how far vertically and horizontally you need to move on the grid to find the desired location. In coordinate geometry, coordinates are used to pinpoint a position. You may have heard of locations being described in terms of latitude and longitude. For instance, Sydney's position on Earth is located at 33.87°S, 151.21°E — this is a spherical coordinate system and is more complex than the Cartesian plane. The Cartesian plane is also useful for plotting graphs of linear relationships.

Linear relationships form part of algebra and are used to model many real-life situations. Things that change at a constant rate over time produce a straight-line graph and are known as a linear relationship. A car travelling at a constant speed, the interest earned by a simple-interest bank account, and a wage based on hours worked are all linear relationships.

Knowledge of linear relationships can help you convert different currencies, such as Australian to US dollars, or temperatures, such as Fahrenheit to Celsius. Being able to graph linear relationships allows you to solve many questions that you might not be able to answer with the coordinates alone. The graph will allow you to find other values that are not given in the coordinates, and will give you a visual representation of the relationship.

1. **MC** If a point with coordinate $(3, 4)$ is translated (moved) 4 units to the right and 7 units down, identify the coordinates of the new position of the point.

   A. $(7, 11)$ **B.** $(10, 8)$ **C.** $(-4, 7)$ **D.** $(7, -3)$

2. Determine the missing value in the table.

x	0	1	2	3
y	5	8		14

3. **MC** From the following options, identify the missing value from the table.

x	2	5	6	11
$y = 3x - 4$	2	11	14	

   A. 22   **B.** 29   **C.** 33   **D.** 37

4. **MC** Which of the following coordinates lie on the line $y = 2x + 1$? Select all that apply.

   A. $(1, 4)$ **B.** $(1, 3)$ **C.** $(3, 10)$ **D.** $(0, 1)$

5. A plumber charges a \$90 call-out fee and \$65 per hour for any job they are asked to carry out. If the plumber's bill comes to , calculate the number of hours they spent on the job.

6. **MC** What does $y = 6x - 3$ mean?
   A. The y-value equals the x-value subtracted by 6.
   B. The y-value equals the x-value multiplied by negative 3 and then multiplied by 6.
   C. The y-value equals the x-value multiplied by 6, then subtracted by 3.
   D. The y-value equals the x-value plus 6, then subtracted by 3.

7. **MC** Which of the following points does not lie on the graph of the equation $y = 3x - 1$?

   A. $(1, 2)$ **B.** $(2, 1)$ **C.** $(3, 8)$ **D.** $(4, 11)$

8. a. Using the equation, $y = x - 3$ to complete the table of values.

x	0	1	2	3	4
y					

b. Graph the equation, $y = x - 3$ using the table in part **a**.

9. Determine the rule for each of the following tables.

a.

x	0	1	2	3	4
y	0	3	6	9	12

b.

x	0	1	2	3	4
y	1	3	5	7	9

10. **MC** Use one of the following words to describe the gradient of a graph that is represented by a vertical line.
   A. Undefined
   B. Positive
   C. Negative
   D. Zero

11. Draw a table of values and graph of these rules on the same Cartesian plane.
   $y = 2x + 1$
   $y = x - 1$
   Describe the relationship between these lines.

12. **MC** Determine which of the following equations are parallel to $y = 3x - 1$. Select all that apply.
   A. $y - 3x = 2$
   B. $6x + 3y = 2$
   C. $3y - 9x = 4$
   D. $3y + 9x = 3$

13. Determine the rule for the equation represented by the following graph.

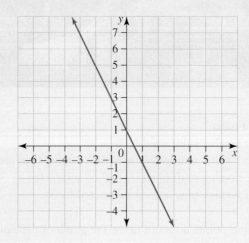

14. Use the graph of $y = -2x + 2$ to solve the linear equation $-2x + 2 = -4$

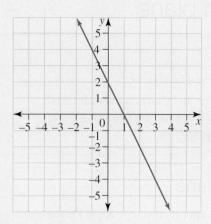

15. While working in the garden, Bill accidentally cut the electricity to his house. He called the power company and they informed him their emergency call-out charge was a \$250 call-out fee, plus \$50 for every 15 minutes of the repairer's time.

a. Calculate how much it would cost Bill to have his electricity restored if the repairer was there for:

   i. $\frac{1}{4}$ of an hour

   ii. $\frac{1}{2}$ of an hour

   iii. $\frac{3}{4}$ of an hour.

b. Determine the rule that satisfies these values, where y is the total cost and x is the time (in quarter-hours).

LESSON
9.2 The Cartesian plane

LEARNING INTENTION

At the end of this lesson you should be able to:

- understand the Cartesian plane and Cartesian coordinates
- determine the coordinates of points on the Cartesian plane
- plot points on the Cartesian plane.

▶ 9.2.1 The Cartesian plane

eles-4500

- The **Cartesian plane** (named after its inventor René Descartes) is a visual means of describing locations on a plane by using two numbers as coordinates.
- The Cartesian plane is formed by two perpendicular lines, which are called the axes. The horizontal axis is called the **x-axis**; the vertical axis is called the **y-axis**.
- The centre of the Cartesian plane (where the x- and y-axis intersect) is called the **origin**.
- Both axes are evenly scaled and numbered, with 0 (zero) placed at the origin. On the x-axis the numbers increase from left to right, while on the y-axis the numbers increase from bottom to top.
- Arrows are placed on the ends of each axis to show that they continue infinitely.

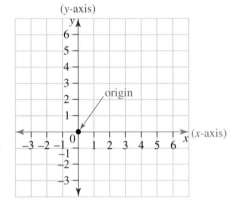

Cartesian coordinates

- To locate any point on the Cartesian plane, we use a pair of numbers called *Cartesian coordinates*. They are written as (x, y), where x refers to the horizontal position of the point and is called 'the x-coordinate' of the point, and y refers to the vertical position of the point and is called 'the y-coordinate' of the point. The coordinates of the origin are $(0, 0)$.
- To locate a point on the Cartesian plane, move along the x-axis to the number indicated by the x-coordinate and then along the y-axis to the number indicated by the y-coordinate. For example, to locate the point with coordinates $(3, 2)$, beginning at the origin, move 3 units right and then 2 units up.

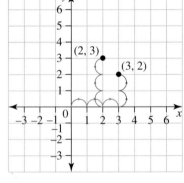

- The coordinates are also called **ordered pairs** because they come in pairs and their order matters. The point $(3, 2)$ is different from $(2, 3)$, as shown on the right.
- *Hint:* To help remember the order in which Cartesian coordinates are measured, think about using a ladder. Remember we must always walk across with our ladder and then climb up it.

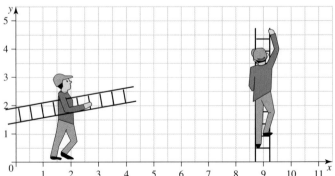

WORKED EXAMPLE 1 Plotting points on a Cartesian plane

Draw a Cartesian plane with axes extending from 0 to 6 units. Mark the following points with a dot, and label them.

a. (4, 2) **b. (0, 5)** **c. (2, 0)** **d.** $\left(2, 3\frac{1}{2}\right)$

THINK

1. Rule up and label the axes.
2. Mark each point.
   a. (4, 2) means start at the origin, move 4 units right and then 2 units up.
   b. (0, 5) means start at the origin, move 0 units right and then 5 units up. It lies on the y-axis.
   c. (2, 0) means start at the origin, move 2 units right and then 0 units up. It lies on the x-axis.
   d. $\left(2, 3\frac{1}{2}\right)$ means start at the origin, move 2 units right and $3\frac{1}{2}$ units up.

WRITE

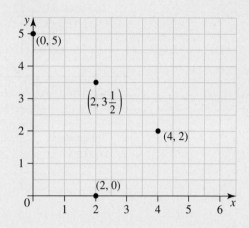

WORKED EXAMPLE 2 Determining the coordinates of points on a Cartesian plane

Determine the Cartesian coordinates for each of the points A, B, C and D.

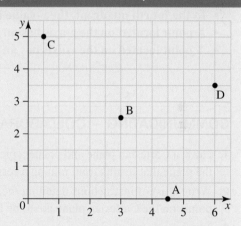

THINK

Point A is $4\frac{1}{2}$ units right and 0 unit up.

Point B is 3 units right and $2\frac{1}{2}$ units up.

Point C is $\frac{1}{2}$ units right and 5 units up.

Point D is 6 units right and $4\frac{1}{2}$ units up.

WRITE

A is at $\left(4\frac{1}{2}, 0\right)$.

B is at $\left(3, 2\frac{1}{2}\right)$.

C is at $\left(\frac{1}{2}, 5\right)$.

D is at $\left(6, 4\frac{1}{2}\right)$.

▶ 9.2.2 Quadrants and axes

eles-4501

- The Cartesian axes extend infinitely in both directions, as represented by the arrows on the axes.
- The axes divide the Cartesian plane into four sections called **quadrants**. The quadrants are numbered in an anti-clockwise direction, starting with the top right corner.

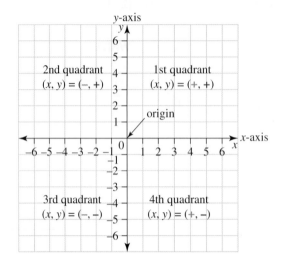

Quadrants of the Cartesian plane

- **In the first quadrant, the x- and y-values are both positive.**
- **In the second quadrant, the x-value is negative and the y-value is positive.**
- **In the third quadrant, both the x- and y-values are negative.**
- **In the fourth quadrant, the x-value is positive and the y-value is negative.**
- **A point located on the y-axis has an x-coordinate of 0.**
- **A point located on the x-axis has an y-coordinate of 0.**

COMMUNICATING — COLLABORATIVE TASK: Creating a picture

1. Draw a Cartesian plane and create a picture of your choice on the plane that results from joining a series of points.
2. Write a list of instructions detailing the order in which the points on the Cartesian plane are to be joined.
3. Test your instructions on a classmate.

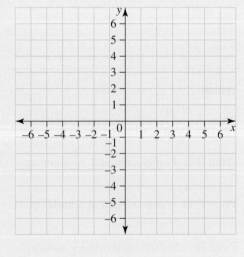

Consider the following points: A (-1, 2), B (2, -4), C (0, -3), D (4, 0), E (-5, -2).

a. **Without plotting the points, state the location of each point on the plane (i. e. the quadrant, or the axis on which it sits).**

b. **Plot these points on the Cartesian plane and confirm your answers to part a.**

THINK	WRITE
a. 1. Consider the signs of the x- and y-coordinates of each point to determine and state its location.	a. 1. A (-1, 2): The x-coordinate is negative and the y-coordinate is positive. A is in the 2nd quadrant. B (2, -4): The x-coordinate is positive and the y-coordinates is negative. B is in the 4th quadrant. C (0, -3): The x-coordinate is zero. C is on the y-axis. D (4, 0): The y-coordinate is zero. D is on the x-axis. E (-5, -2): The x- and y-coordinates are both negative. E is in the 3rd quadrant.
b. 1. Draw a set of axes, ensuring they are long enough to fit all of the values, and label the quadrants. By examining the given coordinates, it is clear that a scale of -5 to 5 on both axes will fit all the points.	b. 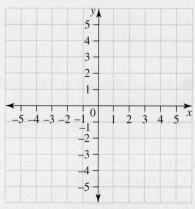
2. Plot the points. Point A is one unit to the left and two units up from the origin; point B is 2 units right and 4 units down from the origin (and so on).	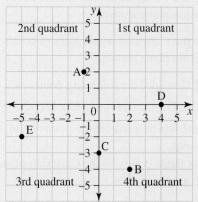
3. State the location of each point and confirm your answers to part a.	b. Point A is in the second quadrant. Point B is in the fourth quadrant. Point C is on the y-axis. Point D is on the x-axis. Point E is in the third quadrant. This confirms the answers to part a.

Exercise 9.2 The Cartesian plane

learn on

| **9.2 Quick quiz** on | **9.2 Exercise** |

Individual pathways

■ PRACTISE	■ CONSOLIDATE	■ MASTER
1, 4, 7, 10, 12, 13, 15, 18, 21	2, 5, 8, 11, 14, 16, 19, 22	3, 6, 9, 17, 20, 23, 24

Fluency

1. **WE1** Draw a Cartesian plane that extends from −6 to 6 on the *x*-axis and −6 to 6 on the *y*-axis, and plot and label the following points.

   a. $A(3, 3)$ b. $B(2, 5)$ c. $C(5, 1)$ d. $D(−1, 4)$

2. Draw a Cartesian plane that extends from −6 to 6 on the *x*-axis and −6 to 6 on the *y*-axis, and plot and label the following points.

   a. $E(−4, 2)$ b. $F(−2, 0)$ c. $G(−2, −3)$ d. $H(−4, −5)$

3. Draw a Cartesian plane and plot and label the following points. (*Hint*: Decide how big the Cartesian plane needs to be before you plot the points.)

   a. $I(0, −3)$ b. $J(1, −2)$ c. $K\left(3, -1\frac{1}{2}\right)$ d. $L\left(4\frac{1}{2}, 0\right)$

4. **WE2** Write the Cartesian coordinates of points A to D marked on the Cartesian plane shown.

5. Write the Cartesian coordinates of points E to H marked on the Cartesian plane shown.

6. Write the Cartesian coordinates of points I to L marked on the Cartesian plane shown.

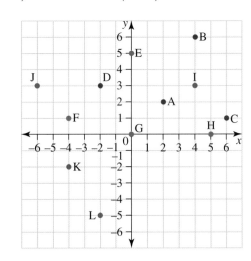

7. **WE3** Consider the following points on a Cartesian plane.

   $A(2, 5), B(−3, 2), C(−1, −5), D(−2, −5)$

   a. State which quadrant the points lie in, or whether they sit on an axis.
   b. Plot these points on the Cartesian plane and confirm your answers to part a.

8. a. Plot the following points on a Cartesian plane.

   $E(−10, 0), F(0, 0), G(−8, 15), H(−9, 24)$

   b. State which quadrant the points lie in, or whether they sit on an axis.

9. a. Plot the following points on a Cartesian plane.

   $I(24, 0), J(−1, 1), K(−7, −1), L(0, −8)$

   b. State which quadrant the points lie in, or whether they sit on an axis.

10. **MC** The point $(3, 4)$ gives the position on the Cartesian plane of:
    Note: There may be more than one correct answer.

    A. 3 on the y-axis, 4 on the x-axis.   B. 3 left, 4 up.
    C. 3 on the x-axis, 4 on the y-axis.   D. 3 right, 4 up.

11. **MC** The point $(-2, 0)$ gives a position on the Cartesian plane of:
    Note: There may be more than one correct answer.

    A. left 2, up 0.   B. left 0, down 2.
    C. -2 on the x-axis, 0 on the y-axis.   D. -2 on the y-axis, 0 on the x-axis.

Understanding

12. Each of the following sets of Cartesian axes (except one) has something wrong with it. Match the mistake in each diagram with one of the sentences in the list below.

    A. The units are not marked evenly.
    B. The y-axis is not vertical.
    C. The axes are labelled incorrectly.
    D. The units are not marked on the axes.
    E. There is nothing wrong.

    a.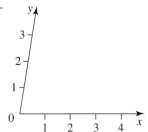

    b.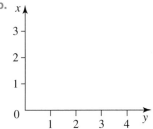

    c.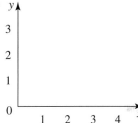

    d.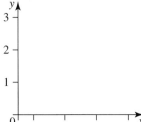

    e.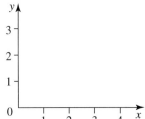

    f.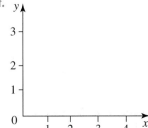

13. On 1-cm graph paper, draw a Cartesian plane with an x-axis from -6 to 6 and a y-axis from -6 to 6. Connect these groups of points.
    START $(4, 6) (-4, 6) (-6, 0) (-4, -6) (4, -6) (6, 0) (4, 6) (-6, 0) (4, -6) (4, 6)$ STOP
    START $(-4, 6) (-4, -6) (6, 0) (-4, 6)$ STOP
    START $(4, 0) (2, 2.5) (-2, 2.5) (-4, 0) (-2, -2.5) (2, -2.5) (4, 0)$ STOP
    Colour the 6 triangles between the star and the hexagon. For example the triangle $(6, 0) (4, 6) (4, 1)$ could be coloured pink. Colour the 6 triangles inside the star. For example $(4, 0) (4, 1) (2, 2.5)$ could be coloured green.

14. Draw a Cartesian plane. Check the following coordinates to find the lowest and highest x- and y-value needed on the axes. Then, follow the steps below to draw a cartoon character.
    START $(6, 7) (7.5, 9) (5, 9) (4.5, 12) (2, 11) (0, 13) (-1.5, 10) (-5, 11) (-5, 8) (-8, 6) (-6, 4) (-8, 2)$
    $(-6, 1) (-7, -2) (-4, -1.5) (-4, -3.5) (-1.5, -3) (-2, -4) (-4, -7) (-5, -8)$ STOP
    START $(-2, -9) (-1, -7) (1, -8) (3, -8) (4, -7.5) (5, -10)$ STOP
    START $(4, -7.5) (3.5, -6) (3.5, -4) (4, -3) (5, -2.5) (5, -2) (4, -1.5) (4, -1) (5, 0.5) (7, 1) (8, 2) (8, 2.5)$
    $(6.5, 3)$ STOP
    START $(4, -2.5) (2, -3) (0.5, -3) (0, -2) (1, -1) (2, -0.5) (3, 0) (7, 1)$ STOP

START (6, 2.5) (6.5, 3) (6.5, 4) (6, −4) (4, 3) STOP
START (6, 7) (5, 7.5) (4, 7) (3, 6) (1, 6) (0, 5) (−1, 4) (0, 2) (1.5, 1.5) (3, 2) (4, 4) (6.5, 4)(7, 5) (7, 6)
(6, 7) STOP
START (4, 4) (4, 5) (3, 6) STOP
START (1, −1) (5, 0) STOP
EYES AT (1, 3) AND (5, 5)
EYELASHES (−1, 4) TO (−2, 4.5) , (0, 5) TO (−0.5, 6) , (1, 6) TO (0.5, 7) , (2, 6) TO (2, 7) , (4, 7) TO
(3.5, 8) , (5, 7.5) TO (5, 8.5) , (6, 7) TO (6.5, 8) , (6.5, 6.5) TO (7, 7)

15. **MC** Consider the following set of points: A (2, 5), B (−4, −12), C (3, −7), D (0, −2), E (−10, 0), F (0, 0),
    G (−8, 15), H (−9, −24), I (18, −18), J (24, 0).
    Identify which of the following statements are true.

    A. Points A and J are in the first quadrant.
    B. Points B and H are in the third quadrant.
    C. Only point I is in the fourth quadrant.
    D. Only one point is in the second quadrant.

16. **MC** Consider the following set of points: A (2, 5), B (−4, −12), C (3, −7), D (0, −2), E (−10, 0), F (0, 0),
    G (−8, 15), H (−9, −24), I (18, −18), J (24, 0).
    Identify which of the following statements are true.

    A. Point F is at the origin.
    B. Point J is not on the same axis as point E.
    C. Point D is two units to the left of point F.
    D. Point C is in the same quadrant as point I.

17. Messages can be sent in code using a grid like the one shown, where the letter B is
    represented by the coordinates (2, 1). Use the diagram to decode the answer to the
    following riddle.
    Q: Where did they put the man who was run over by a steamroller?
    A: (4, 2) (4, 3) (3, 2) (5, 3) (4, 4) (1, 4) (4, 2) (5, 4) (1, 1) (2, 3) (3, 6) (4, 2) (4, 3) (3, 5)
    (1, 1) (3, 4) (4, 1) (4, 4) (4, 4) (4, 2) (4, 5) (4, 4) (5, 1) (2, 5) (5, 1) (4, 3) (5, 1) (4, 2)
    (2, 2) (3, 2) (5, 4) (1, 1) (4, 3) (4, 1) (4, 3) (4, 2) (4, 3) (5,1) (2, 6)

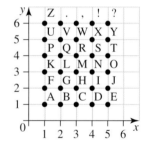

Communicating, reasoning and problem solving

18. A line passes through points A (–5, 6) and B (3, 6).

    a. Draw the line interval AB on a Cartesian plane.
    b. Determine the length of AB.
    c. Plot the point that is in the middle of the line interval AB. Determine the coordinates of the point.
    d. Draw another horizontal line and determine the coordinates of the middle point.
    e. Is there a formula that can be used to calculate the x-coordinate of the middle point of a horizontal line?

19. A line passes through points C (1, 7) and D (1, −5).

    a. Draw the line interval CD on a Cartesian plane.
    b. Determine the length of CD.
    c. Plot the point that is in the middle of the line interval CD. Determine the coordinates of the point.
    d. Draw another vertical line and determine the coordinates of the middle point.
    e. Is there a formula that can be used to calculate the y-coordinate of the middle point of a vertical line?

20. Explain why the x-coordinate must always be written first and the y-coordinate second.

21. A line connects the points $(0, 0)$ and $(5, 5)$ as shown on the Cartesian plane.

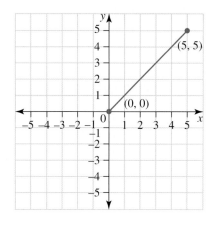

   a. Here is a list of points on the line. Fill in the blanks.

   $(1, \underline{\quad}), (2, \underline{\quad}), (\underline{\quad}, 3), (\underline{\quad}, 4)$

   b. On this line, when $x = \dfrac{1}{2}$, what does y equal?

   c. Imagine that the line is extended so that the points $(0, 0)$ and $(-5, -5)$ are connected. Here is a list of points on the extended line. Fill in the blanks.

   $(-1, \underline{\quad}), (-2, \underline{\quad}), (\underline{\quad}, -3), (\underline{\quad}, -4)$

22. Consider the square ABCD shown.

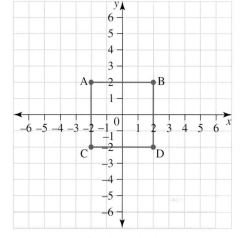

   a. State the coordinates of the vertices of the square ABCD.
   b. Calculate the area of the square ABCD.
   c. Move the points A and B up so that the shape is a rectangle and the area is doubled. Determine the new coordinates of A and B.

23. Consider the square ABCD shown.

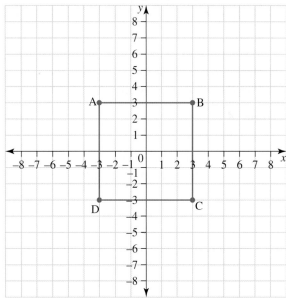

   a. State the coordinates of the vertices of the square ABCD.
   b. Calculate the area of the square ABCD.
   c. Extend sides AB and CD 3 squares to the right and 3 squares to the left.
   d. Calculate the area of the new shape.
   e. In your new diagram, extend sides AD and BC 3 squares up and 3 squares down.
   f. Calculate the area of the new shape.
   g. Compare the three areas calculated in parts **b**, **d** and **f**. Explain the changes of area in relation to the change in side length.

24. Consider the rectangle formed by connecting the points $(3, 2), (9, 2), (9, 5)$ and $(3, 5)$ on a Cartesian plane.

   a. Calculate the area and the perimeter of this shape.
   b. Determine another set of points that have the same area and perimeter.

LESSON
9.3 Linear patterns

LEARNING INTENTION

At the end of this lesson you should be able to:
- plot number patterns on the Cartesian plane
- identify whether number patterns are linear or non-linear.

▶ 9.3.1 Plotting points from a table of values

eles-4502

- Number patterns can be described by rules. For example, the sequence 1, 4, 7, 10, ... increases by 3 from one term to the next. This number pattern can be described by the rule 'start with 1 and add 3 each time'.
- For any number pattern, we can create a table of values relating the value of a term to its position in a number pattern.
- The position in the pattern corresponds to the x-value on the Cartesian plane; the value of the term corresponds to the y-value on the Cartesian plane.
- For the sequence 1, 4, 7, 10, ... the table of values would be:

Position in the pattern (x)	1	2	3	4	...
Value of the term (y)	1	4	7	10	...

- Once a table of values has been constructed, we can plot these values on a Cartesian plane and observe the pattern of the points.

WORKED EXAMPLE 4 Plotting number patterns

A number pattern is formed using the rule 'start with 2 and add 1 each time'.
a. Write the first five terms of the number pattern.
b. Draw up a table of values relating the value of a term to its position in the pattern.
c. Plot the points from your table of values on a Cartesian plane.

THINK	WRITE/DRAW
a. 1. Start with 2 and add 1.	**a.** $2 + 1 = 3$
2. Keep adding 1 to the previous answer until five numbers have been calculated.	$3 + 1 = 4$ $4 + 1 = 5$ $5 + 1 = 6$
3. Write the answer.	The first five numbers are 2, 3, 4, 5 and 6.
b. Draw up a table relating the position in the pattern to the value of the term.	**b.**

Position in the pattern	1	2	3	4	5
Value of the term	2	3	4	5	6

---

c. Plot the points from your table of values one at a time. Start with the first column of numbers, remembering that 'position in the pattern' relates to the *x*-coordinates and 'value of the term' relates to the *y*-coordinates.

c. Points to plot are: $(1, 2), (2, 3), (3, 4), (4, 5)$ and $(5, 6)$

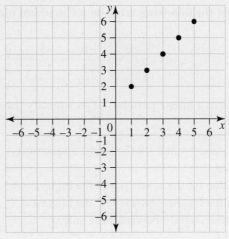

9.3.2 Straight-line patterns

eles-4503

- If the pattern formed by the set of points is a straight line, we refer to it as a **linear pattern**. For example, the points $(1, 2), (2, 3), (3, 4), (4, 5)$ and $(5, 6)$ when plotted will form a linear pattern, as shown in the previous worked example.
- A diagram formed by plotting a set of points on the Cartesian plane is referred to as a **graph**. If the points form a straight line, then the graph is a linear (straight-line) graph.

WORKED EXAMPLE 5 Plotting a set of points

a. Plot the following points on a Cartesian plane. Check the lowest and highest values to help you decide the length of the axes.
   $(-5, 6), (-3, 2), (-2, 0)$ **and** $(1, -6)$
b. Comment on any pattern formed.

THINK

a. Look at the *x*- and *y*-values of the points and draw a Cartesian plane.
The lowest value for the *x*-axis is −5; the highest is 1.
The lowest value for the *y*-axis is −6; the highest is 6.
Extend each axis slightly beyond these values.
Plot each point.

WRITE

a.

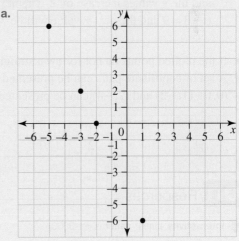

b. Comment on any pattern formed.

b. The pattern is linear because the points form a straight line.

a. Plot the points in the following table on a Cartesian plane.

x	−6	−3	0	3	6
y	−1	1	3	5	7

b. Do the points form a linear pattern? If so, identify the next point in the pattern.

THINK

a. 1. Look at the x- and y-values of the points and draw a Cartesian plane.
The lowest value for the x-axis is −6; the highest is 6.
The lowest value for the y-axis is −1; the highest is 7.
Extend each axis slightly beyond these values.
Plot each point.

WRITE

a.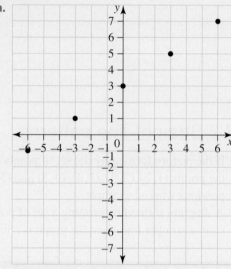

b. 1. Look at the position of the points and answer the question.
Note: The points form a straight line, so we have a linear pattern.

2. Study the pattern and answer the question.
Note: The pattern shows that the x-values increase by 3 and the y-values increase by 2. The next x-value will be $6 + 3 = 9$ and the next y-value will be $7 + 2 = 9$.

b. Yes, the points do form a linear pattern.

The next point in the pattern is $(9, 9)$.

- If the plotted points do not follow a straight line, we refer to this as a **non-linear** pattern.
- A set of coordinate pairs can be presented in the form of a table. For example, the points $(−2, 4)$, $(−1, 2)$, $(0, 0)$, $(1, −2)$ and $(2, −4)$ can be presented as shown.

x	−2	−1	0	1	2
y	4	2	0	−2	−4

 Resources

 Interactivity Linear patterns (int-3833)

9.3 Quick quiz on	9.3 Exercise

Individual pathways

■ PRACTISE	■ CONSOLIDATE	■ MASTER
1, 2, 4, 6, 8, 11, 15, 19	3, 7, 9, 12, 16, 17, 20	5, 10, 13, 14, 18, 21

Fluency

1. **WE4** A number pattern is formed using the rule 'start with 9 and subtract 2 each time'.

   a. Write down the first five terms of the number pattern.
   b. Draw up a table of values relating the value of a term to its position in the pattern.
   c. Plot the points from your table of values on a Cartesian plane.

2. **WE5** a. Plot the following points on a Cartesian plane. (Check the lowest and highest values to help you decide what scale to mark on the axes.)

       i. $(-2, -3), (-1, -2), (0, -1), (1, 0)$ and $(2, 1)$ ii. $(-2, 0), (-1, 1), (0, 2), (1, 3)$ and $(2, 4)$
       iii. $(-2, -4), (-1, -2), (0, 0), (1, 2)$ and $(2, 4)$

   b. Comment on any pattern formed.

3. a. Plot the following points on a Cartesian plane. (Check the lowest and highest values to help you decide what scale to mark on the axes.)

       i. $(-2, -5), (-1, -2), (0, 1), (1, 4)$ and $(2, 7)$ ii. $(-2, 2), (-1, 1), (0, 0), (1, -1)$ and $(2, -2)$
       iii. $(-2, 0), (-1, -1), (0, -2), (1, -3)$ and $(2, -4)$

   b. Comment on any pattern formed.

4. Plot the following points on a Cartesian plane.

   a.

x	-2	-1	0	1	2
y	1	2	3	4	5

   b.

x	-2	-1	0	1	2
y	-2	-1	0	1	2

   c.

x	-2	-1	0	1	2
y	-7	-4	-1	2	5

5. Plot the following points on a Cartesian plane.

   a.

x	-2	-1	0	1	2
y	3	2	1	0	-1

   b.

x	-2	-1	0	1	2
y	-1	-0.5	0	0.5	1

   c.

x	-2	-1	0	1	2
y	4	2	0	-2	-4

Understanding

6. **WE6** **a.** Plot the following points on a Cartesian plane.

    i. $(-3, -3), (-2, -1), (-1, 1), (0, 3), (1, 5), (2, 7)$ and $(3, 9)$
    ii. $(-3, -5), (-2, -3), (-1, 0), (0, 1), (1, 4), (2, 5)$ and $(3, 7)$

  b. Do the points form a linear pattern? If so, identify the next point in the pattern.

7. **a.** Plot the following points on a Cartesian plane.

    i.

x	-2	-1	0	1	2
y	3	-1	-2	-3	-4

    ii.

x	-2	-1	0	1	2
y	-6	-3	0	3	6

  b. Do the points form a linear pattern? If so, identify the next point in the pattern.

8. Consider the number pattern $9, 6, 3, ...$

  a. Complete a table of values for the first five terms.
  b. Describe the number pattern in words by relating the value of a term to its position in the pattern.
  c. Describe the number pattern in algebra, with x representing the position in the pattern and y representing the value of a term.
  d. Represent the relationship on a Cartesian plane.

9. Consider the number pattern $1, 5.5, 10, ...$

  a. Complete a table of values for the first five terms.
  b. Describe the number pattern in words by relating the value of a term to its position in the pattern.
  c. Describe the number pattern in algebra, with x representing the position in the pattern and y representing the value of a term.
  d. Represent the relationship on a Cartesian plane.

10. Consider the number pattern generated when you start with -4 and add 3 each time.

  a. Complete a table of values for the first five terms.
  b. Describe the number pattern in words by relating the value of a term to its position in the pattern.
  c. Describe the number pattern in algebra, with x representing the position in the pattern and y representing the value of a term.
  d. Represent the relationship on a Cartesian plane.

11. **MC** The next point in the linear pattern made by $(-2, 0), (-1, 1), (0, 2), (1, 3)$ and $(2, 4)$ is:

  A. $(5, 3)$ **B.** $(-3, -5)$ **C.** $(3, -5)$ **D.** $(3, 5)$

12. **MC** The next point in the linear pattern made by $(-2, 9), (-1, 8), (0, 7), (1, 6)$ and $(2, 5)$ is:

  A. $(-3, 8)$ **B.** $(3, 4)$ **C.** $(3, 6)$ **D.** $(4, 3)$

13. **MC** The next point in the linear pattern made by $(-2, -18), (-1, -14), (0, -10), (1, -6)$ and $(2, -2)$ is:

  A. $(-3, 3)$ **B.** $(-3, -20)$ **C.** $(3, 2)$ **D.** $(3, 6)$

14. **MC** Identify which of the following sets of points would make a linear pattern.
    Note: There may be more than one correct answer.

  A. $(-2, -1), (-1, -2), (0, -3), (1, -4)$ and $(2, -5)$ **B.** $(-2, 12), (-1, 10), (0, 8), (1, 6)$ and $(2, 4)$
  C. $(-2, -1), (-1, 0), (0, 1), (1, -1)$ and $(2, 0)$ **D.** $(-2, -5), (-1, 0), (0, 4), (1, 5)$ and $(2, 8)$

Communicating, reasoning and problem solving

15. By looking at a graph of a number pattern, explain how you can tell whether the number pattern is increasing or decreasing.

16. A student starts the 'Get fit' plan shown in the table.

Day (d)	Distance (D)
Monday (1)	1 km
Tuesday (2)	1.25 km
Wednesday (3)	1.5 km
Thursday (4)	
Friday (5)	
Saturday (6)	
Sunday (7)	

    a. Complete the pattern for the 'Get fit' plan to show the distances run for 7 days.
    b. Determine whether the increasing distance represents a linear pattern.
    c. Determine the constant amount by which the student increases their run each day.
    d. Calculate the distance run on the 10th day.

17. Explain whether it is possible to say whether or not a set of points will form a straight line without actually plotting the points.

18. The time taken for your teacher to write a Maths test is 12 minutes per question plus a 10-minute rest after each lot of 5 questions.

    a. Calculate how much time it takes your teacher to write:
       i. one question   **ii.** two questions   **iii.** six questions.
    b. Generate a table of values for the number of questions written, q, versus the total time taken, t, for up to 20 questions.
    c. Determine whether this relationship represents a linear pattern.

19. This table of values shows the total amount of water wasted by a dripping tap.

Number of days	1	2	3	4	5
Total amount of water wasted (L)	6	12	18		30

    a. Identify the missing number in the table.
    b. State how much water is wasted each day.
    c. If the tap drips for 7 days, calculate the total amount of water wasted.
    d. If the tap drips for 20 days, calculate the total amount of water wasted.

20. The latest version of a mobile phone operating system is scheduled to be available for download on 1 January. An expert software engineer claims there are 400 bugs in this release; however, the manufacturer claims that it can fix 29 bugs each month. Determine when the software will be bug-free.

21. Consider the relationship $\dfrac{1}{y} = \dfrac{1}{x} + 1$. The equation can be rearranged so that the value of y can be calculated easily. This is shown below.

$$\frac{1}{y} = \frac{1}{x} + 1$$
$$\frac{1}{y} = \frac{1}{x} + \frac{x}{x}$$
$$\frac{1}{y} = \frac{1+x}{x}$$
$$y = \frac{x}{1+x}$$

    a. Generate a table of values for x versus y, for $x = -3$ to 3.
    b. Plot a graph to show the points contained in the table of values.
    c. Use your plot to confirm whether this relationship is linear.

LESSON
9.4 Linear graphs

LEARNING INTENTION

At the end of this lesson you should be able to:
- plot linear graphs on the Cartesian plane
- determine the rule of the equation by completing a table of values
- determine the rule of the equation from a graph
- determine the sign of the gradient.

▶ 9.4.1 Plotting linear graphs

eles-4504

- A **linear graph** is a straight-line graph defined by a linear relationship.
- Each point on a linear graph is an ordered pair (x, y), which represents a coordinate that satisfies the linear relationship.
- The straight line of a linear graph is continuous, meaning that all points on the linear graph satisfy the linear relationship. This means that there is an infinite number of ordered pairs that satisfy any given linear relationship.

Plotting linear graphs

- **To plot a linear graph whose equation is given, follow these steps.**
  1. **Create a table of values first.**
     - **Draw a table.**
     - **Select some x-values. For example: $x = -2, -1, 0, 1, 2, \ldots$**
     - **Substitute the selected x-values into the rule to find the corresponding y-values.**
  2. **Draw a Cartesian plane.**
  3. **Plot the points from the table and join them with the straight line. (Extend the straight line in both directions past the points you have plotted.)**
  4. **Label the graph.**

WORKED EXAMPLE 7 Plotting linear graphs

For $y = 2x + 1$, draw a table of values, plot the graph and label the line.

THINK	WRITE/DRAW
1. Write the rule.	$y = 2x + 1$

2. Draw a table and choose simple x-values.

x	−2	−1	0	1	2
y					

3. Use the rule to find the y-values and enter them in the table.
When $x = -2$, $y = 2 \times -2 + 1 = -3$.
When $x = -1$, $y = 2 \times -1 + 1 = -1$.
When $x = 0$, $y = 2 \times 0 + 1 = 1$.
When $x = 1$, $y = 2 \times 1 + 1 = 3$.
When $x = 2$, $y = 2 \times 2 + 1 = 5$.

x	−2	−1	0	1	2
y	−3	−1	1	3	5

4. Draw a Cartesian plane and plot the points.
5. Draw a line joining the points that goes past the points at each end.
   Insert arrows at both ends of the line to indicate that the pattern continues.

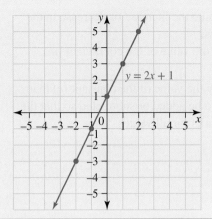

- The straight line representing a linear graph can be extended past the given points in both directions forever, meaning that there is an infinite number of ordered pairs that satisfy any given linear relationship.

COMMUNICATING — COLLABORATIVE TASK: Reading coordinates from a linear graph

1. Use a table of values to plot the graph of $y = 2x - 1$.
2. Compare your results with your classmates. Are there any points that you noted that they didn't? Discuss your results.

▶ 9.4.2 Determining the rule from a table

eles-6213

- We can find the rule for a linear equation by examining the table of values.
  For example, for the following table of values, by observing the pattern we can easily determine that the rule is $y = 3x$. Some patterns can not be so easily found, such as in Worked Example 8.

x	0	1	2	3	4	5
y	0	3	6	9	12	15

- Rules for linear equations are generally written in the form $y = mx + c$, where:
  - m (the coefficient value in front of x) is the value that y changes by as the value of x increase by 1
  - c is the constant that is added or subtracted so the rule works for all values.

WORKED EXAMPLE 8 Determining the rule from a table

Determine the rule for the following tables of values.

a.

x	−1	0	1	2	3
y	−5	−1	3	7	11

b.

x	2	3	4	5	6
y	8	5	2	−1	−4

THINK

a. 1. The general form of a linear equation is $y = mx + c$. The coefficient (m) is the increase in y as x increases by 1.

WRITE

a. x-values: $-1, 0, 1, 2, 3$
The x-values increase by 1 each time.

x	−1	0	1	2	3
y	−5	−1	3	7	11

  +4   +4   +4   +4

2. Determine the increase or decrease in the *y*-values.
This is the value of the coefficient (*m*).

The *y*-values increase by 4.
The value of the coefficient (*m*) is 4.
That is, $m = 4$.
This means the rule is in the form $y = 4x + ?$

3. Identify the value of the constant (*c*) that needs to be added or subtracted so that the rule works for all values.

Input any *x*- and *y*-values to determine the missing constant (*c*) value:
When $x = -1$ and $y = -5$:
$-5 = 4(-1) + ?$
$-5 = -4 + ?$
The missing value is -1.
Note: Check another *x*- and *y*-value pair mentally to ensure it works.

4. Write the answer.

The rule for this equation is $y = 4x - 1$.

b. 1. The coefficient (*m*) is the increase in *y* as *x* increases by 1.

b. *x*-values: $2, 3, 4, 5, 6$
The *x*-values increase by 1 each time.

2. Determine the increase or decrease in the *y*-values.
This is the value of the coefficient (*m*).

x	2	3	4	5	6
y	8	5	2	−1	−4

        −3   −3   −3   −3

The *y*-values decrease by 3.
The coefficient (*m*) is −3. That is, $m = 3$.
This means that the rule is in the form $y = -3x + ?$

3. Identify the value of the constant (*c*) that needs to be added or subtracted so that the rule works for all values.

Input any *x*- and *y*-values to determine the missing constant (*c*) value:
When $x = 2$ and $y = 8$:
$8 = -3(2) + ?$
$8 = -6 + ?$
The missing value is 14.
Note: Check another *x* and *y* value pair mentally to ensure it works.

4. Write the answer.

The rule for this equation is $y = -3x + 14$.

- In Worked Example 8a, you will notice that the value of the constant (*c*) can be also be seen when $x = 0$ in the table.
- In Worked example 8b, you will notice that the pattern of a decrease by 3 when *x* increases by 1 can be used to determine the value of *y* when $x = 0$. This is the value of *c*.

x	0	1	2	3	4
y	14	11	8	5	2

       −3   −3   −3   −3

- *c* is also called the *y*-intercept, as it is the *y*-value when the line intersects the *y*-axis (when $x = 0$).

▶ 9.4.3 Determining the rule from a graph

eles-6212

- A rule can be determined for a graph by first creating a table of values containing points on the graph.
- As found in Worked Example 8, the coefficient (*m*) is found when the *x*-values increase by 1, so record these values in your table.

Determine the rule for the equation represented by the graph shown.

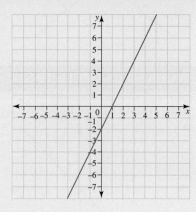

THINK	WRITE

1. Draw a table of values of points from the graph so that the value of x increases by 1.

x	-2	-1	0	1	2
y	-6	-4	-2	0	2

$+2 \quad +2 \quad +2 \quad +2$

2. Determine the increase or decrease in the y-values. This is the value of the coefficient (m).

The y-values increase by 2 as the x-values increase by 1.
This means the rule is in the form $y = 2x + ?$

3. Identify the value of the constant (c) that needs to be added or subtracted so that the rule works for all values.

Input any x- and y-values to determine the missing constant (c) value:
When $x = 0$ and $y = -2$:
$$y = 2x + ?$$
$$-2 = 2(0) + ?$$
$$-2 = 0 + ?$$
The missing value is -2.

4. Write the answer.

The rule for this equation is $y = 2x - 2$.

▶ 9.4.4 The gradient

eles-6214

- The slope or gradient of a line is the rate of change in y over the change in x.
- When rules for linear equations are written in the form $y = mx + c$, the gradient is the constant m. The gradient is constant throughout the line
- Depending on which way a line slopes, the gradient of a straight line could be:
  - a positive number (think of walking uphill)
  - a negative number (think of walking downhill)
  - zero (think of walking along flat horizontal ground)
  - undefined (think of trying to walk at a 90° angle — can't be done!).

Sign of the gradient

- If the line *slopes upward* from left to right (i.e. it *rises*), the gradient is *positive*.
- If the line *slopes downward* from left to right (i.e. it *falls*), the gradient is *negative*.
- If the line is horizontal, there is no slope; hence, the value of the gradient is zero.
- If the vertical, we say that its gradient is **infinite** or **undefined**.

gradient is positive   gradient is negative   gradient is zero   gradient is undefined

---

WORKED EXAMPLE 10 Stating the sign of the gradient from a graph

State whether these lines have a positive, negative, zero or undefined gradient.

a.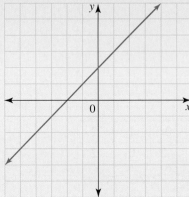

b.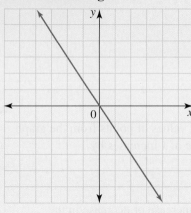

c.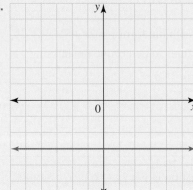

d.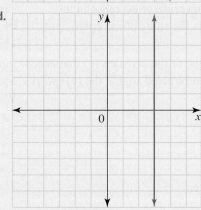

THINK

a   A line that rises from left to right, /, has a positive gradient.

b.   A line that drops from left to right, \, has negative gradient.

c.   A horizontal line, —, has a zero gradient.

d.   A vertical line, |, has an undefined gradient.

WRITE

a.   Positive gradient

b.   Negative gradient

c.   Zero gradient

d.   Undefined gradient

---

 on Resources

   Interactivity Plotting linear graphs (int-3834)

▶ 9.4.5 Graphing using digital tools

eles-6215

- There are many digital technologies that can be used to graph linear relationships.
- The Desmos Graphing Calculator is a free graphing tool that can be found on the internet.
- Other commonly used digital technologies include Microsoft Excel and other graphing calculators.
- Digital technologies can help identify important features and patterns in graphs.
- Depending on the digital technology used, the steps involved to produce a linear graph may vary slightly.

This graph of $y = 5x - 2$ was created using Excel.

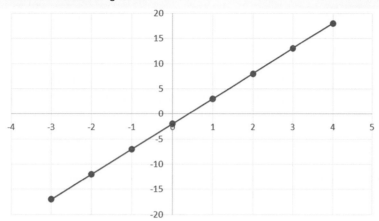

Sketching a linear graph using the Desmos Graphing Calculator

- Most graphing calculators have an entry (or input) box to type in the equation of the line you wish to sketch.
- When using the Desmos Graphing Calculator tool, you can simply type $y = 3x + 3$ into the input box to produce its graph.

This graph of $y = 3x + 3$ was created using Desmos.

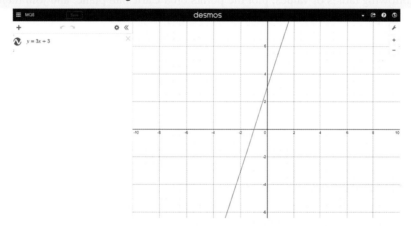

COMMUNICATING — COLLABORATIVE TASK: Digital tools

Use the Desmos Graphing Calculator tool to sketch the following lines on the same set of axes.
a. $y = 5x$
b. $y = 2x - 4$
c. $y = x + 6$
d. $y = 9 - 4x$
e. $y = 2x - 6$
f. $2x + 3y = -6$

In groups, discuss the following questions.
1. What is special about the graphs of $y = 2x - 4$ and $y = 2x - 6$?
2. Where do the graphs of $y = x + 6$ and $2x + 3y = -6$ cross each other?
3. Which equations have a negative gradient?
4. Which line is the steepest?
5. Compare the gradients of the equations. Would you be able to determine the steepest line just by comparing the gradients? How?

Exercise 9.4 Linear graphs

9.4 Quick quiz on	9.4 Exercise

Individual pathways

■ PRACTISE	■ CONSOLIDATE	■ MASTER
1, 6, 7, 13, 17	2, 4, 8, 9, 10, 11, 12, 14, 18, 19, 21	3, 5, 15, 16, 20, 22

Fluency

1. **WE7** Complete the following tables of values, plot the points on a Cartesian plane, and join them to make a linear graph. Label the graphs with the rules.

a. Rule: $y = x + 3$

x	−2	−1	0	1	2
y	1	2	3		

b. Rule: $y = x - 5$

x	−2	−1	0	1	2
y		−6			−3

c. Rule: $y = 5x$

x	−2	−1	0	1	2
y	−10		0		

d. Rule: $y = 2x + 4$

x	−2	−1	0	1	2
y	0			6	

2. Complete the following tables of values, plot the points on a Cartesian plane, and join them to make a linear graph. Label the graphs with the rules.

a. Rule: $y = 3x + 2$

x	−2	−1	0	1	2
y					

b. Rule: $y = 2x - 2$

x	−2	−1	0	1	2
y					

c. Rule: $y = 4x - 3$

x	−2	−1	0	1	2
y					

d. Rule: $y = -3x + 2$

x	−2	−1	0	1	2
y					

3. **MC** a. What does $y = 3x + 4$ mean?
   A. The y-value equals the x-value with 3 added and then multiplied by 4.
   B. The y-value equals the x-value multiplied by 3 and with 4 added.
   C. The x-value equals the y-value times 3, with 4 added.
   D. The y-value equals 4 times the x-value divided by 3.

   b. A table of values shows:
   A. a rule.
   B. coordinates.
   C. a linear graph.
   D. an axis.

4. Draw a table of values and plot the graph for each of the following rules. Label each graph.

   a. $y = x + 2$
   b. $y = x - 4$
   c. $y = x - 1$
   d. $y = x + 5$
   e. $y = 3x$
   f. $y = 7x$

5. Draw a table of values and plot the graph for each of the following rules. Label each graph.

   a. $y = 4x + 1$
   b. $y = 2x - 3$
   c. $y = 3x - 5$
   d. $x = -\dfrac{1}{2}y$
   e. $6x + y = 2$
   f. $5x + y - 4 = 0$

6. **WE10** State whether each of the following lines has a positive, negative, zero or undefined gradient.

a.
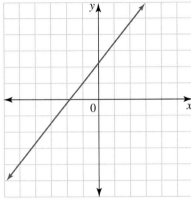

b.
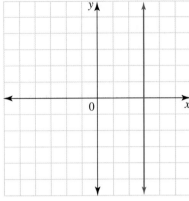

c.
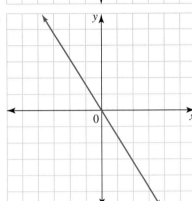

d.
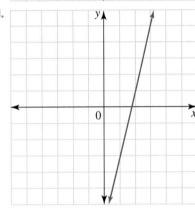

e.
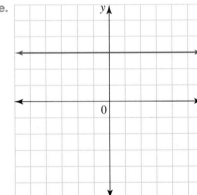

f.
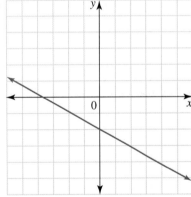

Understanding

7. Plot a graph of the following rules from the tables of values provided. Label the graphs, then copy and complete the sentences.

a. $y = 4$

x	−2	−1	0	1	2
y	4	4	4	4	4

For the rule $y = 4$, the y-value of all coordinates is _____.

b. $y = 1$

x	−2	−1	0	1	2
y	1	1	1	1	1

For the rule $y = 1$, the y-value of all coordinates is _____.

c. $y = -2$

x	−2	−1	0	1	2
y	−2	−2	−2	−2	−2

For the rule $y = -2$, the y-value of all coordinates is _____.

d. $y = -5$

x	−2	−1	0	1	2
y	−5	−5	−5	−5	−5

For the rule $y = -5$, the y-value of all coordinates is _____.

8. Draw a table of values and plot the graph for each of the following rules.

   a. $y = 3$ b. $y = 2$ c. $y = -2$ d. $y = -4$

9. Plot the graph of each of the following rules from the table of values provided. Label the graph, then copy and complete the sentence.

   a. $x = 1$

x	1	1	1	1	1
y	−2	−1	0	1	2

For the rule $x = 1$, the x-value of all coordinates is _____.

   b. $x = 3$

x	3	3	3	3	3
y	−2	−1	0	1	2

For the rule $x = 3$, the x-value of all coordinates is _____.

   c. $x = -2$

x	−2	−2	−2	−2	−2
y	−2	−1	0	1	2

For the rule $x = -2$, the x-value of all coordinates is _____.

   d. $x = -7$

x	−7	−7	−7	−7	−7
y	−2	−1	0	1	2

For the rule $x = -7$, the x-value of all coordinates is _____.

10. Draw a table of values, then plot and label the graph for each of the following.

   a. $x = 2$ b. $x = 5$ c. $x = -5$ d. $x = 0$

11. **WE8** Determine the rule for each of the following tables of values.

   a.

x	0	1	2	3	4
y	4	6	8	10	12

   b.

x	−1	0	1	2	3
y	7	10	13	16	19

   c.

x	2	3	4	5	6
y	−3	−7	−11	−15	−19

12. **WE9** Determine the rule for the equations represented by the following linear graphs

   a.

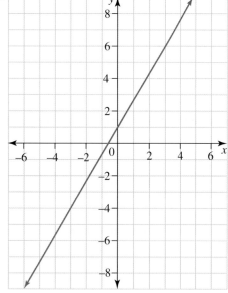

   b.

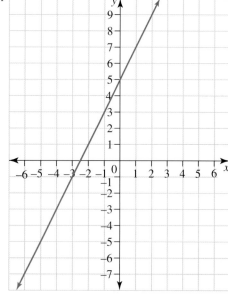

c.

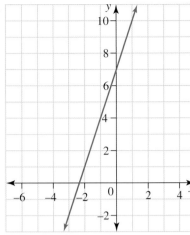

d.

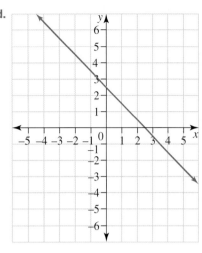

Communicating, reasoning and problem solving

13. Draw a table of values and graph each of these rules on the same Cartesian plane.

    a. $y = 2x$ **b.** $y = 2x - 1$ **c.** $y = 2x + 1$

    Describe the lines that are formed.

14. Draw a table of values and graph each of these rules on the same Cartesian plane.

    a. $y = 3x + 1$ **b.** $y = -2x + 1$

    What do you notice?

15. Draw a table of values and graph each of these rules on the same Cartesian plane.

    a. $y = -x$ **b.** $y = x + 2$

    What do you notice?

16. The monthly cost in dollars, C, of renting a mobile phone is given by the equation $C = 15 + 0.5x$, where x is the call time in minutes.

    a. Plot the graph of the equation.
    b. If July's bill was $100, calculate your call time.
    c. If August's bill was only $50, calculate your call time.

17. **a.** Create a table of values that shows the cost of lengths of fabric in the photo from 0 to 8 metres.
    b. Plot a graph showing the cost of 0 to 8 metres of fabric.
    c. Is this a linear relationship? Explain.

18. When you make an international phone call, you usually pay a flagfall (charge for just making the call) and then you are charged for how long you talk.

    One phone company offers 25 cents flagfall and 35 cents per 30 seconds.

    a. Create a table of values that shows the cost of a call between 0 and 5 minutes (in 30-second blocks).
    b. Plot a graph showing the cost of a phone call between 0 and 5 minutes.

19. Chris's fridge is not working. He called a repair company, and they are sending someone to repair the fridge. Company charges a $55 call-out fee, plus $45 for every $\frac{1}{2}$ hour the repairer is there.

a. Copy and complete the table to show how much it could cost Chris to have his fridge repaired.

Time (hours)	0	$\frac{1}{2}$	1	$1\frac{1}{2}$	2	$2\frac{1}{2}$	3
Cost ($)	55	100	145				

b. Draw a graph of this information. Place *time* on the *x*-axis and *cost* on the *y*-axis.
c. Is there a linear relationship between cost and time?
d. Determine the gradient (*m*) and the *y*-intercept (*c*) of the graph.
e. Determine the rule for this graph using $y = mx + c$.
f. Use your equation to calculate the cost if the repairer takes 4 hours to fix Chris's fridge.

20. Kyle was very bored on his holidays and decided to measure how much the grass in his backyard grew every day for one week. His results are shown in the table.

Day number	0	1	2	3	4	5	6	7
Height of grass (mm)	10	12	14	16	18	20	22	24

a. Kyle knew his dad would want the grass cut as soon as it was 2.5 cm (25 mm) long. On which day would this occur?
b. Plot the points from the table on a Cartesian plane, putting *days* on the *x*-axis and *height* on the *y*-axis.
c. Do the points form a linear graph?
d. Determine the gradient and *y*-intercept of the graph.
e. Develop an equation for the height of the grass by filling in the blanks.

Height = _____ × no. of days + _____ or $h =$ _____ $d +$ _____

f. Use your answer to part e to calculate:
  i. the height of the grass after 14 days if it is not cut
  ii. how long it would take for the grass to grow 50 mm if it is not cut.

21. a. Complete the tables and determine the equation for each of the following tables of values.
    b. Plot the linear graphs on separate Cartesian planes.

i.

x	−2	−1	0	1	2	3	4
y	−3		−2		−1	−0.5	0

ii.

x	0	10	20	30	40	50
y	5	7	9		13	

22. a. Use the equation $y = 2x - 5$ to complete the table.

x				2			
y	−13	−9	−5		3	7	11

b. Using the completed table of values, plot the linear graph.
c. Explain what you did differently to determine the values for this table.

LESSON
9.5 Solving equations graphically

LEARNING INTENTION

At the end of this lesson you should be able to:
- apply graphs of linear relationships to solve linear equations.

▶ 9.5.1 Using linear graphs to solve linear equations

eles-4511

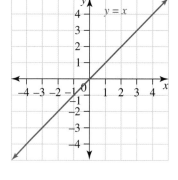

- The graph of an equation shows all of the points (x, y) that are solutions to the equation.
- The graph $y = x$ contains all of the points at which the x-coordinate is equal to the y-coordinate, such as $(0, 0)$, $(1, 1)$, $(-2.5, -2.5)$.
- To solve a linear equation, we can determine the x-value that corresponds to the required y-value.
- For example, to solve the equation $2x + 3 = 5$, we can plot the graph of $y = 2x + 3$ and determine which x-value corresponds to a y-value of 5.

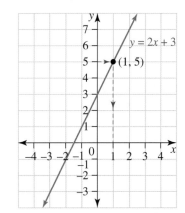

- From the graph we can see that $2x + 3$ is equal to 5 when $x = 1$. Therefore, $x = 1$ is the solution to the equation $2x + 3 = 5$.
- All points on the line satisfy the equation $y = 2x + 3$.

WORKED EXAMPLE 11 Solving linear equations graphically

Use the graph of the equation $y = 2x + 5$ to solve the linear equation $2x + 5 = 15$.

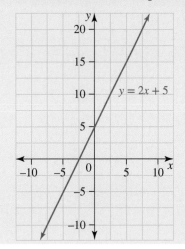

THINK	**WRITE/DRAW**
1. Rule a horizontal line (pink) at $y = 15$. This is the right-hand side of the original equation. This line meets the graph at point A. Rule a vertical line (purple) from point A to the x-axis. The line meets the x-axis at 5.	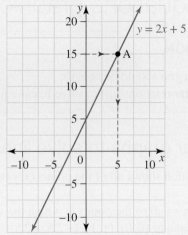
2. When $x = 5$, $2x + 5 = 15$.	The solution to the linear equation $2x + 5 = 15$ is $x = 5$.

9.5.2 Solving equations using the point of intersection

eles-4512

- Another method of solving linear equations is by using the point of intersection.
- If two graphs intersect at a point, that point is the solution to both equations.
- Any two linear graphs intersect at a point unless they are parallel.
- This method is essentially the same as the method used in Worked example 11 if you consider the two graphs to be $y = 2x + 5$ and $y = 4 - x$.
- This method can be used to solve equations that have unknowns on both sides. Simply draw both equations on the same set of axes and determine their point of intersection.
- For example, to solve $2x - 5 = 4 - x$, find the intersection of the graphs $y = 2x - 5$ and $y = 4 - x$.

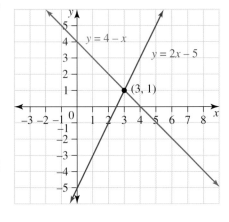

- Looking at the graph, we can see that the intersection occurs when $x = 3$, so the solution to the equation $2x - 5 = 4 - x$ is $x = 3$.

Use the graphs of $y = x - 7$ and $y = -3x + 5$ to solve the equation $-3x + 5 = x - 7$.

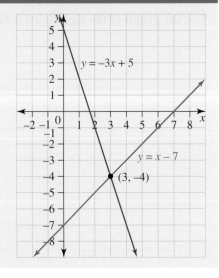

THINK	WRITE
1. The solution is given by the intersection of the two graphs.	The point of intersection is $(3, -4)$.
2. The x-value of the point of intersection is the solution.	The solution to the equation $-3x + 5 = x - 7$ is $x = 3$.

WORKED EXAMPLE 13 Solving equations graphically with technology

Use a digital technology of your choice to solve the equation $2x + 3 = 3x - 5$ graphically.

THINK

1. We can let both sides of the equation equal y, which gives us two equations:
$y = 2x + 3$ and $y = 3x - 5$.
Sketch the graphs of $y = 2x + 3$ and $y = 3x - 5$ on the same set of axes using a digital technology of your choice.

WRITE/DRAW

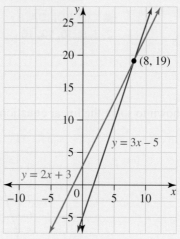

2. Locate the point of intersection of the two lines. This gives the solution.

The point of intersection is $(8, 19)$.

3. The x-value of the point of intersection is the solution.
Check the solution by substituting $x = 8$ into $2x + 3 = 3x - 5$.

Substituting $x = 8$ into $2x + 3 = 3x - 5$ gives:
$$2(8) + 3 = 3(5) - 5$$
$$16 + 3 = 24 - 5$$
$$19 = 19$$

Since LHS = RHS, we can confirm that $x = 8$ is the correct solution to the problem.

4. State the solution.

The solution is $x = 8$.

Here is an example of using Desmos for worked example 13.

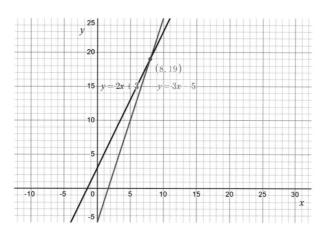

Exercise 9.5 Solving equations graphically

learn on

9.5 Quick quiz on	9.5 Exercise

Individual pathways

■ PRACTISE	■ CONSOLIDATE	■ MASTER
1, 4, 6, 8, 12, 15	2, 7, 9, 10, 13, 16	3, 5, 11, 14, 17

Fluency

1. **WE12** Use the graph shown to solve the linear equation $3x + 5 = 8$.

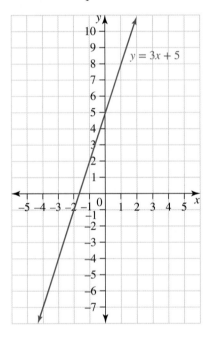

2. Use the graph shown to solve the linear equation $-5x + 6 = -9$.

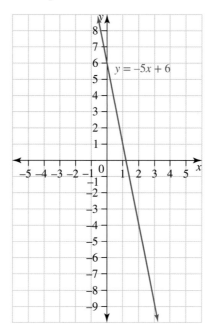

$y = -5x + 6$

3. Use the graph shown to solve the linear equation $0.5x - 1 = -4$.

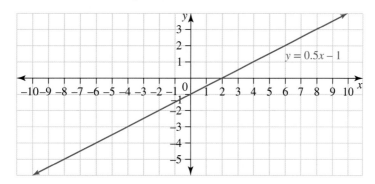

$y = 0.5x - 1$

4. Use the graph shown to solve the linear equation $-2x + 3 = 0$.

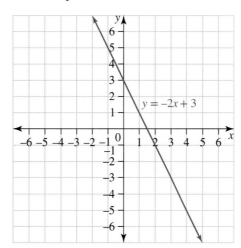

$y = -2x + 3$

5. Use the graph shown to solve the linear equation $6x + 1 = 0$.

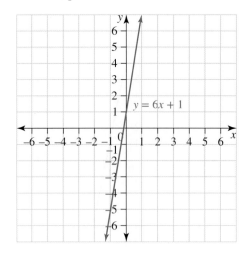

6. **WE12** Use the graphs to solve the equation $2x = 3x - 1$.

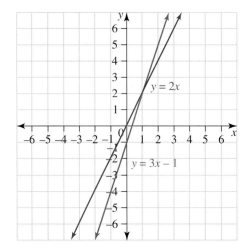

7. Use the graphs to solve the equation $-6x - 9 = \frac{1}{2}x + 4$.

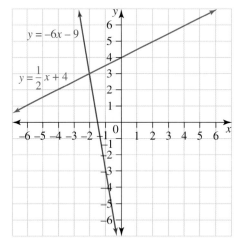

Understanding

8. For each of the following graphs, determine the coordinates of the point of intersection.

a.

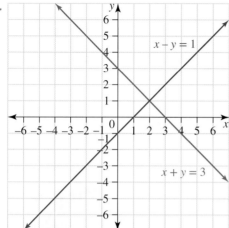

b.

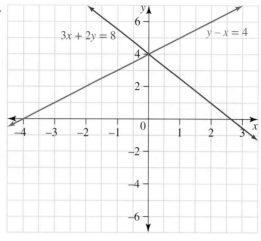

9. For each of the following graphs, determine the coordinates of the point of intersection.

a.

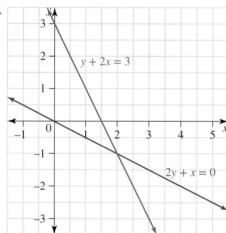

b.

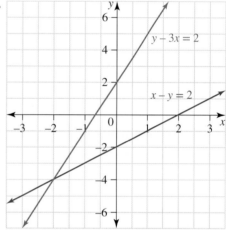

10. **WE12** Use a digital technology of your choice to solve the equation $3x - 7 = -2x + 3$ graphically.

11. Use a digital technology of your choice to solve the equation $\dfrac{x}{3} + 1 = 2 - \dfrac{2x}{5}$ graphically.

12. A triangle is formed by the graphs of $x + y = 13$, $x + 2y = 9$ and $2x + y = 9$. Determine the vertices of the triangle by identifying the coordinates of the points at which each pair of lines cross.

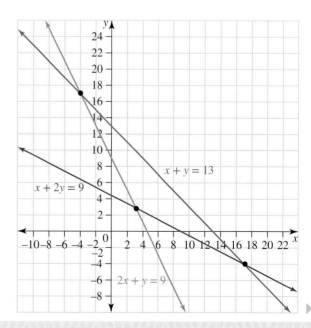

13. The photo shows two friends during a race. The graph shows the distance in kilometres covered by the two friends in a given time in hours. At what time in the race was the photo taken?

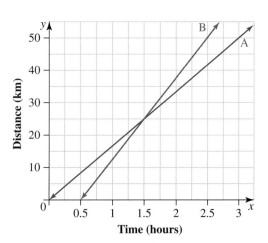

14. You are given some money for your birthday and decide to save a certain amount per week to buy a bike. This information is represented in the equation $A = 17.5w + 110$, where A is the amount in dollars and w is the number of weeks.

    a. Use a digital technology of your choice to plot the graph of the equation.
    b. Use the graph to determine how much you will have saved after 10 weeks.
    c. State how much money you received for your birthday.
    d. Calculate how many weeks it would take for you to save a total of $670.
    e. If the bike costs $800, calculate how long it will take you to save the full amount at this rate.

Communicating, reasoning and problem solving

15. In the list of linear equations below, there are three lines that do not cross each other.

$$y = 3x - 1 \qquad y = 2x \qquad y = 3x + 2$$
$$y = 4x + 1 \qquad y = x - 1 \qquad y = 3x$$

    a. Use digital technology to graph these equations. Identify which three lines do not cross each other.
    b. Looking at the equations in your answer to part a, describe what they have in common.

16. Heidi knows that two lines have a point of intersection at the coordinate $(1, 5)$. The rules for the lines are:
Rule 1: $y = 3x + 2$
Rule 2: $y = -2x + ?$
Use digital technology and trial and error to complete rule 2.

17. A fire truck and an ambulance have been called to an accident scene. The ambulance starts at the hospital; the equation of the distance it travels with respect to time is $d = 1.8t$. The fire truck leaves the garage and travels following the equation $d = 1.5t + 3.6$. In both equations, d is the distance from the hospital in kilometres, and t represents the time in minutes.

    a. Use a digital technology of your choice to plot the two graphs on the same set of axes, with d on the vertical axis and t on the horizontal axis.
    b. Calculate how far from the hospital the ambulance was after 5 minutes.
    c. If the two vehicles met at the scene of the accident, determine how long it took them to arrive.
    d. Determine the distance the two vehicles travelled to the scene of the accident.

LESSON
9.6 Review

9.6.1 Topic summary

The Cartesian plane

- The Cartesian plane consists of two axes: the x-axis (horizontal, with equation $y = 0$) and the y-axis (vertical, with equation $x = 0$).
- Cartesian coordinates are written as (x, y).
- The axes are divided into four quadrants.
  - A point $(+x, +y)$ is in the 1st quadrant.
  - A point $(-x, +y)$ is in the 2nd quadrant.
  - A point $(-x, -y)$ is in the 3rd quadrant.
  - A point $(+x, -y)$ is in the 4th quadrant.
- $(0, y)$ is on the y-axis and $(x, 0)$ is on the x-axis.

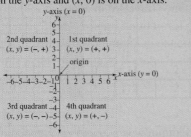

LINEAR RELATIONSHIPS

Plotting linear patterns

- Points that form a straight line when plotted have a linear pattern.
- The set of points is referred to as a linear graph.

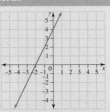

Sketching linear graphs

- Each point is an ordered pair (x, y).
- From the equation, e.g. $y = 2x + 1$:
  - draw up a table of values

x	−2	−1	0	1	2
y	−3	−1	1	3	5

  - draw a Cartesian plane
  - plot the points and connect them with a straight line.
  - Rules for linear equations are generally written in the form $y = mx + c$, where m and c are constants.
  - m is the gradient, the rate of change of y over x, and c is the y-intercept, the y-value when $x = 0$ and the line intersects the y-axis.

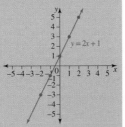

Solving equations graphically

- Equations can be solved graphically on the Cartesian plane.
- For example, to solve the equation $2x + 3 = 5$, we can plot the graph of $y = 2x + 3$ and determine which x-value corresponds to a y-value of 5.

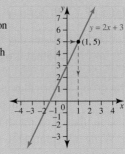

Gradient

- Gradient is a measure of the steepness of a straight line graph.
- The gradient is also called the slope.
- The gradient can be positive, negative, zero or undefined, as shown below.

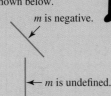

m is positive.

m is negative.

m is 0.

m is undefined.

Graphing with technology

- Graphing technology can be used to sketch linear graphs and find solutions to equations.
- The point of intersection of two graphs can be calculated.
- Some popular graphing technologies include Desmos, GeoGebra or Excel.

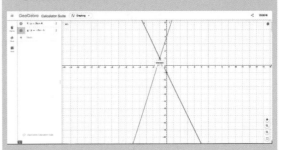

9.6.2 Project

Choosing the right hire car

A group of tourists have just arrived at Sydney Airport and are investigating hire car deals. They decide to study the different options offered by Bonza Car Rentals.

Option 1 $60 per day unlimited kilometres

Option 2 $30 per day and 25 cents/km

Option 3 $40 per day up to 100 km and then an additional 35 cents per kilometre over 100 km

The group knows that on their first day they will be visiting the local attractions close to Sydney, so they will not be driving much.

1. Calculate how much each option would cost if the total distance travelled in a day was 90 km.
2. Write an equation to show the cost of hiring a car for a day for options 1 and 2. Use C to represent the total cost of hiring a car for a day and x to represent the distance in kilometres travelled in a day.
3. Use digital technology to plot the graphs of the three options on the set of axes provided, showing the cost of hiring a car for a day to travel 200 km.

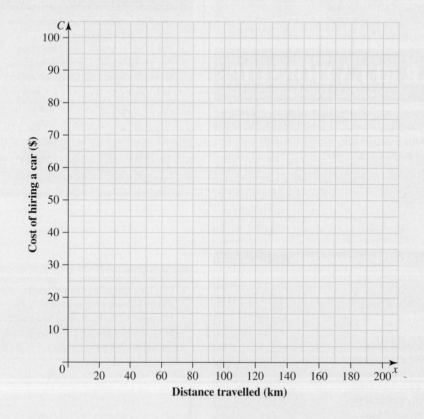

4. Examine the graphs of the three options carefully. Write a brief statement to explain the costs associated with each option over 200 km.

5. Calculate how much it costs to travel 200 km in one day with each option.

The group decides to drive to Melbourne and to spend 2 days travelling there, stopping overnight in Canberra. They plan to stay for 4 days in Melbourne and return to Sydney on the seventh day, driving the whole distance in one day and returning the car to Bonza Car Rentals.

6. In their 4 days in Melbourne, the group travelled a total of 180 km. What was the total distance travelled on this trip?

7. Calculate how much it would cost to hire a car for each option to cover this trip. (Option 3 gives 100 km 'free travel' per day. For a 7-day hire, you get 700 km 'free'.)

8. The group took option 1. Explain whether this was the best deal.

9. Investigate some different companies' rates for car hire. Use digital technology to draw graphs to represent their rates and present your findings to the class.

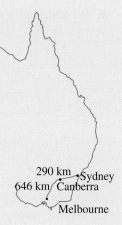

290 km • Sydney
646 km / Canberra
• Melbourne

 on Resources

Interactivities Crossword (int-2640)
Sudoku puzzle (int-3193)

Exercise 9.6 Review questions

learnon

Fluency

1. Use the graph to answer the following.
   a. Identify who watches the most television.
   b. Identify who is the youngest of the three.
   c. Does the youngest watch the least amount of television?

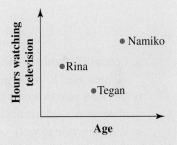

2. Draw a Cartesian plane and plot the following points.
   a. A (1, 4) b. B (5, 3)
   c. C (0, 2) d. D (−2, 5)

3. Draw a Cartesian plane and plot the following points.
   a. E (−4, 1) b. F (−5, 0) c. G (−6, −6) d. H (−5, −4)

4. Draw a Cartesian plane and plot the following points.
   a. I (0, −5) b. J (2, −1) c. K (2, −3) d. L (2, 0)

5. Write down the coordinates of the points A to H marked on the Cartesian plane shown.

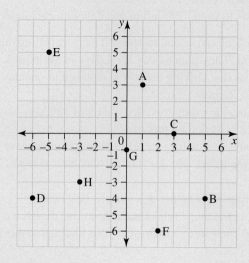

6. Plot the following points on a Cartesian plane.
   a. $(-3, -1), (-2, 0), (-1, 1), (0, 2), (1, 3), (2, 4)$ and $(3, 5)$
   b. $(-3, -12), (-2, -9), (-1, -6), (0, -3), (1, 0), (2, 3)$ and $(3, 6)$

Understanding

7. Follow the instructions for the pattern of shapes shown.

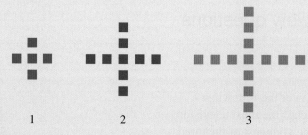

1   2   3

   a. Construct a table to show the relationship between the number of each figure and the number of squares used to construct it.
   b. Devise a rule in words that describes the pattern relating the number of each figure and the number of squares used to construct it.
   c. Use your rule to work out the number of squares required to construct a figure made up of 20 such shapes.

8. Consider the equation $y = x - 2$.
   a. Construct a table of values.   b. Plot the graph on a Cartesian plane.

9. Consider the equation $y = x + 5$.
   a. Construct a table of values.   b. Plot the graph on a Cartesian plane.

10. Consider the equation $4x - y - 2 = 0$.
   a. Construct a table of values.   b. Plot the graph on a Cartesian plane.

11. Use digital technology to graph the line $y = 3x - 8$ and use this graph to solve the equation $3x - 8 = 19$.

12. Use the graphs of $y = x + 7$ and $y = -2x - 2$ to solve the equation $x + 7 = -2x - 2$.

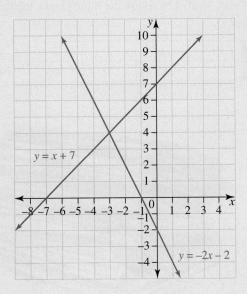

13. Graph the following using a digital technology of your choice.

   a. $y = 3x - 5$ **b.** $y = -2x + 4$

Communicating, reasoning and problem solving

14. Lena and Alex have set up savings accounts. Each month they write down their savings. If the trend continues, evaluate who will be the first to save $300.

Lena's account:

t (months)	0	1	2
A ($)	100	120	140

Alex's account:

t (months)	0	1	2
A ($)	150	160	170

15. Fast Track company is fitting internet cable in the neighbourhood. It takes $1\frac{1}{2}$ hours to install $300\,\text{m}$ of cable. In 2 hours, $450\,\text{m}$ can be installed. If this is a linear relationship, calculate how much can be laid in 4 hours.

16. a. For the equation $y = 2x + 4$, complete the table of values.

x	-3	-2	-1	0	1	2	3
y							

   b. Draw the graph of $y = 2x + 4$ by plotting the points from the table

17. Chris has the newspaper delivered 7 days a week. He saves his newspapers for recycling. Over a month, the newspaper pile grows very high. The following table shows the height of the newspaper recycling pile at the end of each week.

Time (weeks)	0	1	2	3	4	5	6
Height (cm)		35	70	105	140		

a. Complete the table.
b. Draw a set of axes showing height on the vertical axis and time on the horizontal axis. Plot the information from your table on the axes and join the points to form a linear graph.
c. Determine the rule for the graph.
d. Determine the height of the pile after 20 weeks.

18. Lara sells computers and is paid $300 per week plus $20 for every computer she sells.
a. Draw a table to show how much money Lara would be paid if she sold between 0 and 10 computers per week.
b. Plot the points in the table on a Cartesian plane.
c. Identify whether the points form a linear graph.
d. Determine an equation for the graph.
e. If Lara sold 25 computers in a week, calculate how much money she would be paid.

19. James and his sister are going for a bike ride. They know they can ride 25 km in an hour (on average).
a. Complete the following table.

Time (hours)	0	25			100	
Distance (km)			2	3		5

b. Plot these points on a Cartesian plane.
c. Identify whether the points form a linear graph.
d. Complete the following equation.

Distance = _____ × time + _____ or

$d =$ _____ $t +$ _____

e. Use the equation you found in part d to work out how far they would have ridden after 7 hours.
f. If they leave home at 9:00 am and arrive home at 5:00 pm, determine how far James and his sister would have ridden.

20. As Rachel was driving her new car, she kept watch on her petrol usage. The graph shows how the amount of petrol has changed.

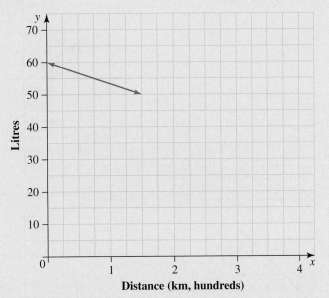

a. If this pattern continues, determine how much petrol there will be after 300 km.

b. The equation for this graph would be:

Number of litres remaining = _____ − _____ × number of km travelled (in hundreds)

This could be shortened to $l =$ _____ − _____k.

c. Using the equation from part **b**, calculate the amount of petrol left when $k = 5$.

d. Calculate how far Rachel will have travelled when she has used 60 litres of petrol.

21. Two yachts are sailing in Sydney harbour. At the end of a very windy day, rain begins to fall and visibility is heavily reduced. One yacht is travelling in the direction given by the equation $y = x + 6$ and the other in the direction given by $y = -2x + 3$.

a. Decide whether the yachts are likely to crash. (Assume both yachts are travelling towards the shore). Graph both lines using digital technology to justify your answer.

b. If they are likely to crash, state the coordinates of the point where they will meet.

c. Explain how you found your answer.

To test your understanding and knowledge of this topic, go to your learnON title at www.jacplus.com.au and complete the **post-test**.

Answers

Topic 9 Linear relationships

9.1 Pre-test

1. D
2. 11
3. B
4. B and D
5. 3
6. C
7. B
8. a.

x	0	1	2	3	4
y	−3	−2	−1	0	1

b.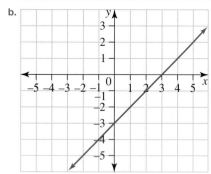

9. a. $y = 3x$ b. $y = 2x + 1$
10. A
11. B
12. A and C
13. $y = -2x + 1$
14. $x = 3$
15. a. i. $300   ii. $350   iii. $400

    b. $y = 50x + 250$

9.2 The Cartesian plane

1. A, B, C, D

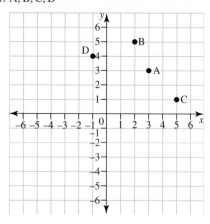

2. E, F, G, H

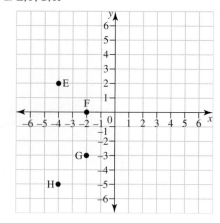

3. I, J, K, L

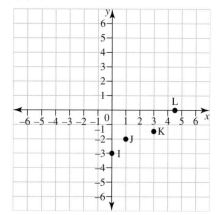

4. A (2, 2), B (4, 6), C (6, 1), D (−2, 3)
5. E (0, 5), F (−4, 1), G (0, 0), H (5, 0)
6. I (4, 3), J (−6, 3), K (−4, −2), L (−2, −5)
7. a. Point A is in quadrant 1; B is in quadrant 2; C and D are in quadrant 3.

b.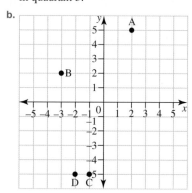

The plotting of the points on the Cartesian plane confirms the answer to part a.

8. a.

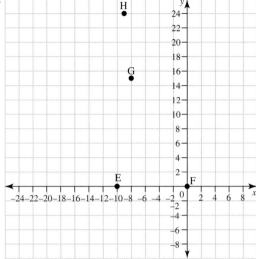

The graph $y = x$ contains all of the points that have the x-coordinate equal to the y-coordinate. For example, $(0, 0), (1, 1), (-2.5, -2.5)$

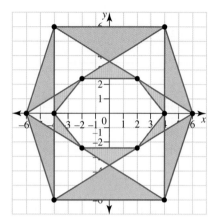

b. Point E sits on the x-axis; F sits on both the x- and y-axis (the origin); G and H are in quadrant 2.

9. a.

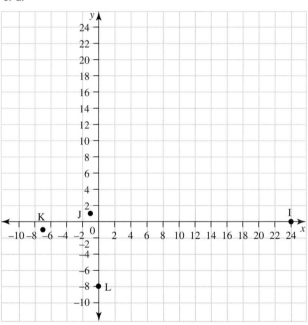

b. Point I sits on the x-axis; J is in quadrant 2; K is in quadrant 3; L sits on the y-axis.

10. C and D

11. A and C

12. a. B   **b.** C   **c.** D
    d. A   **e.** E   **f.** A

13. The graph of an equation shows all of the points (x, y) that are solutions to the equation.

14.

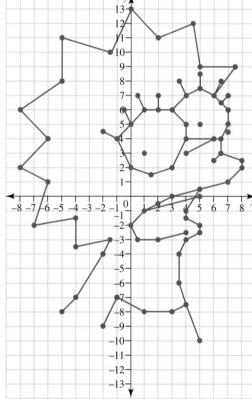

15. B and D

16. A and D

17. In hospital, in wards six, seven, eight and nine

18. a.

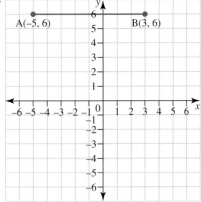

b. 8 units

c. $(-1, 6)$

d. Sample responses can be found in the worked solutions in the online resources.

e. $\dfrac{x_1 + x_2}{2}$

19. a.

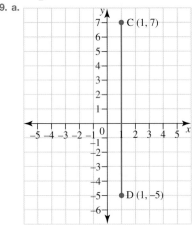

b. 12 units

c. $(1, 1)$

d. Sample responses can be found in the worked solutions in the online resources.

e. $\dfrac{y_1 + y_2}{2}$

20. It is customary for the x-coordinate to be written first. This ensures that anyone reading a coordinate pair will know the correct location. If the x- and y-coordinates were reversed, the location would be incorrect.

21. a. $(1, 1), (2, 2), (3, 3), (4, 4)$

b. $y = \dfrac{1}{2}$

c. $(-1, -1), (-2, -2), (-3, -3), (-4, -4)$

22. a. $A(-2, 2), B(2, 2), C(-2-2), D(2, -2)$

b. 16

c. $A(-2, 6)$ and $B(2, 6)$

23. a. $A(-3, 3), B(3, 3), C(3, -3), D(-3, -3)$

b. 36 square units

c.

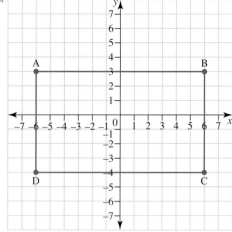

d. 72 square units

e.

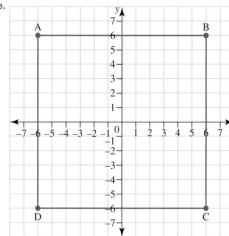

f. 144 square units

g. The length doubled, so the area doubled. When both the length and the width doubled, the area became four times bigger.

24. a. Area is 18 square units; perimeter is 18 units.

b. Sample responses can be found in the worked solutions in the online resources.

9.3 Linear patterns

1. a. 9, 7, 5, 3, 1

b.

Position in pattern	1	2	3	4	5
Value of term	9	7	5	3	1

c.

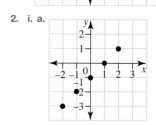

2. i. a.

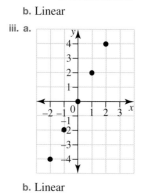

b. Linear

ii. a.

b. Linear

iii. a.

b. Linear

3. i. a.

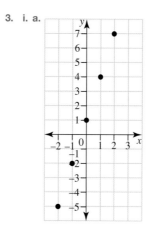

b. Linear

ii. a.

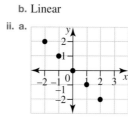

b. Linear

iii. a.

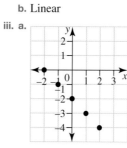

b. Linear

4. a.

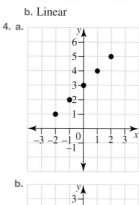

b.

c.

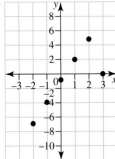

5. a.

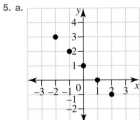

b.

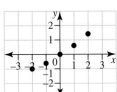

c.

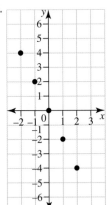

6. i. a.

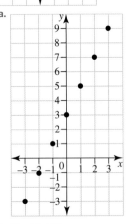

b. Yes (4, 11)

ii. a.

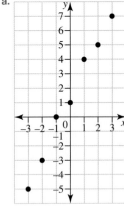

b. No

7. i. a.

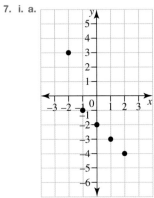

b. No

ii. a.

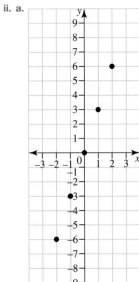

b. Yes (3, 9)

8. a.

Position in pattern	1	2	3	4	5
Value of term	9	6	3	0	−3

b. Value of term = 12 − position in pattern × 3

c. $y = 12 - 3x$

d.

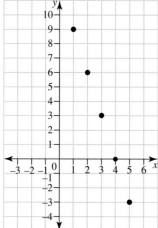

9. a.

Position in pattern	1	2	3	4	5
Value of term	1	5.5	10	14.5	19

b. Value of term = position in pattern × 4.5 − 3.5

c. $y = 4.5x - 3.5$

d.

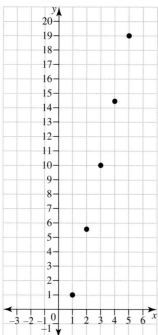

10. a.

Position in pattern	1	2	3	4	5
Value of term	−4	−1	2	5	8

b. Value of term = position in pattern × 3 − 7

c. $y = 3x - 7$

d.

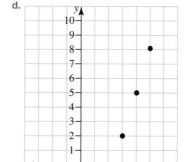

11. D

12. B

13. C

14. A, B

15. If the points of a number pattern go from the bottom left to the top right of a number grid, the pattern is increasing. If the points go from the top left to the bottom right, the pattern is decreasing.

16. a.

Day (d)	Distance (D)
Monday (1)	1 km
Tuesday (2)	1.25 km
Wednesday (3)	1.5 km
Thursday (4)	1.75 km
Friday (5)	2 km
Saturday (6)	2.25 km
Sunday (7)	2.5 km

b. Linear

c. 0.25 km per day

d. 3.25 km

17. Yes, if the points form a linear pattern, then they will form a straight line when plotted.

18. a.   **i.** 12 minutes

    ii. 24 minutes

    iii. 82 minutes

b. See the table at the foot of the page.*

c. Not linear

19. a. 24   **b.** 6 L   **c.** 42 L   **d.** 120 L

20. 22 February of the following year

21. a. See the table at the foot of the page.*

b.

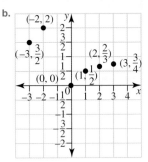

c. Not linear

9.4 Linear graphs

1. a.

x	−2	−1	0	1	2
y	1	2	3	4	5

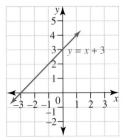

b.

x	−2	−1	0	1	2
y	−7	−6	−5	−4	−3

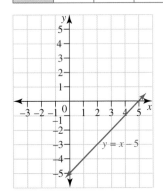

c.

x	−2	−1	0	1	2
y	−10	−5	0	5	10

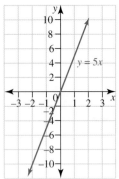

d.

x	−2	−1	0	1	2
y	0	2	4	6	8

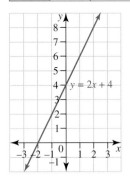

2. a.

x	−2	−1	0	1	2
y	−4	−1	2	5	8

---

*18. b.

q	0	1	2	3	4	5	6	7	8	9	10
t	0	12	14	36	48	60	82	94	106	118	130
q	11	12	13	14	15	16	17	18	19	20	
t	152	164	176	188	200	222	234	246	258	270	

*21. a.

x	−3	−2	−1	0	1	2	3
y	$\dfrac{3}{2}$	2	Undefined	0	$\dfrac{1}{2}$	$\dfrac{2}{3}$	$\dfrac{3}{4}$

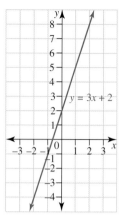

y = 3x + 2

b.

x	−2	−1	0	1	2
y	−6	−4	−2	0	2

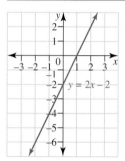

y = 2x − 2

c.

x	−2	−1	0	1	2
y	−11	−7	−3	1	5

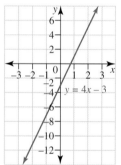

y = 4x − 3

d.

x	−2	−1	0	1	2
y	8	5	2	−1	−4

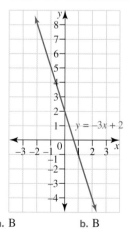

y = −3x + 2

3. a. B **b.** B

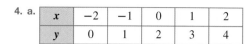

x	−2	−1	0	1	2
y	0	1	2	3	4

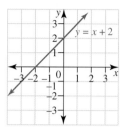

y = x + 2

b.

x	−2	−1	0	1	2
y	−6	−5	−4	−3	−2

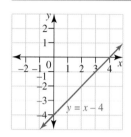

y = x − 4

c.

x	−2	−1	0	1	2
y	−3	−2	−1	0	1

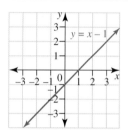

y = x − 1

d.

x	−2	−1	0	1	2
y	3	4	5	6	7

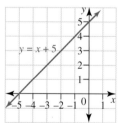

y = x + 5

e.

x	−2	−1	0	1	2
y	−6	−3	0	3	6

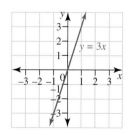

y = 3x

f.

x	−2	−1	0	1	2
y	−14	−7	0	7	14

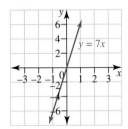

5. a.

x	−2	−1	0	1	2
y	−7	−3	1	5	9

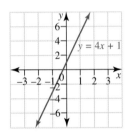

b.

x	−2	−1	0	1	2
y	−7	−5	−3	−1	1

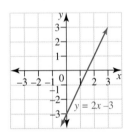

c.

x	−2	−1	0	1	2
y	−11	−8	−5	−2	1

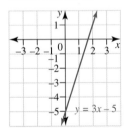

d.

x	−2	−1	0	1	2
y	4	2	0	−2	−4

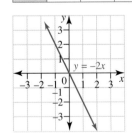

e.

x	−2	−1	0	1	2
y	14	8	2	−4	−10

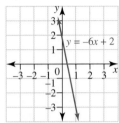

f.

x	−2	−1	0	1	2
y	14	9	4	−1	−6

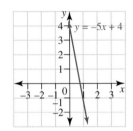

6. a. positive   **b.** undefined   **c.** negative

    d. positive   **e.** zero   **f.** negative

7. a. 4

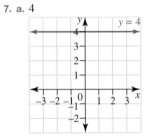

b. 1

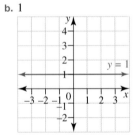

c. −2

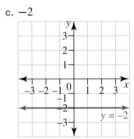

d. −5

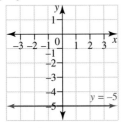

8. a.

x	−2	−1	0	1	2
y	3	3	3	3	3

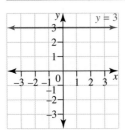

b.

x	−2	−1	0	1	2
y	2	2	2	2	2

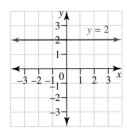

c.

x	−2	−1	0	1	2
y	−2	−2	−2	−2	−2

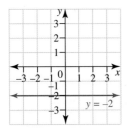

d.

x	−2	−1	0	1	2
y	−4	−4	−4	−4	−4

9. a. 1

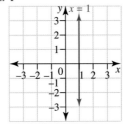

b. 3

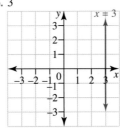

c. −2

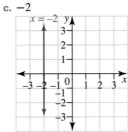

d. −7

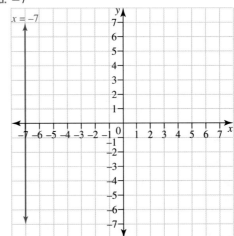

10. a.

x	2	2	2	2	2
y	−2	−1	0	1	2

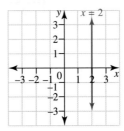

b.

x	5	5	5	5	5
y	−2	−1	0	1	2

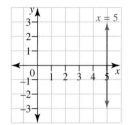

c.

x	−5	−5	−5	−5	−5
y	−2	−1	0	1	2

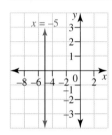

d.

x	0	0	0	0	0
y	−2	−1	0	1	2

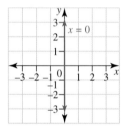

11. a. $y = 2x + 4$ **b.** $y = 3x + 10$ **c.** $y = -4x + 5$

12. a. $y = \dfrac{5}{3}x + 3$ **b.** $y = 2x + 5$ **c.** $y = -3x + 7$

d. $y = -x + 2.5$

13. a.

x	−2	−1	0	1	2
y	−4	−2	0	2	4

b.

x	−2	−1	0	1	2
y	−5	−3	−1	1	3

c.

x	−2	−1	0	1	2
y	−3	−1	1	3	5

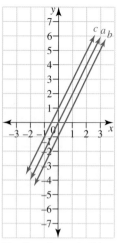

The lines formed all have the same positive gradient; that is, they are parallel.

14. a.

x	−2	−1	0	1	2
y	−5	−2	1	4	7

b.

x	−2	−1	0	1	2
y	5	3	1	−1	−3

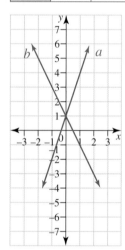

The lines meet at (0, 1).

15. a.

x	−2	−1	0	1	2
y	2	1	0	−1	−2

b.

x	−2	−1	0	1	2
y	0	1	2	3	4

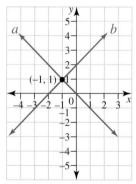

The lines meet at (−1, 1) and are perpendicular.

16. a.

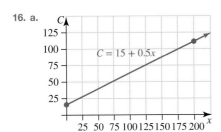

$$C = 15 + 0.5x$$

b. 170 minutes

c. 70 minutes

17. a. See the table at the foot of the page.*

b. This is a linear relationship as the graph is a straight line.

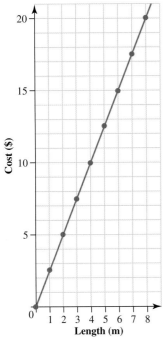

18. a. See the table at the foot of the page.*

b.

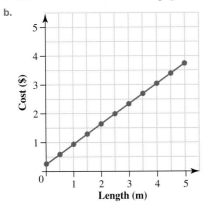

19. a.

Time (hours)	0	$\frac{1}{2}$	1	$1\frac{1}{2}$	2	$2\frac{1}{2}$	3
Cost ($)	55	100	145	190	235	280	325

b.

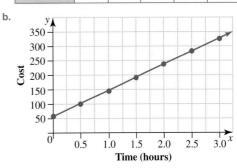

c. Yes

d. $m = 90$; $c = 55$

e. $y = 90x + 55$

f. $C = \$415$

20. a. Day 8

b.

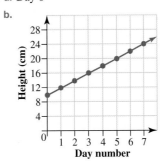

c. Yes

d. $m = 2$, $c = 10$

e. $h = 2d + 10$

f. i. $h = 38\,mm$

ii. 20 days

21. i. a.

x	-2	-1	0	1	2	3	4
y	-3	-2.5	-2	-1.5	-1	-0.5	0

b.

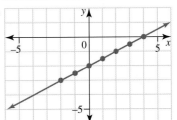

---

*17. a.

Length (m)	0	1	2	3	4	5	6	7	8
Cost ($)	0	2.50	5.00	7.50	10.00	12.50	15.00	17.50	20.00

*18. a.

Time (min)	0	0.5	1.0	1.5	2.0	2.5	3.0	3.5	4.0	4.5	5.0
Cost ($)	0.25	0.60	0.95	1.30	1.65	2.00	2.35	2.70	3.05	3.40	3.75

ii. a.

x	0	10	20	30	40	50
y	5	7	9	11	13	15

b.

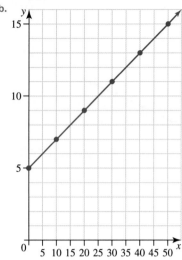

22. a.

x	−4	−2	0	2	4	6	8
y	−13	−9	−5	−1	3	7	11

b.

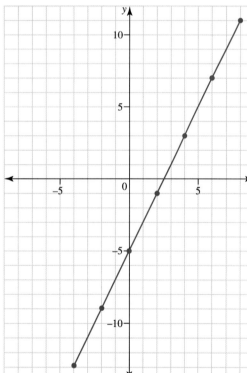

c. Look at the pattern of y-values to find the missing y-value. Use the y-values to calculate the x-values.

9.5 Solving equations graphically

1. $x = 1$

2. $x = 3$

3. $x = -6$

4. $x = \dfrac{3}{2}$

5. $x = -\dfrac{1}{6}$

6. $x = 1$

7. $x = -2$

8. a. $(2, 1)$

   b. $(0, 4)$

9. a. $(2, -1)$

   b. $(-2, -4)$

10. $x = 2$

11. $x = \dfrac{15}{11}$

12. $(3, 3), (-4, 17), (17, -4)$

13. Approximately 1.5 hours

14. a.

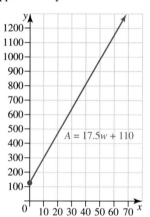

   b. $285

   c. $110

   d. 32 weeks

   e. 40 weeks

15. a. $y = 3x - 1$, $y = 3x + 2$ and $y = 3x$

    b. They all have a gradient of 3.

16. $y = -2x + 7$

17. a.

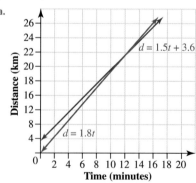

   b. 9 km

   c. 12 min

   d. 21.6 km

Project

1. Option 1: $60; Option 2: $52.50; Option 3: $40
2. Option 1: $C = \$60$; Option 2: $C = \$(30 + 0.25x)$
3. See the graph at the bottom of the page.*
4. Option 1 starts off the most expensive but, by the end of the 200 km, it is the least expensive, remaining at $60 for the day. Option 2 looks the least expensive at the start, but after travelling 200 km, it is the most expensive at $80. Option 3's cost is $40 up to 100 km travelled, but it increases to $75 by the end of 200 km of travel.
5. Option 1: $60; Option 2: $80; Option 3: $75
6. 2052 km
7. Option 1: $420; Option 2: $723; Option 3: $753.20
8. Yes
9. Responses will vary depending upon student research.

9.6 Review questions

1. a. Namiko   b. Rina   c. No

2.

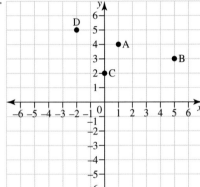

3.

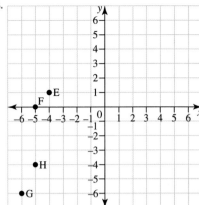

4.

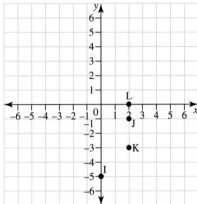

5. A$(1, 3)$, B$(5, -4)$, C$(3, 0)$, D$(-6, -4)$, E$(-5, 5)$, F$(2, -6)$, G$(0, -1)$, H$(-3, -3)$

6. a.

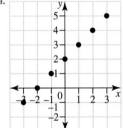

*3.

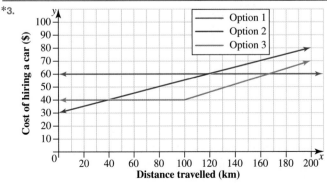

b.

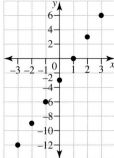

7. a.

Number of figure	1	2	3
Number of squares	5	9	13

b. Number of squares = number of figure × 4 + 1

c. 81

8. a.

x	−2	−1	0	1	2
y	−4	−3	−2	−1	0

b.

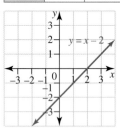

9. a.

x	−2	−1	0	1	2
f(x)	3	4	5	6	7

b.

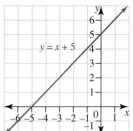

10. a.

x	−2	−1	0	1	2
f(x)	−10	−6	−2	2	6

b.

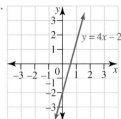

11. $x = 9$

12. $x = -3$

13. a.

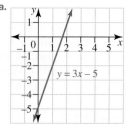

b.

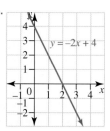

14. Lena (it will take her 10 months, while Alex will need 15 months)

15. 1200 m

16. a.

x	−3	−2	−1	0	1	2	3
y	−2	0	2	4	6	8	8

b.

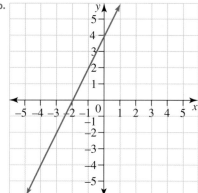

17. a. See the table at the bottom of the page.*

b.

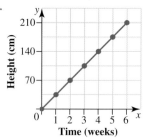

c. $y = 35x + 0$ or $y = 35x$

d. Height = 700 cm

---

*17. a.

Time (weeks)	0	1	2	3	4	5	6
Height (cm)	0	35	70	105	140	175	210

18. a. See the table at the bottom of the page.*

b.

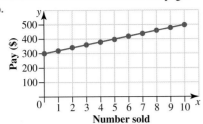

c. Yes

d. $P = 20n + 300$

e. $800

19. a. See the table at the bottom of the page. *

b.

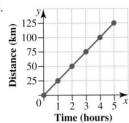

c. Yes

d. Distance $= 25 \times$ time $+ 0$ or $d = 25t + 0$ or $d = 25t$

e. 175 km

f. 200 km

20. a. 40 litres

b. Number of litres remaining $= 60 - \dfrac{20}{3} \times$ distance in km travelled

$l = 60 - \dfrac{20}{3}k$ (l is the number of litres; k is the distance in hundreds of kilometres travelled)

c. $26\dfrac{2}{3}$ litres

d. 900 km

21. a. Yes, they could crash, as the lines are not parallel.

b. They would meet at the point $(-1, 5)$.

c. By drawing the graphs of the two equations and finding the intersection point

---

*18. a.

Number sold (n)	0	1	2	3	4	5	6	7	8	9	10
Pay ($)	300	320	340	360	380	400	420	440	460	480	500

*19. a.

Distance (km)	0	25	50	75	100	125
Time (hours)	0	1	2	3	4	5

10 Data

LESSON
10.1 Overview

Why learn this?

We see statistics and data everywhere. We read them in newspapers, hear them on the TV and see them quoted on social media. It is very important that we can understand and interpret the data that we see. If you read an article that uses the mean or median or shows a graph to describe data, you need to be able to make sense of the information. Imagine you hear on the TV that the mean age of a social media user is 40 years. If you have no idea what the mean is and what it tells us, then this information will not tell you anything. This topic will help you to understand and make sense of statistical data and graphs.

Statistics are used by most professions. Sporting organisations use statistics to analyse game data, advertising agencies to try to make us buy certain products, the media in reporting, economists to analyse share markets, and organisations to describe their performance. In any occupation, you would need to read and interpret statistics at some point, so it is important you can do this easily.

1. Andy asked each person in his class what their favourite sport was.
   State the type of data that Andy collected.

2. Consider the following set of data.

   $$3, 3, 3, 3, 4, 5, 9, 11, 13$$

   Calculate the:
   a. mode
   b. range
   c. median.

3. **MC** From the following list, select all the questions that should not be used in a questionnaire about school uniforms.
   A. The present uniform is pretty ugly. Don't you agree we should change it?
   B. To which age group do you belong?
   C. Do you like the skirt, tie and blazer?
   D. How do you feel about the price of the uniform?

4. **MC** A sample of people were asked about the number of hours per week they spend exercising.

   The data was put into a grouped frequency table with class intervals of:
   $$0-<4$$
   $$4-<8$$
   $$8-<12 \text{ and so on.}$$

   Which kind of graph would be the best choice for presenting this data?
   A. Bar chart
   B. Histogram
   C. Stem-and-leaf plot
   D. Dot plot

5. A group of friends recorded how many hours they each spent watching television in a week. Their results are shown in the following table.

Value	Frequency
0	5
1	7
2	1
3	1

   a. Calculate the mean for the data, correct to 2 decimal places.
   b. Identify the median for this data.

6. **MC** This graph shows the number of children per household in a survey of 20 households.

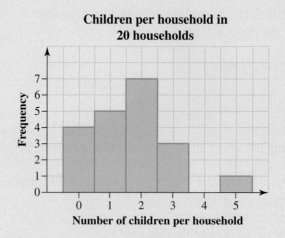

**Children per household in
20 households**

Frequency

Number of children per household

Select the statement that is false.
A. The modal number of children in a household is 2.
B. There are 20 children in total.
C. The range of the number of children is 5.
D. There are 33 children in total.

7. Calculate the mean, $\bar{x}$, for this set of data:

$$3, 3, 3, 3, 4, 5, 7$$

8. Consider the following data:

$$0, 8, 9, 10, 10, 12, 13, 13, 14$$

Determine which measure of centre would be most appropriate in this situation.

9. **MC** The mean, mode, median and range are calculated for the following data:

$$2, 7, 8, 9, 10, 10, 12, 13, 14$$

The value of 2 is considered an outlier (or an anomaly) and is rejected. The measures are recalculated. From the following, select the true statement after the recalculation.
A. The mode changes.
B. The mean changes.
C. The range remains unchanged.
D. The median remains unchanged.

10. Consider the following frequency distribution table.

Value	Frequency
0	8
1	7
2	4
3	2

What is the mean of this data?

11. A group of 22 Year 8 students measured their heights and recorded them in a grouped frequency table.

Height (cm)	Frequency
150– < 155	2
155– < 160	3
160– < 165	4
165– < 170	7
170– < 175	5
175– < 180	1

Note: The interval 150– < 155 in the table means that numbers from 150 up to 155 (but not including 155) are contained within this group.
Calculate the mean height for the group of Year 8 students, correct to 1 decimal place.

12. **MC** From the following list, choose which methods of data collection can be used for primary data. Select all that apply.
A. Survey
B. Questionnaire
C. Observation
D. Reading a report

13. The following values represent the number of hours worked in a week by 10 workers in a store:

$$34, 44, 28, 38, 36, 39, 10, 24, 28, 36$$

If all of these values were doubled, determine the effect this would have on the mean.

14. Determine the sequence of six numbers with a median of 5, a mode of 5, a range of 2 and a mean of 4.5. Write the numbers from lowest to highest.

15. A data set containing 7 pieces of data has the highest value of q and the lowest of p. The mean is m. If another value $(m + 2)$ is added to the data set, determine by how much the value of the mean would increase.

LESSON
10.2 Data collection methods (Foundational)

LEARNING INTENTION

At the end of this lesson you should be able to:
- understand the difference between populations and samples
- understand a variety of sampling methods
- determine sample size
- understand bias
- understand data collection methods.

▶ 10.2.1 Population and samples

eles-6216

- In statistics, a **population** refers to all the members of a particular group being considered in a research study. That is, a population is the entire set about which we want to draw conclusions.
- A **sample** is a subset or group of members selected from the population. This sample information is used to make inferences about the population.

- When data is collected for analysis, consideration needs to be given to whether the data represents the population or a sample.
- If data is collected about the number of students at a school who ate at the canteen on a particular day, the *population* would be every student at the school and a *sample* could be one class or one year level.
- When a sample of data is collected, it is important to ensure that the sample is indicative of the population and is selected at random.
- If the sample is not large enough and is not selected at random, it may provide data that is **biased** towards one particular group of people.

WORKED EXAMPLE 1 Random selections

A school with 750 students is surveying 25 randomly selected students to determine which sport is the most popular.
a. Who makes up the population and how many people are in it?
b. Who makes up the sample and how many people are in it?

THINK	WRITE
a. The population is made up of everyone who could possibly be asked this question. That would be every student at the school.	a. The population is made up of every student at this school, so that is 750 people.
b. The sample is made up of people randomly selected to take part in the survey.	b. The sample is the number of students selected, so that is 25 people.

▶ 10.2.2 Sampling methods

eles-6217

- Usually populations are too large for researchers to attempt to survey all of their members. **Sampling methods** are methods used to select members from the population to be in a statistical study.
- It is important to have a group of people who will participate in a survey and be able to represent the whole target population. This group is called a sample. Determining the right kind and number of participants to be in a sample group is one of the first steps in collecting data.
- Before you begin to select a sample, you first need to define your target population. For example, if your goal is to know the effectiveness of a product or service, then the target population should be the customers who have utilised it.
- In this topic, we will study four different types of sampling methods (**simple random sampling**, **stratified sampling**, **systematic sampling**, **self-selected sampling**, also known as voluntary sampling).

Sampling method	Description	Examples
Simple random sampling	Each member of the population has an equal chance of selection. This is a simple method and is easy to apply when small populations are involved. It is free of bias.	A Tattslotto draw — a sample of 6 numbers is randomly generated from a population of 45, with each number having an equal chance of being selected.

Systematic sampling	This technique requires the first member to be selected at random as a starting point. There is then a gap or interval between each further selection. A sampling interval can be calculated using $I = \dfrac{N}{n}$, where N is the population size and n is the sample size. This method is only practical when the population of interest is small and accessible enough for any member to be selected. A potential problem is that the period of the sampling may exaggerate or hide a periodic pattern in the population.	Every 20th item on a production line is tested for defects and quality. The starting point is item number 5, so the sample selected would be the 5th item, the 25th, the 45th, … Every 10th person who enters a particular store is selected, after a person has been selected at random as the starting point. Occupants in every 5th house in a street are selected, after a house has been selected at random as a starting point.
Stratified sampling	The population is divided into groups called strata, based on chosen characteristics, and samples are selected from each group. Examples of strata are states, ages, sex, religion, marital status and academic ability. An advantage is that information can be obtained on each stratum as well as the population as a whole.	A national survey is conducted. The population is divided into groups based on geography — north, east, south and west. Within each stratum, respondents are randomly selected.
Self-selected sampling	A voluntary sample is made up of people who self-select into the survey. The sample can often be biased, as the people who volunteer tend to have a strong interest in the main topic of the survey. The sample tends to over-represent individuals who have strong opinions.	A news channel on TV asks viewers to participate in an online poll. The sample is chosen by the viewers.

WORKED EXAMPLE 2 Determining sample size

Calculate the number of female students and male students required to be part of a sample of 25 students if the student population is 652 with 317 male students and 335 female students.

THINK	WRITE
1. State the formula for determining the sample size.	Sample size for each subgroup $= \dfrac{\text{subgroup size}}{\text{population size}} \times \text{sample size}$
2. Calculate the sample size for each subgroup.	Sample size for females $= \dfrac{335}{652} \times 25$ $= 13$ Sample size for males $= \dfrac{317}{652} \times 25$ $= 12$
3. State the answer.	The sample should contain 13 female students and 12 male students.

⊙ 10.2.3 Bias

eles-6218

- Generalising from a sample that is too small may lead to conclusions about a larger population that lack credibility. However, there is no need to sample every element in a population to make credible, reliable conclusions. Providing that a sufficiently large sample size has been drawn (as discussed below), a sample can provide a clear and accurate picture of a data set. However, it is important to try to eliminate bias when choosing your sampling method.
- Bias can be introduced in sampling by:
  - selecting a sample that is too small and not representative of the bigger population
  - relying on samples made up of volunteer respondents
  - sampling from select groups within a population, without including the same proportion from all the groups in the population
  - sampling from what is readily available
  - selecting a sample that is not generated randomly.
- A good sample is representative. If a sample is not randomly selected, it will be biased in some way and the data may not be representative of the entire population. The bias that results from an unrepresentative sample is called **selection bias**. Some common examples of selection bias are:
  - **undercoverage** — this occurs when some members of the population are inadequately represented in the sample.
  - **non-response bias** — individuals chosen for the sample are unwilling or unable to participate in the survey. Non-response bias typically relates to questionnaire or survey studies. It occurs when the group of study participants that responds to a survey is different in some way from the group that does not respond to the survey. This difference leads to survey sample results being skewed away from the true population result.
  - **voluntary response bias** — this occurs when sample members are self-selected volunteers.

⊙ 10.2.4 Determining the sample size

eles-6219

- Once you have identified the target population, you have to decide the number of participants in the sample. This is called the **sample size**.
- A sample size must be sufficiently large. As a general rule, the sample size should be at least $\sqrt{N}$, where N is the size of the population.
- If a sample size is too small, the data obtained is likely to be less reliable than that obtained from larger samples.

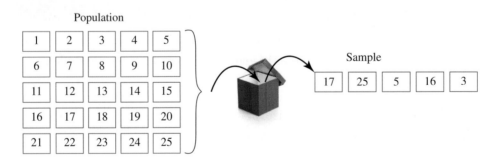

Calculators or computer software make the process of selecting a sample a lot easier by using a **random number generator**. An Excel worksheet generates random numbers using the **RAND( )** or **RANDBETWEEN (a,b)** command.

The **RAND( )** command generates a random number between 0 and 1. Depending on the size of the sample, these generated numbers have to be multiplied by n, where n is the size of the population.

▲	A	B	C	D	E
1	0.433821				
2	0.78391				
3	0.547901				
4	0.892612				
5	0.390466				
6	0.264742				
7	0.690003				
8	0.070899				
9	0.876409				
10	0.976012				

RANDBETWEEN(a,b) generates a random number from **a** to **b**. For the population of Year 12 Mathematics Standard 2 students, the sample of five students can be generated by using RANDBETWEEN$(1, 25)$.

▲	A	B	C	D	E	F
1	9					
2	3					
3	25					
4	3					
5	17					

If the same number appears twice, a new number will need to be generated in its place to achieve five different numbers.

WORKED EXAMPLE 3 Sampling

Select a sample of 12 days between 1 December and 31 January:
a. 'by hand'
b. using a random number generator.

THINK	WRITE
a. 1. Write every member of the population on a piece of paper.	a. There are 31 days in December and 31 days in January. This will require 62 pieces of paper to write down each day. 1/12, 2/12, 30/01, … , 31/01 *Note:* Alternatively, each day could be assigned a number, in order, from 1 to 62.
2. Fold all papers and put them in a box.	Ensure that the papers are folded properly so that no number can be seen.
3. Select the required sample. A sample of 12 days is required, so randomly choose 12 papers from the box.	Sample: $15, 9, 41, 12, 1, 7, 36, 13, 26, 5, 50, 48$
4. Convert the numbers into the data represented.	15 represents 15/12, 9 represents 9/12, 41 represents 10/01, 12 represents 12/12, 1 represents 1/12, 7 represents 7/12, 36 represents 5/01, 13 represents 13/12, 26 represents 26/12, 5 represents 5/12, 50 represents 19/01, 48 represents 17/01.
5. State the sample selected.	15/12, 9/12, 10/01, 12/12, 1/12, 7/12, 5/01, 13/12, 26/12, 5/12, 19/01, 17/01.

b. 1. Assign each member of the population a unique number.

b. Assign each day a number, in order, from 1 to 62: $1/12 = 1$, $2/12 = 2$, ..., $31/01 = 62$.

2. Open a new Excel worksheet and use the RANDBETWEEN (1, 62) command to generate the random numbers required.

	A	B	C	D	E	F
1	29					
2	27					
3	17					
4	26					
5	4					
6	45					
7	41					
8	25					
9	10					
10	28					
11	16					
12	8					

Note: Some of the numbers may be repeated. For this reason, more than 12 numbers should be generated. Select the first 12 unique numbers.

3. Convert the numbers into the data represented.

29 represents 29/12, 27 represents 27/12, 17 represents 17/12, 26 represents 26/12, 4 represents 4/12, 45 represents 14/01, 41 represents 10/01, 25 represents 25/12, 10 represents 10/12, 28 represents 28/12, 16 represents 16/12, 8 represents 8/12.

4. State the sample selected.

29/12, 27/12, 17/12, 26/12, 4/12, 14/01, 10/01, 25/12, 10/12, 28/12, 16/12, 8/12.

WORKED EXAMPLE 4 Calculating sample intervals

A factory produces 5000 mobile phones per week. Phones are randomly checked for defects and quality.
a. What type of probability sampling method would be used?
b. What sample size would be appropriate?
c. Calculate the sampling interval.

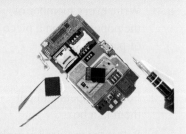

THINK

a. Think about how the data is collected. It would be easy to pick out every 50th member of the population.

b. The sample size should be at least $n = \sqrt{N}$, where N is the size of a population.

WRITE

a. Systematic sampling would be the most appropriate probability sampling method.

b. $N = 5000$
$$n = \sqrt{N}$$
$$= \sqrt{5000}$$
$$\approx 70.71$$
A sample size of 71 would be appropriate.

c. To calculate the sampling interval, use $I = \dfrac{N}{n}$.

c. $I = \dfrac{N}{n}$

$= \dfrac{5000}{71}$

$= 70.42$

≈ 70

Every 70th mobile phone should be selected.

eles-6220

⏵ 10.2.5 Potential flaws in data collection

- To derive conclusions from data, we need to know how the data was collected. There are four methods of data collection.

1. Census

- Sometimes the entire population will be sufficiently small, and the researcher can include the entire population in the study. This type of research is called a **census**, because the data is gathered from every member of the population. For most studies, a census is not practical because of the cost and time required.

2. Sample survey

- A sample survey is a study that obtains data from a subset of a population in order to estimate population attributes. When writing survey questions, care must be taken to avoid language and phrases that may introduce bias. Biased questions are sometimes referred to as 'leading questions' because they are more likely to lead to particular responses.
- Bias can be introduced in a survey by:
  - selecting questions that include several unpopular choices along with one favoured choice
  - phrasing questions positively or negatively
  - limiting the number of options provided when respondents have to make a choice
  - including closed questions without giving respondents the opportunity to give a reason for a particular response
  - relying on a sample of one that reflects a personal opinion, which is often based on limited experiences.

WORKED EXAMPLE 5 Identify whether information was obtained by census or survey

In each of the following, state whether the information was obtained by census or survey.
a. A school uses a roll to count the number of students who are absent each day.
b. Television ratings are determined by 2000 families completing a questionnaire on the programs they watch over a one-week period.
c. A battery manufacturer tests every hundredth battery off the production line.
d. A teacher records exam marks for each student in her class.

THINK	WRITE
a. Every student is counted at roll call each morning.	a. Census
b. Not every family is asked to complete a ratings questionnaire.	b. Survey
c. Not every battery is tested.	c. Survey
d. The marks of every student are recorded.	d. Census

3. Experiment

- An **experiment** is a controlled study in which the researcher attempts to understand cause-and-effect relationships. It is a method of applying treatments to a group and recording the effects. A good group experiment will have two basic elements: a control and a treatment. The control group remains untreated throughout the duration of an experiment. The study group is controlled as the researcher controls how subjects are assigned to groups and which treatment each group receives.

4. Observation

- An **observation** study is a study in which researchers simply collect data based on what is seen and heard. Researchers then make inferences based on the data collected. Researchers should not interfere with the subjects or variables in any way. They cannot add in any extra information. All of the information must be evidence in the observational study.

▶ 10.2.6 Misunderstanding samples and sampling

eles-6221

- Media reports often focus on one person's opinion and do not include data to support a claim. Sometimes data is misunderstood by the media. These reports should require further investigations to determine whether a larger or more appropriate sample reflects the same results.
- Factual information must have integrity, objectivity and accuracy. It is important to recognise that information can be misinterpreted by personal bias, inaccurate statistics, and even by the addition of fictional data.
- Some people and organisations do manipulate information for their own uses. For this reason, always be critical about information that is provided to you.
- Make sure that you know where the information is coming from and find out whether or not the source is credible. Also, try to find out what sampling processes and methods were used to collect the data.

WORKED EXAMPLE 6 Identifying bias

As part of a Year 11 research project, students need to collect data. A group of students put out a message on social media asking for responses to their survey.
a. What type of sampling method is used by the students?
b. Why is the sampling method used probably biased?

THINK	WRITE
a. Deduce what type of sampling method was used from the various sampling methods.	a. The sampling method used was self-selected sampling.
b. Explain the potential issues with the sampling method.	b. Self-selected sampling is a non-probability sampling method. The survey was made up of people who were willing to volunteer to answer the questions. The sample was not randomly generated and is probably not representative of the population.

---

1. Working in pairs, think of a topic on which you would like to gather your classmates' opinions. Some suggestions are: pets, movies, interests, favourite phone app, subjects studied.
2. Think of five questions you would like to ask your classmates about this topic, ensuring that you collect both categorical and numerical data.
3. When writing your questions, you should consider what sorts of answers you want. Do you want questions requiring *yes* and *no* answers, or would you like questions that require people to make a judgement on a scale? Questions should be easy to understand and must be relevant to your topic.

4. Write down how you intend to collect the data, the type of data you expect to collect and how you think it is best to represent this data. Think about the possible responses you may get.
5. As a class, listen to and offer constructive advice on the questions that each pair has developed. Feedback may consist of comments about the type of responses you might expect, how you could collect the data, whether the questions make sense and whether they are relevant to the topic.
6. Once you have received your feedback, you may wish to reword some of your questions and rethink the way you collect the data.
7. Collect and store the data. You will need to use it again during this topic.

on Resources

Interactivities Collecting data (int-3807)
Questionnaires (int-3809)
Planning a questionnaire (int-3810)
Selecting samples (int-3811)
Biased samples (int-3812)

Exercise 10.2 Data collection methods (Foundational)   learn on

10.2 Quick quiz on	10.2 Exercise

Individual pathways

■ PRACTISE	■ CONSOLIDATE	■ MASTER
1, 4, 7, 10, 13, 16, 19, 22, 25	2, 5, 8, 11, 14, 17, 20, 23, 26	3, 6, 9, 12, 15, 18, 21, 24, 27, 28

Fluency

1. **WE1** A company with 1200 employees and offices all over the world conducts a survey to see how happy their employees are with their work environment.
They survey people from offices in London (120 employees), Sydney (180 employees), Milan (45 employees) and Japan (75 employees).

   a. Who makes up the population and how many people are in it?
   b. Who makes up the sample and how many people are in it?

2. A university has 55 000 student enrolments. The university conducts a survey about online access for students. They survey students from the city campus (250 students) and the country campus (45 students).
   a. Who makes up the population and how many people are in it?
   b. Who makes up the sample and how many people are in it?

3. A school has 1240 students. An investigation concerning bell times is being conducted. Fifty students from the school are randomly selected to complete the survey on bell times.
   a. What is the population size?
   b. What is the size of the sample?

4. **WE2** Calculate the number of female and male students required to be a part of a sample of 80 students if the student population is 800 with 350 males and 450 female students.

5. **WE3** Select a sample of 5 students to participate in a debate from a group of 26 students:
   a. 'by hand'
   b. using a random number generator.

6. A small business has 29 employees. The owner of the business has decided to survey 8 employees on their opinion about the length of lunch breaks.
   Select the required sample:
   a. 'by hand'
   b. using a random number generator.

7. **WE4** A clothing manufacturer produces 2000 shirts per week. Shirts are randomly checked for defects and quality.
   a. What type of probability sampling method would be used?
   b. What sample size would be appropriate?
   c. Calculate the sampling interval.
   d. From the sample, 5 shirts were found to be defective in one week.
      Estimate the total number of shirts each week that are defective.

Understanding

8. **MC** Interviewing all members of a given population is called:
   A. a sample
   B. a Gallup poll
   C. a census
   D. a Nielsen audit

9. **MC** The best sample is one that is:
   A. a systematic sample
   B. convenient
   C. representative of the population
   D. purposefully selected

10. **MC** Select the option from the following that is an example of a non-probability sampling method.
    A. Simple random sampling
    B. Stratified sampling
    C. Self-selected sampling
    D. Systematic sampling

11. Zak wants to know what percentage of students at his school have a computer. State which strategy for sampling will be more likely to produce a representative sample and explain your answer.

Strategy 1: Obtain an alphabetised list of names of all students in the school and pick every 10th student on the list to survey.

Strategy 2: Send an email to every student asking them if they have a computer, and count the first 50 surveys that get returned.

12. **MC** Astur randomly selected 10 students from every year level at her school. Select the type of sampling from the following options.
    A. Simple random sampling
    B. Systematic sampling
    C. Stratified sampling
    D. Self-selected sampling

13. **MC** Each student has a student identification number. A careers counsellor generates 50 random student identification numbers on a computer, and those students are asked to take a survey. Select the type of sampling from the following options.
    A. Simple random sampling
    B. Stratified sampling
    C. Self-selected sampling
    D. Systematic sampling

14. **WE5** For each of the following, state whether a census or a survey has been used.
    a. Fifty people at a shopping centre are asked to nominate the supermarket where they do most of their grocery shopping.
    b. To determine the most popular new car on the road, 300 new-car buyers are asked what make and model they purchased.
    c. To determine the most popular new car on the road, the make and model of every new registered car are recorded.
    d. To test the life of a light bulb, every 100th bulb is tested.

15. For each of the following, recommend whether you would use a census or a survey to determine:
    a. the most popular TV program on Sunday night at 8.30 pm
    b. the number of 4-wheel-drive cars sold in a year
    c. the number of cars travelling on a toll road each day
    d. the percentage of defective SIM cards produced by a mobile phone manufacturing company.

16. **WE6** A TV host asks his viewers to visit his website and respond to an online poll.
    a. What type of sampling method is used?
    b. Why is the sampling method used probably biased?

17. **MC** A restaurant leaves comment cards on all of its tables and encourages customers to participate in a brief survey about their overall experience. Select the type of sampling from the following options.
    A. Stratified sampling
    B. Self-selected sampling
    C. Systematic sampling
    D. Simple random sampling

Communicating, reasoning and problem solving

18. Describe a sampling technique that could be used for each of the following.
    a. Three winning tickets are to be selected in an Easter egg raffle.
    b. The New South Wales Department of Tourism wants visitors' opinions of the information facilities that have been set up near the Opera House and Sydney Harbour Bridge.

19. Explain why it is important to consider sample size and randomness when collecting data from a sample of a population.

20. When would it be essential to survey the entire population and not just take a sample?

21. Briefly explain the difference between a census and a sample survey.

22. State one main disadvantage of a telephone survey.

23. State the research strategy that is being used in each of the following situations.

   a. To determine the effect of a new fertiliser on productivity of tomato plants, one group of plants is treated with the new fertiliser while a second group is grown without the treatment.
   b. A sociologist joins a group of homeless people to study the hardships they face in their everyday lives.
   c. A company sends a satisfaction questionnaire to its current customers at the end of the year.

24. For a political survey, 1470 householders were selected at random from the electoral roll and asked whether they would vote for the currently elected political party. In the survey, 520 householders answered 'Yes' to voting for the currently elected political party.

   a. If there are 17 million people in Australia over the age of 18, estimate how many would vote for 'No'. Give your answer to the nearest million.
   b. What percentage of Australians over 18, to the nearest whole number, would vote 'Yes' in your estimation?

25. Do you agree or disagree with the following statement? Explain.
   'I don't trust telephone surveys anymore. More and more individuals — particularly young individuals — do not have a landline. Moreover, these individuals are likely to differ from older individuals on key issues. If we are missing these younger individuals, our survey estimates will be biased.'

26. Some distance education students are enrolled in an online course. Depending on the location of the students, they are allocated to a region. There are 20 regions. In 10 of these regions, students are allocated to one of three tutors; in 7 of these regions students are allocated to one of two tutors; and in the remaining 3 regions, there is a single tutor. There are 10–15 students in each tutor's tutorial group.
   The distance education centre is planning a survey of the students to find out their opinion on the course. Suggest a way of selecting a sample of regions using the stratified sampling method.

27. A hotel manager is undecided about ways of administering a questionnaire. In particular, he is unsure whether to leave questionnaires in the hotel rooms or post them to clients' home addresses, and whether to select clients who book in during a 2-month period or select a proportion of clients who book in during a full year. Discuss which approach you would use and why.

28. An insurance company wishes to obtain customers' views on their satisfaction with the service they received. The company decides to survey callers who telephone its call centre to obtain their views. The call centre receives approximately 400 calls a day.
   If systematic sampling is used to select a sample of 100 callers over a six-day period from Monday to Saturday, estimate n where n represents every nth caller to be selected.

LESSON
10.3 Primary and secondary data (Foundational)

LEARNING INTENTION

At the end of this lesson you should be able to:
- understand the difference between primary and secondary data
- understand various methods of collecting primary data, including observation, measurement and surveys
- understand the source and reliability of secondary data.

▶ 10.3.1 Primary data

eles-4440

- **Primary data** are data that you have collected yourself.
- A variety of methods of collecting primary data is available. These include observation, measurement, survey, experiment or simulation.

Observation

- Observation involves recording the behavioural patterns of people, objects and events in a systematic manner.
- The data can be collected as a disguised observation (respondents are unaware they are being observed) or undisguised observation (the respondents are aware).
- Closed-circuit television (CCTV) cameras are an example of people knowing that their movements are being recorded but not always being aware of where the recording takes place.
- Observations can be in a natural environment (for example in a food hall), or a contrived environment (for example a food-tasting session for a food company).
- Mechanical devices (for example video cameras, CCTV or counting devices across a road) can also be used.

Measurement

- Measurement involves using a measuring device to collect data.
- This generally involves conducting an experiment of some type.
  - The height of everyone in your class can be measured.
  - The mass of all newborn babies can be collected.
  - A pedometer can measure the number of steps the wearer takes.

Surveys

- Surveys are questionnaires designed to interview people. Often the questionnaire requires many rewrites to ensure it is clear and unbiased.
- The interview can be in person — face to face — or by telephone. The advantage of an in-person interview is that you are able to see the reactions of those you are interviewing and explain particular questions, if necessary.

- Until the beginning of the COVID-19 pandemic, email was the most frequently used interviewing tool; however, platforms such as WebEx and Zoom have become more popular as methods of communication. Some of the advantages and disadvantages of an email survey are listed as follows.
- *Advantages:*
  - It can cover a large number of people or organisations.
  - Wide geographic coverage is possible.
  - It avoids embarrassment on the part of the respondent.
  - There is no interviewer bias.
  - The respondent has time to consider responses.
  - It is relatively cheap.
- *Disadvantages:*
  - The questions have to be relatively simple.
  - The response rate is often quite low.
  - The reliability of the answers is questionable.
  - There is no control over who actually completes the questionnaire.
  - Questionnaires may be returned incomplete.

Experiment

- Generally, when conducting an experiment, the data collected is quantitative.
- Particular care should be taken to ensure the experiment is conducted in a manner that would produce similar results if repeated.
- Care must be taken with the recording of results.
- The results must be in a form that can readily be analysed.
- All results need to be recorded, including unusual or unexpected outcomes.

Simulation

- Experiments such as rolling a die, tossing a coin or drawing a card from a deck may be conducted to model real-life situations.
- Simulations occur in areas such as business, engineering, medicine and scientific research.
- Simulations are often used to imitate real-life situations that may be dangerous, impractical or too expensive to explore by other means.

WORKED EXAMPLE 7 Designing a simulation

It is widely believed that there is an equal chance of having a boy or girl with each birth. Although genetics and the history of births in a family may influence the sex of the child, ignore those factors in this question.

a. Design an experiment to simulate the chance of giving birth to a boy or a girl.

b. Describe how your experiment could be conducted to determine the number of children a couple should have, on average, to ensure they have offspring of both sexes.

THINK

a. Use a device that can simulate two outcomes that are equally likely.
A fair coin could be tossed so that a Head represents a boy and a Tail represents a girl.
This could be a random number generator to generate two integers, say 0 (representing a boy) and 1 (representing a girl).

b. 1. Describe how the experiment will be conducted.

2. Display the table of results.

WRITE

a. A fair coin will be tossed, with a Head representing a boy (B) and a Tail representing a girl (G).

b. The experiment will be conducted 50 times, and a record kept of each experiment.
For each experiment, the coin will be tossed until both sexes appear.
This may mean that there could be, for example, 7 trials in an experiment (GGGGGGB) before both sexes are represented.

The table below shows the results of the 50 experiments.

Exp. no.	Results	No. of trials	Exp. no.	Results	No. of trials
1	BG	2	26	GGGB	4
2	GGB	3	27	GGGGB	5
3	BG	2	28	GGGB	4
4	GGGGB	5	29	BG	2
5	BBBBBBG	7	30	BBBG	4
6	GGGB	4	31	BG	2
7	BBG	3	32	GB	2
8	BBG	3	33	GGGB	4
9	BBBBG	5	34	BG	2
10	GB	2	35	GGGGGGB	7
11	BG	2	36	BBBBBBG	7
12	GGGB	4	37	GB	2
13	BBG	3	38	BG	2
14	BBG	3	39	GGB	3
15	GB	2	40	GGGGB	5
16	BG	2	41	BBG	3
17	GGB	3	42	BBBBBG	6
18	GB	2	43	GGB	3
19	GGB	3	44	GGB	3
20	BBBG	4	45	BBBG	4
21	BG	2	46	BBG	3
22	GB	2	47	GGGGGGB	7
23	GGGGB	5	48	BG	2
24	BG	2	49	BBG	3
25	GGGGB	5	50	GGGGGB	6
				Total	175

This table shows that 175 trials were undertaken in 50 experiments where each experiment resulted in both sexes.

3. Determine the average number of children required to produce offspring of both sexes.	Average number of children $= \dfrac{175}{50}$
	$= 3.5$
4. Write a conclusion.	The average number of children a couple should have to reach the goal of having both sexes is 4.

- Before collecting any primary data, it must be clear what data is to be collected.
- A decision must be made as to the method of collection.
- The advantages and disadvantages of the collection method must be acknowledged.
- The reason for the data collection should be clear from the outset.

WORKED EXAMPLE 8 Collecting data

You have been asked to obtain primary data to determine the methods of transport the students at your school use to travel to school.

The data collected is meant to provide support for the student council's proposal for a school bus.

a. State what data should be collected.

b. Outline possible methods that could be used to collect the data.

c. Decide which method you consider to be the best, and discuss its advantages and disadvantages.

THINK	WRITE
a. Outline the various forms of transport available to the students.	a. The modes of transport available to students at the school are car, bus, train, bicycle and walking.
b. Consider all the different ways of collecting the data.	b. Several methods could be used to collect the data. • Stand at the school gate one morning and ask students as they arrive. • Design a questionnaire. • Ask students to write their mode of transport on a piece of paper and then place it in a collection tin.
c. 1. Decide on the best option.	c. The first option (standing at the school gate) is time-consuming, and students could arrive at another entrance. The third option does not seem reliable, as some students may not comply, and other students may place multiple pieces of paper in the collection tin. The second option seems the best of the three.
2. Discuss the advantages and disadvantages.	Advantages of a questionnaire include: • There is a permanent record of responses. • It is not as time-consuming to distribute or collect. • Students can complete it at their leisure. Disadvantages of a questionnaire include: • Students may not return or complete it. • Printing copies could get expensive.

Note: This example does not represent the views of all those collecting such data. It merely serves to challenge students to explore and discuss available options.

- Sometimes the primary data required is not obvious at the outset of the investigation.
- For example, you are asked to investigate the claim:
  Most students do not eat a proper breakfast before school.
  What questions would you ask to prove or refute this claim?

⊳ 10.3.2 Secondary data

eles-4441

- **Secondary data** is data that has already been collected by someone else.
- The data can come from a variety of sources:
  - Books, journals, magazines, company reports
  - Online databases, broadcasts, videos
  - Government sources — the Australian Bureau of Statistics (ABS) provides a wealth of statistical data
  - General business sources — academic institutions, stockbroking firms, sporting clubs
  - Media — newspapers, TV reports
- Secondary data sources often provide data that would not be possible for an individual to collect.
- Data can be qualitative or quantitative — that is, categorical or numerical.
- The accuracy and reliability of data sometimes needs to be questioned, depending on its source.
- The age of the data should always be considered.
- It is important to learn the skills necessary to critically analyse secondary data.

WORKED EXAMPLE 9 Understanding data

Bigbite advertise the energy and fat content of some of the sandwiches on their menus.

a. Determine the information that can be gained from this data.
b. Bigbite advertise that they have a range of sandwiches with less than 6 grams of fat. Comment on this claim.
c. This could be the starting point of a statistical investigation. Discuss how you could proceed from here.
d. Investigations are not conducted simply for the sake of investigating. Suggest some aims for investigating further.

BIGBITE

Bigbite fresh sandwiches	Energy (kJ)	Fat (g)	Sat. fat (g)
Roasted vegetable	900	3.0	1.0
Ham	1100	6.0	1.4
Turkey	1140	4.8	1.7
BBQ beef	1150	5.0	1.5
Bigbite ribbon	1130	4.8	1.3
Turkey and ham	1250	4.5	1.5
BBQ chicken	1460	4.7	1.2
Chicken tandoori	1110	4.0	1.0
Fresh dessert			
Fruit slices	200	<1	<1

Bigbite sandwiches
Regular sandwiches include white and/or wholemeal bread, salads and meat.
Nutritional value is changed by adding cheese or sauces.

THINK

a. Look at the data to gain as much information as possible.

WRITE

a. The data reveals the following information:
  - Higher energy content of a sandwich does not necessarily mean that the fat content is higher.
  - As the fat content of a sandwich increases, generally the saturated fat content also increases.
  - The addition of some types of protein (ham, turkey, beef, chicken) increases the energy content of the sandwich.
  - The data is only for those sandwiches on white or wholemeal bread with salads and meat.
  - The addition of condiments (sauces) or cheese will alter these figures.
  - A fruit slice has much less energy and fat than a sandwich.

b. Examine the data to discover if there is evidence to support the claim. Make further comment.

b. All but one of the sandwiches displayed have less than 6 grams of fat, so Bigbite's claim is true. It must be remembered that the addition of cheese and sauce to these sandwiches would increase their fat content. Also, if the sandwich was on any bread other than white or wholemeal, the fat content could be higher than 6 grams.

c. Determine the next step in the investigation.

c. Conducting a web search for Bigbite's contact details, or to see whether more nutritional information is posted on their website, would be a good next step.

d. What are some interesting facts that could be revealed through a deeper investigation?

d. Suggested aims for investigating further could be:
- How much extra fat is added to a sandwich by the addition of cheese and/or sauce?
- What difference does a different type of bread make to the fat content of the sandwich?
- Which sandwich contains the highest fat content?
- What is the sugar content of the sandwiches?

DISCUSSION

Discuss some of the difficulties that you may come across with obtaining data from either primary or secondary sources.

Consider where the data would need to be collected from, the reliability of the data, what digital technologies might be needed and anything else that may influence the results.

COMMUNICATING — COLLABORATIVE TASK: Simulations

Working in small groups, design an experiment to simulate the following situation.

A restaurant menu features 4 desserts that are assumed to be equally popular. How many dessert orders must be filled (on average) before the owner can be sure all types will have been ordered?

Carry out the experiment and discuss the results of the experiment with the class.

Discuss whether your answer would change if the menu features 6 desserts, all equally popular.

 Resources

 Interactivity Primary and secondary data (int-3814)

10.3 Quick quiz on	10.3 Exercise

Individual pathways

■ PRACTISE	■ CONSOLIDATE	■ MASTER
1, 4, 8, 11	2, 5, 9, 12	3, 6, 7, 10, 13

Fluency

1. **WE7** Devise an experiment to simulate each of the following situations and specify the device used to represent the outcomes.

   a. A true/false test is used in which answers are randomly distributed.
   b. A casino game is played, with outcomes grouped in colours of either red or black.
   c. Breakfast cereal boxes are bought containing different types of plastic toys.
   d. From a group of six people, one person is to be chosen as the group leader.
   e. A choice is to be made between three main meals on a restaurant's menu, all of which are equally popular.
   f. Five possible holiday destinations are offered by a travel agent; all destinations are equally available and equally priced.

2. **WE8** You have been asked to obtain primary data from students at your school to determine what internet access students have at home.
   The data collected will provide support for opening the computer room for student use at night.

   a. Suggest what data should be collected.
   b. Outline possible methods that could be used to collect this data.
   c. Decide which method you consider to be the best option, and discuss its advantages and disadvantages.

3. **WE9** This label shows the nutritional information of Brand X rolled oats.

Nutrition Information			
Servings Per Package: 25			Serving Size 30 g
	Per Serving 30 g	%DI* Per Serving	Per 100 g
Energy	486 kJ	6%	1620 kJ
Protein	4.3 g	9%	14.3 g
Fat - Total	2.8 g	4%	9.3 g
- Saturated	0.5 g	2%	1.7 g
- Trans	Less than 0.1 g	–	Less than 0.1 g
- Polyunsaturated	1.0 g	–	3.2 g
- Monounsaturated	1.3 g	–	4.4 g
Carbohydrate	16.8 g	5%	56.0 g
- Sugars	0.9 g	1%	3.0 g
Dietary Fibre	3.1 g	10%	10.4 g
Sodium	0.7 mg	0.1%	2 mg

* %DI = Percentage daily intake

   a. State the information gained from this data.
   b. This could be the starting point of a statistical investigation. Discuss how you could proceed from here.
   c. Suggest some aims for investigating further.

4. State which of the following methods could be used to collect primary data.
   Census, observation, newspaper article, journal, online response, DVD, interview, experiment, TV news report

Understanding

5. State which of the five methods below is the most appropriate to use to collect the following primary data.

    Survey, observation, newspaper recordings, measurement, census

    a. Heights of trees along the footpaths of a tree-lined street
    b. Number of buses that transport students to your school in the morning
    c. Sunrise times during summer
    d. Student opinion regarding length of lessons

6. Comment on this claim.

    > We surveyed 100 people to find out how often they eat chocolate.
    > Sixty of these people said they regularly eat chocolate.
    > We then measured the heights of all 100 people.
    > The conclusion:
    > Eating chocolate makes you taller!

7. The following claim has been made regarding secondary data.

    There's a lot more secondary than primary data. It's a lot cheaper and it's easier to acquire.

    Comment on this statement.

Communicating, reasoning and problem solving

8. Pizza King conducted a survey by asking their customers to compare 10 of their pizza varieties with those of their nearby competitors. After receiving and analysing the data, they released an advertising campaign with the headline 'Customers rate our pizzas as 25% better than the rest!'
    The details in the small print revealed that this was based on the survey of their Hawaiian pizzas.
    Explain what was wrong with Pizza King's claim.

9. Addison, a prospective home buyer, wishes to find out the cost of a mortgage from financial institutions. She realises that there are a lot of lenders in the marketplace. Explain how she would collect the necessary information in the form of:

    a. primary data
    b. secondary data.

10. The local Bed Barn was having a sale on selected beds by Sealy and SleepMaker. Four of the beds on sale were:

Sealy Posturepremier	on sale for $1499	a saving of $1000
Sealy Posturepedic	on sale for $2299	a saving of $1600
SleepMaker Casablanca	on sale for $1199	a saving of $800
SleepMaker Umbria	on sale for $2499	a saving of $1800

    The store claimed that all these beds had been discounted by at least 40%. Comment on whether this statement is true, supporting your comments with sound mathematical reasoning.

11. Alessandra has two different data sets. Data set A contains the newborn baby weights of each student in her class after she surveyed each student. Data set B contains the average newborn baby weights for the last twenty years.

    a. Identify which data is primary data.
    b. Identify which data is secondary data.
    c. Explain how you determined which data was primary and which data was secondary.

12. Hashem is planning on running a stall at a fundraiser selling ice-cream. There are 1000 students in his school ranging from Year 7 to Year 12. There are five Year 8 classes, each with 25 students (boys and girls).

Hashem intends to ask a group of 10 students chosen at random from each of these five classes to select their favourite three ice-cream flavours. Hamish is confident that this random sampling method encompassing a total of 50 students should give him an accurate picture of the ice-cream preferences for the school. Is Hamish correct or is he facing a financial disaster? Explain your answer.

13. Kirsty, chief marketing manager of Farmco Cheeses, has decided to run a major TV advertising campaign.
    a. Suggest how she should choose a TV channel and time slot to run her advertisements.
    b. Suggest how she should decide which demographic/age groups to target.
    c. Discuss whether the answer to part **b** has any bearing on the answer to part **a**.

LESSON
10.4 Organising and displaying data

LEARNING INTENTION

At the end of this lesson you should be able to:
- organise data into a frequency table, using class intervals where necessary
- construct a histogram from a frequency table
- use technology to construct a histogram
- create statistical infographics.

▶ 10.4.1 Examining data

eles-4442

- Once collected, data must be organised so that it can be displayed graphically and interpreted.
- When this has been done, any anomalies in the data will be highlighted.
- Anomalies could have occurred because of:
  - recording errors
  - unusual responses.
- Sometimes a decision is made to disregard these anomalies, which are regarded as **outliers**.
- Outliers can greatly affect the results of calculations, as you will see later in the topic.

Frequency tables

- Organising raw data into a frequency table is the first step in allowing us to see patterns (trends) in the data.
- A frequency table lists the values (or scores) of the variable and their frequencies (how often they occur).

WORKED EXAMPLE 10 Constructing a frequency table

In a suburb of 350 houses, a sample of 20 households was surveyed to determine the number of children living in them. The data were collected and recorded as follows:

0, 2, 3, 2, 1, 3, 5, 2, 0, 1, 2, 0, 2, 1, 2, 1, 2, 3, 1, 0

a. Organise the data into a frequency table.
b. Comment on the distribution of the data.
c. Comment on the number of children per household in the suburb.

THINK		WRITE	
a. Draw up a frequency table and complete the entries.		a.	

Children per household	Frequency
0	4
1	5
2	7
3	3
4	0
5	1

b. Look at how the data are distributed.

b. The data value of 5 appears to be an outlier. This is probably not a recording error, but it is not typical of the number of children per household. Most households seem to have 1 or 2 children.

c. Does this sample seem to reflect the population characteristics?

c. The sample is an appropriate size, and would probably reflect the characteristics of the population. It would be reasonably safe to say that most houses in the suburb contained 1 or 2 children.

- Sometimes data can take a large range of values (for example age (0–100)) and listing all possible ages would be tedious. To solve this problem, we group the data into a small number of convenient intervals, called **class intervals**.
  - Class intervals should generally be the same size and be set so that each value belongs to one interval only.
  - Examples of class intervals are $0-<5, 5-<10, 10-<15$ and so on; intervals represent the range of values that a particular group can take.
  - For example, the interval $0-<5$ means numbers from 0 up to 5 (but not including 5) are contained within this group.

WORKED EXAMPLE 11 Constructing a frequency table with class intervals

A sample of 40 people was surveyed about the number of hours per week they spent watching TV.
The results, rounded to the nearest hour, are listed below.

12, 18, 9, 17, 20, 7, 24, 16, 9, 27, 7, 16, 26, 15, 7, 28, 11, 20, 9, 11,
23, 19, 29, 12, 19, 12, 16, 21, 8, 4, 16, 20, 17, 10, 24, 21, 5, 13, 29, 26

a. Organise the data into a frequency table using class intervals of $5-<10, 10-<15$ and so on. Show the midpoint or class centre of each class interval.
b. Comment on the distribution of the data.

THINK

a. 1. Draw up a frequency table with three columns: class interval (hours of TV), midpoint or class centre, and frequency.

2. The midpoint is calculated by adding the two extremes of the class interval and dividing by 2. For example, the midpoint of the first class interval is $\dfrac{5+10}{2} = 7.5$.

3. Systematically go through the list, determine how many times each score occurs and enter the information into the frequency column.

b. Look at how the data is distributed.

WRITE

a.

Hours of TV	Midpoint	Frequency
5–<10	7.5	9
10–<15	12.5	7
15–<20	17.5	10
20–<25	22.5	8
25–<30	27.5	6

b. The TV viewing times are fairly evenly distributed, with the most frequent class interval being 15–<20 hours per week.

⏵ 10.4.2 Histograms

eles-4444

- **Histograms** are used for displaying grouped discrete or continuous numerical data and can be used to highlight trends and distributions.
- Histograms display data that has been summarised in a frequency table.
- A histogram has the following characteristics:
  1. The vertical axis (*y*-axis) is used to represent the frequency of each item.
  2. No gaps are left between columns.
  3. A space measuring a half-column width is sometimes placed between the vertical axis and the first column of the histogram if the first bar does not start at zero.

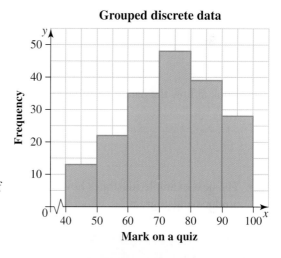
Grouped discrete data

- Before constructing a histogram, identify the smallest and largest values for both axes to help choose an appropriate scale.
- Both axes must be labelled. The vertical axis is labelled 'Frequency'.
- When presenting grouped data graphically, we generally label the horizontal axis (score) with the class interval.

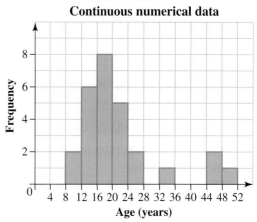
Continuous numerical data

Consider the grouped frequency table created in Worked example 11.
a. **Display the data as a histogram.**
b. **Comment on the shape of the graph.**

Hours of TV	Frequency
5– < 10	9
10– < 15	7
15– < 20	10
20– < 25	8
25– < 30	6

THINK

a. 1. Rule a set of axes on graph paper. Give the graph a title. Label the horizontal axis 'Number of hours of television watched' and the vertical axis 'Frequency'.

   2. Leaving a $\frac{1}{2}$ unit space at the beginning, draw the first column so that it starts at 5 and ends at the beginning of the next interval, which is 10. The height of this first column should be 9. Repeat this technique for the other scores.

WRITE/DRAW

a.
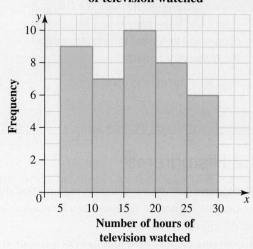

Histogram of hours of television watched

b. Look at how the data is distributed.

b. The number of hours of TV watching is fairly consistent throughout the week. A maximum number of people watch about 15 to 20 hours per week.

• The next example illustrates how to deal with a large amount of data, organise it into class intervals, then produce a histogram.

The following data are the results of testing the lives (in hours) of 100 torch batteries.

20, 31, 42, 49, 46, 36, 42, 25, 28, 37, 48, 49, 45, 35, 25, 42, 30, 23, 25, 26,
29, 31, 46, 25, 40, 30, 31, 49, 38, 41, 23, 46, 29, 38, 22, 26, 31, 33, 34, 32,
41, 23, 29, 30, 29, 28, 48, 49, 31, 49, 48, 37, 38, 47, 25, 43, 38, 48, 37, 20,
38, 22, 21, 33, 35, 27, 38, 31, 22, 28, 20, 30, 41, 49, 41, 32, 43, 28, 21, 27,
20, 39, 40, 27, 26, 36, 36, 41, 46, 28, 32, 33, 25, 31, 33, 25, 36, 41, 28, 33

a. **Choose a suitable class interval for the given data and present the results in a frequency distribution table.**
b. **Draw a histogram of the data.**

	THINK		WRITE/DRAW

THINK

a. 1. To choose a suitable size for the class intervals, calculate the range. To determine the range, subtract the smallest value from the largest.

2. Divide the results obtained for the range by 5 and round to the nearest whole number.
Note: A class interval of 5 hours will result in 6 groups.

3. Draw a frequency table and list the class intervals in the first column, beginning with the smallest value.
Note: The class interval 20– < 25 includes hours ranging from and including 20 to less than 25.

4. Systematically go through the data and determine the frequency of each class interval.

5. Calculate the total of the frequency column.

WRITE/DRAW

a. Range = largest value − smallest value
$$= 49 - 20$$
$$= 29$$

Number of class intervals: $\dfrac{29}{5} = 5.8$
$$= 6$$

Lifetime (hours)	Tally	Frequency (f)
20– < 25	ЖЖ ||	12
25– < 30	ЖЖ ЖЖ ЖЖ ЖЖ |||	23
30– < 35	ЖЖ ЖЖ ЖЖ ЖЖ	20
35– < 40	ЖЖ ЖЖ ЖЖ		16
40– < 45	ЖЖ ЖЖ |||	13
45– < 50	ЖЖ ЖЖ ЖЖ		16
	Total	100

b. 1. Rule and label a set of axes on graph paper. Give the graph a title.

2. Add scales to the horizontal and vertical axes. *Note:* Leave a half interval at the beginning and end of the horizontal axis.

3. Draw in the first column so that it starts at 20 and finishes at 25 and reaches a vertical height of 12 units.

4. Repeat step 3 for each of the other scores.

b.

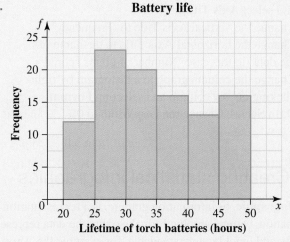

▶ 10.4.3 Using a spreadsheet to draw a histogram

eles-4445

- Spreadsheets such as Excel can also be used to easily tabulate and graph data.
- Enter the data values or class intervals into the first column and frequencies into the second column, as shown.

	A	B
1	Number of points scored	Frequency
2	0– < 5	11
3	5– < 10	8
4	10– < 15	6
5	15– < 20	6
6	20– < 25	4
7	25– < 30	2
8	30– < 35	1

- To construct a histogram, follow the steps outlined below:
  1. Highlight all the cells in your table containing data.
  2. Click on the **Insert** tab at the top of the screen.
  3. Select the **Clustered Column** graph from the Charts section.
  4. Next to the column graph click on the + symbol and select **Axis Titles**.
  5. Change the vertical Axis Title to 'Frequency'.
  6. Change the horizontal Axis Title to the name of the data (in this case 'Number of points scored').
  7. Change the column graph to a histogram by right-clicking on any column and selecting **Format Data Series**. Change the Gap Width to 0%.

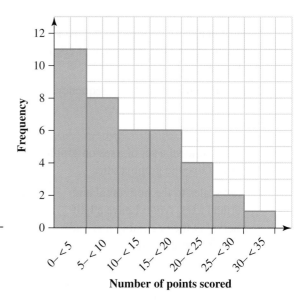

▶ 10.4.4 Creating statistical infographics

eles-6222

- An **infographic**, or information graphic, is a representation of information, using a graphic format, so that the data represented is easily and quickly understandable, for example the Aboriginal and Torres Strait Islander Guide to Healthy Eating shown.
- There are different types of infographics, such as timeline infographics, map infographics, process infographics and 'how-to' infographics.
- Infographics can include text but are generally not text heavy.
- Infographics are effective practical tools used everywhere, and you are more than likely already extremely familiar with them.

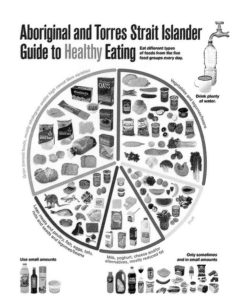

pH scale

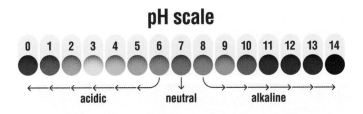

- A **statistical infographic** is a way to represent data in a visually appealing manner, using appropriate graphical illustrations such as pie charts, histograms and bar graphs.
- Statistical infographics are commonly used when communicating data to the general public, for example during the COVID-19 pandemic.

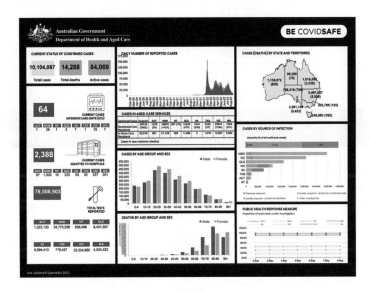

DISCUSSION

Observe the following statistical infographic on mortality rates due to selected causes for those aged 10–19 according to sex (based on data from 2019)

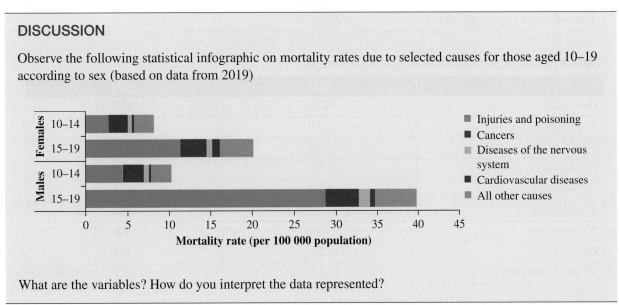

What are the variables? How do you interpret the data represented?

COMMUNICATING — COLLABORATIVE TASK: Creating a statistical infographic

Working in small groups, select one of the following tables and use the data it contains to create a statistical infographic. Justify the choice of the graphical representation you used.

TABLE 1 First Nations Australians by state and territory (data from the 2022 ABS Census)

	Count (no.)	% as a proportion of Australia	% as a proportion of state/territory
New South Wales	278 043	34.2	3.4
Victoria	65 646	8.1	1.0
Queensland	237 303	29.2	4.6
South Australia	42 562	5.2	2.4
Western Australia	88 693	10.9	3.3
Tasmania	30 186	3.7	5.4
Northern Territory	61 115	7.5	26.3
Australian Capital Territory	8 949	1.1	2.0
Australia	812 728	100.0	3.2

TABLE 2 First Nations Australians in capital city areas (data from the 2022 ABS Census)

Capital city	Person count within capital city	% of state or territory
Sydney	90 939	32.7
Melbourne	32 952	50.2
Brisbane	76 942	32.4
Adelaide	23 761	55.8
Perth	42 083	47.4
Hobart	11 216	37.2
Darwin	14 539	23.8
Canberra	8 908	99.5
Total living in capital city areas	301 330	37.1

TABLE 3 First Nations Australians by age (in percentage)

	2011	2021
0–4 years	12.3	10.6
5–14 years	23.6	22.1
15–24 years	19.3	18.5
25–34 years	13.1	14.5
35–44 years	12.3	10.8
45–54 years	9.7	10.1
55–64 years	5.9	7.6
65–74 years	2.6	4.2
75 years and over	1.2	1.7

Source: All data is from the Australian Bureau of Statistics website.

DISCUSSION

Statistical infographics are useful tools but the data they represent can be misunderstood, especially when the choice of graphical representation is not suitable. The data can also voluntarily be presented in a way that will mislead inattentive people. Consider the data in the following table.

Water used per capita and per country in 2009		
Country	Daily water used per capita (litres)	Yearly water used per country (10^9 m^3)
China	1165	598.1
United States	3794	444.3
India	1689	761.0
Australia	1821	16.13
Japan	1736	81.45

A first student has created the following infographic to represent this data.

Who are the biggest daily water consumers?

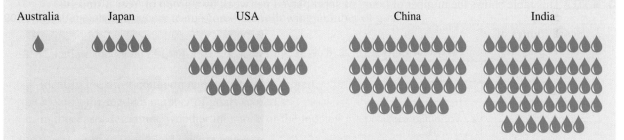

One of their classmates affirms that the infographic is misleading, and produces the following infographic instead.

Which country consumes the most water each year?

Which visual representation do you think best represent the data? How would you represent this data? Explain your choice.

 **Resources**

Interactivities Frequency tables (int-3816)
Column graphs (int-3817)

Exercise 10.4 Organising and displaying data

learn on

10.4 Quick quiz on	10.4 Exercise

Individual pathways

■ PRACTISE	■ CONSOLIDATE	■ MASTER
1, 3, 6, 9, 12	2, 5, 8, 10, 13	4, 7, 11, 14

Fluency

1. **WE10** In a suburb of roughly 1500 houses, a random sample of 40 households was surveyed to calculate the number of children living in each.
   The data were collected and recorded as follows.

   0, 3, 2, 4, 1, 2, 3, 2, 2, 2, 2, 1, 3, 4, 5, 2, 3, 1, 1, 1,
   0, 0, 2, 3, 4, 1, 3, 4, 2, 2, 0, 1, 2, 3, 2, 0, 2, 4, 5, 1

   a. Organise the data into a frequency table.
   b. Comment on the distribution of the data.
   c. Comment on the number of children per household in the suburb.

2. A quality control officer selected 25 boxes of smart watches at random from a production line.
   She tested every single smart watch and displayed the number of defective smart watches in each box as follows:

   1, 3, 2, 5, 2, 2, 1, 5, 2, 1, 2, 4, 3, 0, 5, 3, 2, 1, 3, 2, 1, 3, 4, 2, 1

   a. Comment on the sample.
   b. Organise the data into a frequency table.
   c. Comment on the distribution of the data.
   d. Comment on the population of smart watches.

3. **WE12** This table shows the number of hours of sport played per week by a group of Year 8 students.

Score (hours of sport played)	Frequency (f)
1– < 2	3
2– < 3	8
3– < 4	10
4– < 5	12
5– < 6	16
6– < 7	8
7– < 8	7
Total	64

   a. Draw a histogram to display the data.
   b. Comment on the shape of the graph.
   c. Discuss whether you feel this sample reflects the sporting habits of Year 8 students generally.

4. A block of houses in a suburb was surveyed to determine the size of each house (in m^2). The results are shown in the following table.

Size of house (m^2)	Frequency
100–<150	13
150–<200	18
200–<250	19
250–<300	17
300–<350	14
350–<400	11
Total	92

a. Draw a histogram to display the data.
b. Comment on the shape of the graph.
c. Discuss whether you feel this sample reflects the size of the houses in the suburb.

5. Forty people joined a weight-loss program. Their mass (in kg) was recorded at the beginning of the program and is shown in the frequency table.

a. Draw a histogram to display the data.
b. Comment on the shape of the graph.
c. Discuss whether you feel this sample reflects the mass of people in the community.

Class interval	Frequency
60–<70	2
70–<80	5
80–<90	9
90–<100	12
100–<110	7
110–<120	3
120–<130	2
Total	40

Understanding

6. **WE11&13** Forty people in a shopping centre were asked about the number of hours per week they spent watching TV. The result of the survey is shown as follows.

> 10, 13, 7, 12, 16, 11, 6, 14, 6, 11, 5, 14, 12, 8, 27, 17, 13, 8, 14, 10,
> 13, 7, 15, 10, 16, 8, 18, 14, 21, 28, 9, 12, 11, 13, 9, 13, 29, 5, 24, 11

a. Organise the data into class intervals of 5–<10 hours, and so on, and draw up a frequency table.
b. Draw a histogram to display the data.
c. Comment on the shape of the graph.
d. Discuss whether you feel this sample reflects the TV-viewing habits of the community.

7. The number of hours of sleep during school weeknights for a Year 8 class are recorded below.

$$6, \quad 9, \quad 7, \quad 8, \quad 7, \quad 8\frac{1}{2}, \quad 6\frac{1}{2}, \quad 8, \quad 7\frac{1}{2}, \quad 7\frac{1}{2},$$

$$8, \quad 8\frac{1}{2}, \quad 6\frac{1}{2}, \quad 8, \quad 8, \quad 7, \quad 7\frac{1}{2}, \quad 8, \quad 9, \quad 8$$

a. Organise the data into suitable class intervals and display it as a frequency table.
b. Display the data as a histogram.
c. Comment on the sleeping habits of the Year 8 students.
d. Discuss whether you feel these sample results reflect those of Year 8 students generally.

8. The amount of pocket money (in dollars) available to a random sample of 13-year-olds each week was found to be as shown below.

$$10, 15, 5, 4, 8, 10, 4, 15, 5, 6, 10, 6, 5, 10, 8, 10, 5, 10, 10, 6$$

a. Organise the data into class intervals of 0–<5, 5–<10 dollars, and so on, and display it as a frequency table.
b. Display the data as a histogram.
c. Comment on the shape of the histogram.
d. Discuss whether you feel these sample results reflect those of 13-year-olds generally.

Communicating, reasoning and problem solving

9. Show that the midpoint for the interval 12.5–<13.2 is 12.85.

10. The following histogram shows ages of patients treated by a doctor during a shift.

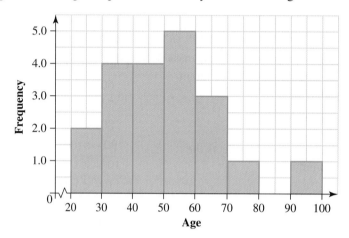

Complete a frequency table for the histogram.

11. The following data gives the results of testing the lives (in hours) of 100 torch batteries.

25,	36,	30,	34,	21,	40,	36,	46,	29,	38,
20,	41,	34,	45,	25,	40,	31,	39,	24,	45,
27,	44,	23,	35,	47,	49,	20,	37,	43,	26,
35,	28,	48,	30,	20,	36,	41,	26,	32,	42,
21,	31,	45,	42,	26,	37,	33,	24,	45,	38,
36,	43,	21,	34,	38,	35,	28,	41,	30,	22,
29,	32,	39,	25,	44,	21,	35,	38,	41,	35,
30,	23,	37,	43,	33,	34,	28,	39,	22,	31,
35,	42,	38,	27,	36,	46,	28,	34,	37,	29,
24,	30,	39,	44,	31,	24,	36,	28,	47,	21

a. Choose a suitable class interval for the given data, and present the results in a frequency distribution table.
b. Draw a histogram of the data.
c. Comment on the trends shown by the histogram.
d. Discuss whether you feel these results reflect those of the battery population.

12. A building company recorded the number of weekends during which their tradespeople needed to work over the course of one year, as shown in the table.

Class interval	Frequency
1–4	4
5–8	6
9–12	13
13–16	11
17–20	5
21–24	7
25–28	10

a. Identify how many tradespeople work at the company.
b. Determine the most common number of weekends worked.
c. Is it possible to determine the maximum number of weekends worked by a tradesperson? Explain.

13. The following data were collected on the number of times people go to the cinema per month.

4, 5, 7, 9, 1, 2, 5, 2, 4, 8, 3, 6, 2, 3, 8, 1, 1, 4,
5, 3, 3, 6, 1, 2, 7, 1, 3, 2, 2, 4, 10, 0, 1, 3, 4, 6

a. Organise the data into class intervals of 0–2, 3–5 and so on, and display it as a frequency table.
b. Draw a histogram to represent the data.
c. Determine how many people go to the cinema fewer than three times a month.
d. Determine how many people go to the cinema at least three times a month.
e. Is it reasonable to draw conclusions about the whole population based on this sample? Give reasons for your answer.

14. A random sample of 30 students in Year 9 undertook a survey to investigate the heights of Year 9 students. These were their measured heights (in cm).

146, 163, 156, 168, 159, 170, 152, 174, 156, 163, 157, 161, 178, 151, 148,
167, 162, 157, 166, 154, 150, 166, 160, 155, 164, 157, 171, 168, 158, 162

a. Organise the data into class intervals of 145– < 150 cm and so on, and display it as a frequency distribution table.
b. Draw a histogram displaying the data.
c. Reorganise the class intervals into 145– < 148 cm and so on, and construct a new frequency distribution table.
d. Draw a new histogram displaying the data in part c.
e. Comment on the similarities and differences between the two histograms.

LESSON
10.5 Measures of centre and spread

LEARNING INTENTION

At the end of this lesson you should be able to:
- determine the mean, median and mode of a set of data
- determine the range of a set of data
- identify possible outliers in a set of data and understand how they affect the mean, median and range.

▶ 10.5.1 Mean

eles-4446

- The **measures of centre** or **measures of location** give some idea of the average or middle of the data set.
- The two types of measures of centre used in interpreting data are the **mean** and **median**.
- The mean is the average of a set of scores, and is denoted by the symbol $\bar{x}$ (pronounced x bar).
- The mean can only be calculated for a set of numerical data.
- The value of the mean of a set of data is often not one of the given scores.

Determining the mean

To calculate the mean (or average) of a set of data, use the following formula.

$$\bar{x} = \frac{\text{sum of the data values (or scores)}}{\text{total number of data values}}$$

WORKED EXAMPLE 14 Calculating the mean

Jan's basketball scores were 18, 24, 20, 22, 14 and 12. Calculate his mean score, correct to 1 decimal place.

THINK	WRITE
1. Calculate the sum of the basketball scores.	Sum of scores $= 18 + 24 + 20 + 22 + 14 + 12$ $= 110$
2. Count the number of basketball scores.	Total number of scores $= 6$
3. Define the rule for the mean.	Mean $= \dfrac{\text{sum of the data values}}{\text{total number of data values}}$
4. Substitute the known values into the rule.	$\bar{x} = \dfrac{110}{6}$
5. Evaluate, rounding to 1 decimal place. *Note:* Jan's typical (or average) score per game is 18.3 points.	$= 18.333\,33\ldots$ $= 18.3$

- Sometimes calculations need to be performed from a frequency distribution table.
- Calculating the mean from a frequency table requires a different process than that used for raw data.

Determining the mean from a frequency table

To calculate the mean (or average) from a frequency table, use the following formula.

$$\overline{x} = \frac{\text{total of (frequency} \times \text{score) column}}{\text{total of frequency column}}$$

WORKED EXAMPLE 15 Calculating the mean from a frequency table

Calculate the mean of the frequency distribution data given below correct to 1 decimal place.

Score (x)	Frequency (f)
1	3
2	2
3	4
4	0
5	5

THINK

1. Copy and complete the frequency table and include an extra column called frequency $\times$ score ($f \times x$).

2. Enter the information into the third column. The score of 1 occurred 3 times. Therefore, $f \times x = 3 \times 1 = 3$.
   The score of 2 occurred 2 times. Therefore, $f \times x = 2 \times 2 = 4$.
   Continue this process for each pair of data.

3. Determine the total of the 'Frequency' column. This shows how many scores there are altogether.

4. Determine the total of the 'Frequency $\times$ score' column. This shows the sum of the values of all the scores.

5. Define the rule for the mean.

6. Substitute the known values into the rule.

7. Evaluate the answer to 1 decimal place.
   Note: The typical (or average) value of the set of data is 3.1.

WRITE

Score (x)	Frequency (f)	Frequency $\times$ score ($f \times x$)
1	3	$3 \times 1 = 3$
2	2	$2 \times 2 = 4$
3	4	$4 \times 3 = 12$
4	0	$0 \times 4 = 0$
5	5	$5 \times 5 = 25$
Total	14	44

$$\text{Mean} = \frac{\text{total of frequency} \times \text{score column}}{\text{total of frequency column}}$$

$$\overline{x} = \frac{44}{14}$$
$$= 3.142\,857\ldots$$
$$= 3.1$$

10.5.2 Median

eles-4447

- The median is the middle value if the data values are placed in numerical (ascending) order.
- The median can only be calculated for a set of numerical data.

Determining the median

The following formula determines the *position* of the median value of a set of scores in numerical order.

$$\text{Location of median} = \left(\frac{n+1}{2}\right)\text{th score, in a set of } n \text{ scores}$$

Note: This formula does not determine the median value. It simply locates its position in the data set.

- For sets of data containing an odd number of scores, the median will be one of the actual scores; for sets with an even number of scores, the median will be positioned halfway between the two middle scores.

WORKED EXAMPLE 16 Determining the median

Determine the median of each of the following sets of scores.

a. **10, 8, 11, 5, 17** b. **9, 3, 2, 6, 3, 5, 9, 8**

THINK

a. 1. Arrange the values in numerical (ascending) order.

2. Select the middle value.
Note: There is an odd number of scores: 5. Hence, the third value is the middle number or median. Alternatively, the rule $\frac{n+1}{2}$, where $n = 5$, gives the position of the median. The location of the median is $\left(\frac{5+1}{2} = 3\right)$; that is, the 3rd score.

3. Write the answer.

b. 1. Arrange the values in ascending order.

2. Select the two middle values.
Note: There is an even number of scores: 8. Hence, the fourth and fifth values are the middle numbers, or median. Again, the rule $\frac{n+1}{2}$ could be used to locate the position of the median.

3. Obtain the average of the two middle values (the fourth and fifth values).

4. Write the answer.

WRITE

a. 5, 8, 10, 11, 17

5, 8, ⑩, 11, 17

The median of the scores is 10.

b. 2, 3, 3, ⑤, ⑥, 8, 9, 9

$$\text{Location of median} = \frac{n+1}{2}$$
$$= \frac{8+1}{2}$$
$$= \frac{9}{2}$$
$$= 4.5\text{th value}$$
(i.e. between the fourth and fifth values)

$$\text{Median} = \frac{5+6}{2}$$
$$= \frac{11}{2}$$
$$= 5\frac{1}{2} \text{ (or 5.5)}$$

The median of the scores is $5\frac{1}{2}$ or 5.5.

▶ 10.5.3 Mode

eles-4448

- The mode is the most common score or the score with the highest frequency in a set of data. It is *not* considered to be a measure of centre.
- The mode measures the clustering of scores.
- Some sets of scores have more than one mode or no mode at all. There is no mode when all values occur an equal number of times.
- The mode can be calculated for both numerical and categorical data.

Determining the mode

- **The mode is the most common score or the score with the highest frequency.**
- **Some data sets have one unique mode, more than one mode or no mode at all.**

WORKED EXAMPLE 17 Determining the mode

Determine the mode of each of the following sets of scores.

a. **5, 7, 9, 8, 5, 8, 5, 6**   b. **10, 8, 11, 5, 17**   c. **9, 3, 2, 6, 3, 5, 9, 8**

THINK	WRITE
a. 1. Look at the set of data and circle any values that have been repeated.	a. ⑤, 7, 9, ⑧, ⑤, ⑧, ⑤, 6
2. Choose the values that have been repeated the most.	The number 5 occurs three times.
3. Write the answer.	The mode for the given set of values is 5.
b. 1. Look at the set of data and circle any values that have been repeated.	b. 10, 8, 11, 5, 17 No values have been repeated.
2. Answer the question. *Note:* No mode is not the same as a mode that equals 0.	The following set of data has no mode, since none of the scores has the highest frequency. Each number occurs only once.
c. 1. Look at the set of data and circle any values that have been repeated.	c. ⑨, ③, 2, 6, ③, 5, ⑨, 8
2. Choose the values that have been repeated the most.	The number 3 occurs twice. The number 9 occurs twice.
3. Write the answer.	The modes for the given set of values are 3 and 9.

WORKED EXAMPLE 18 Calculating measures of centre

The data from a survey asking people how many times per week they purchased takeaway coffee from a cafe are shown below.

2, 9, 11, 8, 5, 5, 5, 8, 7, 4, 5, 3

Use the data to calculate each of the following.

a. **The mean number of coffees purchased per week**
b. **The median number of coffees purchased per week**
c. **The modal number of coffees purchased per week**

▶

THINK	WRITE
a. 1. To calculate the mean, add all the values in the data set and divide by the total number of data values. There are 12 values in the data set.	**a.** $\bar{x} = \dfrac{\text{sum of all the values}}{\text{total number of values}}$ $= \dfrac{2+9+11+8+5+5+5+8+7+4+5+3}{12}$ $= \dfrac{72}{12}$ $= 6$
2. Write the answer.	The mean number of coffees purchased per week is 6.
b. 1. The median is the value in the middle position. There are 12 values in the data set, so the middle position is between the 6th and 7th values.	**b.** Location of median $= \dfrac{n+1}{2}$ $= \dfrac{12+1}{2}$ $= \dfrac{13}{2}$ $= 6.5$
2. Arrange the data set in order from lowest to highest. The 6th value is 5. The 7th value is 5.	2, 3, 4, 5, 5, ⑤, ⑤, 7, 8, 8, 9, 11 Median $= \dfrac{5+5}{2}$ $= \dfrac{10}{2}$ $= 5$
3. Write the answer.	The median number of coffees purchased per week is 5.
c. The mode is the most common value in the data set. The most common value is 5.	**c.** The modal number of coffees purchased per week is 5.

- If the shape of a distribution for a set of data is symmetrical, then the mean and median values will be the same. This implies that the average value and the middle score will be the same.

▶ 10.5.4 Measures of spread

eles-4449

- In analysing a set of scores, it is helpful to see not only how the scores tend to cluster, or how the middle of the set looks, but also how they spread or scatter.
- For example, two classes may have the same average mark, but the spread of scores may differ considerably.
- The **range** of a set of scores is the difference between the highest and lowest scores.
- The range is a **measure of spread**.

> **Determining the range**
>
> **To determine the range of a set of data, use the following formula.**
>
> **Range = highest score − lowest score**

- The range can only be calculated for a set of numerical data.

Calculate the range of the following sets of data.

a. 7, 3, 5, 2, 1, 6, 9, 8

b.

Score (x)	Frequency (f)
7	1
8	3
9	5
10	2

THINK

a. 1. Obtain the highest and lowest values.

2. Define the range.

3. Substitute the known values into the rule.

4. Evaluate.

5. Write the answer.

b. 1. Obtain the highest and lowest values.
   Note: Consider the values (scores) only, not the frequencies.

2. Define the range.

3. Substitute the known values into the rule.

4. Evaluate.

5. Write the answer.

WRITE

a. Highest value $= 9$
   Lowest value $= 1$
   Range $=$ highest value $-$ lowest value
   $= 9 - 1$
   $= 8$
   The set of values has a range of 8.

b. Highest value $= 10$
   Lowest value $= 7$
   Range $=$ highest value $-$ lowest value
   $= 10 - 7$
   $= 3$
   The frequency distribution table data has a range of 3.

- Although the range identifies both the lowest and highest scores, it does not provide information on how the data is spread out between those values.
- In most cases, the spread of data between the lowest and highest scores is not uniform.

COMMUNICATING — COLLABORATIVE TASK: Analyse this!

Equipment: Data collected from the survey in *Collaborative Task: Designing a survey*, from lesson 10.2.

1. a. For your collected data, calculate the mean, median and mode for each of the questions you asked.
   b. Are these statistics appropriate for the type of data you collected in each question? Think about what these values mean for your data and whether the values you are achieving are appropriate for the type of data you have collected.
2. Choose an appropriate visual representation for the data you collected for each question. What do you need to take into consideration before selecting the visual representation?
3. a. Select the data from one question to represent as grouped data using class intervals.
   b. Calculate the mean, median and modal class for the group data.
   c. Compare the values of mean, median and mode for the grouped data with those of the ungrouped data. What do you notice? Suggest a reason for anything you noticed.
4. As a class, discuss any similarities and differences you found between the statistics for your grouped and ungrouped data.

10.5.5 Clusters, gaps and outliers

eles-4467

- Clusters, gaps and outliers in the data set can be seen in histograms.
- A cluster is a grouping of data points that are close together.
- Consider the following two histograms.

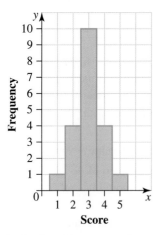
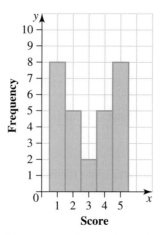

- In the first histogram, the mean, median and mode of the data set is 3 and the data is clustered around the mean.
- In the second histogram, the data is not clustered and there are two modes — one at either end of the distribution. The mean and median of this data set are also 3, but the modes are now 1 and 5.
- Data can also be clustered at either the lower end or the upper end of a distribution, as shown in the following histograms.

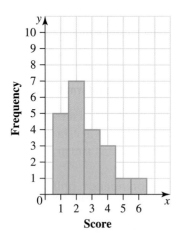
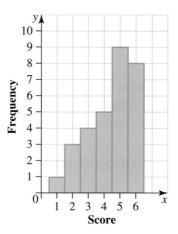

- Data values that do not follow the general pattern of the distribution could be classified as **outliers**.
- Looking at this histogram, we can see that data value 15 does not follow the general pattern of the distribution and is possibly an outlier.
- Clusters, gaps and outliers can also be identified in other data representations, including stem-and-leaf plots and dot plots.

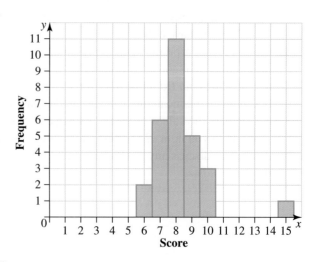

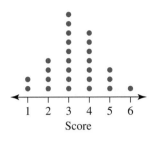

Key: 1|2 = 12

Stem	Leaf
1	2 4 6
2	0 1 3 7
3	2 3 4 4 6 7 8
4	0 1 5
5	
6	
7	8

The effect of outliers on measures of centre and spread

- The presence of one or more outliers may have a considerable effect on the measures of centre and spread of a particular set of data.

WORKED EXAMPLE 20 Determining the effect of an outlier

A netball team scored the following points in 10 games:
$$17, 23, 31, 19, 50, 29, 16, 23, 30, 32$$
a. Calculate the mean, median, mode and range of the team scores.
b. The following Saturday, the regular goal shooter was ill, and Lauren, who plays in a higher division, was asked to play. The team's score for that game was 200.
   Recalculate the mean, median, mode and range after 11 games.
c. Comment on the similarities and differences between the two sets of summary statistics.

THINK	WRITE
a. 1. Rearrange the scores in numerical order.	a. 16, 17, 19, 23, 23, 29, 30, 31, 32, 50
2. Calculate the mean.	$\bar{x} = \dfrac{16 + 17 + 19 + 23 + 23 + 29 + 30 + 31 + 32 + 50}{10} = 27$
3. Calculate the median.	Median $= \dfrac{23 + 29}{2} = 26$
4. Calculate the mode.	23 appears the most times.
5. Calculate the range.	Range $= 50 - 16 = 34$
6. Write the answer.	Mean $= 27$
	Median $= 26$
	Mode $= 23$
	Range $= 34$
b. 1. Rearrange the 11 scores in numerical order.	b. 16, 17, 19, 23, 23, 29, 30, 31, 32, 50, 200
2. Calculate the mean.	$\dfrac{16 + 17 + 19 + 23 + 23 + 29 + 30 + 31 + 32 + 50 + 200}{11} = 43$
3. Calculate the median.	The middle value of the data set is 29.
4. Calculate the mode.	23 appears the most times.
5. Calculate the range.	$200 - 16 = 184$
6. Write the answers.	Mean $= 43$
	Median $= 29$
	Mode $= 23$
	Range $= 184$

| c. Compare the two sets of summary statistics. | c. The inclusion of an extreme value or outlier has dramatically increased the mean and the range of the data, marginally increased the median and left the mode unchanged. |

Note: The important point to learn from Worked example 20 is that when a set of data includes extreme values, the mean may not be truly representative of the data.

Which measure of centre is most useful?

- It is important to know which measure of centre will be most useful in a given situation.
  - The mean is appropriate when no extreme values or outliers distort the information.
  - The median is appropriate when outliers are present.
  - The mode is appropriate when the most common result is significant.

 Resources

 Video eLesson Mean and median (eles-1905)

Interactivities Mean (int-3818)
Median (int-3819)
Mode (int-3820)
Range (int-3822)
Outliers (int-3821)

Exercise 10.5 Measures of centre and spread

learn on

| 10.5 Quick quiz on | 10.5 Exercise |

Individual pathways

| ■ PRACTISE | ■ CONSOLIDATE | ■ MASTER |
| 1, 2, 6, 11, 12, 13, 17, 21, 23, 26, 27, 30 | 3, 4, 8, 9, 15, 18, 19, 22, 24, 28, 31 | 5, 7, 10, 14, 16, 20, 25, 29, 32, 33 |

Fluency

1. **WE14** Mai's basketball scores were 28, 25, 29, 30, 27 and 22. Calculate her mean score correct to 1 decimal place.

2. Calculate the mean (average) of each set of the following scores.
   Give the answers correct to 2 decimal places.

   a. 1, 2, 3, 4, 7, 9
   b. 2, 7, 8, 10, 6, 9, 11, 4, 9
   c. 3, 27, 14, 0, 2, 104, 36, 19, 77, 81
   d. 4, 8.4, 6.6, 7.0, 7.5, 8.0, 6.9

3. Francesca's soccer team has the following goals record this season:

$$2, 0, 1, 3, 1, 2, 4, 0, 2, 3$$

   a. State the total number of goals the team has scored.
   b. State the number of games the team has played.
   c. Calculate the team's average score.

4. **MC** Matteo's athletics coach timed 5 consecutive 200-metre training runs. He recorded times of 25.1, 23.9, 24.8, 24.5 and 27.3 seconds. His mean 200-metre time (in seconds) is:

   A. 24.60   B. 25.20
   C. 25.12   D. 25.42

5. Two Year 8 groups did the same Mathematics test. Their results out of 10 were:

   Group A: 5, 8, 7, 9, 6, 7, 8, 5, 4, 2
   Group B: 5, 6, 4, 5, 9, 7, 8, 8, 9, 7

   Determine which group had the highest mean.

6. **WE15** Calculate the mean of this frequency distribution, correct to 2 decimal places.

Score (x)	Frequency (f)
1	4
2	3
3	6
4	1
5	0

7. Calculate the mean of this frequency distribution, correct to 2 decimal places.

Score (x)	Frequency (f)
6	2
7	8
8	3
9	6
10	2

8. **WE16a** Calculate the median of the following scores.

   a. 5, 5, 7, 12, 13
   b. 28, 13, 17, 21, 18, 17, 14

9. **WE16b** Calculate the median of each of the following sets of scores.

   a. 52, 46, 52, 48, 52, 48
   b. 1.5, 1.7, 2.0, 1.8, 1.5, 1.7, 1.8, 1.9

10. **WE17** For each set of scores in questions **8** and **9**, state the mode.

Questions **11** and **12** refer to the following set of scores.

$$1, 1, 1, 4, 4, 5, 5, 6, 3, 3, 7, 6, 5, 4, 6, 2, 1, 8$$

11. **MC** The median of the given scores is:

A. 1
B. 4.5
C. 4
D. 5

12. **MC** The mode of the given scores is:

A. 5
B. 4
C. 3
D. 1

13. **WE19a** Calculate the range of the following scores.

a. 5, 5, 7, 12, 13
b. 28, 13, 17, 21, 18, 17, 14
c. 2, 52, 46, 52, 48, 52, 48
d. 4, 1.5, 1.7, 2.0, 1.8, 1.5, 1.7, 1.8, 1.9

14. **WE19b** Determine the range of the following sets of data.

a.

Score (x)	Frequency (f)
6	1
7	5
8	10
9	7
10	3

b.

Score (x)	Frequency (f)
1	7
2	9
3	6
4	8
5	10
6	10

c.

Score (x)	Frequency (f)
5	1
10	5
15	10
20	7

d.

Score (x)	Frequency (f)
110	2
111	2
112	2
113	3
114	3

15. Determine the range of each of the following data sets.

a. Key: $1\,|\,8 = 18$

Stem	Leaf
1	1 2 7 8 9
2	2 8
3	1 3 7 9
4	0 1 2 6

b. Key: $24\,|\,7 = 247$

Stem	Leaf
24	2 7
25	2 4 6 6 8
26	0 1 3 5 9
28	5 6 6 8

c. Key: $17\,|\,4 = 174$

Stem	Leaf
15	6 2 4
16	8 6 1 3 9
17	0 2 1 8 6 7 3 4
18	4 1 5 2 7 1

16. Calculate the mean, median, mode and range of the data shown in the following histogram.

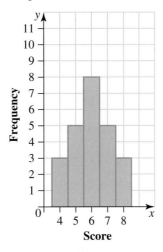

Understanding

17. A third Year 8 group had the following results in the same test as in question **5**:

Group C: 5, 7, 8, 4, 6, 8, 5, 9, 8

a. Calculate the average score of this group, correct to 1 decimal place.
b. Determine the score that a tenth student (who was originally absent) would need to achieve to bring this group's average to 7.

18. A survey of the number of occupants in each house in a street gave the following data:

2, 5, 1, 6, 2, 3, 2, 4, 1, 2, 0, 2, 3, 2, 4, 5, 4, 2, 3, 4

Prepare a frequency distribution table with an $f \times x$ column and use it to calculate the mean number of people per household.

19. The mean of 5 scores is 7.2.

a. Calculate the sum of the scores.
b. If four of the scores are 9, 8, 7 and 5, determine the fifth.

20. Over 10 matches, a soccer team scored the following number of goals:

2, 3, 1, 0, 4, 5, 2, 3, 3, 4

a. Identify the most common number of goals scored.
b. Identify the median number of goals scored.
c. In this case, determine whether the mode or the median give a score that shows a typical performance.

21. **WE18** The following scores represent the number of muesli bars sold in a school canteen each day over two weeks:

$$54, 64, 51, 58, 56, 59, 10, 34, 48, 56$$

a. Calculate the mean.
b. Calculate the median.
c. Calculate the mode.
d. Of the mean, median and mode, explain which best represents a typical day's sales at the school canteen.

22. A small business pays the following annual wages (in thousands of dollars) to its employees:

$$18, \ 18, \ 18, \ 18, \ 26, \ 26, \ 26, \ 40, \ 80$$

a. Identify the mode of the distribution.
b. Identify the median wage.
c. Calculate the mean wage.
d. Explain which measure you would expect the employees' union to use in wage negotiations.
e. Discuss which measure the boss might use in such negotiations.

23. The following histogram shows the distribution of a set of scores.

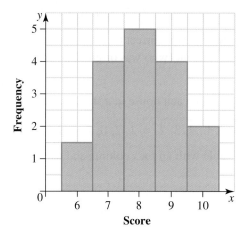

a. Identify the mode.
b. Determine whether there are any outliers in the data set. If so, calculate their value(s).

24. Consider the following distribution. Determine the modal class.

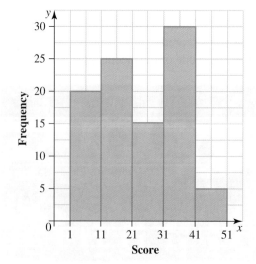

25. **MC** Select the option from the following that shows data that is clustered.

A.

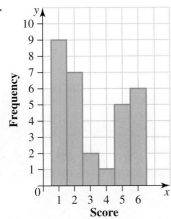

B.

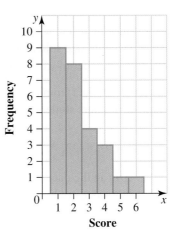

C.

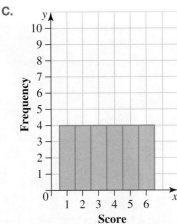

D.

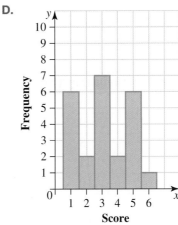

26. **WE20** A rugby team scored the following numbers of points in 10 games:

$$24, 18, 33, 29, 22, 16, 38, 30, 26, 30$$

a. Calculate the mean, median, mode and range of the team scores.
b. The following Saturday, the team played a side who had lost their previous 35 games and who were on the bottom of the ladder. The team's score for that game was 86. Recalculate the mean, median, mode and range after 11 games.
c. Comment on the similarities and differences between the two sets of summary statistics.

Communicating, reasoning and problem solving

27. These data show the number of hours Year 8 students used a computer in a particular week:

$$5, 3, 6, 7, 3, 5, 2, 5, 2, 3, 6, 7$$

a. Calculate the mean, median and mode.
b. Comment on the value of the mean compared with the median.
c. Comment on the value of the mean compared with the mode.
d. Explain which is the best indicator of the centre of the data set.

28. Determine which measure of centre is most appropriate to use in the following situations. Explain your answers.

   a. Analysing property values in different suburbs of a capital city
   b. Determining the average shoe size sold at a department store
   c. Determining the average number of tries scored over a season of rugby league

29. The following scores represent data from an online survey asking about the average number of hours students spent exercising each week:

$$3, 5, 1, 4, 0, 8, 23, 4, 2, 0, 2, 6$$

   a. Identify the potential error in the data set.
   b. Explain whether this data value could have been a genuine outlier.

30. Calculate the mean number of books presented in this frequency table, using the midpoint of each interval as the x-value for the interval.

Number of books (x)	Frequency (f)
1–15	3
16–30	9
31–45	8
46–60	11
61–75	10
76–90	14
91–105	15
106–120	18

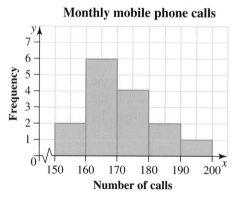

31. Identify the mean, median and mode in the following paragraph:
It was an amazing game of cricket today. The winning team hit more sixes than any other number of runs. This meant that, even though the middle value of runs per over was 3, the sixes brought the average up to about five runs per over. What an incredible game!

32. The mean of 5 different test scores is 15. Evaluate the largest and smallest possible test scores, given that the median is 12. All test scores are whole numbers. Justify your answer.

33. Evaluate the mean number of calls made on mobile phones in the month shown in the graph, using the midpoint of each interval to represent the number of phone calls per month.

Monthly mobile phone calls

LESSON
10.6 Analysing data

10.6.1 Describing data sets

eles-6223

- Data sets can be described as uniform, when they have no mode, **unimodal** when they have one mode, **bimodal** when they have two modes, and **multimodal** when they have multiple modes.
- The shape and distribution of a data set can be **symmetrical**, **negatively skewed** or **positively skewed**.
- For a unimodal data set, the **skewness** is a measure of the asymmetry of the data set.
- If a unimodal data set is symmetrical, its skewness is zero. This happens when the mean, the median and the mode have the same value.
- If the mean and the median of a unimodal data set are less than its mode, then the data set is asymmetrical and negatively skewed.
- If the mean and the median of a unimodal data set are greater than its mode, then the data set is asymmetrical and positively skewed.

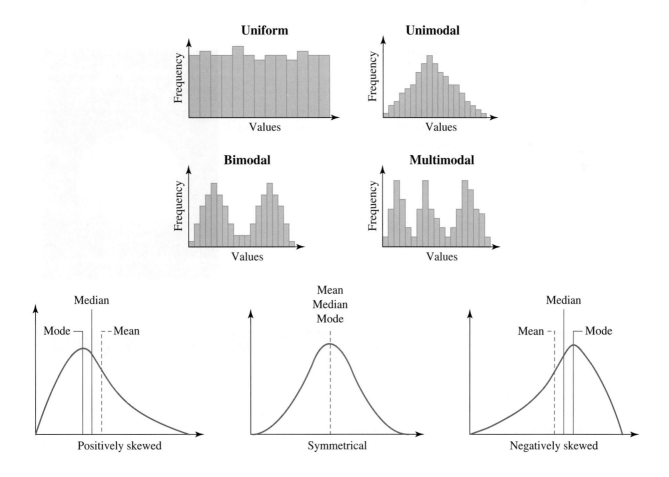

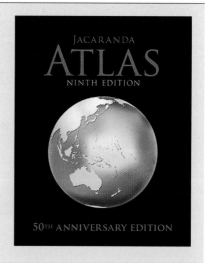

▶ 10.6.2 Analysing data sets

eles-4450

- The mean, median, mode and range are collectively known as **summary statistics** and play an important role in analysing data.
- To analyse a data set:
  - calculate the measures of centre — mean and median
  - determine the mode
  - calculate the spread — range
  - construct frequency tables and histograms.
- Remember:
  - the data comes from surveys of samples of the population
  - what each statistical measure gives.

Statistical measures	Definition and purpose
Mode	The most common score or category. It tells us nothing about the rest of the data. Data may have no mode, one mode or more than one mode.
Median	The score in the exact middle of the values placed in numerical order. It provides information about the centre of the distribution. It tells us nothing about the rest of the data. It is unaffected by exceptionally large or small scores (outliers).
Mean	The typical (or average) score expected. It can be calculated as the sum of all the scores divided by the number of scores and is affected by exceptionally large or small scores (outliers).
Range	The difference between the highest score and the lowest score. It shows how far the scores are spread apart. It is particularly useful when combined with the mean or the median. It is affected by outliers.

ACTIVITY: Analysing data from other subjects

Look for data from one of your other school texts.

For example, your Geography text or atlas may contain information about sustainable food production or population statistics. Calculate and compare sample summary statistics for the data you have found.

WORKED EXAMPLE 21 Identifying the statistical measure

Explain which statistical measure is referred to in these statements.

a. **The majority of people surveyed prefer Activ-8 sports drink.**
b. **The ages of fans at the Rolling Stones concert varied from 8 to 80.**
c. **The average Australian family has 2.5 children.**

THINK	WRITE
a. 1. Write the statement and highlight the keyword(s).	**a.** The **majority** of people surveyed prefer Activ-8 sports drink.
2. Relate the highlighted word to one of the statistical measures.	Majority implies most, which refers to the mode.
3. Answer the question.	This statement refers to the mode.
b. 1. Write the statement and highlight the keyword(s).	**b.** The ages of fans at the Rolling Stones concert **varied** from 8 to 80.
2. Relate the highlighted word to one of the statistical measures.	The statement refers to the range of fans' ages at the concert.
3. Answer the question.	This statement refers to the range.
c. 1. Write the statement and highlight the keyword(s).	**c.** The **average** Australian family has 2.5 children.
2. Relate the highlighted word to one of the statistical measures.	The statement deals with surveying the population (census) and finding out how many children are in each family.
3. Answer the question.	This statement refers to the mean.

▶ ## 10.6.3 Using a sample to predict the properties of a population

eles-4451

- Once survey data has been collated and analysed, the data set can be used to predict the characteristics of the population from which it was taken. Consider this example.

WORKED EXAMPLE 22 Making predictions

The 153 students in Year 8 all sat for a 10-question multiple-choice practice test for an upcoming exam. A random sample of the results of 42 of the students gave this distribution.

Score (x)	Frequency (f)
1	2
2	3
3	6
4	7
5	11
6	8
7	4
8	0
9	0
10	1

a. **Calculate the mean mark, correct to 1 decimal place.**
b. **Determine the median mark.**
c. **Give the modal mark.**
d. **Determine which measure of centre best represents the data.**
e. **Comment on any prediction about the properties of the population from this sample.**

▶

THINK

a. 1. Add a third column called $f \times x$. Multiply the frequency by its corresponding mark in each row to complete the column.

2. Determine the totals of the frequency column and the $f \times x$ column.

3. Define the rule for the mean.

4. Substitute the known values into the rule and evaluate, giving your answer correct to 1 decimal place.

b. The median is the score in the middle; that is, the $\left(\dfrac{42+1}{2}\right)$th score — the average of the 21st and 22nd score. Add a cumulative frequency column to the original frequency distribution. (Find the total frequency at that point for each mark.) Look in the cumulative frequency column to see where the 21st and 22nd scores lie.

c. The modal mark is the one that occurs most frequently. Look for the one with the highest frequency.

WRITE

a.

Mark (x)	Frequency (f)	$f \times x$
1	2	2
2	3	6
3	6	18
4	7	28
5	11	55
6	8	48
7	4	28
8	0	0
9	0	0
10	1	10
Total	42	195

$$\text{Mean} = \frac{\text{total of } (f \times x) \text{ column}}{\text{total of frequency column}}$$

$$= \frac{195}{42}$$

$$= 4.6$$

b.

Mark (x)	Frequency (f)	Cumulative frequency
1	2	2
2	3	$2+3=5$
3	6	$5+6=11$
4	7	$11+7=18$
5	11	$18+11=29$
6	8	$29+8=37$
7	4	$37+4=41$
8	0	$41+0=41$
9	0	$41+0=41$
10	1	$41+1=42$
Total	42	42

The 21st and 22nd scores are both 5. The median is 5.

c. The mode is 5.

d. Compare the results for the mean, median and mode. Look for similarities and differences.

d. The mean is 4.6, and the median is 5. It seems that either of these measures would be appropriate to use as a measure of centre of the data.

However, check the mark of 10 as it could be a possible outlier. When the mark of 10 is disregarded, the mean is calculated to be $185 \div 41 = 4.5$.

Since the value of the mean did not change significantly after removing the mark of 10, we can be safe in concluding that the mark of 10 is not an outlier.

So, the mean or median could be used as a measure of centre of the data.

e. Consider whether these results from the sample would reflect those of the population.

e. It seems likely that these results would reflect those of the whole population. The sample is random and of sufficient size. The one perfect score of 10 indicates that there would be a few students with full marks, and at least half the students passed the test.

- It is important to note that summary statistics may vary from sample to sample even though they are taken from the same population.

  For example, if you collect data on heights of students and one sample consisted only of boys and a second sample only of girls, the statistical measures would vary significantly.

Using a spreadsheet to calculate summary statistics

- Spreadsheets such as Excel can calculate statistical measures.
- Enter the data values into a column as shown in column B, rows 2 to 9, in the spreadsheet shown.
- To calculate the mean, use the formula '=AVERAGE(' and then select all of the cells containing your data. Close the brackets and the mean will be calculated.
  Type '=**AVERAGE(B2:B9)**' into cell B11, then press ENTER.
- To calculate the median, use the formula '=MEDIAN(' and then select all of the cells containing your data. Close the brackets and the median will be calculated.
  Type '=**MEDIAN(B2:B9)**' into cell B12, then press ENTER.
- To calculate the mode, use the formula '=MODE(' and select all of the cells containing your data. Close the brackets and the mode will be calculated.
  Note: If there is more than one mode, this method will only display one of the modes, so double check the data set.
  Type '=**MODE(B2:B9)**' into cell B13, then press ENTER.
- To calculate the range, use the formula '=MAX(' and then select all of the cells containing your data. Close the brackets and type '−MIN(' and again select all of the cells containing your data. Close the brackets and the range will be calculated.
  Type '=**MAX(B2:B9)−MIN(B2:B9)**' into cell B14, then press ENTER.

	A	B
1		Scores
2		5
3		10
4		7
5		5
6		8
7		12
8		6
9		11
10		
11	Mean	8
12	Median	7.5
13	Mode	5
14	Range	7

Exercise 10.6 Analysing data

learnon

| 10.6 Quick quiz on | 10.6 Exercise |

Individual pathways

| ■ PRACTISE | ■ CONSOLIDATE | ■ MASTER |
| 1, 2, 4, 6, 11, 14 | 3, 5, 8, 12, 15 | 7, 9, 10, 13, 16 |

Fluency

1. **WE21** Explain which statistical measure is referred to in these statements.
   a. There was a 15 °C temperature variation during the day.
   b. Most often you have to pay $79.95 for those sports shoes.
   c. The average Australian worker earns about $1659 per week.
   d. A middle-income family consisting of 2 adults and 2 children earns about $116 600 per annum.

2. **WE22** The following frequency table shows the results of a random sample of 15 students (from a class of 30) who sat for a 10-question multiple-choice test.

Score (x)	Frequency (f)
4	1
5	2
6	5
7	4
8	3
Total	15

   a. Calculate the mean mark.
   b. Determine the median mark.
   c. Give the modal mark.
   d. Which measure of centre best represents the data?
   e. Comment on any prediction of properties of the population from this sample.

3. Consider the following frequency distribution tables.

a.

Score (x)	Frequency (f)
1	4
2	3
3	2
4	1
5	0

b.

Score (x)	Frequency (f)
6	2
7	8
8	3
9	4
10	2

For each one:

i. calculate the mean score to 1 decimal place
ii. determine the median score
iii. identify the modal score
iv. indicate which measure of centre best describes the distribution.

4. Consider the following stem plots.

a. Key: $1\,|\,0 = 10$

Stem	Leaf
1	0 2
2	1 3 3 5
3	
4	4

b. Key: $10\,|\,0 = 100$

Stem	Leaf
10	0
11	0 2 2 2
12	0 4 6 6
13	3

For each one:

i. calculate the mean score to 1 decimal place
ii. determine the median score
iii. identify the modal score
iv. indicate which measure of centre best describes the distribution.

5. Consider the following dot plots.

a.

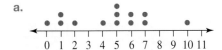

b.

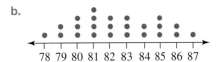

For each one:

i. calculate the mean score to 1 decimal place
ii. determine the median score
iii. identify the modal score
iv. indicate which measure of centre best describes the distribution.

Understanding

6. Three different samples looked at the average yearly incomes of NSW households.
The figures given are in thousands of dollars (for example $97 000) and have been rounded to the nearest whole number.

Sample 1: 97, 135, 52, 106, 189, 158, 70, 81, 122, 69
Sample 2: 102, 131, 85, 204, 77, 85, 114, 90, 111, 126
Sample 3: 66, 89, 110, 90, 173, 77, 129, 166, 256, 98

a. Calculate the mean, median, mode and range for each sample.
b. Compare the summary statistics for the three different samples.

7. A survey of the number of people living in each house in a street produced the following data:

$$2, 5, 1, 6, 2, 3, 2, 1, 4, 3, 4, 3, 1, 2, 2, 0, 2, 4$$

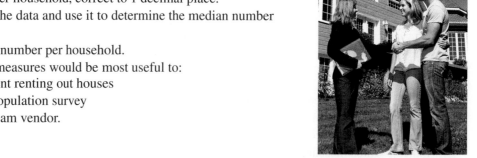

   a. Display the data as a frequency table and determine the average (mean) number of people per household, correct to 1 decimal place.
   b. Draw a dot plot of the data and use it to determine the median number per household.
   c. Identify the modal number per household.
   d. State which of the measures would be most useful to:
      i. a real-estate agent renting out houses
      ii. a government population survey
      iii. a mobile ice-cream vendor.

8. The contents of 20 packets of matches were counted after random selection. The following numbers were obtained:

$$138, 139, 139, 141, 137, 140, 137, 141, 139, 142,$$
$$140, 141, 141, 139, 141, 138, 139, 140, 141, 138$$

   a. Construct a frequency distribution table for the data.
   b. Determine the mode, median and mean of the distribution. Give answers correct to 1 decimal place where necessary.
   c. Comment on which of the three measures best supports the manufacturer's claim that there are 140 matches per box.

9. A class of 26 students had a median mark of 54 in Mathematics; however, no-one actually obtained this result.
   a. Explain how this is possible.
   b. Explain how many students must have scored below 54.

10. A soccer team had averaged 2.6 goals per match after 5 matches. After their sixth match, the average had dropped to 2.5. How many goals did they score in that latest match?

Communicating, reasoning and problem solving

11. A tyre manufacturer selects 48 tyres at random from the production line for testing. The total distance travelled during the safe life of each tyre is shown in the table.

Distance in km (×1000)	46	50	52	56	78	82
Number of tyres	4	12	16	10	4	2

   a. Calculate the mean, median and mode.
   b. Discuss which measure best describes 'average' tyre life. Explain your answer.
   c. Recalculate the mean with the 6 longest-lasting tyres removed. By how much is it lowered?
   d. If you selected a tyre at random, determine the distance it would be most likely to last.
   e. In a production run of 10 000 tyres, determine how many could be expected to last for a maximum of 50 000 km.
   f. As the manufacturer, explain for what distance you would be prepared to guarantee your tyres.

12. Read the following paragraph and explain what statistics are represented and what they mean.

It's been an exciting day at the races today. There was a record fast time of 38 seconds, and also a record slow time of 4 minutes and 52 seconds. We had an unbelievable number of people who ran the race in exactly 1 minute. Despite this, the average time was well over 2 minutes, due to the injury of a few runners.

13. If you take more than one sample from the same population, explain why the summary statistics will vary from sample to sample.

14. The following graph displays the movement of the price of BankSave shares over a 30-day period.

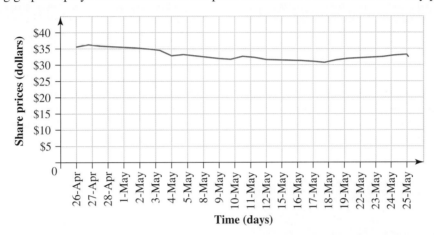

The closing price of the shares after 22 days of trading, rounded to the nearest dollar, were:

$$35, 36, 36, 35, 35, 34, 33, 33, 32, 32, 33,$$
$$32, 32, 31, 31, 31, 32, 32, 33, 33, 33, 33$$

Using the rounded amounts, calculate the mean (correct to 2 decimal places), median, mode and range of the share prices.

15. At a preview cinema session, the ages of the viewers were recorded and displayed in a stem-and-leaf plot. For this data, evaluate the:

   a. range
   b. mean
   c. median
   d. mode.

Key: 1 | 6 = 16 years

Stem	Leaf
1	5 6 7 7 8 9 9
2	1 2 4 8 8
3	0 1 1 1 5
4	2 3
5	3

16. The number of goals a netballer scored in the 12 games of a season was as follows:

$$1, 1, 1, 1, 2, 2, 2, 3, 3, 3, 8, 12$$

A local newspaper reporter asked the netballer what her average was for the season.

   a. Explain which measure of centre (mean, median or mode) the netballer should give the reporter as her 'average' so that the value of the average is as high as possible.
   b. Explain which measure of centre you would choose to best describe the 'average' number of goals the netballer scored each game.

LESSON
10.7 Review

10.7.1 Topic summary

Collecting data

- Data can be collected via:
  - observation
  - survey
  - experiment.
- A census is a survey of the population.
- Due to time and cost, samples are often surveyed.
- Samples should be randomly selected and the size $= \sqrt{\text{population size}}$.

Organising data

- Data can be displayed in many ways, such as:
  - frequency tables
  - histograms
  - spreadsheets.
- Where the range of values is large, the data may be grouped in class intervals, such as 5 or 10.
- Outliers (data points that differ significantly from the others) may be excluded from the analysis.

DATA

Primary and secondary data

- Primary data is data you have collected.
- Methods of collecting primary data include observation, measurement, survey, experiment and simulation.
- Secondary data is data that has been collected by someone else.
- Sources of secondary data include magazines, journals, videos, television and websites.

Measures of spread

- The range is a measure of spread. It is the difference between the highest and lowest values in the data set, including outliers.

  Range = highest value − lowest value
- The range is greatly affected by outliers.

Describing data sets

- Data sets can be described as uniform, unimodal, bimodal or multimodal.
- A unimodal data set can be symmetrical, positively skewed or negatively skewed.

Measures of centre

- The measures of centre are the mean and median.
- The mean, $\bar{x}$, is the average of the data set.

$$\bar{x} = \frac{\text{sum of data values}}{\text{total number of data values}}$$

Or, if the data is from a frequency table:

$$\bar{x} = \frac{\text{total of (frequency} \times \text{score) column}}{\text{total of frequency column}}$$

- The median is the middle value of the data set.

$$\text{Location of median} = \left(\frac{n+1}{2}\right) \text{th value}$$

- The mode is the most common value or value with the highest frequency.
- The mean is affected by outliers, whereas the median is not.

10.7.2 Project

Analysing the English language

The English alphabet contains 26 letters that combine to form words. Have you ever wondered why some letters appear more often than others?

Scrabble® is a game that allows players to form interlocking words in a crossword style. The words are formed using lettered tiles that carry a numerical value. Players compete against each other to form words of the highest score until all the tiles have been used.

The following table displays the letter distribution and the value of each letter tile in the game of Scrabble.

Letter	Number of tiles	Letter score	Letter	Number of tiles	Letter score	Letter	Number of tiles	Letter score	Letter	Number of tiles	Letter score
A	9	1	B	2	3	C	2	3	D	4	2
E	12	1	F	2	4	G	3	2	H	2	4
I	9	1	J	1	8	K	1	5	L	4	1
M	2	3	N	6	1	O	8	1	P	2	3
Q	1	10	R	6	1	S	4	1	T	6	1
U	4	1	V	2	4	W	2	4	X	1	8
Y	2	4	Z	1	10	Blank	2	0			

1. State which letters are most common in Scrabble and which are least common.
2. What do you notice about the relationship between the number of tiles used for each letter and their letter score?

This passage about newspapers and magazines is taken from an English textbook.

Who reads news?

N**ewspapers and magazines** are produced for different categories of readers. These are known as the 'target audience'. Particular groups of readers buy certain types of newspapers and magazines because these publications have content which interests them as a reader. Magazines and newspapers cover different types of news; for example, a local paper contains information on local issues whereas a magazine like *Movie* has all the latest news on films and film stars. General newspapers like *The Age* and *The Australian* try to appeal to a broader audience by including sections on various topics such as sport, business and property.

3. Complete a frequency table for the distribution of letters in this passage.
4. How do the results from this frequency table compare to the Scrabble game's frequency table?

Knowing which letters in the English language are most common can be very helpful when we are trying to solve coded messages.

The following paragraph is written in code.

> Tnlmktebt bl ehvtmxw bg max lhnmaxkg axfbliaxkx tgw bl ftwx ni hy lbq lmtmxl tgw mph mxkkbmhkbxl. Max vtibmte hy Tnlmktebt bl Vtguxkkt.

5. Study the coded message carefully. Use an appropriate method to decode the message.
6. Explain the strategies you used to decode the message.

 Resources

Interactivities Crossword (int-2760)
  Sudoku puzzle (int-3191)

Exercise 10.7 Review questions

learn_{on}

Fluency

1. For each of the following statistical investigations, state whether a census or a survey has been used.
   a. The average price of petrol in Sydney was estimated by averaging the price at 40 petrol stations.
   b. The Australian Bureau of Statistics has every household in Australia complete an online questionnaire or information form every 5 years.
   c. The performance of a cricketer is measured by looking at his performance in every match he has played.
   d. Public opinion on an issue is sought by a telephone poll of 2000 homes.

2. MC Identify which of the following is an example of a census.
   A. A newspaper conducts an opinion poll of 2000 people.
   B. A product survey of 1000 homes is carried out to determine what brand of washing powder is used.
   C. Every 200th jar of Vegemite is tested to see if it is the correct mass.
   D. A federal election is held.

3. MC Identify which of the following is an example of a random sample.
   A. The first 50 students who arrive at school take a survey.
   B. Fifty students' names are drawn from a hat, and those drawn take the survey.
   C. Ten students from each year level at a school are asked to complete a survey.
   D. One class at a school is asked to complete a survey.

Understanding

4. Discuss how bias can be introduced into statistics through:
   a. questionnaire design
   b. sample selection.

5. Explain how you can determine an appropriate sample size from a population of known size.

6. A number of people were asked to rate a movie on a scale of 0 to 5.
   Here are their scores:

   $$1, 0, 2, 1, 0, 0, 1, 0, 2, 3, 0, 0, 1, 0, 1, 2, 5, 3, 1, 0$$

   a. Sort the data into a frequency distribution table.
   b. Determine the mode.
   c. Identify the median.
   d. Calculate the range.

7. Weekly earnings from casual work performed by a sample of 50 high school students were rounded to
   the nearest dollar, as follows:

205,	189,	216,	224,	227,	194,	232,	178,	228,	198,
227,	223,	235,	221,	194,	230,	213,	226,	241,	220,
179,	235,	186,	208,	194,	208,	223,	238,	226,	234,
219,	219,	197,	225,	216,	249,	228,	186,	229,	232,
217,	197,	208,	217,	231,	234,	214,	204,	228,	214

   a. Organise the data into a frequency distribution with class intervals $170{-}<180, 180{-}<190$
      and so on.
   b. Display the data as a histogram.
   c. Describe the data set.

8. Calculate the mean of the following scores: 1, 2, 2, 2, 3, 3, 5, 4 and 6.

9. The mean of 10 scores was 5.5. Nine of the scores were 4, 5, 6, 8, 2, 3, 4, 6 and 9.
   Calculate the 10th score.

10. Consider the following distribution table.

Score (x)	Frequency (f)
2	3
3	2
5	8
6	2

    Determine the:
    a. mean   b. mode   c. median   d. range.

11. **a.** Determine the mode of the following values: 3, 2, 6, 5, 9, 8, 1, 7. Explain your answer.
    b. Determine the median of the following values: 10, 6, 1, 9, 8, 5, 17, 3.
    c. Calculate the range of the following values: 1, 6, 15, 7, 21, 8, 41, 7.

Communicating, reasoning and problem solving

12. Consider the following distribution table.

Score (x)	Frequency (f)
1.5	10
2.0	20
2.5	8
3.0	5
3.5	6

   a. Calculate the mean score.
   b. Determine the median score.
   c. Give the modal score.
   d. Indicate which measure of centre best describes the distribution.

13. Consider the following stem plot.

   Key: 6.1 | 8 = 6.18

Stem	Leaf
6.1	8 8 9
6.2	0 5 6 8
6.3	0 1 2 4 4 4

   a. Calculate the mean score.
   b. Determine the median score.
   c. Give the modal score.
   d. Indicate which measure of centre best describes the distribution.

14. Study this dot plot to answer the following questions.

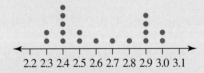

   2.2 2.3 2.4 2.5 2.6 2.7 2.8 2.9 3.0 3.1

   a. Calculate the mean score (correct to 2 decimal places).
   b. Determine the median score.
   c. Give the modal score.
   d. Indicate which measure of centre best describes the distribution.

15. A frozen goods section manager recorded the following sales of chickens by size during a sample week.

$$
\begin{array}{cccccccccc}
16, & 14, & 13, & 12, & 15, & 14, & 13, & 11, & 12, & 14, \\
14, & 16, & 15, & 13, & 11, & 12, & 14, & 13, & 15, & 17, \\
13, & 12, & 14, & 16, & 13, & 11, & 15, & 14, & 12, & 11, \\
15, & 12, & 13, & 12, & 12, & 15, & 13, & 11, & 11, & 13, \\
16, & 13, & 12, & 15, & 17, & 13, & 14, & 16, & 12, & 15
\end{array}
$$

a. Construct a frequency distribution table showing x, f and $f \times x$ columns. You may include a tally column if you wish.
b. Identify the mode of the distribution.
c. Calculate the mean and median sizes of the chickens sold.
d. Determine which size the manager should order most. Explain.
e. Calculate the range of sizes.
f. Calculate what percentage of total sales are in the 12–14 size group.

16. The following table displays the results of the number of pieces of mail delivered in a week to a number of homes.

Number of pieces of mail	Frequency
0	7
1	25
2	34
3	11
4	8
5	2
6	4
7	5
8	3
9	1

a. Determine the most common number of pieces of mail delivered.
b. Calculate the mean number of pieces of mail delivered.
c. Calculate the range.
d. Explain what this shows about the mail delivery service to these homes.

To test your understanding and knowledge of this topic, go to your learnON title at www.jacplus.com.au and complete the **post-test**.

Answers

Topic 10 Data

10.1 Pre-test

1. Primary data
2. a. Mode $= 3$ b. Range $= 10$ c. Median $= 4$
3. A and C
4. B
5. a. Mean $= 0.86$ b. Median $= 1$
6. B
7. 4
8. Median
9. B and D
10. Mean $= 1$
11. 165.5 cm
12. A, B and C
13. The mean would double.
14. 3, 4, 5, 5, 5, 5
15. $\frac{1}{4}$

10.2 Data collection methods (Foundational)

1. a. The population is the company's 1200 employees.
   b. The sample consists of 120 employees from London, 180 employees from Melbourne, 45 employees from Milan and 75 employees from Japan. The total sample is 420 employees.
2. a. The population is the university's total student enrolment, which is 55 000.
   b. The sample is made up of 250 city campus students and 45 country campus students.
   The total sample is 295 students.
3. a. 1240   b. 50
4. Male $= 35$
   Female $= 45$
5. a, b. Any sample with 5 different numbers between 1 and 26.
6. a, b. Any sample with 8 different numbers between 1 and 29.
7. a. Systematic sampling
   b. 45
   c. 44
   d. 227
8. C
9. C
10. C
11. Strategy 1
12. C
13. A
14. a. Survey   b. Survey
    c. Census   d. Survey
15. a. Survey   b. Census
    c. Census   d. Survey

16. a. Self-selected sampling
    b. Answers will vary. The sample is made up of self-selection with people volunteering to respond. People who volunteer to respond will tend to have strong opinions. This can mean over-representation and may cause bias.
17. B
18. a. Random sampling   b. Self-selected sampling
19. Answers will vary. A sample size should be sufficiently large and random. It should not be biased. The sample should be representative of the population. A sample size that is too small is less reliable.
20. Answers will vary but could include when the population is sufficiently small and all the members can be included, or when a census is conducted.
21. Answers will vary. In a census, everyone in the population is intended to be included. In a sample survey, only a subset of the population is included.
22. Answers will vary. The person surveyed could respond multiple times, and you don't know whether the person responding fits the criteria of the survey.
23. a. Experiment   b. Observation   c. Sample survey
24. a. 11 million   b. 35%
25. Agree. Answers will vary. Most households do not have landlines. Younger individuals only have mobile phones. If younger people are not well represented in the surveys, the survey samples will be biased.
26. Answers will vary but could include:
    Stratified sampling — the students can be divided into 20 groups based on the 20 regions. Students will be randomly selected from each group.
27. Answers will vary. A sample answer is given below.
    It would be best for the hotel manager to post surveys to clients' homes and select a sample based on clients who book during a full year. This will give better representation and randomness, as leaving a questionnaire in a room will allow for self-selection. Also, clients who have grievances and want their opinions considered will be the majority who complete on site; this will cause bias. A two-month period will not give a complete representation of the hotel experience compared to a full year, in which clients are staying in the hotel in different seasons.
28. 24

10.3 Primary and secondary data (Foundational)

1. These are examples of simulations that could be conducted.
   a. A coin could be flipped (Heads representing 'True' and Tails representing 'False').
   b. A coin could be flipped (Heads representing 'red' and Tails representing 'black').
   c. Spinner with 4 equal sectors (each sector representing a different toy)
   d. Roll a die (each face represents a particular person).
   e. Spinner with 3 equal sectors (each one representing a particular meal)
   f. Spinner with 5 equal sectors (each one representing a particular destination)

2. a. Some possible suggestions include:
   Which students have internet access at home?
   Do the students need access at night?
   What hours would be suitable?
   How many would use this facility?

   b. Answers could include a survey or online questionnaire.

   c. Sample responses can be found in the worked solutions in the online resources.

3. Sample responses can be found in the worked solutions in the online resources.

4. Some possible suggestions include:
   Census, interview, observation, online response, experiment

5. a. Measurement

   b. Observation

   c. Newspaper recordings

   d. Survey

6. The claim is false. It is not a logical deduction.

7. Sample responses can be found in the worked solutions in the online resources.

8. Pizza King's advertising campaign was misleading as it sounds as though all of their pizzas were rated 25% better than their competitors'. Only one of the 10 varieties was rated 25% better.

9. Some possible suggestions include:
   Primary data: Addison could visit each of the institutions in turn either in person or online.
   Secondary data: She could seek advice from friends and colleagues. She could enlist the help of a mortgage broker or similar professional.

10. Sealy Posturepremier: 40% off $\left(\dfrac{1000}{2499} \times 100\% \right)$

    Sealy Posturepedic: 41% off $\left(\dfrac{1600}{3899} \times 100\% \right)$

    SleepMaker Casablanca: 40% off $\left(\dfrac{800}{1999} \times 100\% \right)$

    SleepMaker Umbria: 42% off $\left(\dfrac{1800}{4299} \times 100\% \right)$

    True; the discount is at least 40% off these beds.

11. a. Data set A

    b. Data set B

    c. Primary data is data that is collected by the researcher, and secondary data is data that has been collected by another source. Data set A was collected by Hannah, whereas data set B would have been found by another source.

12. A sample from one year group only will not give an indication of the preferences of the whole school. Hamish should ask students from all year levels.

13. Some possible suggestions include:

    a. The most-watched channel and the time slot with the highest viewing rating

    b. Conduct a survey in a supermarket to establish the typical customer for her product.

c. Yes. If the typical customer was a five-year-old child, there would be little point in advertising during the 6 o'clock news, even though that time slot may have the highest rating on the most-watched channel.

10.4 Organising and displaying data

1. a.

x	Frequency
0	5
1	8
2	13
3	7
4	5
5	2
Total	40

   b. The data are distributed fairly evenly around 2 children per household, and there appear to be no outliers. The graph clearly shows 2 children per household is the most common.

   c. The sample is a random one, and of sufficient size; we can be confident that the suburb also exhibits these same properties.

2. a. The sample is a random one, so it seems to be a reliable reflection of the population of smart watches.

   b.

x	Frequency
0	1
1	6
2	8
3	5
4	2
5	3
Total	25

   c. There was only 1 box with no defective smart watches. Most boxes had only 1, 2 or 3 defective smart watches, while 3 boxes were found to have 5 defective smart watches.

   d. Since the sample was randomly selected, it seems to be a reliable reflection of the characteristics of the population. It would be reasonably safe to say that most boxes would have only 1, 2 or 3 defective smart watches.

3. a.

Hours of sport played by Year 8 students

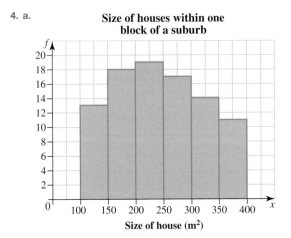

b. The graph rises steadily to a maximum, then falls away sharply at the upper end of the data.

c. This is likely to be a true reflection of the sporting habits of Year 8 students. Some do a minimum of only 1 hour per week, quite a few do 2, 3 or 4 hours per week, with the maximum number doing 5 hours per week. The committed sports players would put in 6 or 7 hours per week.

4. a.

Size of houses within one block of a suburb

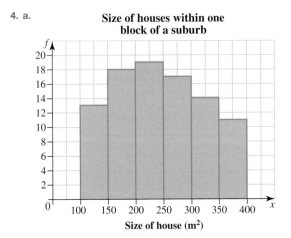

b. The graph is roughly symmetrical, rising to a maximum at the 200-m² to 250-m² size, then decreasing slowly.

c. Since this is one block of houses in the suburb, it is not a random sample. It is common for houses in a block of a suburb to be of similar style. For this reason, we could not say it reflects house sizes in the whole suburb.

5. a.

Mass of people joining a weight-loss program

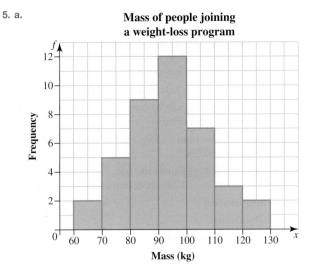

b. The graph is quite symmetrical, rising to a maximum at the 90-kg to 100-kg mass, then decreasing more rapidly to the 130-kg mass.

c. This would most likely not reflect the masses of people in the community because these are people who have enrolled in a program to lose weight.

6. a.

Class interval	Tally	Frequency
5– < 10	JHT JHT	11
10– < 15	JHT JHT JHT	19
15– < 20	JHT	5
20– < 25	||	2
25– < 30	|||	3
	Total	40

b.

Hours of TV watched per week

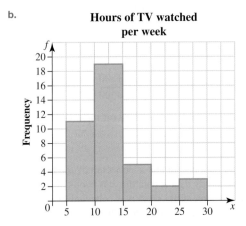

c. The graph is heavily weighted towards the lower end of the scale, with most people watching fewer than 15 hours of TV per week. There were 3 people who watched almost 30 hours of TV per week.

d. Since these people were interviewed in a shopping centre, the sample is not a random one. It could not therefore be taken to reflect the viewing habits of the community.

7. a.

Hours of sleep	Frequency
6–<7	3
7–<8	6
8–<9	9
9–<10	2
Total	20

b.

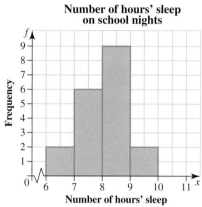

Number of hours' sleep on school nights

c. The histogram peaks sharply at the 8 hours' sleep mark, indicating that generally, Year 8 students get 8 hours of sleep per night during the week. Some get less, and a few get more.

d. It seems likely that these sample results would reflect the sleeping habits of Year 8 students generally.

8. a.

Pocket money ($)	Frequency
0–<5	2
5–<10	9
10–<15	7
15–<20	2
Total	20

b.

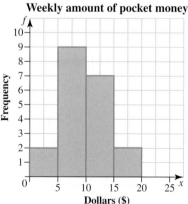

Weekly amount of pocket money

c. The histogram shows no general trend. The maximum is at $10, indicating that a popular amount of pocket money is $10 per week. Quite a few receive less than this, with only 2 receiving more.

d. Since this is a random sample, it is quite likely that these results reflect the general population of 13-year-olds when it comes to pocket money.

9. $\dfrac{12.5 + 13.2}{2} = 12.85$

10.

Class interval	Frequency
20–<30	2
30–<40	4
40–<50	4
50–<60	5
60–<70	3
70–<80	1
80–<90	0
90–<100	1

11. a.

Lifetime (hours)	Frequency
20–<25	16
25–<30	16
30–<35	18
35–<40	25
40–<45	15
45–<50	10
Total	100

b.

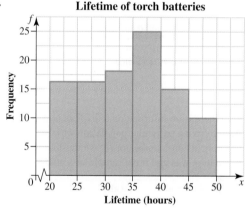

Lifetime of torch batteries

c. The histogram shows that the majority of torch batteries last for about 40 hours. A few last longer than this.

d. It seems reasonable that the torch battery population would display a similar trend.

12. a. 56

b. 9–12

c. No, it is not possible to determine the maximum number of weekends worked. However, the highest interval is 25–28. Therefore, the most weekends worked may have been 28; but we cannot be certain whether any tradesperson actually worked 28 weekends.

13. a. Answers will depend on class intervals chosen. An example is given.

Number of visits to the cinema	Tally	Frequency (f)
0–2	卌 卌 II	13
3–5	卌 卌 III	14
6–8	卌 卌 II	7
9–11	II	2
	Total	36

b.

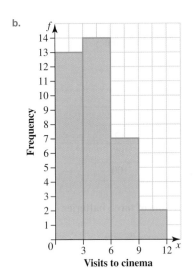

c. 13

d. 23

e. No. There is no information to explain how the sample of people from a particular population was obtained. Before any conclusions can be drawn, we must know what the population was and if the sample was random.

14. a.

Class interval	Tally	Frequency
145–<150	II	2
150–<155	IIII	4
155–<160	卌 II	8
160–<165	卌 II	7
165–<170	卌	5
170–<175	III	3
175–<180	I	1
	Total	30

b.

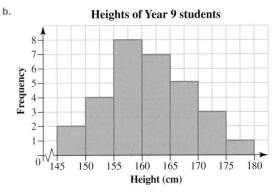

Heights of Year 9 students

c.

Class interval	Tally	Frequency
145–<148	I	1
148–<151	II	2
151–<154	II	2
154–<157	IIII	4
157–<160	卌	5
160–<163	IIII	4
163–<166	III	3
166–<169	卌	5
169–<172	II	2
172–<175	I	1
175–<178	—	0
178–<181	I	1
	Total	30

d. See the graph at the foot of the page.*

e. The two histograms represent the same data set, but appear to be quite different. The first histogram appears to be roughly symmetrical, with a maximum number of students having a height of 155–160 cm. The second histogram has two modes, with the most common height for students being 157–160 cm, or 166–169 cm. It illustrates the fact that the interpretation of a histogram displaying grouped data is dependent on the class interval used.

10.5 Measures of centre and spread

1. 26.8

2. a. 4.33   **b.** 7.33   **c.** 36.30   **d.** 6.91

3. a. 18   **b.** 10   **c.** 1.8 goals

4. C

5. Group A: 6.1; Group B: 6.8. Group B has a higher mean.

*14. d.

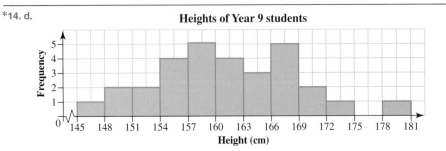

Heights of Year 9 students

6. 2.29

7. 7.90

8. a. 7   b. 17

9. a. 50   b. 1.75

10. Question 8:
   a. 5
   b. 17
   Question 9:
   a. 52
   b. 1.5, 1.7, 1.8

11. C

12. D

13. a. 8   b. 15   c. 50   d. 2.5

14. a. 4   b. 5   c. 15   d. 4

15. a. 35   b. 46   c. 35

16. Mean = 6; median = 6; mode = 6; range = 4

17. a. 6.7   b. 10

18.

x	f	$f \times x$
0	1	0
1	2	2
2	7	14
3	3	9
4	4	16
5	2	10
6	1	6
Total	20	57

Mean = 2.85

19. a. 36   b. 7

20. a. 3   b. 3   c. Both

21. a. 49   b. 55
   c. 56   d. Median

22. a. $18 000
   b. $26 000
   c. $30 000
   d. Mean (the highest value)
   e. Mode (the lowest value)

23. a. 8   b. No

24. 31–40

25. B

26. a. Mean = 26.6; median = 27.5; mode = 30; range = 22
   b. Mean = 32; median = 29; mode = 30; range = 70
   c. The inclusion of an outlier dramatically increased the range and significantly increased the mean of the data. The median was marginally increased while the mode was unchanged.

27. a. Mean = 4.5; median = 5; modes = 3 and 5
   b. The median is higher than the mean.
   c. The mean is between the two modal values.
   d. As there are no outliers, the mean is probably the best indicator of the centre of the data set.

28. a. The median, as there are likely to be some outliers which will significantly affect the mean.
   b. The mode, as this will be the most popular shoe size and will be a number that is easy to interpret.
   c. The mean, as this will be the average scored in a game.

29. a. The potential error is the data value of 23.
   b. This value could have been a genuine outlier if the particular student was a multi-sport athlete who trained for a number of hours each day.

30. 73.45

31. Mean = 5; median = 3; mode = 6

32. Highest score is 49 and lowest score is 0.
   The scores would be 0, 1, 12, 13, 49.

33. 171

10.6 Analysing data

1. a. Range   b. Mode
   c. Mean   d. Median

2. a. 6.4
   b. 6
   c. 6
   d. They are all quite close, so any would do.
   e. Since these were the results of half the class, and the sample was random, it seems likely that the population results would be similar.

3. a. i. 2   ii. 2
      iii. 1   iv. Mean or median
   b. i. 7.8   ii. 7
      iii. 7   iv. Any of the three

4. a. i. 22.6   ii. 23
      iii. 23   iv. Any of the three
   b. i. 117.5   ii. 116
      iii. 112   iv. Mean

5. a. i. 4.5   ii. 5
      iii. 5   iv. Any of the three
   b. i. 82.4   ii. 82
      iii. 81   iv. Mean or median

6. a. Sample 1: mean = 107.9; median = 101.5; no mode; range = 137
   Sample 2: mean = 112.5; median = 106.5; mode = 85; range = 127
   Sample 3: mean = 125.4; median = 104; no mode; range = 190
   b. All three median values lie close together. Sample 3 has a significantly larger mean and range than the other two samples, possibly caused by the data value 256, which appears to be an outlier. Sample 2 is the only sample that has a mode, so modes cannot be compared.

Score (x)	Frequency (f)
0	1
1	3
2	6
3	3
4	3
5	1
6	1
	$n = 18$

$\bar{x} = 2.6$

b.

Number of people per household

Median $= 2$

c. 2

d. i. Median   ii. Mean   iii. Mode

8. a.

Score (x)	Frequency (f)
137	2
138	3
139	5
140	3
141	6
142	1
	$n = 20$

b. 141, 139.5, 139.6

c. Mean

9. a. The median was calculated by taking the average of the 2 middle scores.

b. 13

10. 2

11. a. 55 250 km, 52 000 km, 52 000 km

b. The mean, as the data did not appear to have any outliers.

c. 51 810 km; it is reduced by 3440 km.

d. 52 000 km

e. 3333

f. 50 000 km; 92% last that distance or more.

12. The range was 4 minutes 14 seconds (fastest time of 38 seconds; slowest time of 4 minutes 52 seconds). The most common time (mode) was 1 minute. The average (mean) time was more than 2 minutes.

13. The data values in each sample will vary, so the sample statistics taken from these samples will also vary. Samples from the same population may have different summary statistics due to natural variation within the population, the presence of outliers within samples and some subsets of the population not being represented in all

samples. By taking sufficiently large random samples, these variations can be reduced.

14. Mean $= 33.05$; median $= 33$; mode $= 33$; range $= 5$

15. a. 38 years   b. 27 years
    c. 26 years   d. 31 years

16. a. The mean (3.25). The median is 2 and the mode is 1.

b. Sample responses can be found in the worked solutions in the online resources.

Project

1. The letters E, A, I and O are the most common; the letters J, K, Q, X and Z are least common.

2. The more tiles used for a letter, the lower the score. The fewer tiles used for a letter, the higher the score.

3. The frequency distribution table includes the letters in the passage heading.

Letter	Frequency	Letter	Frequency
A	61	O	31
B	6	P	23
C	18	Q	0
D	18	R	39
E	73	S	50
F	13	T	34
G	10	U	14
H	14	V	4
I	31	W	11
J	0	X	1
K	3	Y	6
L	18	Z	4
M	10		
N	41		

4. Results from both frequency tables show letters J, K, Q, X and Z to be the least common. However, while results from the scrabble frequency table show letters A, E, I and O to be the most common, the results from the textbook frequency table show letters A, E, I, N, O, R, S and T to be most common.

5. Australia is located in the southern hemisphere and is made up of six states and two territories. The capital of Australia is Canberra.

6. Common strategies involve looking for common letters and frequently used two- and three-lettered words.

10.7 Review questions

1. a. Survey   b. Census
   c. Census   d. Survey

2. D

3. B

4. Sample responses can be found in the worked solutions in the online resources.

5. The appropriate sample size is $\sqrt{\text{population size}}$.

6. a.

Video rating	Frequency
0	8
1	6
2	3
3	2
4	0
5	1
Total	20

b. 0

c. 1

d. 5

7. a.

Class interval	Frequency (f)
170– < 180	2
180– < 190	3
190– < 200	6
200– < 210	5
210– < 220	9
220– < 230	14
230– < 240	9
240– < 250	2
	$n = 50$

b. See the graph at the foot of the page.*

c. The data set is unimodal and negatively skewed.

8. 3.1

9. 8

10. a. 4.3　　　b. 5　　　c. 5　　　d. 4

11. a. There is no mode since none of the values occurs more than once.

b. 7

c. 40

12. a. 2.3

b. 2.0

c. 2.0

d. Mode

13. a. 6.27

b. 6.28

c. 6.34

d. Mean or median

14. a. 2.63

b. 2.55

c. 2.4

d. Mean

15. a.

x	f	$f \times x$
11	6	66
12	10	120
13	11	143
14	8	112
15	8	120
16	5	80
17	2	34
Total	50	675

b. 13

c. 13.5, 13

d. 13, as this is the most frequently sold size.

e. 6

f. 58%

16. a. 2

b. 2.6

c. 9

d. Most homes get up to about 3 pieces of mail. Some do get more.

---

*7. b.

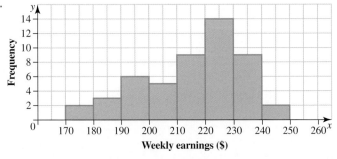

11 Probability

LESSON
11.1 Overview

Why learn this?

Probability is a topic in mathematics that measures how likely something is to happen. It looks at the chance of an event occurring. Probability can be measured with a number between 0 and 1. They can be expressed as fractions, decimals or percentages and can be described using words such as impossible, not likely, even chance, highly likely and certain. Think about the question 'What is the chance that the sun will rise tomorrow?' The answer could be 100% or 1 or *certain*. Understanding probability is a vital skill that allows you to understand and consider risks and make decisions accordingly. If the weather report shows an 80% chance of rain, you might consider taking an umbrella with you, whereas if the report shows a 10% chance of rain, you might not. Probability is widely used to describe everyday events. It is used to describe the chance of a sporting team winning, the chance of the weather being sunny on Christmas Day, and the chance of you winning at a board game.

You can also use probability to determine which insurance policy may best suit you based on the chance of your needing to use it. For instance, you may decide to insure your mobile phone because you think you are likely to lose or damage your phone. The ability to understand and analyse probabilities to inform your decisions will be vital throughout your life.

1. **MC** If the success of an event was said to be 'unlikely', select the percentage from the list that could match the probability of success of the event.
   - **A.** 25%
   - **B.** 0%
   - **C.** 50%
   - **D.** 70%

2. A bag contains six counters, five of which are blue and the sixth green. A counter is taken from the bag. Calculate the probability that it is not green. Write this probability as a fraction in simplest form.

3. **MC** When spinning a nine-sided spinner numbered 1 to 9, select the approximate probability of landing on an even number.
   - **A.** 0.44
   - **B.** 0.5
   - **C.** 0.55
   - **D.** 0.11

4. **MC** Identify how many different possible outcomes there are when rolling a 6-sided die once.
   - **A.** 2
   - **B.** 36
   - **C.** 12
   - **D.** 6

5. **MC** Two regular (6-sided) dice, one red and one blue, are rolled at the same time. Select the probability of scoring a total of 10 when the uppermost faces of the dice are added together.
   - **A.** $\dfrac{5}{6}$
   - **B.** $\dfrac{15}{18}$
   - **C.** $\dfrac{1}{4}$
   - **D.** $\dfrac{1}{12}$

6. **MC** When rolling a standard 6-sided die, determine which **two** of the following events are complementary.
   - **A.** Rolling a 4 or a 5
   - **B.** Rolling a prime number
   - **C.** Rolling a factor of 6
   - **D.** Rolling a multiple of 2

7. **MC** Select the probability that best describes the likelihood of a *Spinosaurus* walking into your classroom.
   - **A.** Certain
   - **B.** Unlikely
   - **C.** Even chance
   - **D.** Impossible

8. Auntie Kerry owns a house in Coffs Harbour. Every week, a total of 3 kookaburras, 2 galahs and 6 fantails visit her. Calculate the probability that the first bird to visit Auntie Kerry in a week is a galah, if all the birds have an equal chance of visiting her first.

9. A bag contains 6 black-coloured marbles, 4 red-coloured marbles, 3 green-coloured marbles and 2 blue-coloured marbles.

   If a marble is drawn at random, calculate the probability that is black or green.

10. Yu Chen is hosting a Halloween party. For the party, 4 friends dress as characters from *Star Wars*, 2 friends dress as pirates, 1 friend dresses as a fairy, and 3 friends don't wear costumes.

    Assuming all the friends have an equal chance of leaving first, determine the probability that a friend *not* dressed as a pirate leaves first.

    Write your answer in simplified fraction form.

11. Shrish is rolling two 6-faced fair dice. He claims that getting two odd numbers and getting zero odd numbers are complementary events. Is he right? Explain your answer.

▶

12. **MC** Among the following spinners, select the one for which there is not an equal chance of landing on each colour.

A.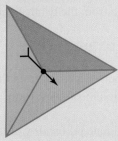

B.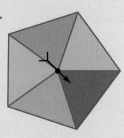

C.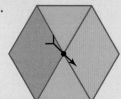

D.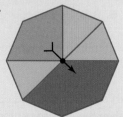

13. Amir and Hussein run a race against each other; Hussein is four times more likely to win this race than Amir. The probability that Amir will win is $\frac{1}{5}$. Explain whether this statement is true or false.

14. David is planning 7 dinners for the week.
    Meal 1: Beef lasagne
    Meal 2: Roast chicken
    Meal 3: Thai green curry with chicken
    Meal 4: Beef stew
    Meal 5: Fish tacos
    Meal 6: Fish and chips
    Meal 7: Vegetarian quiche
    Assuming all meals have an equal chance of being cooked on Monday, evaluate the probability of David cooking either a beef or vegetarian meal.

15. Dr Sofia Samaras is running a microbiology experiment in which she will attempt to culture different species of bacteria. The table shows a list of the bacteria she is hoping to grow and their level of risk to humans.
    Assume all bacteria have an equal chance of growing first.

Bacterial genus	Risk
Geobacter	Low
Escherichia	High
Pseudomonas	Moderate
Campylobacter	High
Desulfobacter	Low
Halomonas	Low
Giardia	High

    a. Evaluate the probability of a high-risk bacteria growing first.
    b. Determine the probability of either a high- or low-risk bacteria growing first.

LESSON
11.2 Observed probability

LEARNING INTENTION

At the end of this lesson you should be able to:
- calculate the relative frequency of an event
- determine the expected value of occurrences of an event.

▶ 11.2.1 Probability and relative frequency

eles-4474

- **Experiments** are performed to provide data, which can then be used to forecast the outcome of similar events in the future.
- An experiment that is performed in the same way each time is called a **trial**.
- An **outcome** is a particular result of a trial.
- A **favourable outcome** is one that we are looking for.
- An **event** is the set of favourable outcomes in each trial.
- The **relative frequency** of an event occurring is the observed probability of it occurring.

Relative frequency or experimental probability

The relative frequency (or observed probability) of an event is:

$$\text{relative frequency} = \frac{\text{number of times the event occurs (frequency)}}{\text{total number of trails}}$$

WORKED EXAMPLE 1 Calculating relative frequencies of an event

The table shows the results of a fair coin that was tossed 20 times. Calculate the relative frequency of:
a. Heads
b. Tails.

Event	Frequency
Heads	8
Tails	12
Total	20

THINK	WRITE
a. 1. Write the frequency of the number of Heads and the total number of trials (tosses).	a. Frequency of Heads $= 8$ Total number of tosses $= 20$
2. Write the rule for the relative frequency.	Relative frequency $= \dfrac{\text{frequency of Heads}}{\text{total number of tosses}}$
3. Substitute the known values into the rule.	Relative frequency of Heads $= \dfrac{8}{20}$
4. Evaluate and simplify if possible.	$= \dfrac{2}{5}$ (or 0.4)
5. Write the answer.	The relative frequency of Heads is $\dfrac{2}{5}$.

			b.	Frequency of Tails $= 12$

b. 1. Write the frequency of the number of Tails and the total number of trials (tosses).

    b. Frequency of Tails $= 12$
        Total number of tosses $= 20$

2. Write the rule for the relative frequency.

$$\text{Relative frequency} = \frac{\text{frequency of Tails}}{\text{total number of tosses}}$$

3. Substitute the known values into the rule.

$$\text{Relative frequency of Tails} = \frac{12}{20}$$

4. Evaluate and simplify if possible.

$$= \frac{3}{5} \ (\text{or } 0.6)$$

5. Write the answer.

The relative frequency of Tails is $\frac{3}{5}$.

- In probability, the **expected value** is the average value of an experiment over many repetitions — it is the number of times we would typically expect something to happen.

  For example, if the chances of a football team winning their match is $\frac{1}{4}$, then we would expect that out of 100 matches they would win 25 of them.

Expected value

$$\textbf{Expected value} = \textbf{relative frequency} \times \textbf{number of trials}$$

WORKED EXAMPLE 2 Calculating relative frequency and the expected value

Forty people picked at random were asked where they were born. The results were coded as follows.

Place of birth:

1. Melbourne

2. Elsewhere in Victoria

3. Interstate

4. Overseas

Responses:

1, 3, 2, 1, 1, 4, 3, 1, 2, 1, 2, 1, 3, 4, 1, 2, 3, 1, 3, 4,

4, 3, 2, 1, 2, 3, 1, 4, 1, 2, 3, 4, 1, 2, 3, 1, 1, 4, 2, 3

a. Organise the data into a frequency table.

b. Calculate the relative frequency of each category as a fraction and a decimal.

c. Determine the total of the relative frequencies.

d. If a person is selected at random, identify where they are most likely to have been born.

e. From a group of 100 people, decide how many people you would expect to have been born overseas.

THINK

a. 1. Draw a table with 3 columns. The column headings are in the order of Score, Tally and Frequency.

2. Enter the codes 1, 2, 3 and 4 into the score column.

3. Place a stroke into the tally column each time a code is recorded.

    Note: ЖГ represents a score of five.

WRITE

a.

Score	Tally	Frequency
Code 1	ЖГ ЖГ IIII	14
Code 2	ЖГ IIII	9
Code 3	ЖГ ЖГ	10
Code 4	ЖГ II	7
	Total	40

4. Count the number of strokes corresponding to each code and record them in the frequency column.

5. Add the total of the frequency column.

b. 1. Write the rule for the relative frequency.

b. Relative frequency $= \dfrac{\text{frequency of category}}{\text{total number of people}}$

Category 1: People born in Melbourne

    2. Substitute the known values into the rule for each category.

    3. Evaluate and simplify where possible. Write the answer.

Relative frequency $= \dfrac{14}{40}$

$= \dfrac{7}{20}$ or 0.35

Category 2: People born elsewhere in Victoria

Relative frequency $= \dfrac{9}{40}$ or 0.255

Category 3: People born interstate

Relative frequency $= \dfrac{10}{40}$

$= \dfrac{1}{4}$ or 0.25

Category 4: People born overseas

Relative frequency $= \dfrac{7}{40}$ or 0.175

c. 1. Add each of the relative frequency values.

c. Total $= \dfrac{7}{20} + \dfrac{9}{40} + \dfrac{1}{4} + \dfrac{7}{40}$

$= 0.35 + 0.225 + 0.25 + 0.175$

$= 1$

    2. Write the answer.

The relative frequencies sum to a total of 1.

d. 1. Using the results from part **b**, obtain the code that corresponds to the largest frequency.
Note: A person selected at random is *most* likely to have been born in the place with the largest frequency.

d. Melbourne (code 1) corresponds to the largest frequency.

    2. Write the answer.

A person selected at random is most likely to have been born in Melbourne.

e. 1. Write the relative frequency of people born overseas and the number of people in the sample.

e. Relative frequency (overseas) $= \dfrac{7}{40}$

Number of people in the sample $= 100$

    2. Write the rule for the expected value.
Note: Of the 100 people, $\dfrac{7}{40}$ or 0.175 would be expected to be born overseas.

Expected value $=$ relative frequency $\times$ number of people

3. Substitute the known values into the rule.

$$\text{Expected value} = \frac{7}{40} \times 100$$
$$= \frac{700}{40}$$

4. Evaluate.

$$= 17.5$$

5. Round the value to the nearest whole number.
   Note: We are dealing with people. Therefore, the answer must be represented by a whole number.

$$\approx 18$$

6. Write the answer.

We would expect 18 of the 100 people to be born overseas.

 Resources

 Interactivity Experimental probability (int-3825)

Exercise 11.2 Observed probability

learn**on**

11.2 Quick quiz **on**	11.2 Exercise

Individual pathways

■ PRACTISE	■ CONSOLIDATE	■ MASTER
1, 2, 8, 11, 13, 16	3, 4, 6, 9, 14, 17	5, 7, 10, 12, 15, 18

Fluency

1. **WE1** The table shows the result of tossing a fair coin 150 times. Calculate the relative frequency of:

   a. Heads
   b. Tails.

Event	Frequency
Heads	84
Tails	66
Total	150

2. A fair coin was tossed 300 times. A Head came up 156 times.

   a. Calculate the relative frequency of Heads as a fraction.
   b. Calculate the relative frequency of Tails as a decimal.

3. A die is thrown 50 times, with 6 as the favourable outcome. The 6 came up 7 times. Determine the relative frequency of:

   a. a 6 occurring
   b. a number that is not a 6 (that is, any number other than a 6) occurring.

4. The spinner shown with 3 equal sectors was spun 80 times, with results as shown in the table:

Score	1	2	3
Frequency	29	26	25

i. What fraction of the spins resulted in 3?
ii. What fraction of the spins resulted in 2?
iii. Express the relative frequency of the spins that resulted in 1 as a decimal.

5. **WE2** 100 people picked at random were asked which Olympic event they would most like to see. The results were coded as follows:

Swimming — 1   Athletics — 2   Gymnastics — 3   Rowing — 4

The recorded scores were:

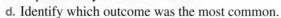

1, 1, 4, 3, 2, 2, 2, 4, 4, 3, 1, 1, 4, 2, 1, 1, 1, 4, 2, 2, 1, 3, 3, 3, 4,
1, 1, 3, 2, 2, 1, 2, 1, 1, 1, 1, 2, 3, 3, 3, 3, 2, 2, 4, 1, 1, 1, 3, 2, 2,
4, 1, 1, 1, 3, 3, 3, 3, 2, 1, 2, 2, 2, 2, 3, 4, 4, 1, 1, 1, 2, 3, 3, 2, 1,
4, 3, 2, 3, 1, 1, 2, 4, 1, 1, 3, 2, 2, 3, 3, 4, 4, 2, 1, 1, 3, 1, 2, 4, 1

a. Organise the data into a frequency table.
b. Calculate the relative frequency of each category as a fraction and a decimal.
c. Calculate the total of the relative frequencies.
d. If an Olympic event is selected at random, identify which event is most likely to be seen.
e. From a group of 850 people, decide how many people you would expect to prefer to watch the gymnastics.

6. The following are results of 20 trials conducted for an experiment involving the 5-sector spinner shown.

1, 4, 2, 5, 3, 4, 5, 3, 2, 5, 1, 3, 2, 4, 2, 1, 4, 3, 3, 2

a. Organise the data into a frequency table.
b. Calculate the relative frequency of each outcome.
c. Determine how many times you would have expected each outcome to appear. Explain how you came to this conclusion.
d. Identify which outcome was the most common.
e. Calculate the total of all the relative frequencies.

7. A card is *randomly* (with no predictable pattern) drawn 60 times from a hand of 5 cards; it is recorded, then returned, and the five cards are reshuffled. The results are shown in the frequency distribution table. For each of the following, determine:
i. the favourable outcomes that make up the event
ii. the relative frequency of these events.

Card	Frequency
3 ♥	13
Q ♦	15
3 ♦	12
3 ♣	9
3 ♠	11

   a. A heart   b. A red card
   c. A 3   d. A spade or a heart
   e. A 3 or a queen   f. The king of spades

Understanding

8. When 60 light bulbs were tested, 3 were found to be faulty.

a. State the relative frequency of faulty bulbs.
b. Calculate the fraction of the bulbs that were not faulty.
c. In a carton of 600 such bulbs, determine how many you would expect to be faulty.

9. The square spinner shown was trialled 40 times and the number it landed on each time was recorded as shown below.

2, 4, 3, 1, 3, 2, 1, 4, 4, 3, 3, 1, 4, 2, 1, 2, 3, 1, 4, 2,
4, 2, 1, 2, 1, 3, 1, 4, 3, 1, 3, 1, 4, 2, 3, 1, 3, 2, 4, 4

a. Determine the relative frequency of each outcome.
b. Organise the data into a frequency table and calculate the actual experimental relative frequency of each number.
c. Calculate the relative frequency of the event *odd number* from the table obtained in part **b**.
d. What outcomes make up the event *prime number*? *Hint:* Remember a prime number has exactly 2 factors: itself and 1.
e. Calculate the relative frequency of the event *prime number* from the table obtained in part **b**.

10. The following table shows the progressive results of a coin-tossing experiment.

	Outcome		Relative frequency	
Number of coin tosses	Heads	Tails	Heads (%)	Tails (%)
10	6	4	60	40
100	54	46	54	46
1000	496		49.6	50.4

a. Complete the missing entry in the table.
b. Comment on what you notice about the relative frequencies for each trial.
c. If we were to repeat the same experiment in the same way, explain whether the results would necessarily be identical to those in the table.

11. **MC** A fair coin was tossed 40 times and it came up Tails 18 times. The relative frequency of Heads was:

A. $\dfrac{9}{11}$ B. $\dfrac{11}{20}$ C. $\dfrac{9}{20}$ D. $\dfrac{20}{11}$

12. **MC** Olga observed that, in 100 games of roulette, red came up 45 times. Out of 20 games on the same wheel, select the relative frequency of red.

A. 4.5   B. $\dfrac{4}{9}$ C. $\dfrac{9}{4}$ D. 9

Communicating, reasoning and problem solving

13. Sadiq has a box of 20 chocolates. The chocolates come in four different flavours: caramel, strawberry, mint and almond. Sadiq recorded the number of each type of chocolate in this table.

a. Determine the frequency of mint.
b. Nico says that the experimental probability of choosing caramel is $\dfrac{5}{20}$, but his friend June says it is $\dfrac{1}{4}$. Explain why they are both correct.

Flavour	Frequency
Caramel	5
Strawberry	7
Mint	
Almond	4
Total	20

14. The game 'rock, paper, scissors' is played all over the world, not just for fun but also for settling disagreements.

The game uses three different hand signs.

Simultaneously, two players 'pound' the fist of one hand into the air three times. On the third time each player displays one of the hand signs. Possible results are shown.

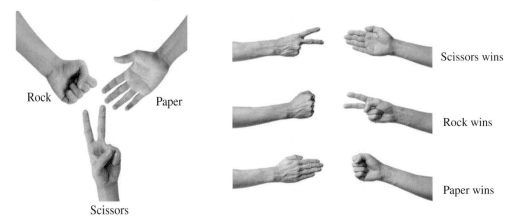

a. Play 20 rounds of 'rock, paper, scissors' with a partner. After each round, record each player's choice and the result in a table like the following one. (Use R for rock, P for paper and S for scissors.)

Round	Player 1	Player 2	Result
1	P	R	Player 1 wins
2	S	R	Player 2 wins
3	S	S	Tie

b. Based on the results of your 20 rounds, calculate the experimental probability of:
   i. you winning
   ii. your partner winning
   iii. a tie.

c. Explain whether playing 'rock, paper, scissors' is a fair way to settle a disagreement.

15. There is a total of 200 green and red marbles in a box. A marble is chosen, its colour is noted, and it is replaced in the box. This experiment is conducted 65 times. Only 13 green marbles are chosen.

   a. Give a reasonable estimate of the number of red marbles in the box. Show all of your working.
   b. If this experiment is conducted n times and g green marbles are chosen, give is a reasonable estimate of the number of red marbles in the box. Show all of your working.

16. The gender of babies in a set of triplets is simulated by flipping 3 coins. If a coin lands Tails up, the baby is a boy. If a coin lands Heads up, the baby is a girl. In the simulation, the trial is repeated 40 times and the following results show the number of Heads obtained in each trial:

$$0, 3, 2, 1, 1, 0, 1, 2, 1, 0, 1, 0, 2, 0, 1, 0, 1, 2, 3, 2,$$
$$1, 3, 0, 2, 1, 2, 0, 3, 1, 3, 0, 1, 0, 1, 3, 2, 2, 1, 2, 1$$

   a. Determine the probability that exactly one of the babies in the set of triplets is female.
   b. Determine the probability that more than one of the babies in the set of triplets is female.

17. A survey of the favourite foods of Year 8 students was conducted, with the following results.

Meal	Tally
Hamburger	45
Fish and chips	31
Macaroni and cheese	30
Lamb souvlaki	25
BBQ pork ribs	21
Cornflakes	17
T-bone steak	14
Banana split	12
Corn on the cob	9
Hot dog	8
Garden salad	8
Veggie burger	7
Smoked salmon	6
Muesli	5
Fruit salad	3

  a. Estimate the probability that macaroni and cheese is the favourite food among Year 8 students.
  b. Determine the probability that a vegetarian dish is the favourite food.
  c. Estimate the probability that a beef dish is the favourite food.

18. A standard deck of 52 playing cards consists of four suits (clubs, diamonds, hearts and spades) as shown in the table (sample space).

Clubs	Diamonds	Hearts	Spades
A♣	A♦	A♥	A♠
2♣	2♦	2♥	2♠
3♣	3♦	3♥	3♠
4♣	4♦	4♥	4♠
5♣	5♦	5♥	5♠
6♣	6♦	6♥	6♠
7♣	7♦	7♥	7♠
8♣	8♦	8♥	8♠
9♣	9♦	9♥	9♠
10♣	10♦	10♥	10♠
J♣	J♦	J♥	J♠
Q♣	Q♦	Q♥	Q♠
K♣	K♦	K♥	K♠

One card is chosen at random. Evaluate the probability that the card is:
  a. a red card
  b. a picture card (jack, queen or king)
  c. an ace
  d. an ace or a heart
  e. an ace and a heart
  f. not a diamond
  g. a club or a 7
  h. neither a heart nor a queen
  i. a card worth 10 (10 and picture cards)
  j. a red card or a picture card.

LESSON
11.3 Probability scale

LEARNING INTENTION

At the end of this lesson you should be able to:
- understand the concept of chance and probabilities lying between 0 and 1
- classify the chance of an event occurring using words such as *certain*, *likely*, *unlikely*, *even chance* or *impossible*
- estimate the probability of an event occurring.

⏵ 11.3.1 Describing and assigning values to the likelihood of events

eles-4475

- **Probability** is defined as the chance of an event occurring.
- A probability scale from 0 to 1 is used to allocate the probability of an event as follows:
- Probabilities may be expressed as fractions, decimals or percentages.

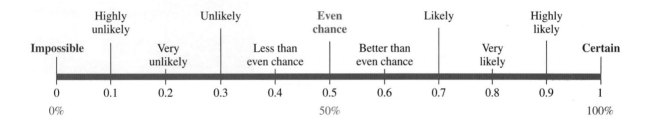

WORKED EXAMPLE 3 Describing the likelihood of an event

Using the words in the probability scale, describe the probability of each of the following events occurring.
a. February follows January.
b. You draw the queen of diamonds from a standard deck of playing cards.
c. You will represent your school in gymnastics at state finals.
d. You roll a standard die and obtain an even number.
e. Every Mathematics student will obtain a score of 99.95% in an examination.

THINK	WRITE
a. This is a true statement. February always follows January.	a. It is *certain* this event will occur.
b. In a standard deck of 52 playing cards, there is only one queen of diamonds. So, you have an extremely slim chance of drawing this particular card.	b. It is *highly unlikely* this event will occur.
c. The chance of a person competing in the state finals is small. However, it could happen.	c. It is *unlikely* this event will occur.

▶

d. There are six possible outcomes when rolling a die, each of which is equally likely. Three of the outcomes are even while three are odd.

d. There is an *even chance* this event will occur.

e. Due to each student having different capabilities, there is an extremely small chance this could occur.

e. It is *highly unlikely* that this event will occur.

- It is important to note that the responses for particular situations, such as part **c** in Worked example 3, are not always straightforward and may differ for each individual. A careful analysis of each event is required before making any predictions about their future occurrences.

WORKED EXAMPLE 4 Estimating probabilities

Assign a fraction to represent the estimated probability of each of the following events occurring.

a. A high tide will be followed by a low tide.
b. Everyone in your class will agree on every matter this year.
c. A tossed coin lands Heads.
d. A standard die is rolled and the number 5 appears uppermost.
e. One of your 15 tickets in a 20-ticket raffle will win.

THINK

a. The tide pattern occurs daily; this event seems *certain*.

b. Total agreement among many people on every subject over a long time is virtually *impossible*.

c. When tossing a coin, there are two equally likely outcomes, a Head or a Tail.

d. When rolling a die, there are six equally likely outcomes: 1, 2, 3, 4, 5, 6.

e. There are 15 chances out of 20 of winning.

WRITE

a. The probability of this event occurring is equal to 1.

b. The probability of this event occurring is equal to 0.

c. The probability of this event occurring is equal to $\dfrac{1}{2}$.

d. The probability of this event occurring is equal to $\dfrac{1}{6}$.

e. The probability of this event occurring is equal to $\dfrac{15}{20}$, which when simplified is equal to $\dfrac{3}{4}$.

 Resources

 Interactivity Probability scale (int-3824)

Exercise 11.3 Probability scale

| 11.3 Quick quiz on | 11.3 Exercise |

Individual pathways

■ PRACTISE	■ CONSOLIDATE	■ MASTER
1, 4, 6, 9, 10, 13	2, 7, 11, 14	3, 5, 8, 12, 15

Fluency

1. **WE3** Using words, describe the probability of each of the following events occurring.

   a. The sun will set today.
   b. Every student in this class will score 100% in the next Mathematics exam.
   c. Your school bus will have a flat tyre tomorrow.
   d. Commercial TV stations will reduce time devoted to ads.
   e. A comet will collide with Earth this year.

2. Using words, describe the probability of each of the following events occurring.

   a. The year 2028 will be a leap year.
   b. You roll a standard die and an 8 appears uppermost.
   c. A tossed coin lands on its edge.
   d. World records will be broken at the next Olympics.
   e. You roll a standard die and an odd number appears uppermost.

3. Using words, describe the probability of each of the following events occurring.

   a. You draw the queen of hearts from a standard deck of playing cards.
   b. You draw a heart or diamond card from a standard deck of playing cards.
   c. One of your 11 tickets in a 20-ticket raffle will win.
   d. A red marble will be drawn from a bag containing 1 white marble and 9 red marbles.
   e. A red marble will be drawn from a bag containing 1 red and 9 white marbles.

4. **WE4** Assign a fraction to represent the estimated probability of each of the following events occurring.

   a. A Head appears uppermost when a coin is tossed.
   b. You draw a red marble from a bag containing 1 white and 9 red marbles.
   c. A standard die shows a 7 when rolled.
   d. You draw a yellow disk from a bag containing 8 yellow disks.
   e. The next baby in a family will be a boy.

5. Assign a fraction to represent the estimated probability of each of the following events occurring.

   a. A standard die will show a 1 or a 2 when rolled.
   b. You draw the king of clubs from a standard deck of playing cards.
   c. One of your 15 tickets in a 20-ticket raffle will win.
   d. A standard die will show a number less than or equal to 5 when rolled.
   e. You draw an ace from a standard deck of playing cards.

Understanding

6. **MC** The probability of Darwin experiencing a white Christmas this year is closest to:

   A. 1   **B.** 0.75   **C.** 0.5   **D.** 0

7. **MC** The word that best describes the probability that a standard die will show a prime number is:

   A. impossible.   **B.** very unlikely.   **C.** even chance.   **D.** very likely.

8. The letters of the word MATHEMATICS are each written on a small piece of card and placed in a bag. If one card is selected from the bag, determine the probability that it is:

   a. a vowel   **b.** a consonant   **c.** the letter M   **d.** the letter C.

Communicating, reasoning and problem solving

9. For the spinner shown, answer the following questions.

   a. Explain whether there is an equal chance of landing on each colour.
   b. List all the possible outcomes.
   c. Determine the probability of each outcome.

10. Discuss some events that would have a probability of occurring of 0, 1 or $\frac{1}{2}$.

11. For the spinner shown, answer the following questions.

    a. Explain whether there is an equal chance of landing on each colour.
    b. List all the possible outcomes.
    c. Determine the probability of each outcome.

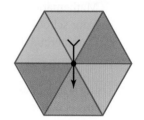

12. For the spinner shown, answer the following questions.

    a. Explain whether there is an equal chance of landing on each colour.
    b. List all the possible outcomes.
    c. Determine the probability of each outcome.
    d. If the spinner is spun 30 times, explain how many times you would expect it to land on green.

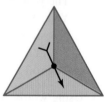

13. All the jelly beans shown are placed in a bag for a simple probability experiment.

    a. Which colour jelly bean is most likely to be randomly selected from the bag? Explain.
    b. Which colour jelly bean is least likely to be randomly selected from the bag? Explain.
    c. Using a horizontal line with the words 'least likely' on the far left and 'most likely' on the far right, place the jelly bean colours along the line.

14. Draw spinners with the following probabilities.

    a. Probability of blue $= \frac{1}{3}$ and probability of white $= \frac{2}{3}$

    b. Probability of blue $= \frac{1}{2}$, probability of white $= \frac{1}{4}$, probability of green $= \frac{1}{8}$ and probability of pink $= \frac{1}{8}$

    c. Probability of blue $= 0.75$ and probability of white $= 0.25$

15. Four different types of coloured card are enclosed in a bag: red, blue, green and yellow. There is a 40% chance of drawing a blue card, and the probability of drawing a red card is 0.25. There are 4 yellow cards. If there is a total of 20 cards in the bag, evaluate:

    a. the probability of drawing a green card
    b. how many green cards are in the bag.

LESSON
11.4 Sample space and theoretical probability

11.4.1 Calculating theoretical probabilities

eles-4476

- **Chance experiments** are performed to provide data, which can then be used to forecast the outcome of similar events in the future.
- An *outcome* is a possible result of a chance experiment.
- A *favourable outcome* is the outcome that we are looking for.
- An *event* is either one or a collection of favourable outcomes.
- Equally likely outcomes have equal probabilities. Consider the following examples:
  - When rolling a regular 6-sided die, rolling an odd number (1, 3 or 5) or rolling an even number (2, 4, 6) are equally likely events, each with a probability of $\frac{1}{2}$.
  - When selecting a marble at random from a jar containing 4 blue marbles, 4 green marbles and 4 red marbles, selecting a blue marble, selecting a red marble and selecting a green marble are equally likely events (4 favourable outcomes out of 12 outcomes), each with a probability of $\frac{1}{3}$.
- The total probabilities of all possible outcomes of an event is 1.
- The **theoretical probability** of a particular event occurring is denoted by the symbol P(event).
- Theoretical probability is also sometimes called expected probability.
- By contrast, observed probability is based on an experiment or trial. It is often called experimental probability.

Theoretical probability (expected probability)

The theoretical probability (or expected probability) of an event is:

$$P(event) = \frac{\text{number of favourable outcomes}}{\text{total number of outcomes}}$$

- The **sample space**, S, is the set of all the possible outcomes.
- If an outcome appears more than once, it is only listed in the sample space once.
  For example, the sample space, S, for the word *mathematics* is {m, a, t, h, e, i, c, s}, **not** {m, m, a, a, t, t, h, e, i, c, s}.

A standard 6-sided die is rolled.
a. **List the sample space for this chance experiment.**
b. **Determine the probability of having the following appear uppermost:**
   i. **4**   ii. **an odd number**   iii. **5 or less.**

THINK	WRITE
a. List all the possible outcomes for the given chance experiment.	a. $S = \{1, 2, 3, 4, 5, 6\}$
b. i. 1. Write the total number of possible outcomes.	b. i. Total number of outcomes $= 6$
2. Write the number of favourable outcomes. *Note:* The favourable outcome is 4.	Number of favourable outcomes $= 1$
3. Write the rule for probability.	$P(event) = \dfrac{\text{number of favourable outcomes}}{\text{total number of outcomes}}$
4. Substitute the known values into the rule and evaluate.	$P(4) = \dfrac{1}{6}$
5. Write the answer.	The probability of 4 appearing uppermost is $\dfrac{1}{6}$.
ii. 1. Write the total number of possible outcomes.	ii. Total number of outcomes $= 6$
2. Write the number of favourable outcomes. The favourable outcomes are 1, 3, 5.	Number of favourable outcomes $= 3$
3. Write the rule for probability.	$P(event) = \dfrac{\text{number of favourable outcomes}}{\text{total number of outcomes}}$
4. Substitute the known values into the rule and simplify.	$P(\text{odd number}) = \dfrac{3}{6}$ $= \dfrac{1}{2}$
5. Write the answer.	The probability of an odd number appearing uppermost is $\dfrac{1}{2}$.
iii. 1. Write the total number of possible outcomes.	iii. Total number of outcomes $= 6$
2. Write the number of favourable outcomes. The favourable outcomes are 1, 2, 3, 4, 5.	Number of favourable outcomes $= 5$
3. Write the rule for probability.	$P(event) = \dfrac{\text{number of favourable outcomes}}{\text{total number of outcomes}}$
4. Substitute the known values into the rule and simplify.	$P(\text{5 or less}) = \dfrac{5}{6}$
5. Write the answer.	The probability of obtaining 5 or less is $\dfrac{5}{6}$.

A card is drawn at random from a standard well-shuffled pack.

Calculate the probability of drawing:

a. a club

b. a king or an ace

c. not a spade.

Express each answer as a fraction and as a percentage.

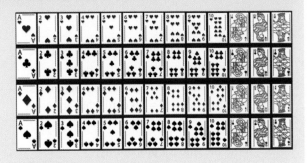

THINK

a. 1. Write the total number of outcomes in the sample space. There are 52 cards in a pack.

  2. Write the number of favourable outcomes. There are 13 cards in each suit.

  3. Write the rule for probability.

  4. Substitute the known values into the rule and simplify.

  5. Convert the fraction to a percentage; that is, multiply by 100%.

  6. Write the answer.

b. 1. Write the total number of outcomes in the sample space.

  2. Write the number of favourable outcomes. There are 4 kings and 4 aces.

  3. Write the rule for probability.

  4. Substitute the known values into the rule and simplify.

  5. Convert the fraction to a percentage, rounded to 1 decimal place.

WRITE

a. Total number of outcomes $= 52$

Number of favourable outcomes $= 13$

$$P(\text{event}) = \frac{\text{number of favourable outcomes}}{\text{total number of outcomes}}$$

$$P(\text{a club}) = \frac{13}{52}$$

$$= \frac{1}{4}$$

$$\text{Percentage} = \frac{1}{4} \times 100\%$$

$$= \frac{100}{4}\%$$

$$= 25\%$$

The probability of drawing a club is $\frac{1}{4}$ or 25%.

b. Total number of outcomes $= 52$

Number of favourable outcomes $= 8$

$$P(\text{event}) = \frac{\text{number of favourable outcomes}}{\text{total number of outcomes}}$$

$$P(\text{a king or an ace}) = \frac{8}{52}$$

$$= \frac{2}{13}$$

$$\text{Percentage} = \frac{2}{13} \times 100\%$$

$$= \frac{200}{13}\%$$

$$\approx 15.4\%$$

| 6. Write the answer. | The probability of drawing a king or an ace is $\frac{2}{13}$ or approximately 15.4%. |

c. 1. Write the total number of outcomes in the sample space.

c. Total number of outcomes $= 52$

2. Write the number of favourable outcomes. There are $52 - 13 = 39$ cards that are *not* a spade.

Number of favourable outcomes $= 39$

3. Write the rule for probability.

$$P(\text{event}) = \frac{\text{number of favourable outcomes}}{\text{total number of outcomes}}$$

4. Substitute the known values into the rule and simplify.

$$P(\text{not a spade}) = \frac{39}{52} = \frac{3}{4}$$

5. Convert the fraction to a percentage.

$$\text{Percentage} = \frac{3}{4} \times 100\%$$
$$= \frac{300}{4}\%$$
$$= 75\%$$

6. Write the answer.

The probability of drawing a card that is not a spade is $\frac{3}{4}$ or 75%.

WORKED EXAMPLE 7 Calculating probability

A shopping centre car park has spaces for 10 buses, 300 cars and 20 motorbikes. If all vehicles have an equal chance of leaving at any time, calculate the probability that the next vehicle to leave will be:
a. a motorbike　　　　**b. a bus or a car**　　　　**c. not a car.**

Express each answer as a fraction.

THINK	WRITE
a. 1. Write the total number of outcomes in the sample space. There are 330 vehicles.	**a.** Total number of outcomes $= 330$
2. Write the number of favourable outcomes. There are 20 motorbikes.	Number of favourable outcomes $= 20$
3. Write the rule for probability.	$P(\text{event}) = \dfrac{\text{number of favourable outcomes}}{\text{number of possible outcomes}}$
4. Substitute the known values into the rule and simplify.	$P(\text{a motorbike}) = \dfrac{20}{330}$ $= \dfrac{2}{33}$
5. Write the answer.	The probability of a motorbike next leaving the car park is $\dfrac{2}{33}$.

b.

1. Write the total number of outcomes in the sample space.
2. Write the number of favourable outcomes. There are 10 buses and 300 cars.
3. Write the rule for probability.
4. Substitute the known values into the rule and simplify.
5. Write the answer.

b. Total number of outcomes $= 330$

Number of favourable outcomes $= 310$

$$P(\text{event}) = \frac{\text{number of favourable outcomes}}{\text{total number of outcomes}}$$

$$P(\text{a bus or a car}) = \frac{310}{330}$$

$$= \frac{31}{33}$$

The probability of a bus or car next leaving the car park is $\frac{31}{33}$.

c.

1. Write the total number of outcomes in the sample space.
2. Write the number of favourable outcomes. There are 10 buses and 20 motorbikes.
3. Write the rule for probability.
4. Substitute the known values into the rule and simplify.
5. Write the answer.

c. Total number of outcomes $= 330$

Number of favourable outcomes $= 30$

$$P(\text{event}) = \frac{\text{number of favourable outcomes}}{\text{total number of outcomes}}$$

$$P(\text{not a car}) = \frac{30}{330}$$

$$= \frac{1}{11}$$

The probability of a vehicle that is not a car next leaving the car park is $\frac{1}{11}$.

COMMUNICATING — COLLABORATIVE TASK: Random number generator

In pairs, use a random number generator to repeat an experiment such as flipping a coin. Count 1 as Heads and 2 as Tails.

1. Conduct the experiment for:
   a. 20 trials
   b. 100 trials
   c. 1000 trials.
2. Compare your results with the rest of the class.
3. Did your results show a 50% outcome of Heads and a 50% outcome of Tails? Why or why not?
4. What did you notice as the trials became larger?

---

on Resources

✦ **Interactivity** Sample spaces and theoretical probability (int-3826)

Exercise 11.4 Sample space and theoretical probability

11.4 Quick quiz on	**11.4 Exercise**

Individual pathways

■ PRACTISE	■ CONSOLIDATE	■ MASTER
1, 4, 7, 8, 12, 15, 16, 17	2, 5, 6, 9, 10, 13, 18, 23, 24	3, 11, 14, 19, 20, 21, 22, 25, 26

Fluency

1. List the sample spaces for these chance experiments.

   a. Tossing a coin
   b. Selecting a vowel from the word *astronaut*
   c. Selecting a day of the week to go to the movies
   d. Drawing a marble from a bag containing 3 reds, 2 whites and 1 black

2. List the sample spaces for these chance experiments.

   a. Rolling a standard 6-sided die
   b. Drawing a picture card from a standard pack of playing cards
   c. Spinning an 8-sector circular spinner numbered from 1 to 8
   d. Selecting even numbers from the first 20 counting numbers

3. List the sample spaces for these chance experiments.

   a. Selecting a piece of fruit from a bowl containing 2 apples, 4 pears, 4 oranges and 4 bananas
   b. Selecting a magazine from a rack containing 3 *Dolly*, 2 *Girlfriend*, 1 *Smash Hits* and 2 *Mathsmag* magazines
   c. Selecting the correct answer from the options A, B, C, D, E on a multiple-choice test
   d. Winning a medal at the Olympic Games

4. **WE5** A standard 6-sided die is rolled.

   a. List the sample space for this chance experiment.
   b. Determine the probability of obtaining the following appearing uppermost:

       i. 6   ii. An even number   iii. At most 4   iv. 1 or 2

5. For a 6-sided die, determine the probabilities of obtaining the following appearing uppermost.

   a. A prime number   b. A number greater than 4
   c. 7   d. A number that is a factor of 60

6. A card is drawn at random from a standard well-shuffled pack. Calculate the probability of drawing:

   a. a red card   b. an 8 or a diamond
   c. an ace   d. a red or a black card.

7. **WE7** A shopping centre car park has spaces for 8 buses, 160 cars and 12 motorbikes. If all vehicles have an equal chance of leaving at any time, determine the probability that the next vehicle to leave will be:

   a. a bus
   b. a car
   c. a motorbike or a bus
   d. not a car.

8. A bag contains marbles coloured as follows: 3 red, 2 black, 1 pink, 2 yellow, 3 green and 3 blue. If a marble is drawn at random, calculate the chance that it is:

   a. red
   b. black
   c. yellow
   d. red or black.

9. A bag contains marbles coloured as follows: 3 red, 2 black, 1 pink, 2 yellow, 3 green and 3 blue. If a marble is drawn at random, calculate the chance that it is:

   a. not blue
   b. red or black or green
   c. white
   d. not pink.

10. A beetle drops onto *one* square of a chessboard. Calculate its chances of landing on a square that is:

    a. black
    b. white
    c. neither black nor white
    d. either black or white.

11. Assuming it is not a leap year, calculate the chance that the next person you meet has their birthday:

    a. next Monday
    b. sometime next week
    c. in September
    d. one particular day next year (assuming it is not a leap year either).

Understanding

12. Consider the spinner shown.

    a. State whether the 4 outcomes are equally likely. Explain your answer.
    b. Determine the probability of the pointer stopping on 1.

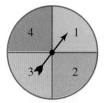

13. Consider the spinner shown.

    a. State whether the 2 outcomes are equally likely. Explain your answer.
    b. Determine the probability of the pointer stopping on 1.

14. Consider the spinner shown.

    a. State whether the 4 outcomes are equally likely. Explain your answer.
    b. Determine the probability of the pointer stopping on 1.

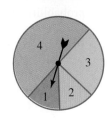

15. Hanna flipped a coin 5 times and each time a Tail showed. Determine the chances of Tails showing on the *sixth* toss.

Communicating, reasoning and problem solving

16. **WE6** A card is drawn at random from a standard well-shuffled pack. (Refer to Worked Example 6 for a representation of a standard 52-card deck.) Determine the probability of drawing:

    a. the king of spades
    b. a 10
    c. a jack or a queen
    d. a club.

    Explain your answer.

17. **a.** Design a circular spinner coloured pink, white, blue, orange and green so that each colour is equally likely to result from any trial.

   b. Determine the angle between each sector in the spinner.

18. **a.** List all the outcomes for tossing a coin once, together with their individual probabilities.

   b. Determine the sum of the probabilities.

19. **a.** List all the outcomes for tossing a coin twice, together with their individual probabilities.

   b. Determine the sum of the probabilities.

20. **a.** Design a circular spinner labelled A, B, C and D so that
   $$P(A) = \frac{1}{4}, P(B) = \frac{1}{3}, P(C) = \frac{1}{6}, P(D) = \frac{1}{4}.$$

   b. Estimate the size of the angles between each sector in the spinner.

21. **a.** Design a spinner with the numerals 1, 2 and 3 so that 3 is twice as likely to occur as either 2 or 1 in any trial.

   b. Determine the size of the angles in each sector or region at the centre of the spinner.

22. **MC** If a circular spinner has three sectors, A, B and C, such that $P(A) = \frac{1}{2}$ and $P(B) = \frac{1}{3}$, then $P(C)$ must be:

   A. $\frac{1}{4}$ **B.** $\frac{2}{5}$ **C.** $\frac{1}{6}$ **D.** $\frac{5}{6}$

23. A fair coin is flipped 3 times. Evaluate the probability of obtaining:

   a. at least two Heads or at least two Tails
   b. exactly two Tails.

24. A fair die is rolled and a fair coin flipped. Calculate the probability of obtaining:

   a. an even number from the die and a Head from the coin
   b. a Tail from the coin
   c. a prime number from the die
   d. a number less than 5 from the die and a Head from the coin.

25. The targets shown are an equilateral triangle, a square and a circle with coloured regions that are also formed from equilateral triangles, squares and circles. If a randomly thrown dart hits each target, determine the probability that the dart hits each target's coloured region.

   a.
   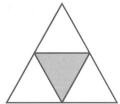
   b.
   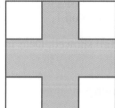
   c.
   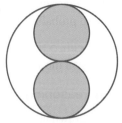

26. Two dice are rolled and the product of the two numbers is found. Evaluate the probability that the product of the two numbers is:

   a. an odd number **b.** a prime number **c.** more than 1 **d.** at most 36.

LESSON
11.5 Complementary events

LEARNING INTENTION

At the end of this lesson you should be able to:
- understand the concept of complementary events
- state the complement of a given event.

▶ 11.5.1 Complementary events

eles-4477

- **Complementary events** are two possible outcomes that have nothing in common. For example, when rolling a regular 6-sided die, the events 'rolling a 6' and 'not rolling a 6' are complementary.
- The diagram shown represents the sample space. If an event is circled in the black rectangular box (the sample space S), the complementary event is every other outcome in the sample space. An event A {6} is shown in the blue circle. Every other outcome {1, 2, 3, 4, 5} is the complementary event, which is shown in pink.
- If an event is denoted by the letter A, its complement is denoted by A'.

Sample space, S

| 1 | 2 | 3 |
| 4 | 5 | ⑥ |

$A = \{6\}$

$A' = \{1, 2, 3, 4, 5\}$

- The probability of rolling a 6 is $\dfrac{1}{6}$ and the probability of not rolling a 6 is $\dfrac{5}{6}$. The probability of an event and its complement is always a total of 1.

- Examples of complementary events include selecting a vowel or selecting a consonant, winning a race or not winning a race, and rolling a prime number or not rolling a prime number.

Complementary events

- **The sum of the probabilities of an event and its complement are:**

$$P(A) + P(A') = 1$$

- **The sum of the probabilities of all outcomes is 1. Thus, it follows that the probabilities of an event and its complement are:**

$$P(A) = 1 - P(A') \text{ and } P(A') = 1 - P(A)$$

WORKED EXAMPLE 8 Stating complementary events

State the complement of each of the following events.
- a. Selecting a red card from a standard deck
- b. Rolling two dice and getting a total greater than 9
- c. Selecting a red marble from a bag containing 50 marbles

THINK	WRITE
a. Selecting a black card will complete the sample space for this chance experiment.	a. The complement of selecting a red card is selecting a black card.
b. When rolling two dice, rolling a total less than 10 will complete the sample space.	b. The complement of rolling a total greater than 9 is rolling a total less than 10.
c. Selecting a marble that is not red is the only way to define the rest of the sample space for this chance experiment.	c. The complement of selecting a red marble in this chance experiment is selecting a marble that is not red.

WORKED EXAMPLE 9 Calculating the probability of complementary events

If a card is drawn from a pack of 52 cards, calculate the probability that the card is not a diamond.

THINK	WRITE
1. Determine the probability of drawing a diamond.	Number of diamonds = 13, number of cards = 52 $$P(\text{event}) = \frac{\text{number of favourable outcomes}}{\text{total number of outcomes}}$$ $$P(\text{diamond}) = \frac{13}{52}$$ $$= \frac{1}{4}$$
2. Write down the rule for obtaining the complement of drawing a diamond: that is, not drawing a diamond.	$P(A') = 1 - P(A)$ $P(\text{not a diamond}) = 1 - P(\text{diamond})$
3. Substitute the known values into the given rule and simplify.	$$= 1 - \frac{1}{4}$$ $$= \frac{3}{4}$$
4. Write the answer.	The probability of drawing a card that is not a diamond is $\frac{3}{4}$.

DISCUSSION

Is there a complementary event for every possible event? Use reasoning to explain your answer. Check with a classmate to see if they agree or disagree with your answer. If they disagree, discuss your reasoning and see if you can come to an agreement.

 Resources

 Interactivity Complementary events (int-3827)

Exercise 11.5 Complementary events

11.5 Quick quiz on	11.5 Exercise

Individual pathways

■ PRACTISE	■ CONSOLIDATE	■ MASTER
1, 4, 6, 8, 10, 14, 18	2, 5, 7, 11, 15, 19	3, 9, 12, 13, 16, 17, 20

Fluency

For questions **1** to **5**, state whether the events are complementary.

1. Having Weet-Bix or Corn Flakes for breakfast

2. Walking or riding your scooter to your friend's house

3. Watching TV or listening to music in the evening

4. Passing or failing your Mathematics test

5. Rolling a number less than 4 or greater than 4 on a die.

6. **WE8** For each of the following, state the complementary event.
   a. Selecting an even-numbered marble from a bag of numbered marbles
   b. Selecting a vowel from the letters of the alphabet
   c. Tossing a coin that lands Heads
   d. Rolling a die and getting a number less than 3

7. State the complementary event for each of the following events.
   a. Rolling two dice and getting a total less than 12
   b. Selecting a diamond from a deck of cards
   c. Selecting an 'e' from the letters of the English alphabet
   d. Selecting a blue marble from a bag of marbles

8. **WE9** If a card is drawn from a pack of 52 cards, calculate the probability that the card is not a queen.

9. **MC** Identify which option is not a pair of complementary events.
   A. Travelling to school by bus or travelling to school by car
   B. Drawing a red card from a pack of 52 playing cards or drawing a black card from a pack of 52 playing cards
   C. Obtaining an even number or obtaining an odd number on a six-sided die
   D. Choosing a black square or choosing a white square on a chessboard

Understanding

10. When a six-sided die is rolled 3 times, the probability of getting 3 sixes is $\frac{1}{216}$. Calculate the probability of not getting 3 sixes.

11. Eight athletes compete in a 100-metre race. The probability that the athlete in lane 1 will win is $\frac{1}{5}$.

    Determine the probability that one of the other athletes wins.
    (Assume that there are no dead heats.)

12. A pencil case has 4 red pens, 3 blue pens and 5 black pens.
    If a pen is drawn randomly from the pencil case, determine:
    a. P(drawing a blue pen)
    b. P(not drawing a blue pen)
    c. P(drawing a red or a black pen)
    d. P(drawing neither a red nor a black pen).

13. Holty is tossing two coins. He claims that getting two Heads and getting zero Heads are complementary events. Is he right? Explain your answer.

Communicating, reasoning and problem solving

14. In a bag there are 4 red cubes and 7 green cubes. If Clementine picks a cube at random, determine the probability that it is not:
    a. red or green
    b. red
    c. green.

15. In a hand of n cards there are r red cards. All cards are either red or black. I choose a card at random. Explain what the probability is that the card is:
    a. red or black
    b. red
    c. black
    d. not black.

16. Explain why the probability of an event and the probability of the complement of the event always sum to 1.

17. In a bag there are 4 red cubes and 7 green cubes. Clementine picks a cube at random, looks at it and notes that it is red. Without putting it back, she picks a second cube from the bag. Evaluate the probability that it is not green. Show your working.

18. Determine the following complementary probabilities.
    a. The probability it will rain today is $\frac{1}{4}$. Evaluate the probability that it will not rain today.
    b. The probability that you eat a sandwich for lunch is 80%. Determine the probability that you don't eat a sandwich for lunch.
    c. The probability that Olga's phone runs out of battery before she gets home is 0.3. Evaluate the probability that her phone doesn't run out of battery before she gets home.

19. There are 100 tickets being sold for the school raffle.

a. If Jeff buys one ticket, calculate his probability of winning.

b. If Jeff buys one ticket, determine the probability of someone else winning.

c. The teachers have a probability of $\frac{1}{5}$ of winning the raffle. Determine how many tickets were bought by the teachers.

d. The parents have a 50% chance of winning the raffle. Evaluate how many tickets were bought by the parents.

20. There are three cyclists in a road race. Cyclist A is twice as likely to win as cyclist B and three times as likely to win as cyclist C. Evaluate the probability that:

a. cyclist B wins

b. cyclist A does not win.

LESSON
11.6 Review

11.6.1 Topic summary

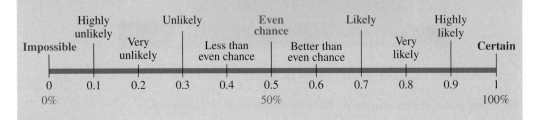

Understanding probability

- The probability of an event is the chance of that event occurring.
- Probabilities can be expressed as fractions, decimals or percentages.
- $0 \leq$ probability of an event ≤ 1, where 1 is certain and 0 is impossible.

| Impossible | | Highly unlikely | Very unlikely | | Unlikely | | Less than even chance | | Even chance | | Better than even chance | | Likely | | Very likely | | Highly likely | | Certain |

0   0.1   0.2   0.3   0.4   0.5   0.6   0.7   0.8   0.9   1
0%   50%   100%

Complementary events

- The complement of an event, A, is the opposite of that event and is denoted by A'.
  e.g. If A = rolling a 2, then A = not rolling a 2.
  $$P(A) + P(A') = 1$$
- If there are only two possible outcomes and there is no chance of both outcomes occurring at the same time, then the events are complementary events.
- Events A and A' are also said to be mutually exclusive.

PROBABILITY

Theoretical probability

- A *favourable outcome* is one that we are looking for.
- *Equally likely outcomes* have an equal chance of occurring.
- The theoretical probability of an event is:
  $$P(event) = \frac{number\ of\ favourable\ outcomes}{total\ number\ of\ outcomes}$$
- The sample space, S, is the set of all possible outcomes.

Observed probability (experimental probability)

- The **observed probability** of an event is determined by the results of repeated trials of the experiment.
- $$P(event) = \frac{number\ of\ successful\ trials}{total\ number\ of\ trials}$$
  e.g. If a team has won 42 out of their last 70 matches, the experimental probability that they will win a match is $\frac{42}{70} = \frac{3}{5} = 60\%$.
- The observed probability of an event becomes more accurate as the number of trials increases, and approaches the theoretical probability.

11.6.2 Project

In a spin

When dealing with events involving chance, we try to predict what the outcome will be. Some events have an even chance of occurring, whereas others have little or no chance of occurring. If the outcomes for events are equally likely, we can predict how often each outcome will appear.

However, will our predictions always be exact?

Spinners are often used to help calculate the chance of an event occurring. There are many different types of spinners, and one with 6 equal sections is shown here. A paperclip is flicked around the pencil placed at the centre.

The instructions below will enable you to construct a spinner to use in a probability exercise.
- Draw a circle with a radius of 8 cm onto a piece of cardboard using a pair of compasses. Using a ruler, draw a line to indicate the diameter of the circle.
- Keep the compasses open at the same width. Place marks on the circumference on both sides of the points where the diameter meets the circle.
- Join the marks around the circumference to produce your hexagon. Draw a line with a ruler to join the opposite corners and cut out the hexagon.
- Number the sections 1 to 6 or colour each section in a different colour.

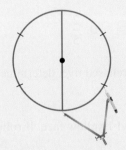
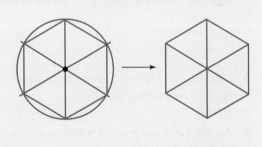

1. Flick the paperclip around the pencil 120 times. Record each outcome in a copy of the table below.

Score	Tally	Frequency
1		
2		
3		
4		
5		
6		
	Total	

Note: If the paperclip lands on a line, re-spin the spinner.

2. Are the outcomes of each number in your spinner equally likely? Explain your answer.
3. What would be the theoretical probability of spinning a 6?
4. Based on the results you obtained from your spinner, list the relative frequency of each outcome on the spinner.
5. How many times would you expect the paperclip to land on each number when you perform 120 spins?
6. How close were your results to the expected results?
Combine your results with your classmates' results.
7. Design a new frequency table for the class results.
8. How do the relative frequencies of the pooled class results compare with your results? Are they closer to the results you expected?
9. If time permits, continue spinning the spinner and pooling your results with the class. Investigate the results obtained as you increase the number of trials for the experiment.

 Resources

 Interactivities Crossword (int-2638)
Sudoku puzzle (int-3192)

Exercise 11.6 Review questions

learnon

Fluency

1. **MC** The chance of getting a 5 on the spinner shown is:

   A. $\dfrac{1}{3}$ **B.** $\dfrac{1}{5}$ **C.** $\dfrac{1}{6}$ **D.** $\dfrac{2}{5}$

   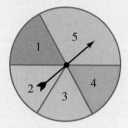

2. One thousand tickets were sold for a raffle. If you purchased five, determine your chances of winning the raffle.

3. A cube has 2 red faces, 2 white faces, 1 green face and 1 yellow face. If rolled, calculate the probability of the top face showing:
   a. red
   b. yellow
   c. not red
   d. green or white.

4. On each of the 5 weekdays, Monday to Friday, garbage is collected. Calculate the chances that garbage is collected at an address chosen at random on:
   a. Wednesday
   b. Thursday or Friday
   c. Sunday
   d. a day other than Monday.

5. A multiple-choice question has as alternatives A, B, C, D and E. Only one is correct.
   a. Determine the probability of guessing:
      i. the right answer
      ii. the wrong answer.
   b. Determine the total sum of the above probabilities.

Understanding

6. Give the probability of each of the numbers in the circular spinner shown.

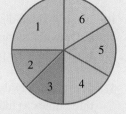

7. State whether each of the following pairs of events is complementary or not. Explain your answer.
   a. Having milk in your coffee or not having milk in your coffee
   b. Going overseas for your holiday or holidaying in Australia
   c. Catching a train to work or catching a bus to work

8. The probability that a traffic light is green is $\frac{3}{7}$.

   Give the probability that the traffic light is not green.

9. State the complement of each of the following events.
   a. Tossing a coin and it landing Tails
   b. Winning a race in which you run
   c. Answering a question correctly
   d. Selecting a black marble from a bag of marbles
   e. Selecting a number less than 20

10. Study the spinner shown.
    a. List the sample space.
    b. Are all outcomes equally likely? Explain your answer.
    c. In theory, determine the chances of spinning:
       i. a 2
       ii. a 6.

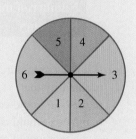

11. a. Design a circular spinner with the numbers 1 to 4 so that:

    $$P(1) = \frac{1}{2}, \ P(2) = \frac{1}{4}, \ P(3) = \frac{1}{8}, \ P(4) = \frac{1}{8}$$

    b. Evaluate the size of the angle in each sector.

Communicating, reasoning and problem solving

12. A letter is chosen at random from the letters in the word PROBABILITY. What is the probability that the letter is:
    a. B
    b. not B
    c. a vowel
    d. not a vowel?

13. During her latest shopping trip, Karen's mother stocked up on a supply of icy poles for the summer. In one box there are 3 icy poles of each of orange, raspberry, lime and cola flavours. Karen likes everything but lime.

    a. Determine the probability that when Karen reaches into the box randomly, she will select a flavour that she likes.
    b. Two days later, 2 raspberry, 1 cola, and 1 lime have been eaten by various members of the family. Explain whether Karen's chances of randomly choosing a lime icy pole have gone up, gone down or stayed the same.

14. In a certain school band there are 6 girls and some boys. A student is selected at random from this group. Calculate the number of boys in the group if the probability that a girl is selected is $\frac{1}{4}$.

15. Jar A contains 5 black, 3 white and 2 coloured marbles. Jar B contains 6 black, 4 white and 4 coloured marbles. Determine which jar are you more likely to draw a black marble from.

16. A fair 4-sided die in the shape of a regular tetrahedron is tossed twice.
    a. Draw the sample space.
    b. Calculate the probability that the two numbers that are thrown are both even.

17. There are only three swimmers in the 100-m freestyle. Swimmer A is twice as likely to win as B and three times as likely to win as C. Determine the probability that B or C wins.

18. James has two dice. One is a regular die. James has altered the other by replacing the 1 with a 6 (so the altered die has two sixes and no 1). He rolls the two dice together.
    Determine the probability that a double is rolled when the two dice are thrown.

19. A box of matches claims on its cover to contain 100 matches. A survey of 200 boxes established the following results.

Number of matches	95	96	97	98	99	100	101	102	103	104
Frequency	1	13	14	17	27	55	30	16	13	14

If you were to purchase a box of these matches, what is the probability that:
a. the box would contain 100 matches
b. the box would contain at least 100 matches
c. the box would contain more than 100 matches
d. the box would contain no more than 100 matches?

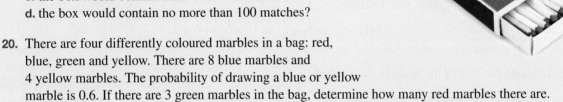

20. There are four differently coloured marbles in a bag: red, blue, green and yellow. There are 8 blue marbles and 4 yellow marbles. The probability of drawing a blue or yellow marble is 0.6. If there are 3 green marbles in the bag, determine how many red marbles there are.

21. a. In a jar, there are 600 red balls, 400 green balls, and an unknown number of yellow balls. If the probability of selecting a green ball is $\frac{1}{5}$, how many yellow balls are in the jar?

    b. In another jar there are an unknown number of balls, N, with 20 of them green. The other colours contained in the jar are red, yellow and blue, with P(red or yellow) $= \frac{1}{2}$, P(red or green) $= \frac{1}{4}$ and P(blue) $= \frac{1}{3}$. Determine the number of red, yellow and blue balls in the jar.

22. A box contains two coins. One is a double-headed coin, and the other is a normal coin with Heads on one side and Tails on the other. You draw one of the coins from a box and look at one of the sides. It is Heads. What is the probability that the other side shows Heads also?

on To test your understanding and knowledge of this topic, go to your learnON title at www.jacplus.com.au and complete the **post-test**.

Answers

Topic 11 Probability

11.1 Pre-test

1. A

2. $\dfrac{5}{6}$

3. A

4. D

5. D

6. A and C

7. D

8. $\dfrac{2}{11}$

9. $\dfrac{3}{5}$

10. $\dfrac{4}{5}$

11. Shrish is incorrect: the two events are not complementary. Shrish did not consider the fact that getting only one odd number is also a possible outcome (he could roll two odd numbers, two even numbers, or one of each).

12. D

13. True

14. $\dfrac{3}{7}$

15. a. $\dfrac{3}{7}$ b. $\dfrac{6}{7}$

11.2 Observed probability

1. a. $\dfrac{14}{25}$ (0.56)   b. $\dfrac{11}{25}$ (0.44)

2. a. $\dfrac{13}{25}$ b. 0.48

3. a. $\dfrac{7}{50}$ (0.14)   b. $\dfrac{43}{50}$ (0.86)

4. a. $\dfrac{5}{16}$ b. $\dfrac{13}{40}$ c. 0.3625

5. a.

Score	Frequency
1	34
2	27
3	24
4	15
Total	100

b. Swimming $= \dfrac{34}{100}$ (0.34)

Athletics $= \dfrac{27}{100}$ (0.27)

Gymnastics $= \dfrac{24}{100}$ (0.24)

Rowing $= \dfrac{15}{100}$ (0.15)

c. 1

d. Swimming

e. 204

6. a.

Score	Frequency
1	3
2	5
3	5
4	4
5	3
Total	20

b. $1 \to \dfrac{3}{20}$ (0.15)

$2 \to \dfrac{1}{4}$ (0.25)

$3 \to \dfrac{1}{4}$ (0.25)

$4 \to \dfrac{1}{5}$ (0.2)

$5 \to \dfrac{3}{20}$ (0.15)

c. 4
There are five possible outcomes, and each has an equal chance of occurring. Therefore, in 20 trials each outcome would be expected to occur 4 times.

d. 2 and 3

e. 1

7. a. i. 3 of hearts   ii. $\dfrac{13}{60}$

b. i. 3 of hearts, queen of diamonds and 3 of diamonds

ii. $\dfrac{2}{3}$

c. i. 3 of each suit   ii. $\dfrac{3}{4}$

d. i. 3 of both spades and hearts

ii. $\dfrac{2}{5}$

e. i. All cards drawn   ii. 1

f. i. None of the cards drawn

ii. 0

8. a. $\dfrac{1}{20}$ (0.05)   b. $\dfrac{19}{20}$ (0.95)   c. 30

9. a. $\dfrac{1}{4}$

b.

Score	Frequency
1	11
2	9
3	10
4	10
Total	40

c. $\dfrac{21}{40}$

d. $2, 3$

e. $\dfrac{19}{40}$

10. a. 504.

   b. The greater the number of trials, the closer the results come to what we would expect; that is, there is a relative frequency of 50% for each event.

   c. The results being identical would be highly unlikely, as you would normally expect each trial to be different from another trial.

11. B

12. D

13. a. 4   **b.** $\dfrac{1}{4} = \dfrac{5}{20}$

14. Sample responses can be found in the worked solutions in the online resources.

15. a. 160   **b.** $200\left(1 - \dfrac{g}{n}\right)$

16. a. $\dfrac{7}{20}$ **b.** $\dfrac{2}{5}$

17. a. $\dfrac{30}{241}$ **b.** $\dfrac{91}{241}$ **c.** $\dfrac{59}{241}$

18. a. $\dfrac{1}{2}$ **b.** $\dfrac{3}{13}$ **c.** $\dfrac{1}{13}$ **d.** $\dfrac{4}{13}$ **e.** $\dfrac{1}{52}$
    f. $\dfrac{3}{4}$ **g.** $\dfrac{4}{13}$ **h.** $\dfrac{9}{13}$ **i.** $\dfrac{4}{13}$ **j.** $\dfrac{8}{13}$

11.3 Probability scale

1. a. Certain   **b.** Highly unlikely
   c. Highly unlikely   **d.** Highly unlikely
   e. Highly unlikely

2. a. Certain   **b.** Impossible
   c. Highly unlikely   **d.** Highly likely
   e. Even chance

3. a. Highly unlikely

   b. Even chance

   c. Better than even chance

   d. Highly likely

   e. Highly unlikely

4. a. $\dfrac{1}{2}$ **b.** $\dfrac{9}{10}$ **c.** 0

   d. 1   **e.** $\dfrac{1}{2}$

5. a. $\dfrac{1}{3}$ **b.** $\dfrac{1}{52}$ **c.** $\dfrac{3}{4}$

   d. $\dfrac{5}{6}$ **e.** $\dfrac{1}{13}$

6. D

7. C

8. a. $\dfrac{4}{11}$ **b.** $\dfrac{7}{11}$ **c.** $\dfrac{2}{11}$ **d.** $\dfrac{1}{11}$

9. a. Yes

   b. Blue; green; purple; orange; pink

   c. Blue: $\dfrac{1}{5}$; green: $\dfrac{1}{5}$; purple: $\dfrac{1}{5}$; orange: $\dfrac{1}{5}$; pink: $\dfrac{1}{5}$

10. There are many events that have probabilities of $0, 1$ or $\dfrac{1}{2}$. An example of each is given below.

Rolling a 7 on a 6-sided die is impossible, so it has a probability of 0.

Rolling a number that is less than 10 on a 6-sided die is certain, so it has a probability of 1.

Rolling an odd number on a 6-sided die has a probability of $\dfrac{3}{6} = \dfrac{1}{2}$.

11. a. No

   b. Blue, green, pink, orange

   c. Blue: $\dfrac{1}{3}$; green: $\dfrac{1}{6}$; pink: $\dfrac{1}{3}$; orange: $\dfrac{1}{6}$

12. a. Yes

   b. Blue, pink, green

   c. Blue: $\dfrac{1}{3}$; pink: $\dfrac{1}{3}$; green: $\dfrac{1}{3}$

   d. 10

13. a. Blue. There are more blue jelly beans than those of any other colour.

   b. Yellow. There are fewer yellow jelly beans than those of any other colour.

   c.

Yellow	Red	Green	Blue
Least likely			Most likely

14. Answers will vary. Some examples are shown.

   a.   **b.**   **c.**

15. a. The probability of a green card is 0.15.

   b. There are 3 green cards in the bag.

11.4 Sample space and theoretical probability

1. a. {Heads, Tails}

   b. {a, o, u}

   c. {Monday, Tuesday, Wednesday, Thursday, Friday, Saturday, Sunday}

   d. {R, W, B}

2. a. $\{1, 2, 3, 4, 5, 6\}$

   b. {king of hearts, diamonds, clubs, spades; queen of hearts, jack of hearts}

   c. $\{1, 2, 3, 4, 5, 6, 7, 8\}$

   d. $\{2, 4, 6, 8, 10, 12, 14, 16, 18, 20\}$

3. a. {apple, pear, orange, banana}

   b. {*Dolly, Girlfriend, Smash Hits, Mathsmag*}

   c. {A, B, C, D, E}

   d. {gold, silver, bronze}

4. a. {1, 2, 3, 4, 5, 6}

  b. i. $\frac{1}{6}$　　**ii.** $\frac{1}{2}$　　**iii.** $\frac{2}{3}$　　**iv.** $\frac{1}{3}$

5. a. $\frac{1}{2}$　　**b.** $\frac{1}{3}$　　**c.** 0　　**d.** 1

6. a. $\frac{1}{2}$, 50%　　　　**b.** $\frac{4}{13}$, 30.8%

  c. $\frac{1}{13}$, 7.7%　　　　**d.** 1, 100%

7. a. $\frac{2}{45}$　　**b.** $\frac{8}{9}$　　**c.** $\frac{1}{9}$　　**d.** $\frac{1}{9}$

8. a. $\frac{3}{14}$　　**b.** $\frac{1}{7}$　　**c.** $\frac{1}{7}$　　**d.** $\frac{5}{14}$

9. a. $\frac{11}{14}$　　**b.** $\frac{4}{7}$　　**c.** 0　　**d.** $\frac{13}{14}$

10. a. $\frac{1}{2}$　　**b.** $\frac{1}{2}$　　**c.** 0　　**d.** 1

11. a. $\frac{1}{365}$　　　　**b.** $\frac{7}{365}$

  c. $\frac{30}{365} = \frac{6}{73}$　　　　**d.** $\frac{1}{365}$

12. a. Yes, equal sectors　　**b.** $\frac{1}{4}$

13. a. No, sector 1 occupies　　**b.** $\frac{2}{3}$
  a larger area.

14. a. No, sector 1 occupies the smallest area.

  b. $\frac{1}{8}$

15. $\frac{1}{2}$

16. a. $\frac{1}{52}$, 1.9%　　　　**b.** $\frac{1}{13}$, 7.7%

  c. $\frac{2}{13}$, 15.4%　　　　**d.** $\frac{1}{4}$, 25%

17. a.

  b. Each sector has an angle of 72° at the centre of the spinner.

18. a. Heads: $\frac{1}{2}$; Tails: $\frac{1}{2}$　　**b.** 1

19. a. HH: $\frac{1}{4}$; HT: $\frac{1}{4}$; TH: $\frac{1}{4}$; TT: $\frac{1}{4}$

  b. 1

20. a.

  b. Sectors A and D: 90°, sector B: 120° and sector C: 60°

21. a.

  b. Sectors 1 and 2 have angles of 90° at the centre of the spinner, and sector 3 has an angle of 180°.

22. C

23. a. 1　　　　　　　　　**b.** $\frac{3}{8}$

24. a. $\frac{1}{4}$　　**b.** $\frac{1}{2}$　　**c.** $\frac{1}{2}$　　**d.** $\frac{1}{3}$

25. a. $\frac{1}{4}$　　**b.** $\frac{5}{9}$　　**c.** $\frac{1}{2}$

26. a. $\frac{1}{4}$　　**b.** $\frac{1}{6}$　　**c.** $\frac{35}{36}$　　**d.** 1

11.5 Complementary events

1. Not complementary, as there are other things that you could have for breakfast

2. Not complementary, as there are other ways of travelling to your friend's house

3. Not complementary, as there are other things that you could be doing

4. Complementary, as this covers all possible outcomes

5. Not complementary, as neither case covers the possibility of rolling a 4

6. a. Selecting an odd number

  b. Selecting a consonant

  c. The coin landing Tails

  d. Getting a number greater than 2

7. a. Getting a total of 12

  b. Not selecting a diamond

  c. Not selecting an E

  d. Not selecting a blue marble

8. $\frac{12}{13}$

9. A

10. $\frac{215}{216}$

11. $\frac{4}{5}$

12. a. $\frac{1}{4}$　　**b.** $\frac{3}{4}$　　**c.** $\frac{3}{4}$　　**d.** $\frac{1}{4}$

13. No, the two events are not complementary, as the sum of their probabilities does not equal one. Getting one Head is also an outcome.

14. a. 0　　**b.** $\frac{7}{11}$　　**c.** $\frac{4}{11}$

15. a. 1　　**b.** $\frac{r}{n}$　　**c.** $\frac{n-r}{n}$　　**d.** $\frac{r}{n}$

16. If an event has only 2 possible outcomes that have nothing in common, they are complementary events. If the probability of the event occurring (P) is added to the probability of the event not occurring ($P\prime$, i.e. the complement), the result will be a probability of 1. The event will either occur or not occur.

17. $\dfrac{3}{10}$

18. a. $\dfrac{3}{4}$ **b.** 20% **c.** 0.7

19. a. $\dfrac{1}{100}$ **b.** $\dfrac{99}{100}$ **c.** 20

  d. 50

20. a. $\dfrac{3}{11}$ **b.** $\dfrac{5}{11}$

Project

1. Sample response:

Score	Tally	Frequency
1	卌 卌 卌 卌 卌 ‖‖	28
2	卌 卌 卌 ‖‖	18
3	卌 卌 卌	15
4	卌 卌 卌 卌 ‖	22
5	卌 卌 卌 卌 ‖‖‖	24
6	卌 卌 ‖‖	13
	Total	120

2. They are equally likely because each section is the same size.

3. $\dfrac{1}{6}$

4. Sample response:

Section 1: $\dfrac{28}{120} = \dfrac{7}{30}$

Section 2: $\dfrac{18}{120} = \dfrac{3}{20}$

Section 3: $\dfrac{15}{120} = \dfrac{3}{24}$

Section 4: $\dfrac{22}{120} = \dfrac{11}{60}$

Section 5: $\dfrac{24}{120} = \dfrac{3}{15}$

Section 6: $\dfrac{13}{120}$

5. 20

6. Sample response:
There is a large variance from the expected results of 20 per section.

7. Sample response:

Score	Frequency
1	513
2	491
3	484
4	520
5	503
6	489
Total	3000

8. Sample response:
The class results were closer to the expected value (500) than the individual results.

9. Sample response:
As the number of trials increases, the data collected seems to approach closer to the expected results.

11.6 Review questions

1. A

2. $\dfrac{1}{200}$

3. a. $\dfrac{1}{3}$ **b.** $\dfrac{1}{6}$ **c.** $\dfrac{2}{3}$ **d.** $\dfrac{1}{2}$

4. a. $\dfrac{1}{5}$ **b.** $\dfrac{2}{5}$ **c.** 0 **d.** $\dfrac{4}{5}$

5. a. i. $\dfrac{1}{5}$ **ii.** $\dfrac{4}{5}$

  b. 1

6. Sector 1: $\dfrac{1}{4}$ Sector 2: $\dfrac{1}{8}$ Sector 3: $\dfrac{1}{8}$

  Sector 4: $\dfrac{1}{6}$ Sector 5: $\dfrac{1}{6}$ Sector 6: $\dfrac{1}{6}$

7. a. Complementary, as all possible outcomes are covered
  b. Complementary, as all possible outcomes are covered
  c. Not complementary, as there are other means to travel to work

8. $\dfrac{4}{7}$

9. a. The coin landing Heads
  b. Losing the race
  c. Answering the question incorrectly
  d. Not selecting a black marble
  e. Selecting a number greater than 19

10. a. $\{1, 2, 3, 4, 5, 6\}$
  b. No, because the sectors are of varying angle size. 3 and 6 have sectors that are double the size of others; therefore, they will have a larger probability.
  c. i. $\dfrac{1}{8}$ **ii.** $\dfrac{1}{4}$

11. a.

b. Sector 1: 180°   Sector 2: 90°

c. Sector 3: 45°   Sector 4: 45°

12. a. $\dfrac{2}{11}$ **b.** $\dfrac{9}{11}$

c. $\dfrac{4}{11}$ **d.** $\dfrac{7}{11}$

13. a. $\dfrac{3}{4}$

b. Stayed the same. Recalculating the probability gives $\dfrac{3}{4}$ again.

14. 18 boys

15. Jar A

16. a. {(1, 1), (1, 2), (1, 3), (1, 4), (2, 1), (2, 2), (2, 3), (2, 4), (3, 1), (3, 2), (3, 3), (3, 4), (4, 1), (4, 2), (4, 3), (4, 4)}

b. $\dfrac{1}{4}$

17. $\dfrac{5}{11}$

18. $\dfrac{1}{6}$

19. a. $\dfrac{11}{40}$ **b.** $\dfrac{16}{25}$

c. $\dfrac{73}{200}$ **d.** $\dfrac{127}{200}$

20. 5

21. a. 1000 yellow balls

b. 10 red balls, 50 yellow balls and 40 blue balls

22. $\dfrac{2}{3}$

12 Right-angled triangles (Pythagoras' theorem)

LESSON
12.1 Overview

Why learn this?

Pythagoras was a famous mathematician and Greek philosopher who lived about 2500 years ago. He is particularly well known for investigating right-angled triangles and proving that there is a special relationship between the lengths of the three sides. Think about where right-angled triangles are used and where it might be helpful to know whether a particular angle is a right angle or not. Think about angles in architecture, construction, navigation, design and woodwork. In all these fields, it is important that people know how to calculate right angles.

It might not always be possible to measure angles using a measuring device such as a protractor, so understanding the theorem relating to side lengths will be helpful here. Being able to apply Pythagoras' theorem will allow you to determine whether an angle is a right angle just from measuring the three side lengths of the triangle. Pythagoras' theorem is one of the great geometrical theorems and you'll explore his findings in this topic.

Hey students! Bring these pages to life online

▶ Watch videos

Engage with interactivities

A+ Answer questions and check solutions

Find all this and MORE in jacPLUS

Reading content and rich media, including interactivities and videos for every concept

Extra learning resources

Differentiated question sets

Questions with immediate feedback, and fully worked solutions to help students get unstuck

1. Evaluate $\sqrt{4^2 + 3^2}$.

2. **a.** Write the decimal 14.3875 correct to 2 decimal places.
   b. Write the decimal 14.3875 correct to 3 significant figures.

3. **MC** A right-angled triangle has side lengths of 12 cm, 13 cm and 5 cm. Determine which of these lengths is the hypotenuse of the triangle.
   A. 12 cm   **B.** 13 cm   **C.** 5 cm   **D.** 169 cm

4. **MC** A right-angled triangle has a base measuring 6 cm and perpendicular height measuring 11 cm.

   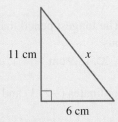

   Select which of the following is the length of the hypotenuse x written in exact form.
   A. 12.5 cm   **B.** 33 cm   **C.** 157 cm   **D.** $\sqrt{157}$ cm

5. **MC** A ladder measuring 2.4 m in length leans up against a wall. The foot of the ladder is 60 cm from the base of the wall. Calculate how high up the wall the ladder will reach.
   A. 2.5 m   **B.** 2.4 m   **C.** 2.3 m   **D.** 2.2 m

6. A bushwalker starts at point A and walks north 2.5 km to point B. At point B, she turns east and walks another 2.1 km to point C. She then walks directly from point C to point A. Calculate how far she has walked in total, giving your answer to 1 decimal place.

7. A square has a diagonal 12 cm in length. Calculate the side length of the square. Write the answer correct to 1 decimal place.

8. Calculate the perimeter of the trapezium shown.

   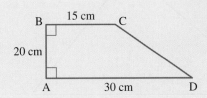

9. Calculate the length of x to 1 decimal place.

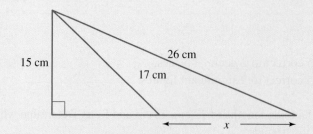

10. The smallest number of a Pythagorean triad is 11. Determine the middle number and the largest number. Write the middle number first.

11. **MC** A box is a cube with a side length of 8 cm. The longest pencil that fits is placed in the box. Select the length of that pencil from the following options.

    A. 8 cm   **B.** 11.3 cm   **C.** 13.9 cm   **D.** 16 cm

12. Determine the length of the line that joins the coordinates $(-2, 1)$ and $(5, 8)$. Write your answer to 1 decimal place.

13. **MC** A square-based pyramid structure is made from wire. The square base has side lengths of 10 cm and the vertical distance from the base of the structure to the vertex is 10 cm. The amount of wire required to make the pyramid is closest to:

    A. 80 cm   **B.** 83 cm   **C.** 86 cm   **D.** 89 cm

14. The total surface area of a cone is found by using the formula $A = \pi r^2 + \pi rs$, where r is the radius of the base circle and s is the slant height of the cone (that is, the distance from the vertex to a point on the circumference of the base).
Evaluate the total surface area of the cone shown.

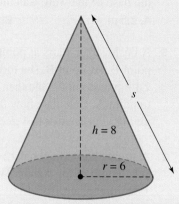

15. A right-angled triangle is drawn with the hypotenuse (60 cm) as the base of the triangle. A vertical line is drawn from the vertex to the base of the triangle. This vertical line creates two further right-angled triangles and splits the base into the ratio 2 : 3.

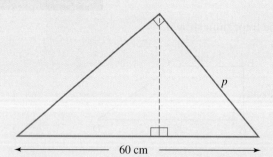

Evaluate the length of the shorter side, p, in the original right-angled triangle to the nearest whole number.

LESSON
12.2 Pythagoras' theorem

LEARNING INTENTION

At the end of this lesson you should be able to:
- understand that the hypotenuse is the longest side of a right-angled triangle
- understand Pythagoras' theorem and how it describes the relationship between the side lengths of right-angled triangles
- apply Pythagoras' theorem to determine the length of the hypotenuse.

▶ 12.2.1 Right-angled triangles

eles-4535

- A **right-angled triangle** contains a 90° angle (right angle: ⌐_).
- In all right-angled triangles, the longest side is always opposite the right angle.
- The longest side is called the **hypotenuse**.

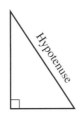 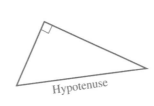 

WORKED EXAMPLE 1 Identifying the hypotenuse side of a right-angled triangle

For the right-angled triangles shown, state which side is the hypotenuse.

a.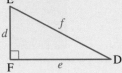

b.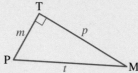

THINK	**WRITE/DRAW**
a. The hypotenuse is always opposite the right angle.	a. Side *f* is opposite the right angle. Therefore, side *f* is the hypotenuse. 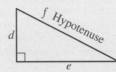
b. The hypotenuse is opposite the right angle.	b. Side *t* is opposite the right angle. Therefore, side *t* is the hypotenuse. 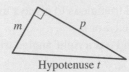

COMMUNICATING — COLLABORATIVE TASK: The relationship between right-angled triangles and squares

For this activity you will work with a partner, and you will need graph paper, coloured pencils, glue and scissors.

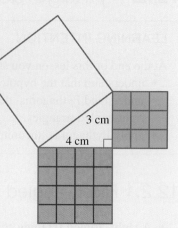

1. On a sheet of graph paper, draw a right-angled triangle with a base of 4 cm and a height of 3 cm.
2. Carefully draw a square on each of the three sides of the triangle. Mark a grid on each so that the square on the base is divided into 16 small squares, while the square on the height is divided into 9 small squares.
3. Colour the square on the base and the square on the height in different colours, as shown, so that you can still see the grid lines.
4. Carefully cut out the two coloured squares from the triangle.
5. Now stick the larger of the coloured squares on the uncoloured square of the triangle (the square on the hypotenuse).
6. Using the grid lines as a guide, cut the smaller square up and fit it on the remaining space. The two coloured squares should have exactly covered the third square.
7. Comment on what you notice about the hypotenuse and the other two sides of the right-angled triangle.

▶ 12.2.2 Pythagoras' theorem

eles-4536

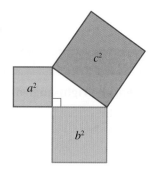

- A theorem is the statement of a mathematical truth.
- The Greek mathematician Pythagoras (c. 582–500 BCE) is credited as the first to describe the relationship now known as Pythagoras' theorem.
- **Pythagoras' theorem** states that for any right-angled triangle, the square of the length of the hypotenuse is equal to the sum of the squares of the lengths of the two shorter sides.

Pythagoras' theorem

For the right-angled triangle shown:

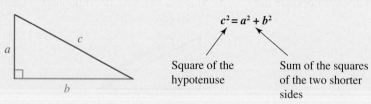

$$c^2 = a^2 + b^2$$

Square of the hypotenuse

Sum of the squares of the two shorter sides

- The key to labelling a right-angled triangle is to always first label the hypotenuse, c. It makes no difference which of the shorter sides is labelled a or b.
- A triangle with side lengths a, b and c is a right-angled triangle if $c^2 = a^2 + b^2$.

WORKED EXAMPLE 2 Determining whether a triangle is right-angled

Determine which of the following triangles are not right-angled triangles.

a.

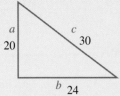

b. **A triangle with side lengths 3 cm, 4 cm and 5 cm**

THINK	WRITE
a. 1. For a right-angled triangle, Pythagoras' theorem will be true.	a. $c^2 = a^2 + b^2$
2. Identify the values of a, b and c. The longest side is c. Substitute the values into the left- and right-hand sides of Pythagoras' theorem to see whether the values make the equation true.	$\begin{aligned} \text{LHS} &= 30^2 \\ &= 900 \end{aligned}$ $\quad$ $\begin{aligned} \text{RHS} &= 20^2 + 24^2 \\ &= 976 \end{aligned}$ $\quad$ $\text{RHS} \neq \text{LHS}$
3. Pythagoras' theorem is not true for this triangle. Write the answer.	The triangle is not a right-angled triangle.
b. 1. Identify the values of a, b and c. The largest value is 5 cm; therefore, this is the hypotenuse, c. It does not matter which of the other sides are a or b.	b. Side lengths: $c = 5\,\text{cm}$ $\qquad\qquad\qquad\quad a = 3\,\text{cm}$ $\qquad\qquad\qquad\quad b = 4\,\text{cm}$
2. Substitute the values into the left- and right-hand sides of Pythagoras' theorem to see whether the values make the equation true.	$c^2 = a^2 + b^2$ $\begin{aligned} \text{LHS} &= 5^2 \\ &= 25 \end{aligned}$ $\quad$ $\begin{aligned} \text{RHS} &= 3^2 + 4^2 \\ &= 25 \end{aligned}$ $\quad$ $\text{RHS} \neq \text{LHS}$
3. Pythagoras' theorem is true for these values. Write the answer.	The triangle is a right-angled triangle.

- Calculations with Pythagoras' theorem often result in a number under the root sign $\left(\sqrt{}\right)$, called the radicand, that is not a square number. In such cases, answers may be left unsimplified. This is called exact (surd) form. A calculator may be used to find an approximate answer, which is usually then given to a specified number of decimal places.
- The square root of a number can be a positive or a negative number, but when we are talking about length it is always positive.
- The $\approx$ symbol means *is approximately equal to*.

WORKED EXAMPLE 3 Determining the length of the hypotenuse

For the triangle shown, determine the length of the hypotenuse, x.

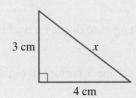

THINK	WRITE/DRAW
1. Copy the diagram and label the sides a, b and c. Remember to label the hypotenuse as c.	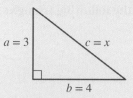
2. Write the Pythagoras' theorem.	$c^2 = a^2 + b^2$
3. Substitute the values of a, b and c into this rule and simplify.	$x^2 = 3^2 + 4^2$ $= 9 + 16$ $= 25$
4. Calculate x by taking the square root of 25.	$x = \sqrt{25}$
5. Write the answer.	$= 5\,\text{cm}$

WORKED EXAMPLE 4 Applying Pythagoras' theorem

There is a fire on top of a building. A child needs to be rescued from a window that is 12 metres above ground level.

If the rescue ladder can be placed no closer than 5 metres from the foot of the building, determine the shortest ladder, in metres, needed for the rescue.

THINK	WRITE/DRAW
1. Draw a diagram and label the sides a, b and c. Remember to label the hypotenuse as c.	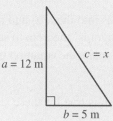
2. Write the Pythagoras' theorem.	$c^2 = 12^2 + 5^2$
3. Substitute the values of a, b and c into this rule and simplify.	$x^2 = 12^2 + 5^2$ $= 144 + 25$ $= 169$
4. Calculate x by taking the square root of 169.	$x = \sqrt{169}$ $x = 13$
5. Write the answer in a sentence.	The shortest ladder needs to be 13 metres long.

Digital technology

Newer scientific calculators can display answers in both decimal and exact form.

In most cases, your calculator's default setting is to display a decimal answer.

For example, type $\sqrt{14}$ and press ENTER (or the = button) to obtain the decimal approximation.

By changing the settings (or mode) it may be possible to display answers as exact answers. The name of this setting differs between calculator brands.

For the TI-30XB calculator shown, change the mode to MATHPRINT.

Being able to change your calculator's settings can be useful when needing to change between a decimal and exact answer.

on Resources

Interactivity Finding the hypotenuse (int-3844)

Exercise 12.2 Pythagoras' theorem

learn on

12.2 Quick quiz on	12.2 Exercise

Individual pathways

■ PRACTISE	■ CONSOLIDATE	■ MASTER
1, 3, 4, 7, 11, 12, 15	5, 8, 9, 13, 16	2, 6, 10, 14, 17, 18

Fluency

1. **WE1** For the right-angled triangles shown, state which side is the hypotenuse.

   a.    b.   c.   d.

2. **WE2** Determine which of the following are right-angled triangles.

   a.

   b.

   c.

   d.

   e. A triangle with side lengths of 22.5 km, 54 km and 58.5 km

   f.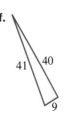

   g. A triangle with side lengths of 53 mm, 185 mm and 104 mm

   h.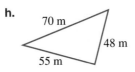

3. **WE3** For the following triangles, calculate the length of the hypotenuse, x, correct to 1 decimal place (where necessary).

a.
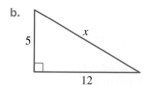

b.
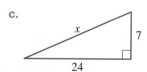

c.
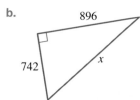

4. For the following triangles, calculate the length of the hypotenuse, x, correct to 1 decimal place (where necessary).

a.
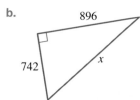

b.
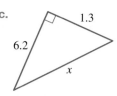

c.
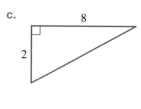

5. For each of the following triangles, determine the length of the hypotenuse. Leave your answers in exact form.

a.
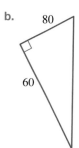

b.
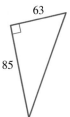

c.
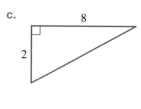

6. For each of the following triangles, determine the length of the hypotenuse. Leave your answers in exact form.

a.
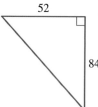

b.
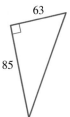

c.
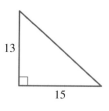

Understanding

7. Determine the lengths, correct to 2 decimal places, of the diagonals of squares that have side lengths of:

   a. 12 cm

   b. 20 mm

   c. 4.9 cm.

8. A right-angled triangle has a base of 5 cm and a perpendicular height of 11 cm. Determine the length of the hypotenuse. Leave your answer in exact form.

9. Determine the lengths, correct to 2 decimal places, of the diagonals of rectangles whose sides are:

   a. 10 cm and 8 cm

   b. 620 cm and 400 cm

   c. 17 cm and 3 cm.

10. An isosceles triangle has a base of 30 cm and a perpendicular height of 10 cm. Determine the length of the two equal sides of the isosceles triangle. Give your answer correct to 2 decimal places.

11. **WE4** A ladder leans against a vertical wall. The foot of the ladder is 1.2 m from the wall, and the top of the ladder reaches 4.5 m up the wall. Calculate the length of the ladder. Give your answer correct to 2 decimal places.

Communicating, reasoning and problem solving

12. Sanjay is building a chicken coop. The frame of the coop is going to be a right-angled triangle, which will look like the diagram below.

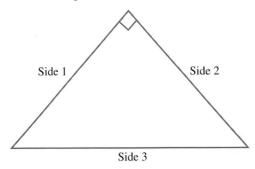

a. Determine which side of this section is the longest.
b. If side 1 and side 2 are each 1 m long, determine how long side 3 is. Give your answer to 1 decimal place and include units.
c. Sanjay plans to buy one long piece of wood, then cut the wood to make the three sides. Estimate the length of the piece of wood that Sanjay needs to buy.

13. A ladder rests against a vertical wall, with the top of the ladder 7 m above the ground. If the bottom of the ladder was moved 1 m further away from the foot of the wall, the top of the ladder would rest against the foot of the wall. Evaluate the length of the ladder, l.

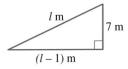

14. A smaller square is drawn inside a large square, as shown in the diagram. Use Pythagoras' theorem to determine the side length and hence the area of the smaller square.

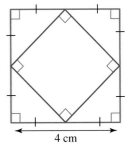

15. a. Calculate the volume of the prism shown.
    b. Determine the length of the sloping edge of the cross-section.

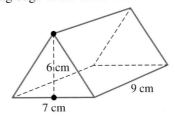

16. A right-angled triangle has a perpendicular height of 17.2 cm, and a base that is half the height. Determine the length of the hypotenuse, correct to 2 decimal places.

17. Wally is installing a watering system in his garden. The pipe is to go all around the edge of the rectangular garden and have a branch running diagonally across the garden. The garden measures 5 m by 7.2 m. If the pipe costs $2.40 per metre (or part thereof), evaluate the total cost of the pipe. Show your working.

18. A right-angled isosceles triangle has an area of 200 cm². Evaluate the area of the semicircle that sits on the hypotenuse.

LESSON
12.3 Calculating shorter side lengths

LEARNING INTENTION

At the end of this lesson you should be able to:
- determine the lengths of the shorter sides of a right-angled triangle.

▶ 12.3.1 Lengths of sides of right-angled triangles
eles-4537

- If the lengths of two sides in a right-angled triangle are known, the third side can be calculated using Pythagoras' theorem.

WORKED EXAMPLE 5 Determining the shorter side length

Determine the length of the unmarked side of the triangle shown.
Leave your answer in exact form.

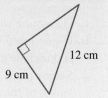

THINK	WRITE/DRAW
1. Copy the diagram and label the sides a, b and c. Remember to label the hypotenuse as c.	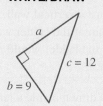
2. Write Pythagoras' theorem.	$c^2 = a^2 + b^2$
3. Substitute the values of a, b and c into this rule and simplify.	$12^2 = a^2 + 9^2$ $144 = a^2 + 81$ $144 - 81 = a^2$ $63 = a^2$
4. Determine the exact value of a by taking the square root of 63.	$a = \sqrt{63}$ cm

---

WORKED EXAMPLE 6 Applying Pythagoras' theorem in real contexts

A ladder that is 4.5 m long leans against a vertical wall. The foot of the ladder is 1.2 m from the wall.

Calculate how far up the wall the ladder reaches. Give your answer correct to 2 decimal places.

THINK	WRITE/DRAW
1. Draw a diagram and label the sides a, b and c. Remember to label the hypotenuse as c.	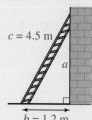

---

2. Write Pythagoras' theorem.

$$c^2 = a^2 + b^2$$

3. Substitute the values of a, b and c into this rule and simplify.

$$4.5^2 = a^2 + 1.2^2$$
$$20.25 = a^2 + 1.44$$
$$20.25 - 1.44 = a^2$$
$$18.81 = a^2$$

4. Calculate a by taking the square root of 18.81. Round to 2 decimal places.

$$a = \sqrt{18.81}$$
$$\approx 4.34 \, \text{m}$$

5. Write the answer.

The ladder will reach a height of 4.34 m up the wall.

DISCUSSION

When taking the square root of a number, why is it not feasible to use the negative solution when solving problems involving Pythagoras' theorem?

 Resources

 Interactivity Finding the shorter side (int-3845)

Exercise 12.3 Calculating shorter side lengths

learn on

12.3 Quick quiz on	**12.3 Exercise**

Individual pathways

■ PRACTISE	■ CONSOLIDATE	■ MASTER
1, 3, 5, 6, 11, 12, 15	2, 7, 9, 13, 16	4, 8, 10, 14, 17

Fluency

1. **WE5** Determine the length of the unmarked side in each of the following triangles. Leave your answer in exact form.

a.

10
8

b.
14
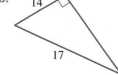
17

c.
32
84

2. Determine the value of the pronumeral in each of the following triangles, correct to 2 decimal places.

a.
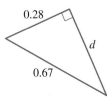
0.28
0.67
d

b.
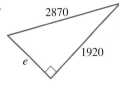
2870
1920
e

c.
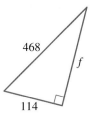
468
114
f

3. Determine the length of the unmarked side in each of the following triangles. Leave your answer in exact form.

a.

382
457

b.
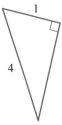
1
4

c.
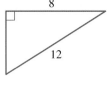
8
12

4. Determine the value of the pronumeral in each of the following triangles, correct to 2 decimal places.

a.
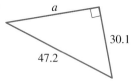
a
30.1
47.2

b.
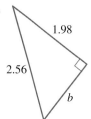
1.98
2.56
b

c.
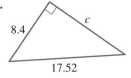
c
8.4
17.52

Understanding

For questions 5 to 15, give your answers correct to 2 decimal places.

5. The diagonal of the rectangular sign shown is 34 cm.
   If the height of this sign is 25 cm, what is the width?

6. The diagonal of a rectangle is 120 cm. One side has a length of 70 cm.
   Determine:
   a. the length of the other side
   b. the perimeter of the rectangle
   c. the area of the rectangle.

7. An equilateral triangle has sides of length 20 cm. Determine the height of the triangle.

8. The roundabout sign shown is in the form of an equilateral triangle.
   Determine the height of the sign and hence its area.

9. **WE6** A ladder that is 7 metres long leans up against a vertical wall. The
   top of the ladder reaches 6.5 m up the wall. Calculate how far from the
   wall is the foot of the ladder.

10. A tent pole that is 1.5 m high is to be supported by ropes attached to the
    top. Each rope is 2 m long. Calculate how far from the base of the pole the
    rope can be pegged.

11. The size of a rectangular television screen is given by the length of its diagonal. The television shown has a 92-cm screen. Determine the length of the shorter side. Show your working.

80 cm

Communicating, reasoning and problem solving

12. You are helping your friend answer some maths questions. This is what your friend has written:

$$c^2 = a^2 + b^2 \qquad \text{Line 1}$$
$$10^2 = a^2 + 4^2 \qquad \text{Line 2}$$
$$100 = a^2 + 8 \qquad \text{Line 3}$$

a. i. Determine the error made by your friend in line 3.
   ii. Rewrite line 3 so that it is correct.
   iii. Evaluate the value of a, giving your answer to 1 decimal place. Show your working.

For another question, this is what your friend has written:

$$c^2 = a^2 + b^2 \qquad \text{Line 1}$$
$$5^2 = a^2 + 4^2 \qquad \text{Line 2}$$
$$25 = a^2 + 16 \qquad \text{Line 3}$$
$$25 + 16 = a^2 \qquad \text{Line 4}$$

b. i. Determine the error made by your friend in line 4.
   ii. Rewrite line 4 so that it is correct.
   iii. Evaluate the value of a. Show your working.

13. Penny is building the roof for a new house. The roof has a gable end in the form of an isosceles triangle, with a base of 6 m and sloping sides of 7.5 m. She decides to put 5 evenly spaced vertical strips of wood as decoration on the gable, as shown. Determine how many metres of this decorative wood she needs. Show your working.

7.5 m   7.5 m
6 m

14. Ben's dog Macca has wandered onto a frozen pond and is too frightened to walk back. Ben estimates that the dog is 3.5 m from the edge of the pond. He finds a plank, 4 m long, and thinks he can use it to rescue Macca. The pond is surrounded by a bank that is 1 m high. Ben uses the plank to make a ramp for Macca to walk up. Explain whether he will be able to rescue his dog.

15. A kite is attached to a string 15 m long. Sam holds the end of the string 1 m above the ground, and the horizontal distance of the kite from Sam is 8 m, as shown.
    Evaluate how far the kite is above the ground. Show your working.

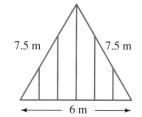
15 m
8 m
1 m

16. An art student is trying to hang her newest painting on an existing hook in a wall. She leans a 1.2-m ladder against the wall so that the distance between the foot of the ladder and the wall is 80 cm.

a. Draw a sketch showing the ladder leaning against the wall.
b. Determine how far the ladder reaches up the wall.
c. The student climbs the ladder to check whether she can reach the hook from the step at the very top of the ladder. Once she extends her arm, the distance from her feet to her fingertips is 1.7 m. If the hook is 2.5 m above the floor, determine whether the student will reach it from the top step.

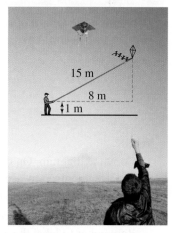

17. a. A rectangle has two sides of length 3 cm and diagonals of length 5 cm. Evaluate the length of the other two sides.
    b. It is known that the diagonals of this rectangle bisect each other at the point of intersection. Using this fact, verify (using appropriate calculations) whether the diagonals are perpendicular to each other.

LESSON
12.4 Applying Pythagoras' theorem

LEARNING INTENTION

At the end of this lesson you should be able to:
- apply Pythagoras' theorem in familiar and unfamiliar contexts.

▶ 12.4.1 Applying Pythagoras' theorem to composite shapes

eles-4538

- A **composite shape** can be defined as a shape that can be made from smaller, more recognisable shapes.

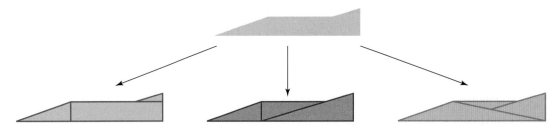

- Dividing a composite shape into simpler shapes creates shapes that have known properties. For example, to calculate the value of x in the trapezium shown, a vertical line can be added to create a right-angled triangle and a rectangle. The length of x can be found using Pythagoras' theorem.

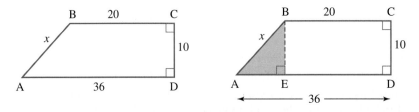

WORKED EXAMPLE 7 Determining the side length of a trapezium

Determine the length of side x in the figure shown. Give your answer correct to 2 decimal places.

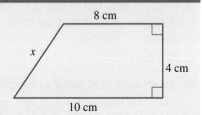

THINK

1. Copy the diagram. On the diagram, create a right-angled triangle and use the given measurements to work out the lengths of the two sides of the triangle needed to determine the hypotenuse.

2. Label the sides of your right-angled triangle as a, b and c. Remember to label the hypotenuse as c.

3. Check that all measurements are in the same units. They are the same.

WRITE/DRAW

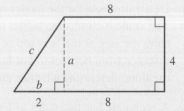

---

4. Write Pythagoras' theorem.

$$c^2 = a^2 + b^2$$

5. Substitute the values of a, b and c into this rule and simplify.

$$x^2 = 4^2 + 2^2$$
$$= 16 + 4$$
$$= 20$$

6. Evaluate x by taking the square root of 20. Round your answer correct to 2 decimal places.

$$x = \sqrt{20}$$
$$\approx 4.47 \, \text{cm}$$

WORKED EXAMPLE 8 Solving triangles

For the diagram shown, determine the lengths of the sides marked x and y correct to 2 decimal places.

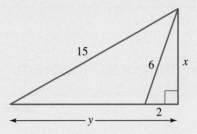

THINK

1. Identify and draw any right-angled triangles contained in the diagram. Label their sides.

2. To determine an unknown side in a right-angled triangle, we need to know two sides, so evaluate x first.

WRITE/DRAW

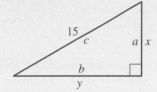

3. For the triangle containing x, write down Pythagoras' theorem, substitute the values of a and b into this rule, transpose to determine x and simplify.

$$c^2 = a^2 + b^2$$
$$6^2 = x^2 + 2^2$$
$$36 = x^2 + 4$$
$$x^2 = 36 - 4$$
$$= 32$$
$$x = \sqrt{32}$$
$$\approx 5.66$$

4. We now know two sides of the other triangle because we can substitute $x = 5.66$.

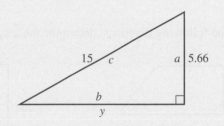

5. For the triangle containing y, write down Pythagoras' theorem.
Substitute the values of c and a into this rule and simplify.

$$c^2 = a^2 + b^2$$
$$15^2 = x^2 + y^2$$
$$225 = \sqrt{32}^2 + y^2$$
$$y^2 = 225 - 32$$
$$= 193$$
$$y = \sqrt{193}$$
$$\approx 13.89$$

 Resources

 Interactivity Composite shapes (int-3847)

Exercise 12.4 Applying Pythagoras' theorem

learn

12.4 Quick quiz on

12.4 Exercise

Individual pathways

■ PRACTISE	■ CONSOLIDATE	■ MASTER
1, 4, 7, 8, 10, 14, 17	2, 5, 9, 11, 13, 15, 18	3, 6, 12, 16, 19, 20

Where appropriate, give answers correct to 2 decimal places.

Fluency

1. **WE7** Determine the length of the side x in the figure shown.

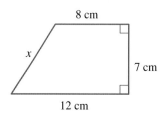

2. For the following diagrams, determine the length of the sides marked x.

a.
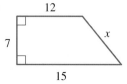

b.
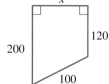

c.
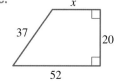

3. **WE8** For each of the following diagrams, determine the length of the sides marked x and y.

a.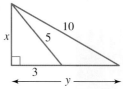

b.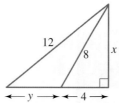

c.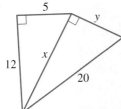

d.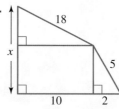

4. **MC** The length of the diagonal of the rectangle shown is:

A. 9.7
B. 19
C. 13.9
D. 12.2

5. **MC** The area of the rectangle shown is:

A. 12.1
B. 84.9
C. 98
D. 169.7

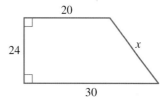

6. **MC** The value of x in this shape is:

A. 24
B. 26
C. 38.4
D. 10

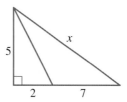

7. **MC** Select the correct value of x in this figure.

A. 5.4
B. 8.6
C. 10.1
D. 10.3

Understanding

8. A feature wall in a garden is in the shape of a trapezium, with parallel sides of 6.5 m and 4.7 m. The wall is 3.2 m high. It is to have fairy lights around the perimeter (except for the base). Calculate how many metres of lighting are required.

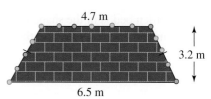

9. Jess paddles a canoe 1700 m to the west, then 450 m south, and then 900 m to the east. They then stop for a rest. Calculate how far they are from her starting point.

10. In a European city, two buildings, 10 m and 18 m high, are directly opposite each other on either side of a street that is 6 m wide (as shown in the figure). Determine the distance between the tops of the two buildings.

11. A yacht race starts and finishes at A and consists of 6 legs: AB, BC, CA, AD, DC, CA, in that order, as shown in the figure. If AB = 4 km, BC = 3 km and CD = 3 km, determine:

    a. AC
    b. AD
    c. the total length of the race.

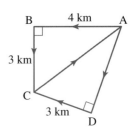

12. A painter uses a trestle to stand on in order to paint a ceiling. The trestle consists of 2 stepladders connected by a 4-m-long plank. The inner feet of the 2 stepladders are 3 m apart, and each ladder has sloping sides of 2.5 m. Calculate how far the plank is above the ground.

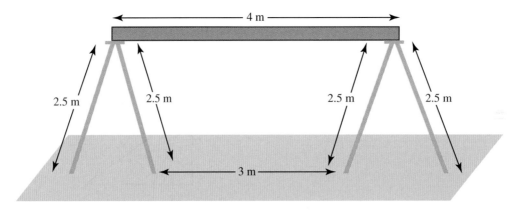

13. A garden bed is in the shape of a right-angled trapezium, with the sloping edge of 2.0 m and parallel sides of 3.2 m and 4.8 m. Determine the width of the garden and hence the area.

Communicating, reasoning and problem solving

14. A rectangular gate is 3.2 m long and 1.6 m high, and consists of three horizontal beams and five vertical beams, as shown in the diagram. Each section is braced with diagonals. Calculate how much timber is needed for the gate.

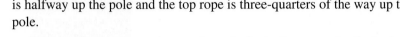

15. A music festival is coming to town. A huge tent is being set up as shown, with poles that are 4 m high. Ropes are needed to secure the poles. The diagram below shows how one set of ropes is attached. The bottom rope is halfway up the pole and the top rope is three-quarters of the way up the pole.

    a. Calculate how many metres up the pole the bottom rope is attached.
    b. Determine how many metres up the pole the top rope is attached.
    c. Determine the length, correct to 2 decimal places, of:
       i. the bottom rope
       ii. the top rope.
    d. Each pole needs to be secured by four (identical) sets of rope. Evaluate how much rope is needed to secure one pole. Show your working.

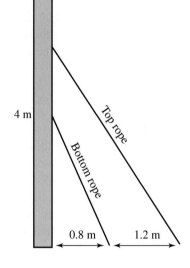

16. Evaluate the distance AB in the following plan of a paddock. Distances are in metres.

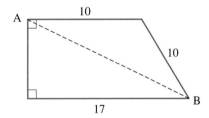

17. Consider the following two shapes, labelled shape 1 and shape 2.

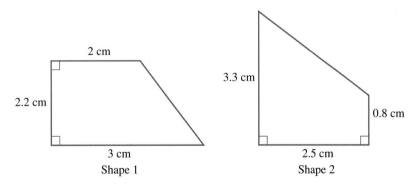

    Answer the following questions, giving your answers correct to 2 decimal places.

    a. Determine the length of the missing side in shape 1.
    b. Calculate the perimeter of shape 1.
    c. Determine the length of the missing side in shape 2.
    d. Calculate the perimeter of shape 2.
    e. Determine which shape has a larger perimeter. Show your working.

18. The front door shown is 1 m wide and 2.2 m high and has four identical glass panels, each 76 cm long and 15 cm wide.

  a. Calculate the total area of the glass panels.
  b. The door is to be painted inside and outside. Evaluate the total area to be painted.
  c. Two coats of paint are needed on each side of the door. If the paint is sold in 1-L tins at $24.95 per litre and each litre covers 8 m² of the surface, determine the total cost of painting the door.

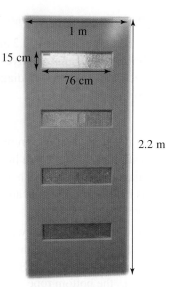

19. An equilateral triangle is drawn on each of the four sides of a square, with the final figure resembling a four-pointed star. If the sides of the square are x cm long, evaluate the area of the complete figure.

20. The leadlight panel shown depicts a sunrise over the mountains. The mountain is represented by a green triangle 45 cm high. The yellow sun is represented by a section of a circle with an 18-cm radius. There are 10 yellow sunrays in the shape of isosceles triangles with a base of 3 cm and a height of 12 cm. The sky is blue.
Evaluate the area of the leadlight panel made of:

  a. green glass
  b. yellow glass
  c. blue glass.

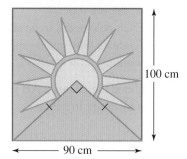

LESSON
12.5 Pythagorean triads

LEARNING INTENTION

At the end of this lesson you should be able to:
  • understand the concept of Pythagorean triads
  • solve Pythagorean triads.

⊙ 12.5.1 Pythagorean triads

eles-4539

  • A **Pythagorean triad** is a group of any three whole numbers that satisfy Pythagoras' theorem. For example, {5, 12, 13} and {7, 24, 25} are Pythagorean triads.

$$13^2 = 5^2 + 12^2$$
$$25^2 = 24^2 + 7^2$$

  • Pythagorean triads are useful when solving problems using Pythagoras' theorem. If two known side lengths in a triangle belong to a triad, the length of the third side can be stated without performing any calculations. Some well-known Pythagorean triads are {3, 4, 5}, {5, 12, 13}, {8, 15, 17} and {7, 24, 25}.
  • Pythagorean triads are written in ascending order. For example, the Pythagorean triad {3, 4, 5} should not be written as {4, 3, 5}.

WORKED EXAMPLE 9 Determining Pythagorean triads

Determine whether the following sets of numbers are Pythagorean triads.
a. {9, 10, 14}   **b. {33, 56, 65}**

THINK	WRITE
a. 1. Pythagorean triads satisfy Pythagoras' theorem. Substitute the values into the equation $c^2 = a^2 + b^2$ and determine whether the equation is true. Remember, c is the longest side.	**a.** $c^2 = a^2 + b^2$ LHS $= c^2$ RHS $= a^2 + b^2$ $= 14^2$ $= 9^2 + 10^2$ $= 196$ $= 81 + 100$ $= 181$
2. State your conclusion.	Since LHS $\neq$ RHS, the set {9, 10, 14} is not a Pythagorean triad.
b. 1. Pythagorean triads satisfy Pythagoras' theorem. Substitute the values into the equation $c^2 = a^2 + b^2$ and determine whether the equation is true. Remember, c is the longest side.	**b.** $c^2 = a^2 + b^2$ LHS $= 65^2$ RHS $= 33^2 + 56^2$ $= 4225$ $= 1089^2 + 3136^2$ $= 4225$
2. State your conclusion.	Since LHS $=$ RHS, the set {33, 56, 65} is a Pythagorean triad.

- If each term in a triad is multiplied by the same number, the result is also a triad. For example, if we multiply each number in {5, 12, 13} by 2, the result {10, 24, 26} is also a triad.
- Builders and gardeners use multiples of the Pythagorean triad {3, 4, 5} to ensure that walls and floors are at right angles.
- The list of Pythagorean triads is endless. Below is a list of all the triads with smallest lengths less than 100.

{3, 4, 5}	{5, 12, 13}	{7, 24, 25}	{8, 15, 17}	{9, 12, 15}	{9, 40, 41}
{11, 60, 61}	{12, 35, 37}	{13, 84, 85}	{15, 20, 25}	{15, 36, 39}	{15, 112, 113}
{16, 63, 65}	{17, 144, 145}	{19, 180, 181}	{20, 21, 29}	{20, 99, 101}	{21, 28, 35}
{21, 72, 75}	{21, 220, 221}	{23, 264, 265}	{24, 45, 51}	{24, 143, 145}	{25, 60, 65}
{27, 36, 45}	{27, 120, 123}	{28, 45, 53}	{28, 195, 197}	{32, 255, 257}	{33, 44, 55}
{33, 56, 65}	{33, 180, 183}	{35, 84, 91}	{35, 120, 125}	{36, 77, 85}	{36, 105, 111}
{39, 52, 65}	{39, 80, 89}	{39, 252, 255}	{40, 75, 85}	{44, 117, 125}	{45, 60, 75}
{45, 108, 117}	{45, 200, 205}	{48, 55, 73}	{48, 189, 195}	{49, 168, 175}	{51, 68, 85}
{51, 140, 149}	{52, 165, 173}	{55, 132, 143}	{55, 300, 305}	{56, 105, 119}	{57, 76, 95}
{57, 176, 185}	{60, 63, 87}	{60, 91, 109}	{60, 175, 185}	{60, 221, 229}	{60, 297, 303}
{63, 84, 105}	{63, 216, 225}	{63, 280, 287}	{65, 72, 97}	{65, 156, 169}	{68, 285, 293}
{69, 92, 115}	{69, 260, 269}	{72, 135, 153}	{75, 100, 125}	{75, 180, 195}	{77, 264, 275}
{81, 108, 135}	{84, 135, 159}	{84, 187, 205}	{84, 245, 259}	{85, 132, 157}	{85, 204, 221}
{87, 116, 145}	{88, 105, 137}	{88, 165, 187}	{93, 124, 155}	{95, 168, 193}	{95, 228, 247}
{96, 247, 265}	{99, 132, 165}	{99, 168, 195}	{100, 105, 145}		

You can find more Pythagorean triads on the Internet.

a. i. Form a new Pythagorean triad from the known triad {7, 24, 25}.
ii. Use substitution to show that the new triad satisfies Pythagoras' theorem.
b. Evaluate x, given that the three numbers {32, x, 68} form a Pythagorean triad.

THINK	WRITE
a. i. 1. If each term in a triad is multiplied by the same number, the result is also a triad. Choose a number to multiply each value in the triad by.	**a.** $7 \times 3 = 21$ $24 \times 3 = 72$ $25 \times 3 = 75$
2. Write the new triad.	{21, 72, 75} is a Pythagorean triad.

ii. 1. Pythagorean triads satisfy Pythagoras' theorem. Substitute the values into the equation $c^2 = a^2 + b^2$ and determine whether the equation is true. Remember, c is the longest side.	$c^2 = a^2 + b^2$
	LHS $= c^2$ RHS $= 21^2 + 72^2$
	    $= 75^2$ $= 441 + 5184$
	    $= 5625$ $= 5625$

2. State your conclusion.	Since LHS $=$ RHS, the set {21, 72, 75} is a Pythagorean triad.
b. 1. Pythagorean triads satisfy Pythagoras' theorem. Substitute the values into the equation $c^2 = a^2 + b^2$.	**b.** $c^2 = a^2 + b^2$ $68^2 = 32^2 + x^2$
2. Solve for x.	$4624 = 1024 + x^2$ $3600 = x^2$ $x = \sqrt{3600}$ $= \pm 60$
3. State the answer. The answer is positive, since the values given were positive.	$x = 60$

Resources

Interactivity Pythagorean triads (int-3848)

Exercise 12.5 Pythagorean triads

learn on

12.5 Quick quiz on	12.5 Exercise

Individual pathways

■ PRACTISE	■ CONSOLIDATE	■ MASTER
1, 2, 6, 9, 12, 15	3, 5, 8, 10, 13, 16	4, 7, 11, 14, 17

Fluency

1. Use Pythagoras' theorem to identify which of these triangles are right-angled.

   a. 6, 8, 10   **b.** 5, 12, 13   **c.** 4, 5, 6

2. Use Pythagoras' theorem to identify which of these triangles are right-angled.

    a. 24, 7, 25   b. 16, 20, 12   c. 14, 16, 30

3. **WE9** Determine whether the following sets of numbers are Pythagorean triads.

    a. {2, 5, 6}   b. {7, 10, 12}   c. {18, 24, 30}

4. Determine whether the following sets of numbers are Pythagorean triads.

    a. {30, 72, 78}   b. {8, 13, 15}   c. {30, 40, 50}

5. **MC** The smallest number of a Pythagorean triad is 9. The middle and third numbers respectively are:

    A. 41, 40   B. 10, 11   C. 11, 10   D. 40, 41

6. **MC** The smallest number of a Pythagorean triad is 11. The middle and third numbers respectively are:

    A. 20, 21   B. 30, 31   C. 60, 61   D. 50, 51

7. **MC** The smallest number of a Pythagorean triad is 13. The middle and third numbers respectively are:

    A. 54, 55   B. 64, 65   C. 74, 75   D. 84, 85

8. **MC** The smallest number of a Pythagorean triad is 29. The middle and third numbers respectively are:

    A. 420, 421   B. 520, 521   C. 620, 621   D. 720, 721

Understanding

9. **WE10** a. Form three new Pythagorean triads from the known triad {5, 12, 13}.
    b. Use substitution to show that the new triads satisfy Pythagoras' theorem.
    c. Evaluate x, given that the three numbers {x, 64, 136} form a Pythagorean triad.

10. If 32, x, 68 is a Pythagorean triad, determine the value of x. Show your working.

11. Construct a Pythagorean triad in which the smallest number is 33. Explain how you reached the solution.

Communicating, reasoning and problem solving

12. Georgios says that {1.5, 4, 2.5} is a Pythagorean triad, while Susan says that it is not. Explain who is correct, showing full working.

13. Discuss some strategies you could use to memorise as many common triads as possible.

14. Prove that $(a^2 - b^2)$, $2ab$, $(a^2 + b^2)$ form a Pythagorean triad.

15. a. If $(p - q)$, p, $(p + q)$ form a Pythagorean triad, determine the relationship between p and q.
    b. If $p = 8$, determine the Pythagorean triad.

16. A cable is stretched between two vertical poles that are installed 12 m apart on level ground. The first pole is 6 m high and the second pole is 11 m high. What is the length, l, of the cable?

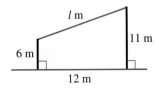

17. The lengths of the sides of a particular triangle are a cm, $2a$ cm and $3a$ cm. Prove that the triad formed is not a Pythagorean triad.

LESSON
12.6 Review

12.6.1 Topic summary

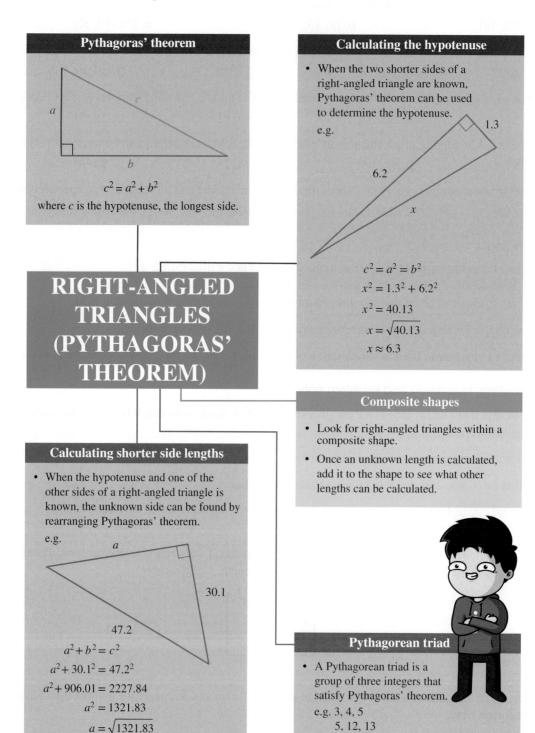

Pythagoras' theorem

$$c^2 = a^2 + b^2$$

where c is the hypotenuse, the longest side.

Calculating the hypotenuse

- When the two shorter sides of a right-angled triangle are known, Pythagoras' theorem can be used to determine the hypotenuse.

  e.g.

  $$c^2 = a^2 = b^2$$
  $$x^2 = 1.3^2 + 6.2^2$$
  $$x^2 = 40.13$$
  $$x = \sqrt{40.13}$$
  $$x \approx 6.3$$

RIGHT-ANGLED TRIANGLES (PYTHAGORAS' THEOREM)

Composite shapes

- Look for right-angled triangles within a composite shape.
- Once an unknown length is calculated, add it to the shape to see what other lengths can be calculated.

Calculating shorter side lengths

- When the hypotenuse and one of the other sides of a right-angled triangle is known, the unknown side can be found by rearranging Pythagoras' theorem.

  e.g.

  $$a^2 + b^2 = c^2$$
  $$a^2 + 30.1^2 = 47.2^2$$
  $$a^2 + 906.01 = 2227.84$$
  $$a^2 = 1321.83$$
  $$a = \sqrt{1321.83}$$
  $$a = 36.4$$

Pythagorean triad

- A Pythagorean triad is a group of three integers that satisfy Pythagoras' theorem.

  e.g. 3, 4, 5
  5, 12, 13
  7, 24, 25

12.6.2 Project

Are these walls at right angles?

Builders often use what is called a 'builder's square' when pegging out the foundations of a building to ensure that adjacent walls are at right angles. Even an error of just 1 or 2 degrees could mean that walls at the opposite end of a building may not end up in line. A builder's square uses the properties of Pythagoras' theorem.

The diagram below shows how a builder's square is constructed. The hypotenuse, c, acts as a brace to keep the two adjacent sides, a and b, in the correct position.

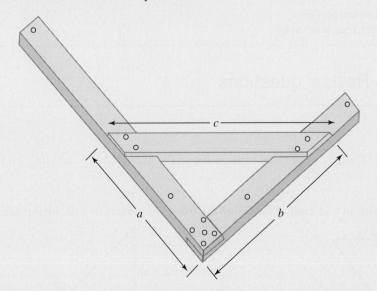

We will investigate the use of the builder's square. Cut two thin strips of paper to represent the arms a and b, as shown in the diagram. Join these two strips at one point with a pin to make the shape of the builder's square. The length of c can be obtained by measuring the distance from the end of a to the end of b. Complete questions 1 to 4 and record your results in the following table.

1. Open the arms so that they make an angle of 90°. Use a protractor to measure the angle. Carefully measure the strips to obtain the values for a, b and c in millimetres. Record your results in the first row of the table.
2. Repeat question **1** with the arms opened up to an angle less than 90°. Complete the second row of the table.
3. Repeat question **2** with the arms opened up to an angle greater than 90°. Complete the third row of the table.
4. Change the lengths of a and b by constructing a new builder's square. Repeat questions **1** to **3** and complete the last three rows of the table.

Length of a	Length of b	Length of c	$a^2 + b^2$	c^2	Angle

5. Consider the last three columns of the table. Using your results, what conclusions can you draw?
6. Construct a new builder's square and open it to an angle that you estimate to be 90°. Take measurements of a, b and c and record them below. What conclusions would you draw regarding your angle estimate? Check by measuring your angle with a protractor.
7. From your investigations, write a paragraph outlining how the measurements of the lengths of a, b and c on a builder's square can be used to determine whether the angle between adjacent walls would be equal to, greater than or less than a right angle.

 Resources

Interactivities Crossword (int-3386)
Sudoku puzzle (int-3194)

Exercise 12.6 Review questions

learn on

Fluency

1. In a right-angled triangle, what is the relationship of the hypotenuse, c, and the two shorter sides, a and b?

2. Calculate the value of x in each of the following triangles, correct to 2 decimal places.

a.

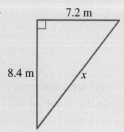

7.2 m

8.4 m

x

b.

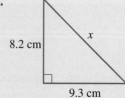

8.2 cm

x

9.3 cm

3. The top of a kitchen table measures 160 cm by 90 cm. A beetle walks diagonally across the table. Calculate how far the beetle walks.

4. A broomstick leans against a wall. The stick is 1.5 m long and reaches 1.2 m up the wall. Calculate the distance between the base of the wall and the bottom of the broom.

5. Calculate the value of x in each of the following triangles. Leave your answers in exact form.

a.
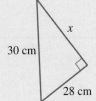

b.
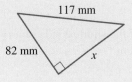

Understanding

6. Calculate how high up a wall a 20-m ladder can reach when it is placed 2 m from the foot of a wall. Give your answer correct to 1 decimal place.

7. A rectangular garden bed measures 3.7 m by 50 cm. Determine the length of the diagonal in cm.

8. A road sign is in the shape of an equilateral triangle with sides of 600 mm. Determine the area of the sign in cm².

9. Calculate x in the figure shown.

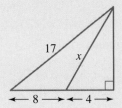

10. Determine the perimeter of the shape shown.

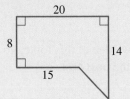

Communicating, reasoning and problem solving

11. Decide whether a triangle with sides 4 cm, 5 cm and 6 cm forms a right-angled triangle.

12. Calculate the length of the hypotenuse of a right-angled triangle with sides of 10 cm and 24 cm.

13. Determine the integer to which the length of the hypotenuse in the triangle shown is closest.

14. Calculate the length of the shortest side of a right-angled triangle with sides of 15 cm and 12 cm.

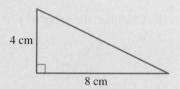

15. Determine the integer to which the length of the shortest side in this triangle is closest.

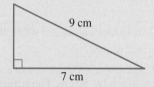

16. Calculate the perimeter of a right-angled triangle with a hypotenuse of 6.5 cm and another side of 2.5 cm.

17. An ironwoman race involves 3 swim legs and a beach run, as shown in the diagram. Determine the total distance covered in the race.

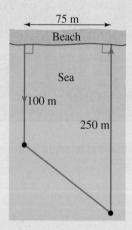

18. Decide which of the following are Pythagorean triads.
    a. 15, 36, 39
    b. 50, 51, 10
    c. 50, 48, 14

 To test your understanding and knowledge of this topic, go to your learnON title at www.jacplus.com.au and complete the **post-test**.

Answers

Topic 12 Pythagoras' theorem

12.1 Pre-test

1. 5
2. a. 14.39   b. 14.4
3. B
4. D
5. C
6. 7.9 km
7. 8.5 cm
8. 90 cm
9. 13.2 cm
10. 60 and 61
11. C
12. 9.9 units
13. D
14. TSA $= 302\,\text{cm}^2$
15. 38 cm

12.2 Pythagoras' theorem

1. a. r b. x c. k d. FU
2. a. Yes b. Yes c. No d. Yes
   e. Yes f. Yes g. No h. No
3. a. $x = 5$ b. $x = 13$ c. $x = 25$
4. a. $x = 21.3$ b. $x = 1163.3$ c. $x = 6.3$
5. a. $\sqrt{16\,781}$ b. 100 c. $\sqrt{68}$
6. a. $\sqrt{9760}$ b. $\sqrt{11\,194}$ c. $\sqrt{394}$
7. a. 16.97 cm b. 28.28 mm c. 6.93 cm
8. $\sqrt{146}$ cm
9. a. 12.81 cm b. 737.83 cm c. 17.26 cm
10. 18.03 cm
11. 4.66 m
12. a. Side 3 b. 1.4 m c. 3.4 m
13. 25 m
14. Side length ≈ 2.83 cm; area $= 8\,\text{cm}^2$
15. a. $189\,\text{cm}^3$ b. 6.95 cm
16. 19.23 cm
17. $79.60
18. $314.16\,\text{cm}^2$

12.3 Calculating shorter side lengths

1. a. 6 b. $\sqrt{93}$ c. $\sqrt{6032}$
2. a. 0.61 b. 2133.19 c. 453.90
3. a. $\sqrt{62\,925}$ b. $\sqrt{15}$ c. $\sqrt{80}$
4. a. 36.36 b. 1.62 c. 15.37
5. 23.04 cm
6. a. 97.47 cm b. 334.94 cm c. $6822.76\,\text{cm}^2$
7. 17.32 cm
8. 65.82 cm; $2501.08\,\text{cm}^2$

9. 2.60 m
10. 1.32 m
11. 45.43 cm
12. a. i. Multiply by 2 instead of squaring.
      ii. $100 = a^2 + 16$
      iii. $a = 9.2$
    b. i. Add instead of subtracting.
      ii. $25 - 16 = a^2$
      iii. $a = 3$
13. 20.62 m
14. Yes
15. 13.69 m
16. a.

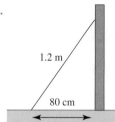

  b. 89.44 cm
  c. Yes, she will reach the hook from the top step.
17. a. 4 cm
  b. They only bisect at perpendicular angles when the rectangle has the same side lengths (i.e. it's a square).

12.4 Applying Pythagoras' theorem

1. 8.06 cm
2. a. $x = 7.62$ b. $x = 60$ c. $x = 20.87$
3. a. $x = 4$, $y = 9.17$ b. $x = 6.93$, $y = 5.80$
  c. $x = 13$, $y = 15.20$ d. $x = 19.55$
4. C
5. B
6. B
7. D
8. 11.35 m
9. 917.88 m
10. 10 m
11. a. 5 km b. 4 km c. 24 km
12. 2.45 m
13. 1.2 m; $4.8\,\text{m}^2$
14. 26.65 m
15. a. 2 m
  b. 3 m
  c. i. 2.15 m   ii. 3.61 m
  d. 23.04 m
16. 18.44 m
17. a. 2.42 cm b. 9.62 cm c. 3.54 cm
  d. 10.14 cm e. Shape 2
18. a. $4560\,\text{cm}^2$ or $0.4560\,\text{m}^2$
  b. $34\,880\,\text{cm}^2$ or $3.488\,\text{m}^2$
  c. The paint job would need 1 tin of paint, costing $24.95. The cost of the paint used would be $21.76.

19. $(1 + \sqrt{3})x^2$

20. a. $2025\,\text{cm}^2$ **b.** $943.41\,\text{cm}^2$ **c.** $6031.59\,\text{cm}^2$

12.5 Pythagorean triads

1. a. Yes   **b.** Yes   **c.** No

2. a. Yes   **b.** Yes   **c.** No

3. a. No   **b.** No   **c.** Yes

4. a. Yes   **b.** No   **c.** Yes

5. D

6. C

7. D

8. A

9. a. Sample responses:
$\{10, 24, 26\}, \{15, 36, 39\}, \{20, 48, 52\}$

    b. Sample response:
$26^2 = 24^2 + 10^2$
$676 = 576 + 100$
$676 = 676$ (proven)

    c. 120

10. 60

11. The following Pythagorean triads can be found:
33, 44, 55
33, 56, 65
33, 180, 183
33, 544, 545

12. Susan is correct because the numbers must be whole numbers.

13. An individual response is required, but here is a sample. All Pythagorean triads are written in the form $\{a, b, c\}$, where c is the hypotenuse. The easiest triad is $\{3, 4, 5\}$, and its multiples are also easy to work out:
$\{3, 4, 5\} \times 2 = \{6, 8, 10\}$ and $\{3, 4, 5\} \times 3 = \{9, 12, 15\}$.
Otherwise, select a few of the more commonly encountered triads and memorise them, for example $\{5, 12, 13\}$, $\{7, 24, 25\}$ and $\{8, 15, 17\}$.

14.
$$\left(a^2 - b^2\right)^2 + (2ab)^2 = a^4 - 2a^2b^2 + b^4 + 4a^2b^2$$
$$= a^4 + 2a^2b^2 + b^4$$
$$= \left(a^2 + b^2\right)^2$$
Therefore, $\left(a^2 - b^2, 2ab, a^2 + b^2\right)$ is a Pythagorean triad.

15. a. $p = 4q$ **b.** 6, 8, 10

16. 13 m

17.
$$(a)^2 + (2a)^2 = a^2 + 4a^2$$
$$= 5a^2$$
$$(3a)^2 = 9a^2$$
Therefore, $(a, 2a, 3a)$ is not a Pythagorean triad.

Project

1. to 4. Individual measurement and data collection required. Your table will need to have the following rows and columns.

Angle size	Length of a	Length of b	Length of c
Angle = 90°			
Angle > 90°			
Angle > 90°			

5. When the angle is equal to 90°, $c^2 = a^2 + b^2$.
For an angle greater than 90°, $c^2 > a^2 + b^2$.
For an angle smaller than 90°, $c^2 < a^2 + b^2$.

6. If $c^2 = a^2 + b^2$, then the angle is equal to 90°. Check by measuring the angle.

7. Personal response required; ensure that you provide a full description of your research findings, relating them to what you have learned about Pythagoras' theorem.

12.6 Review questions

1. $c^2 = a^2 + b^2$

2. a. $11.06\,\text{m}$ **b.** $12.40\,\text{cm}$

3. 150 cm

4. 0.9 m

5. a. $\sqrt{116}\,\text{cm}$ **b.** $\sqrt{6965}\,\text{cm}$

6. 19.9 m

7. 373.36 cm

8. $1558.85\,\text{cm}^2$

9. 12.69

10. 64.81

11. No

12. 26 cm

13. 9 cm

14. 9 cm

15. 6 cm

16. 15 cm

17. 592.71 m

18. a. Yes   **b.** No   **c.** Yes

Semester review 2

The learnON platform is a powerful tool that enables students to complete revision independently and allows teachers to set mixed and spaced practice with ease.

Student self-study

Review the **Course Content** to determine which topics and lessons you studied throughout the year. Notice the green bubbles showing which elements were covered.

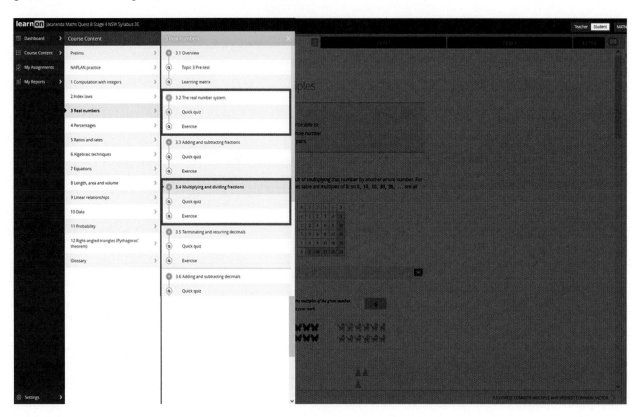

Review your results in **My Reports** and highlight the areas where you may need additional practice.

Use these and other tools to help identify areas of strengths and weakness and target those areas for improvement.

Teachers

It is possible to set questions that span multiple topics. These assignments can be given to individual students, to groups or to the whole class in a few easy steps.

Go to **Menu** and select **Assignments** and then **Create Assignment**. You can select questions from one or many topics simply by ticking the boxes as shown below.

Once your selections are made, you can assign to your whole class or subsets of your class, with individualised start and finish times. You can also share with other teachers.

More instructions and helpful hints are available at www.jacplus.com.au.

GLOSSARY

arc a section of the circumference of a circle

area the amount of flat surface enclosed by a shape; it is measured in square units, such as square metres, m^2, or square kilometres, km^2

Associative Law a number law that refers to the order in which three numbers may be added, subtracted, multiplied or divided, taking two at a time

backtracking the process of working backwards through a flowchart; each step is the inverse operation of the corresponding step in the flowchart. It can be used to solve equations.

base the number or variable that is being multiplied when an expression is written in index form

basic numeral a number; does not include index or exponent notation

biased leaning in a favoured direction

bimodal describes a data set that has two modes

Cartesian plane a coordinate grid formed by an x- and y-axis

census the collection of data in which every member of a target population is surveyed

chance experiment an experiment in which the outcomes are open to chance

chord a straight line joining any two points on the circumference of a circle

circumference the distance around the outside of a circle; it is equal to $2\pi r$ or πD, where $r =$ radius and $D =$ diameter of the circle

class intervals subdivisions of a set of data; for example, students' heights may be grouped into class intervals of 150 cm–154 cm, 155 cm–159 cm

closed describes a question with only one possible solution

coefficient the number part of a term

Commutative Law a number law that refers to the order in which two numbers may be added, subtracted, multiplied or divided

complementary events events that have no common elements and together form a sample space

composite shape a shape made up of more than one basic shape

constant term a term without a pronumeral; that is, a number

cost price the total cost that a business pays for a product, including its production and business overhead costs

cross-multiplication a method of determining whether a pair of ratios are in proportion: if $a : b = c : d$, then $a \times d = c \times b$

data various forms of information

denominator the bottom term of a fraction; it shows the total number of parts the whole has been divided into

diagonal a line that runs from one corner of a closed figure to an opposite corner

diameter the straight-line distance across a circle through its centre

directed numbers numbers that have both size and direction; for example, $+3$ and -7

direction on a number line, the position relative to zero; increasing numbers are to the right of zero and decreasing numbers are to the left of zero

Distributive Law a rule that states that each term inside a pair of brackets is to be multiplied by the term outside the brackets

equivalent fractions fractions that are equal in value; for example, $\dfrac{1}{2} = \dfrac{3}{6}$

estimate an approximate answer when a precise answer is not required

evaluate to examine the context of a problem and apply calculations to find a solution

event a set of favourable outcomes in each trial of a probability experiment

expanding using the Distributive Law to remove the brackets from an expression

expected value the average value of an experiment over many trials, found by multiplying the relative frequency by the number of trials

experiment in probability, the process of performing repeated trials of an activity for the purpose of obtaining data in order to predict the chances of certain things happening; in data collection, a controlled study in which researchers attempt to understand cause-and-effect relationships

exponent *see* **power**

exponent notation *see* **index notation**

expression a group of terms separated by operation symbols and/or brackets. Expressions do not contain equals signs.

favourable outcome the desired result in a probability experiment

gradient a measure of how steep something is; that is, its slope. The gradient of a straight line is given by:

$$m = \frac{\text{vertical distance}}{\text{horizontal distance}}$$

GST Goods and Services Tax; a federal government tax (10%) added to the price of some goods and services

highest common factor (HCF) the largest of the set of factors common to two or more numbers; for example, the HCF of 16 and 24 is 8

histogram a type of column graph in which no gaps are left between columns and each column 'straddles' an *x*-axis score, such that the column starts and finishes halfway between scores. The *x*-axis scale is continuous and usually a half-interval is left before the first column and after the last column.

hypotenuse the longest edge of a right-angled triangle

Identity Law for addition when 0 is added to any number, the original number remains unchanged.

Identity Law for multiplication When any number is multiplied by 1, the original number remains unchanged.

improper fraction a fraction whose numerator is larger than its denominator; for example, $\frac{5}{4}$

index *see* **power**

index notation the short way of writing a number, pronumeral or variable when it is multiplied by itself repeatedly

infinite not finite; never ending; unlimited

infographic a visual representation of information and data combining text, charts, diagrams, videos etc.

integers positive whole numbers, negative whole numbers and zero

Inverse Law for addition When a number is added to its additive inverse, the result is 0.

Inverse Law for multiplication When a number is multiplied by its multiplicative inverse, the result is 1.

irrational describes numbers that cannot be written as fractions

kite a quadrilateral in which two pairs of adjacent sides are equal in length and one pair of opposite angles (those between the sides of unequal length) are equal

like terms terms that contain exactly the same variables (letters); for example, $3ab$ and $7ab$ are like terms, but $5a$ and $6ab$ are not

linear equation an equation in which the pronumeral has an index (power) of 1

linear function a function that is a straight line when drawn

linear graph a graph in which all of the points fall on a straight line

linear pattern a pattern of points that when plotted can be joined to form a straight line

lowest common denominator (LCD) between two or more fractions, the lowest common multiple of the denominators

magnitude size

mean in summary statistics, the sum of all the scores divided by the number of scores. It is also called the average.

measures of centre any of a number of terms used to describe a central value in a data set, for example mean and median

measures of location *see* **measures of centre**

measure of spread any of a number of terms used to describe how the values in a data set are spread or scattered, for example range

median in summary statistics, the middle value if the number of data is odd, or the average of the two middle values if the number of data is even. Data must first be arranged in numerical order.

mixed number a number made up of a whole number and a fraction; for example, $2\frac{3}{4}$

mode the most common score or the score with the highest frequency in a set of data

multimodal describes a data set that has multiple modes

mutually exclusive two events that cannot both occur at the same time

negatively skewed describes a data set in which the mean and the median are less than the mode

non-linear describes a relationship between two variables that does not increase at a constant rate; their graph does not form a straight line

non-response bias a selection bias that occurs when individuals chosen for the sample are unwilling or unable to participate in the survey

numerator the top term of a fraction; it shows how many parts there are

observation a study in which researchers simply collect data based on what is seen and heard

open describes a question with more than one possible answer

ordered pair a pair of coordinates, with the x-coordinate appearing before the y-coordinate

origin the point on a number line where the number 0 sits; the centre of a Cartesian plane, (0, 0), where the x- and y-axes intersect

outcome the particular result of a trial in a probability experiment

outlier a piece of data that is much larger or smaller than the rest of the data

parallelogram a quadrilateral with both pairs of opposite sides parallel to each other; rectangles, squares and rhombuses are parallelograms

per cent out of 100

perimeter the distance around the outside (border) of a shape

population every member or data point under consideration

positively skewed describes a data set in which the mean and the median are greater than the mode

power the number that indicates how many times the base is being multiplied by itself when an expression is written in index form. Also known as an *exponent* or *index*.

primary data data that has been collected first-hand (by you)

prism a solid object with identical parallel ends, and with the same cross-section along its length

probability the likelihood or chance of a particular event (result) occurring:

$$\text{Pr(event)} = \frac{\text{number of favourable outcomes}}{\text{number of possible outcomes}}$$

The probability of an event occurring ranges from 0 (impossible: will not occur) to 1 (certain: will definitely occur).

profit the amount of money made on a sale, calculated by subtracting the costs from the sale price

pronumeral a letter used in place of a number; another name for a variable

proportion equality of two or more ratios

Pythagoras' theorem In any right-angled triangle, the square of the hypotenuse is equal to the sum of the squares of the other two sides; often expressed as $c^2 = a^2 + b^2$.

Pythagorean triad a group of any three whole numbers that satisfy Pythagoras' theorem, for example $\{5, 12, 13\}$ or $\{7, 24, 25\}$

quadrant one of the four sections of a Cartesian plane

quadratic equation an equation in which the pronumeral has an index (power) of 2

quadrilateral a 2-dimensional closed shape formed by four straight sides

questionnaire a set of questions used in a survey

radius a straight line from a circle's centre to any point on its circumference

random following no particular order or pattern. To ensure that they are free from bias, surveys should be as random as possible.

random number generator a method of generating random numbers

range in summary statistics, the difference between the highest and lowest values (scores)

rate a ratio that compares quantities or measurements in different units

rational describes numbers that can be expressed as fractions with non-zero denominators

ratios comparisons of two or more quantities of the same kind

real number a number that belongs to the set of all rational and irrational numbers

recurring decimals numbers that have an infinitely repeating pattern of decimal places

reflection transformation whereby a point or object is reflected in a mirror line

relative frequency the chance of an event happening expressed as a fraction or decimal:

$$\text{relative frequency} = \frac{\text{frequency of an event}}{\text{total number of trials}}$$

rhombus a parallelogram in which all sides are equal and opposite angles are equal

right-angled triangle a triangle that has one of its angles equal to 90° (a right angle)

rounding expressing a number with a certain number of decimal places

rounding down A number ending in 0, 1, 2, 3 or 4 is rounded down.

rounding up A number ending in 5, 6, 7, 8 or 9 is rounded up.

sample part of a whole population

sample size the number of participants in a sample

sample space in probability, the complete set of outcomes or results obtained from an experiment. It is shown as a list enclosed in a pair of braces, { }, and is denoted by the symbols ξ or S.

sampling methods methods used to select members from a population to be in a statistical study

secondary data data that has been collected second-hand (by someone else)

sector a region of a circle bounded by two radii and the arc joining them

segment a section of a circle bounded by a chord and an arc

selection bias bias that results from a sample that is not representative of the population

self-selected sampling a sampling method in which people choose to take part in a survey; also called voluntary sampling

selling price the price of a good or service charged by a business to a customer

simple random sampling a sampling method in which each member of the population has equal change of selection

skewness a measure of the asymmetry of a data set

statistical infographic an infographic that uses pie charts, bar diagrams, histograms, box plots, Venn diagrams etc. to represent a data set

statistics the branch of mathematics that deals with the collection, organisation, display, analysis and interpretation of data, which are usually presented in numerical form

stratified sampling a sampling method in which the population is divided into groups (strata), and samples are selected from each group

substitution the process by which a number replaces a variable in a formula

summary statistics the mean, median, mode and range of a data set

symmetrical describes a data set that has no skewness; the mean, the median and the mode have the same value

systematic sampling a sampling method in which the first member is selected at random and subsequent members are chosen at a fixed interval, for example every 20th member

tangent a straight line that passes through a point on a circle and is perpendicular to the radius

term a group of letters and/or numbers that forms an expression when combined with operation symbols and brackets

terminating decimals decimal numbers that have a fixed number of places; for example, 0.6 and 2.54

theoretical probability the probability of an event based on the number of possible favourable outcomes and the total number of possible outcomes

trapezium a quadrilateral in which one pair of opposite sides is parallel

trial an experiment performed in the same way every time

two-way tables a diagram that represents the relationship between two non-mutually exclusive attributes

undefined a numeric value that cannot be calculated

undercoverage a selection bias that occurs when some members of the population are inadequately represented in a sample

unimodal describes a data set that has one mode

universal set the set containing all the elements specific to a particular problem; denoted by the symbol ξ

variable a letter or symbol in an equation or expression that may take many different values

vinculum the horizontal line used to separate the top of a fraction (the numerator) from the bottom of a fraction (the denominator)

volume the amount of space inside a three-dimensional object; it is measured in cubic units, such as cubic metres, m^3, or cubic kilometres, km^3

voluntary response bias a selection bias that occurs when sample members are self-selected volunteers

***x*-axis** the horizontal axis in a Cartesian plane

***y*-axis** the vertical axis in a Cartesian plane

INDEX